SECOND EDITION

# FLAVOR CHEMISTRY and TECHNOLOGY

SECOND EDITION

# FLAVOR CHEMISTRY and TECHNOLOGY

GARY REINECCIUS

Taylor & Francis
Taylor & Francis Group

Boca Raton London New York Singapore

A CRC title, part of the Taylor & Francis imprint, a member of the Taylor & Francis Group, the academic division of T&F Informa plc.

Published in 2006 by
CRC Press
Taylor & Francis Group
6000 Broken Sound Parkway NW, Suite 300
Boca Raton, FL 33487-2742


Printed in the United States of America on acid-free paper
10 9 8 7 6 5 4 3

International Standard Book Number-10: 1-56676-933-7 (Hardcover)
International Standard Book Number-13: 978-1-56676-933-4 (Hardcover)
Library of Congress Card Number 2005041778



**Library of Congress Cataloging-in-Publication Data**

Reineccius, Gary.
Flavor chemistry and technology – 2nd ed. / Gary Reineccius.
p. cm.
Includes bibliographical references and index.
ISBN 1-56676-933-7
1. Flavor. 2. Flavoring essences. I. Title.

TP372.5.R374 2005
664¢.5—dc22 2005041778

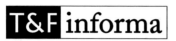

T&F informa

Taylor & Francis Group
is the Academic Division of T&F Informa plc.

**Visit the Taylor & Francis Web site at**
**http://www.taylorandfrancis.com**

**and the CRC Press Web site at**
**http://www.crcpress.com**

# *Dedication*

*To my teachers*

# Preface

This book combines the essentials of both flavor chemistry and flavor technology. Modern flavor chemistry as an academic area of study has existed for over 40 years. The advent of gas chromatography in the early 1960s (rapidly coupled with mass spectrometry) heralded the beginning of this research area. These instruments gave scientists the ability to separate and identify the host of volatiles found in our foods that contribute to the aroma of foods.

Flavor chemistry flourished in the U.S., Europe, and Japan in the late 1960s and early 1970s. Since money was readily available for flavor research, great strides were made in understanding the biosynthetic pathways of flavor formation and the chemical constituents that are important to flavor. Government support for flavor research in the U.S. started to decline in the early 1980s and stabilized at a level that has permitted a limited number of laboratories to exist. Fortunately, flavor research in Europe has continued to be funded, and Europe has become the leader in flavor research. In recent years, the food industry and various commodity groups such as the dairy industry have come to recognize the value of flavor to the consumer and have provided increasing support to the academic community in the U.S. Thus, we find a growth in flavor research once again in the U.S.

Our journey of understanding has taken us from providing long lists of volatile chemicals found in foods (thought to be the way to replicate nature), to studies to determine the key volatile compounds, to studies on food/aroma interactions, flavor release, and most recently, to taste, perception, and cognitive functioning. It is unlikely that this is the final frontier — for we thought we were there nearly 40 years ago! Nature is not readily giving up its secrets to flavor and the pleasure it provides in our daily lives.

Flavor technology is an ancient area of study. Man has searched for a means of making food more pleasurable or palatable since time began. The flavor industry had its beginnings with the search for, and trading of, spices many centuries ago. This searching and trading has been responsible for the discovery of new worlds. Today this industry is nearly invisible to the average consumer since only a few members of the industry sell directly to the consumer: most are ingredient suppliers only known to their client companies who interface with the ultimate consumer.

The flavor industry today has a world market of approximately 16.3 billion dollars (www.Leffingwell.com). The total global sales are only somewhat larger than a single major food company. The flavor industry is comprised of a few large companies and a host of small companies (perhaps 500+). There is little published about today's flavor industry because of its highly competitive and secretive nature. It is difficult and sometimes impossible to protect a flavoring via patents, so flavoring materials, synthetic chemicals, and manufacturing procedures are kept secret to avoid supporting competitive companies.

The author has used the best information available to try to give a view of this industry. Often patents or personal experience have been called on due to a dearth of public information on the industry. While only a sampling of the literature available could be included in many areas, the author has provided abundant referencing to provide direction for the reader. This book is intended to serve as a suitable text for both undergraduate and graduate level courses as well as a reference text for those in academia or industry.

# Acknowledgments

I wish to acknowledge the host of individuals referenced in this text for their efforts to provide understanding to a field that is ever so slowly giving up its secrets. I wish to thank my many teachers for sharing their knowledge and helping me down a career path that has been most challenging and exciting. I have to thank my students also for educating me in their specific research topics — I am certain they taught me more than I taught them. I will forever be in debt to Henry Heath for what he has given to the field and to me personally through his writing. Several sections of this book still reflect his contributions.

# The Author

**Gary Reineccius, Ph.D.,** is a professor in the Department of Food Science and Nutrition at the University of Minnesota. He has been actively involved in flavor research for more than 35 years. During this time he has published over 190 research articles. Dr. Reineccius has spent sabbatical leaves with Fritzsche Dodge and Olcott (New York, flavor creation and production), Nestle (Switzerland, reaction flavors), and most recently Robertet S.A. (France, taste modifiers and manufacturing).

Dr. Reineccius teaches courses in Chemical and Instrumental Analysis of Foods, Food Processing, and Flavor Chemistry and Technology. He has written a college textbook on food flavors with Henry Heath. This was the first textbook in the flavor area that combined both flavor chemistry and technology. Dr. Sara Risch and he edited and were major contributors to two books on flavor encapsulation. He is the editor of the *Source Book of Flavors* and an ACS symposium proceeding titled "Heteroatomic Aroma Compounds."

Dr. Reineccius' achievements have been recognized by several local and international organizations. He is an honorary member of the Society of Flavor Chemists. He has been granted the Palmer Award for his contribution to chromatography by the Minnesota Chromatography Forum. He has received the Distinguished Achievement and Service in Agricultural and Food Chemistry Award from the American Chemical Society and been presented the Stephen S. Chang Award by the Institute of Food Technologists. These are the highest awards given to individuals in the flavor area. He is a Fellow of the American Chemical Society.

He often speaks at public schools and other groups. His favorite lay topics are chocolate (he spent three years researching chocolate flavor for his Ph.D. thesis) and the chemistry of gourmet cooking. From a professional standpoint, his favorite topic is flavor encapsulation. He has been actively engaged in research in this area since 1964.

# Contents

# Part I

## Flavor Chemistry

# 1 An Overview of Flavor Perception

## 1.1 FLAVOR PERCEPTION

In our daily lives, a complete flavor experience depends on the combined responses of our senses and the cognitive processing of these inputs. While flavor per se often is thought of as being limited to olfaction, taste, and the somatosenses (irritation, tactile, and thermal), numerous other sensory inputs are processed by the brain to result in flavor perception ([1]; Figure 1.1). This broad multimodal aspect of flavor perception has only recently been acknowledged and multidisciplinary research directed at its understanding initiated [2]. Historically researchers in both academic and industrial settings have viewed flavor as predominantly aroma with only minor importance given to the contribution of taste and the somatosenses. Current research is proving this to be an unrealistic simplification of human flavor perception.

The very broad nature of flavor perception cannot be addressed in a single chapter and thus the book edited by Taylor and Roberts [3] is recommended for a better appreciation of the overall phenomenon of flavor perception. Limitations in terms of space (and of the author) result in this text focusing discussion on the traditional aspects of flavor, i.e., olfaction, taste, and the somatosenses. These fundamental sensory inputs will be discussed in terms of their functioning in the human to help the reader gain an appreciation of the complexity of flavor perception.

## 1.2 TASTE PERCEPTION

Taste is the combined sensations arising from specialized taste receptor cells located in the mouth. It is primarily limited to the tongue and is broken down into the sensations of sweet, sour, salty, bitter, and umami (the sensation given by the amino acids glutamate, aspartate, and related compounds). Defining taste as being limited to five categories suggests that taste is a simple sensation: this is not true. Within sour, for example, there is the sourness of vinegar (acetic acid), sour milk (lactic acid), lemons (citric acid), apples (malic acid), and wines (tartaric acid). Each of these aspects of sour has a unique sensory character. The same can be said about sweetness, bitterness, and saltiness. How each taste is recognized, specificity by taste cells, and how tastes are coded and interpreted are still largely unknown. Thus, taste is not simple in itself, nor is how it interacts with other sensory properties of food to determine human perception. For excellent comprehensive reviews on this topic, see Lindemann [4], and Rawson and Li [5].

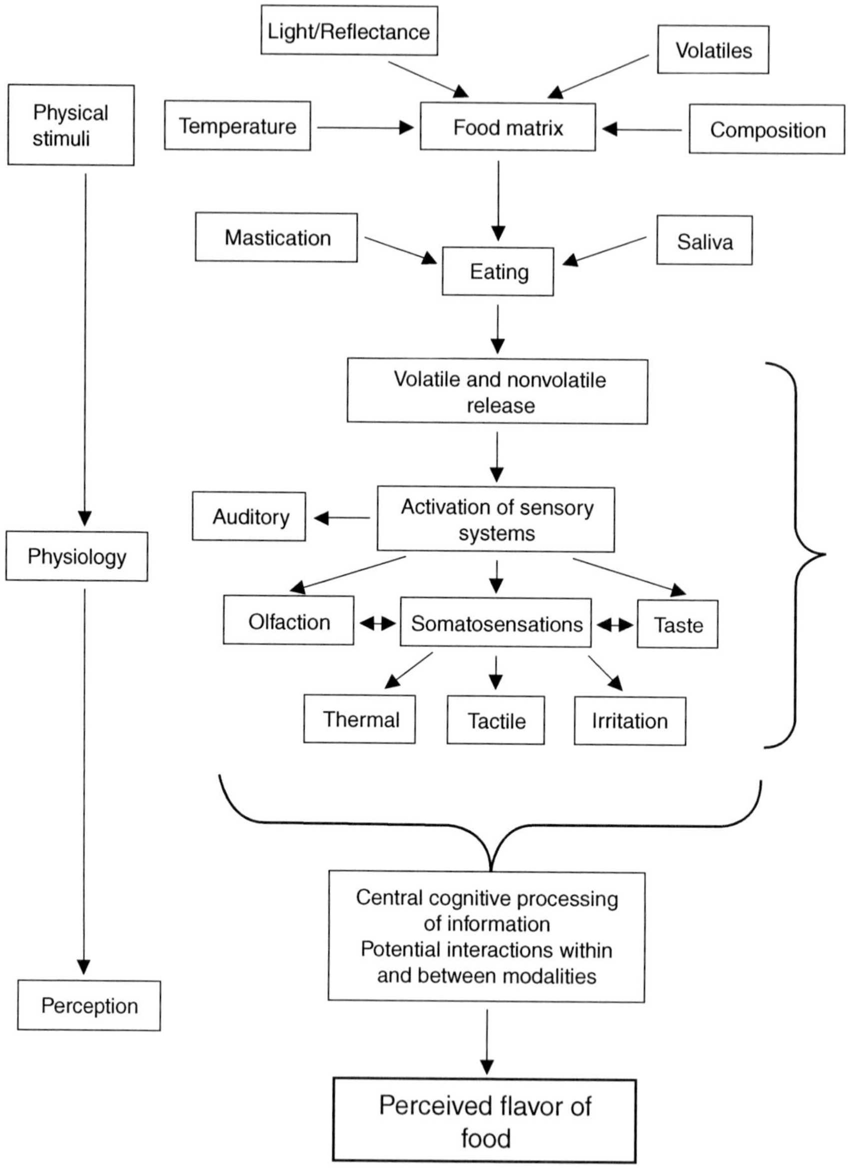


**FIGURE 1.1** Flow diagram of factors that influence flavor perception (From Keast, R.S.J., P.H. Dalton, P.A.S. Breslin, *Flavor Perception*, A.J. Taylor, D. Roberts, Eds., Blackwell Publ., Ames, 2004, p. 228. With permission.)

### 1.2.1 Anatomy of Taste

Tastes are detected by taste buds that are located throughout the oral cavity (tongue, palate, pharynx, larynx, and in the cheeks of infants). The majority of taste buds are located on the tongue within papillae (those little bumps easily seen on the surface of the tongue). The average adult has roughly 10,000 taste buds, children have more, but there exists a large variation within human populations. Damaged taste buds are quickly replaced within 7 to 10 days, and these detectors are maintained throughout life to serve as seekers of nutrients and final protection to the body from potentially harmful materials. However, the ability to taste can decrease or become damaged over time from age, oral infections, gastric reflux (a common cause), repeated scalding, smoking, illness (diabetes mellitus, pernicious anemia), certain drugs, pesticide and metal exposure, head trauma, surgical procedures, and radiation [6].

There are four types of papillae. The most abundant papilla, the filiform, lacks taste buds but is involved in tactile sensation. The three papillae that contain taste buds are the fungiform, foliate, and circumvallate (Figure 1.2). The fungiform, located on the front of the tongue, are mushroom shaped and appear as red spots, contain two to three taste buds, and comprise about 18% of the total taste buds on the tongue. The foliate are leaf-like papilla and appear as small ridges at the back edges of the tongue. There are up to twenty of these ridges and about 600 taste buds on each side, which comprise about 34% of the tongue taste buds. At the back of the tongue are 8 to 12 relatively large circumvallate papillae, each containing about 250 taste buds and makeup the final 48% of the tongue taste buds [7–9].

The taste buds, described as onion-like or navel orange-like structures, are clustered in packs of 20 to 250 depending on the papilla, and each taste bud consists of up to 100 taste cells that represent all five taste sensations. Each taste cell has filament-like structures, called microville, that project through a taste pore located

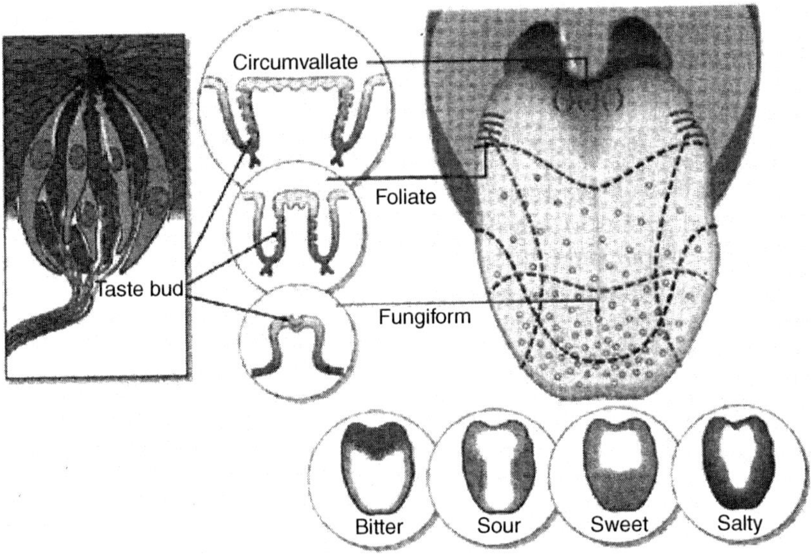


**FIGURE 1.2** Functional anatomy of the human tongue. Diagram of a human tongue highlighting the taste buds and the regional preferences to sweet. sour, bitter, and salty stimuli. While different areas are preferentially responsive to certain taste modalities, there is significant overlap between the regions. (From Hoon, M.A., E. Adler, J. Lindemeier, J.F. Battey, N.J. Ryba, C.S. Zuker, *Cell*, 1999(96): p. 541. With permission.)

at the opening of the taste bud into the papilla and capture tastants. The microville have receptors that involve transmembrane proteins that bind to molecules and ions, which stimulate tastes [7,8].

Historically, there has been a very neat arrangement of taste made on the map of a tongue. Sweetness was mapped on the tip of the tongue, salt-detecting taste buds fell in the center, sour detecting taste buds were labeled on the sides of the tongue, and the bitter tasting taste buds (thought to be the last defense against toxins) were placed at the back of the tongue (closest to the gag response). This orderly arrangement of taste has been depicted inaccurately for a very long time [4]. A more accurate representation of regional preferences for the basic tastes is provided in Figure 1.2 [5].

Over 100 years ago, it was determined that taste cells in the taste buds of different papilla located across the tongue respond to more than one type of stimulus [11]. Although each neuron may respond more strongly to one tastant, it will also respond to unlike taste properties. Also, it is thought that no single taste cell contains receptors for both bitter and sweet. Each taste receptor cell is connected through a network of cellular activities to a sensory neuron that travels to the brain. A single sensory neuron may be connected to several taste cells each within different taste buds [7].

Because taste cell neurons can respond to more than one taste stimuli, it is now proposed that the brain represents different tastes by generating unique patterns of electrochemical activity across a large set of neurons. Stimuli that taste alike give similar electrochemical activity across sets of taste neurons. It is hypothesized that the brain uses a form of pattern recognition to interpret, categorize, and store different taste qualities. These patterns are then interpreted in combination with sight, smell, and other sensory signals, as flavor [7,12,13].

### 1.2.2 Synopses of Tastes

Electrophysiological studies suggest salty and sour tastants permeate the taste cell wall through ion specific channels, but those responsible for sweet, bitter, and umami tastes bind to cell surface receptors (see Figure 1.3 and detailed figure legend for description). The electrochemical changes that signal the brain are ultimately dependent on ion concentration. Taste cells, at rest, have a net negative charge internally and net positive charge externally.

Tastants ultimately depolarize (increase the positive charge within the cell) the charge difference in the taste cell. The final events for perception of each basic taste involve an increase of $Ca^{2+}$ in the taste cell, an electric current is then produced, a transmitter is released, and then an increased firing in the primary afferent nerve signals to the brain. The signals are then relayed via the thalamus to the cortical taste centers, where the information is processed and interpreted [4,15].

For salty taste, sodium chloride (and other salts), the positive ions (e.g., $Na^+$) enter the taste cells through $Na^+$ channels located in the taste cell membrane causing a depolarization of $Ca^{2+}$ (Figure 1.3). $Ca^{2+}$ enters the cell through voltage sensitive cell wall channels. As the interior of the cell becomes more positively charged, a small electric current is produced, a transmitter is released and increased firing in the primary afferent nerve signals the brain, "salty" [16,17].

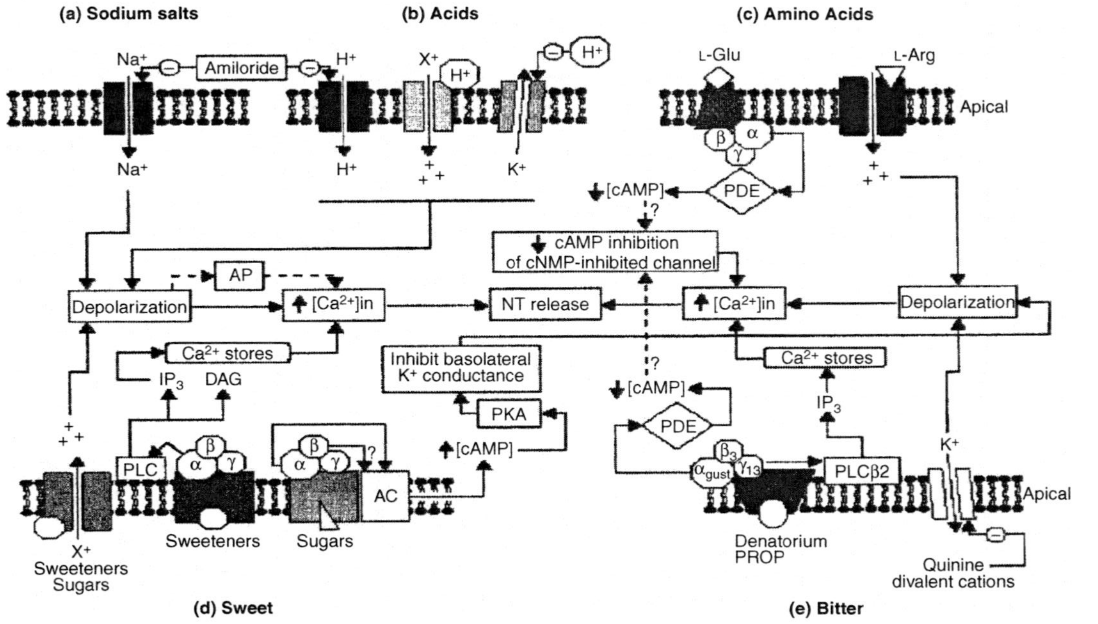


**FIGURE 1.3** Taste is mediated by a diversity of transduction mechanisms. All pathways converge to elicit an increase in intracellular calcium, which triggers neurotransmitter (NT) release. The amiloride-sensitive sodium channel contributes to salt and sour detection. Protons released from sour stimuli are also detected via a nonselective cation channel and via block of an outward K+ conductance. Depolarization results in opening of voltage-gated calcium channels and calcium influx. Glutamate is detected via a G-protein-coupled pathway that may include multiple receptor types. In some species, other amino acids can be detected via ligand-gated ion channels, but the presence of this pathway in humans is not established. Multiple pathways exist for detection of sweet stimuli, including both ligand-activated ion channels (ionotropic) and G-protein-coupled receptor pathways. Bitter compounds are detected via G-protein-coupled receptors activating PLCβ2, leading to release of calcium from intracellular stores. Some bitter compounds and bitter-tasting salts may also act via suppression of K+ efflux. Symbols and abbreviations: α, β, γ refer to G-protein subunits; AC = adenylyl cyclase; AP = action potential; IP3 = inositol trisphosphate; NT = neurotransmitter; PDE = phosphodiesterase; PKA = protein kinase A. (From Gilbertson, T.A., S. Damak, R.F. Margolskee, *Curr. Opinion Neurobiol.*, 2000(10): p. 519. With permission.)

It has been hypothesized for some time that an amiloride molecule (ENaC, amiloride-sensitive epithelial channel) serves as a salt receptor by providing a specific pathway for sodium current into taste cells [16.] ENaC is a hetero-oligomeric complex composed of three homologous subunits, and one of the subunits has been found to be controlled by the hormone aldosterone [18]. It has been observed in salt deficient herbivores, rodents, and humans, that salt sensitivity is increased by an increased flow of aldosterone through the taste buds [19]. This increased flow of aldosterone is a good biological example of adaptive tuning of taste to increase taste sensitivity for salt in an animal deficit in this nutrient. However, in humans, the ENaC channel does not explain our complete sensitivity to salt, which suggests other yet identified taste channels [4]. It is interesting that still today we know little about what would appear to be the simplest of tastes.

Sour taste is proton dependent. Two groups of sour taste receptors have been identified. The first group includes those that are comprised of channels that permit an inward flow of protons from the mouth into taste cells (Figure 1.3b). The molecule ENaC (amiloride-sensitive epithelial channel) functions in this manner. The second category includes those mechanisms that involve $H^+$-gated channels. An example in this category is the apical $K^+$ channel. $H^+$ ions block $K^+$ channels in the taste cell walls. $K^+$ channels are responsible for maintaining cell membrane potential. Once blocked by $H^+$, the cell is depolarized, and $Ca^{2+}$ flows into the cell. In both mechanistic categories, the final event produces an electric current, a transmitter is released and increased firing in the primary afferent nerve signals the brain, "sour" [20,21]. However, these explanations for sour taste transduction are far from complete, and the large number of mechanisms thus far identified show the complexity of this taste perception [4].

Likewise, we have yet to determine the biochemical pathways for perception of sweet, bitter, and umami compounds. Studies to date suggest that there may be more than one mechanism of detection. However, it is now known that initially sweet, bitter, and umami tastants bind to receptor sites located on the taste cell wall membrane (like a lock in key mechanism). The cell coupling of the tastant with a cell surface receptor activates as α-gustducin, a guanine nucleotide-binding protein (G-protein, related to the G-protein transducin, which helps translate light into vision). The relay through α-gustducin initiates tastant dependent reactions that ultimately lead to the increased levels of $Ca^{2+}$ in the cell, and culminates in the firing of the afferent nerve [4,7].

The sweet receptor has been the target of many chemists in the food industry. If a molecular model of the sweet binding site can be identified, high potency sweeteners could be designed from this model. Recently it has been found that sugar and nonsugar sweeteners initially activate sweet-responsive taste receptors called G-protein coupled receptors (GPCRs). Each receptor contains two subunits termed T1R2 and T1R3, which are coupled to α-gustducin [10,22]. Presently the data suggest that sweet tastants activate taste cells through at least two transduction pathways. Sugars are thought to activate adenyl cyclase, elevating intercellular levels of cAMP or cGMP, whereas nonsugar sweeteners activate an alternative $IP_3$ reaction within the same cell. The two pathways may then converge in that elevated cAMP, cGMP, or $IP_3$ produces a PKA-mediated phosphorylation of $K^+$ channels. The flow of $K^+$ is inhibited, and cell depolarization results. $Ca^{2+}$ then enters the cell through activated $Ca^{2+}$ channels, and the electric current is produced [4,10,22].

The relationship between the hormone leptin and sweet taste is of interest. It is thought that leptin, which is secreted by fat cells, is an inherent biological signal used to regulate nutrition and body weight through taste sensitivity. Leptin suppresses insulin secretion by activating ATP-sensitive $K^+$ channels within taste cells. The consequence of this is a blunting of the nerve signals that indicate sweet taste, which in turn, is presumed to make food less enticing. During starvation the production of leptin is decreased, thus increasing sensitivity to sweetness and the desirability of foods [23,24].

Bitter is inherently associated with harmful substances, and this is generally correct. Many organic molecules that originate from plants and interact with the nervous systems of mammals are bitter, including caffeine, nicotine, strychnine, and as well as many pharmaceutical drugs. Unlike other tastes, bitter taste receptor cells are more tuned to respond to specific bitter molecules. That is, they may respond to one type of bitter molecule but not another. Primary transduction of bitter taste is believed to involve a family of about 24 G-protein-coupled receptors, termed T2R that, like sweet taste, are linked to α-gustducin [14,25]. Another pathway is also activated simultaneously involving β and γ subunits of gustducin. The GPCR controlled bitter signals result, which appear to work together, decrease levels of cAMP and cGMP, and the release of a second messenger, inositol-1,4,5-triphosphate ($IP_3$) and diaglycerol. These increases mediate the release of internal stores of $Ca^{2+}$ (extracellular $Ca^{2+}$ is not needed). Once again $Ca^{2+}$ levels are elevated, the cell is depolarized, transmitter is released, and firing of the afferent nerve occurs [26,27]. At this point it is not known why a dual signal is required for bitter taste, perhaps this is a means of sensory amplification [4]. Although the most sensitive responses involve the T2R receptors and gustducin, there are other independent mechanisms in place for perception of bitter tastants [10,15]. For example, quinine and caffeine are known to directly permeate the cell membrane, completely bypassing the G-protein receptor sites [28].

The word *umami* comes from the Japanese term, umami, meaning delicious. This taste was discovered about 100 years ago, and although well established as qualitatively different from sweet, sour, salty and bitter, many fail to recognize umami as the fifth basic taste [29]. Foods that contain L-glutamate such as meat broths (particularly chicken) and aged cheese (e.g., parmesan) give strong umami tastes. Umami taste is also elicited by 5′-ribonucleotides, such as IMP and GMP, and synergism of these nucleotides with L-glutamate occurs. Recently it has been shown that umami tastants are mediated by the metabotropic glutamate receptor (mGluR4). Binding to this receptor activates α-gustducin, which may increase intracellular $Ca^{2+}$ levels. However, there also may exist ionotropic glutamate receptors, associated to ion channels, located on the tongue. When these receptors are activated by umami tastants, nonselective ion channels open and an influx of $Na^+$ and $Ca^{2+}$ ions occur, depolarizing the cell [30,31]. Interestingly, it is found that less of a different tastant is then required to further depolarize the cell and produce transmitter release [31]. This may explain the traditional use of monosodium glutamate to enhance the taste of foods.

## 1.3 CHEMESTHESIS

In addition to the five basic tastes, there are other sensations that are perceived in the mouth. Spicy hot, cooling, and tingle are all responses that result from chemical sensitivity. Chemesthesis is relatively new term for the sensory input formerly called a trigeminal response. In fact a chemical sensitivity response is carried to the brain during eating not only by the trigeminal nerve (anterior oral cavity, and tongue, nasal cavity, face, and parts of the scalp), but also by the glossopharyngeal (posterior tongue and oral pharynx) and vagus (nasal and oral pharynx) nerves, and thus, a more general term is more appropriate. It is believed that the primary evolution of these nerves is to provide a pain response to high temperature or injury.

Chemesthesis responses result from the chemical irritation of nerve systems that sense heat, cold, or pain. Thus, there is a strong influence of temperature on the intensity of these responses.

### 1.3.1 Chemesthetic Responses

Chemesthetic responses (from a flavor standpoint) are most pronounced on the lips, tongue, and olfactory region (when the stimulant is volatile). In the mouth, these neurons are not on the surface of the tissue but are buried below the surface. Thus, response to stimuli is slow in onset and long lasting. One can become painfully aware of this property when eating a "hot" food. Initially, the heat of the food is not apparent but first the lips become inflamed and then the tongue. Just as the sensation was slow in arriving, its departure is equally slow. There are a very limited number of food components, e.g., capsicums, gingers, radish (and other members of this grouping) and mustards, that elicit this response. Recently this sensation has received much more interest within the food and flavor industries due to a consumer desire for more variety in the diet.

On the tongue, the chemesthetic neurons are located in the papillae and are wrapped around the taste buds. The fungiform papillae, though lacking taste cells, possess chemesthetic neurons. These neurons make use of the structure of the taste bud to form a channel to the tongue surface. It has been reported that these neurons outnumber taste receptors nerves three to one [32]. The chemesthetic neurons are similar to taste receptors in that they have chemical specific receptors sites, however, they differ in that the neurons possess a set of other unique receptors. These receptors include mechanoreceptors for tactile response, thermoreceptors that detect temperature change, proprioceptors that detect motion, and nociceptors that mediate pain (comprise the somatosenses as a whole) [33].

Of these receptor sites, it is the distinct subsets of thermoreceptors in combination with nociceptors that give the sensations of heat and cooling by chemical stimulus in the mouth. In mammals it is proposed that a set of ion channels, called transient receptor potential (TRP) channels, are the primary molecular transducers of thermal stimuli. One such molecular transducer is the vanilliod receptor (VR1) channel, which is an ion-gated channel that is activated by temperatures above 43°C and by chemical irritants, such as, capsaicin and acidic pH [34]. Vanilliod receptor, like protein-1 (VRL1), is a structurally related receptor; however, it is activated at extreme heat (53°C) and does not respond to moderate heat, acid, or capsaicin [34]. Recently researchers have identified and cloned cool thermal receptor channels (CMR1 and TRPM8), which are activated by both cooler temperatures and by l-menthol. When expressed together in cloned cells, CMR1 and TR1 provide the cell with defined thermal response thresholds, activated by combinations of chemical and/or temperature stimulants. It is proposed that the TRP channels are the primary transducers of thermal sensation [35,36].

Certain chemicals in spicy, hot foods excite the neurons responsible for sensing heat (Table 1.1). Because the information is carried by the same nerve fibers, nociceptors, that detect both pain and heat, the brain is tricked to perceive thermal heat and thus in addition to pain, often initiates responses such as sweating and face reddening [34].

An active ingredient in hot foods is capsaicin (trans-8-methyl-N-vanilly-6-nonenamide), common to chillies. A molecular receptor has been found in the chemesthetic nerve endings, which respond to capsaicin, as well as, high temperatures (43°C) and local tissue damage. This receptor is an integral membrane protein and has been labeled vanilloid receptor type-1 VR1, as it is the vanilly group of capsaicin that is thought to interact with VR1. Activation of VR1 causes $Ca^{+2}$ to flow into the nerve ending. This initiates nerve impulses that pass to the brain where they are interpreted as a burning pain [34,38].

The perception of mouth "cooling" is thought to occur in much the same way as capsaicin is sensed as being hot, but in this case, certain chemicals stimulate the

**TABLE 1.1**
**Foods that Elicit a Hot, Spicy Sensation**

| Spice | Flavor Contribution | Character-Impact Compounds | Botanical Characteristic |
|---|---|---|---|
| Red pepper | Intensely pungent, biting, hot, sharp | Capsaicin, dihydrocapsaicin, homocapsaicin | *Capsicumfrutescens,* fruit or pods |
| Black pepper | Pungent, hot, biting | Piperine, piperanine, piperylin | *Piper nigrum,* berries from perennial vine |
| Ginger | Aromatic, pungent, biting | Zingiberene β-sesquiphellandrene | *Zingiber officinale,* rhizome of perennial |
| Horseradish | Irritating, pungent, piquant | Sinigrin, gluconasturtin | *Cochlearia armoracia,* rhizome of perennial |
| Mustard | Slightly acrid, pungent | Allyl isothiocynate, sinigrin, sinalbin | *Brassica hirta,* seeds of an annual plant |
| Onion | Marked pungency, bitter | Propyl disulfide thiophene | *Allium cepa,* bulb of biennial plant |
| Garlic | Pungent, sulfurous | Diallyl disulfide | *Allium sativum,* bulb of biennial plant |

Source: From Nagodawithana, T. *Savory flavors*. Esteekay Associates, Milwaukee, 1995, p. 468.

chemesthetic thermoreceptors that register cold temperatures. For example, the cooling sensation of peppermint, wintergreen, and spearmint arise primarily from the component l-menthol. Menthol, unlike capsaicin (essentially nonvolatile), is both volatile and oil soluble. The solubility of this compound allows it permeate the tongue where it activates cooling. The cooling of l-menthol also depends on the volatility of this compound. There is noticeably less of a cooling effect of l-menthol in your mouth if it is closed; however, if you breathe in through your mouth evaporation greatly enhances the cooling sensation. The cooling effect of l-menthol is also concentration dependent. At higher concentrations, l-menthol will stimulate nociceptors, producing an irritating, biting, and even burning effect in the mouth [33,39].

Recently, studies on cloned sensory neurons have provided some understanding of the physiology of the mouth cooling sensation. Both cooler temperatures (between 23–10°C) and l-menthol activate the cloned neuron, TRPM8, and there is evidence that TRPM8 neurons are distinct but related to the heat and pain sensing neurons VR1 and VRL1. Similarly, it was found in another cloned thermoreceptor (cold-and menthol-sensitive receptor, CMR1) that the molecular site for temperature (28–8°C) and l-menthol action is an excitatory ion channel in the trigeminal nerve roots. This demonstrates that menthol elicits a cool sensation by acting as a chemical irritant to a thermally responsive receptor [35]. The nerve impulse activity during temperature cooling and l-menthol application was found to be the same. It has been found that temperature and l-menthol activate an inward ionic current that induces intercellular depolarization [36]. This current is mediated by flux of $Ca^{+}2$ into the cell, and like taste transduction, stimulates the firing of the receptor nerve [40,41]. Menthol is not the only compound to provide cooling. There are many other compounds that are mouth cooling (some newly discovered); these compounds and the relationships to l-menthol are described in Table 1.2 [42].

## 1.3.2 Tactile Response

The chemesthetic neurons also mediate tactile responses. The distinction between a chemical sense and tactile sense often overlaps. For example tannins in foods are a chemical stimuli, but they produce the tactile response of astringency. Tannins give a dry rough feel in the mouth and can cause a tightening effect in the cheeks and facial muscles (puckering). Although tannins are definitely chemicals that give tactile sensations, most connoisseurs of wine would argue astringency is a defining character of wine flavor.

The mechanisms giving rise to astringency are poorly understood. For astringency imparted by tannins in wine, a long-standing popular theory has it that tannins bind to salivary proteins and mucopolysaccharides (the slippery components of saliva), causing them to aggregate or precipitate, thus robbing saliva of its ability to coat and lubricate oral tissues. This causes the rough and dry sensations in the mouth, even when there is fluid in the mouth. However, many acids are often described as more astringent than sour. It has been recently shown that astringency produced by acids is actually caused by promoting interactions between residual salivary phenols and salivary proline rich proteins. Acids without the presence phenols in saliva do not produce astringency.

**TABLE 1.2**
**Examples of Other Cooling Sensate Chemicals**

**3,6-Dihydro-1-(2-hydroxyphenyl)-4-(-3-nitrophenyl)-2(*1H*)-pyrimidinone.** (AKA: Icilin, AG 3 5, CAS No. 36945-98-9, MW 311.3) 200 times more cooling than menthol.[a]

**4-Methyl-3-(1-pyrrolidinyl)-2[5H]-furanone.** 35 times more powerful in the mouth, and 512 times more powerful on the skin than menthol, the active ingredient in mint, and the cooling effect lasts twice as long.[b]

**2-Isopropyl-N,2,3-trimethylbutyramide** (AKA: N,2,3-Trimethyl-2-Isopropyl Butamide, WS-23 by Millennium Chemicals, Inc., CAS No. 51115-67-4, FEMA No. 3804, Molecular Formula: $C_{10}H_{21}NO$, MW 171.29). WS-23 is an almost odorless white powder. It is characterized by a high cooling activity with no side effects such as burning, stinging, or tingling sensations. Typical applications include use as a coolant in medicinal preparations, oral care products, and confectionery products.

**N-Ethyl-p-menthane-3-carboxamide** (AKA: N-Ethyl-5-Methyl-2-(1-Methylethyl)-Cyclohexane carboxamide, WS-3 by Millennium Chemicals, Inc., CAS No. 39711-79-0, FEMA No. 3455, $C_{13}H_{25}NO$, MW 211). Cooling, minty, medicinal. Available in white crystalline form, is almost odorless. Its chief use is as a coolant in medicinal preparations, oral care products, and confectionery products. Five times more cooling than menthol in the mouth.

**p-Menthane-3,8-diols, cis and trans** (PMD38 by Takasago International, $C_9H_{18}O_3$, MW 174.24) have a cooling power approximately 9.5 times that of menthol.[c]

**1,4-Dioxaspiro[4,5]decane-2-menthanol, 9-methyl-6-(1-methylethyl)-** (AKA: 6-Isopropyl-9-methyl-1,4-dioxaspiro(4,5)decane-2-methanol, I-Menthone glycerol ketal, Menthone glycerin acetal, Frescolat MGA by Haarmann & Reimer, FEMA No. 3807, CAS No. 63187-91-7, $C_{13}H_{24}O_3$, MW 228.4). Faint, cool-minty, virtually tasteless, long-lasting cooling effect.

**Propanoic acid, 2-hydroxy-5-methyl-2-(1-methylethyl) cyclohexyl ester, [1R-[1 alpha(R*), 2 beta, 5 alpha]]-** (AKA: 5-methyl-2-(1-methylethyl)-cyclohexyl-2-hydroxypropionate, (l)-Menthyl lactate, l-menthyl lactatic acid-menthyl ester, Frescolat ML by Haarmann & Reimer, CAS No. 59259-38-0, FEMA No. 3748, $C_{13}H_{24}O_3$, MW 228.4) is faintly minty in odor and virtually tasteless, with a pleasant, long-lasting cooling effect.[d]

[a] From Ottinger, H., T. Soldo, T. Hoffmann, *J. Agric. Food Chem.*, 2001(49): p. 5383. With permission.

[b] From Kenmochi, H., T. Akiyama, Y. Yuasa, T. Kobayashi, A. Tachikawa, Method for producing para-menthane-3,8-diol, United States Patent 5959161, 1999, Takasago International Corporation. With permission.

[c] From Grub, H., R. Pelzer, R. Hopp, R. Emberger, H.-J. Bertram, Compositions which have a physiological cooling effect, and active compounds suitable for these compositions, U.S. Patent 5266592, 1993, Haarmann & Reimer GmbH. With permission.

[d] From Maarse, H., C.A. Visscher, L.C. Willemsens, L.M. Nijssen, M.H. Boelens, TNO Nutrition and Food Research, Zeist, The Netherlands, 1994.

*Source:* From Leffingwell, J.C., *Cool without Menthol & Cooler than Menthol.* Leffingwell & Associates, http://www.leffingwell.com/cooler_than_menthol.htm, 2001.

## 1.4 OLFACTION

Olfaction is the sensory component resulting from the interaction of volatile food components with olfactory receptors in the nasal cavity. We generally speak of the aroma or odor of a food. The stimulus for this sensation can be orthonasal, (the odor stimulus enters the olfactory region directly from the nose as one sniffs a food), or retronasal (the odor stimulus enters from the oral cavity as one eats a food).

Aroma is a very complex sensation. While the stimuli available to create taste sensations is limited, more than 7,100 volatile compounds have been identified in foods overall [46–48] each of which may *potentially* contribute to aroma perception, depending upon their concentrations and sensory thresholds. Some of the more complex food aromas, e.g., a thermally processed food such as coffee, may contain over 800 volatile components. Fortunately, the aroma character of most foods can typically be defined by a smaller subset of the total volatile profile.

The human being is exceptionally sensitive to some volatiles (e.g., 2-isobutyl-3-methoxypyrazine has an odor detection threshold in water of 0.002 ppb [49] and 0.015 ppb in wine [50] but insensitive to many other volatiles (e.g., ethanol has an odor threshold of 100,000 ppb in water and a taste threshold of 52,000 ppb in water) [49]. A person's ability to detect odors is also influenced by many other factors such as genetic variability, olfactory fatigue, and naturally occurring and unpredictable factors such as temperature and humidity. The complexity of food aromas and sensitivity required plus the fact that the olfactory system must be able to respond to unknown odorants (it cannot be learned response) make this a most complex phenomenon.

### 1.4.1 Anatomy of Olfaction

It is helpful to understand how odorants reach the olfactory neurons in humans since this mechanism may be partially responsible for determining aspects of odor perception. The olfactory neurons form a neuroepithelium that line protrusions (turbinates) in the lateral walls of the nasal cavity. The turbinates are a series of folds made of bony lateral extensions. In humans, the majority of olfactory epithelium is located on the olfactory cleft which is directly linked to the olfactory bulb in the brain by the cribiform plate (Figure 1.4). Approximately 6 million neurons pass through the cribiform plate to about 8,000 glomeruli in each olfactory bulb. This forms a direct connection between the olfactory receptors and the brain.

In the olfactory epithelium, small appendages protrude outward from the body. At the end of these appendages are located bulb-like structures containing 20 to 30 very fine cilia containing the olfactory receptors and the sensing transduction mechanism. These cilia lie in a layer of mucus covering the tissue that protects the tissue from drying and affords protection of the brain against microbial attack through the open channels where olfactory signals pass to the glomeruli.

### 1.4.2 Odor Receptor Functioning

The initial step in olfaction is the binding of an odorant to an odor binding protein (OBP). This step is essential since most odorants are hydrophobic in nature, and they could not otherwise pass through a polar mucus membrane to reach the olfactory receptors (OR). Thus, odor molecules are bound to the OBP and either simply solubilized by the OBP or perhaps actively transported to the OR by the OBP. Once at the OR, the odorant may be released to interact with the OR or the odorant-OBP complex is sensed by the OR — this has not been determined (Figure 1.5) [52]. The

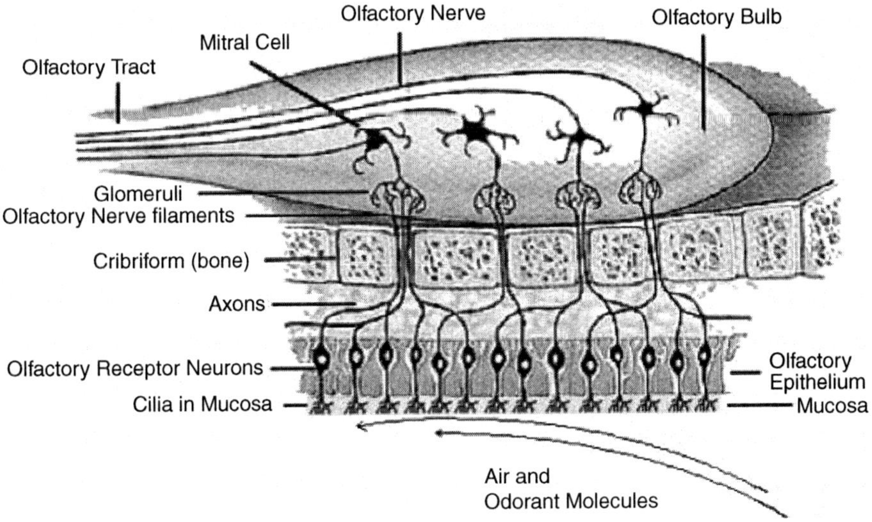


**FIGURE 1.4** Anatomy of human olfactory system. (From Leffingwell J.D., *Olfaction — A Review*. http://www.leffingwell.com/olfaction.htm. 2004. With permission.)

OR is a G-protein-coupled receptor (GPCR), which are a common form of receptors in the body that communicate between the cell and its environment. The GPCR proteins are membrane proteins that have three units, $\alpha$, $\beta$, and $\gamma$, each involved in the communication process. The process involves several enzymatically catalyzed processes that are initiated by an interaction of the odorant molecule with the OR and ultimately produce an influx of $Ca^{2+}$ that depolarizes the cells resulting in an electrical nerve signal that is transmitted to the brain for processing. Several of the reception and transduction steps amplify the signal both in the OR and in the olfactory bulb to result in exceptional sensitivity to the chemical stimuli. Pernollet and Briand [52] have described this process in detail in the figure provided.

### 1.4.3 Signal Encoding

There is little specificity in the interaction of OBP and odorants. The OBPs tend to be reasonably general in their ability to transit odorants. The odor selectivity comes primarily from the ca. 1,000 OR proteins. However OR are not absolutely specific in response but a single OR can generally respond to several odorants which are typically structurally related molecules. Also, different odorants are recognized by a unique combination of OR and the same odorant can be recognized by several different OR [53]. It is likely that different odor receptors may recognize different elements of an odorant thus providing a stereoscopic image of an odorant. The binding strength and perhaps signal intensity may be related to the portion of the molecule being bound. This patterning of OR stimulation by an odorant, or mixture of odorants, results in a coding of odor quality. This coding is transmitted to the brain where pattern recognition takes place. Ultimately, the brain considers all of

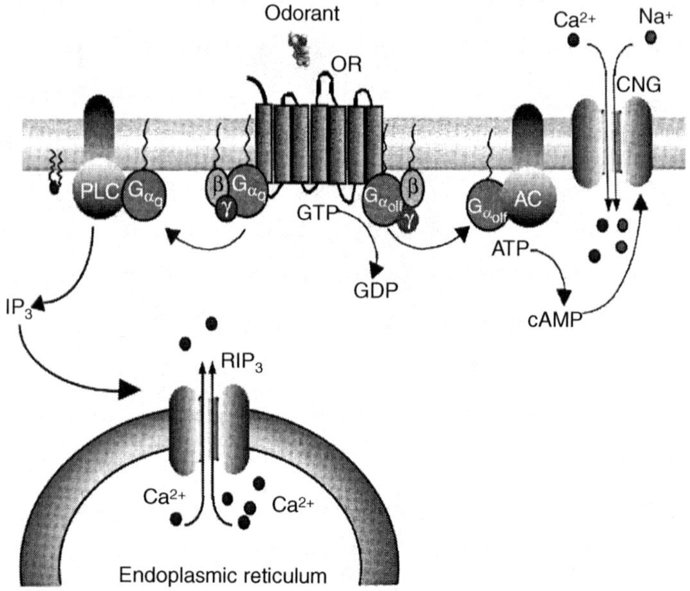


**FIGURE 1.5** Olfactory transduction cascade. Within the cilia of olfactory neurons, a cascade of enzymatic activities transduces odorant binding onto an OR into a calcium influx, which leads to an electrical neuronal signal that can be transmitted to the brain. G-proteins are composed of three distinct subunits called α, β, and γ. Golf$_\alpha$ is the olfactory G$_\alpha$, which, in inactive state, binds guanosine diphosphate (GDP). Upon OR activation, Golf$_\alpha$ binds GTP and activates the enzyme called adenylyl cyclase (AC) that catalyses the formation of 3′,5′-cyclic adenosine-5′-mono-phosphate (cAMP) from adenosine triphosphate (ATP). A cyclic nucleotide-gated (CNG) channel is then opened by cAMP, allowing cation entrance into the cytoplasm. Golf$_\alpha$ hydrolyzes guanosine triphosphate (GTP) into GDP and goes back to inactive state. Alternatively, an inositol-(1,4,5)-triphosphate ($IP_3$) pathway has been described in olfactory neurons. When the Gq α-subunit is activated, it stimulates the enzyme called phospholipase C (PLC) that cleaves phosphatidylinositol-4,5-bisphosphate (PIP) in the cell membrane to release $IP_3$. $IP_3$ diffuses into the cytosol and binds to $IP_3$ receptors ($RIP_3$) located in the endoplasmic reticulum membrane, which permit subsequent $Ca^{2+}$ ion release into the cytoplasm. Arrows indicate stimulatory pathways. (From Pernollet, J.-C. L. Briand, *Flavor Perception*, A.J. Taylor, D.D. Roberts, Eds., Blackwell Publ., Ames, 2004, p. 86. With permission.)

the stimuli associated with eating a food and forms a unique perception of the flavor just experienced.

Pernollet and Rawson [52] have presented a mechanistic rationale for understanding how an odorant can change in odor character as a function of odorant concentration. Figure 1.6 illustrates this, and the legend explains this event in detail. A similar rationale is presented for simply an increase in intensity of an odorant with concentration (no change in odor character): perhaps the combination of OR responses at higher concentrations does not result in a change in character but only an increase in perceived intensity.

Rawson and Li [5] and Schaefer et al. [54,55] point out that there is no simple relationship between the number of odorants in a mixture and the number of GL

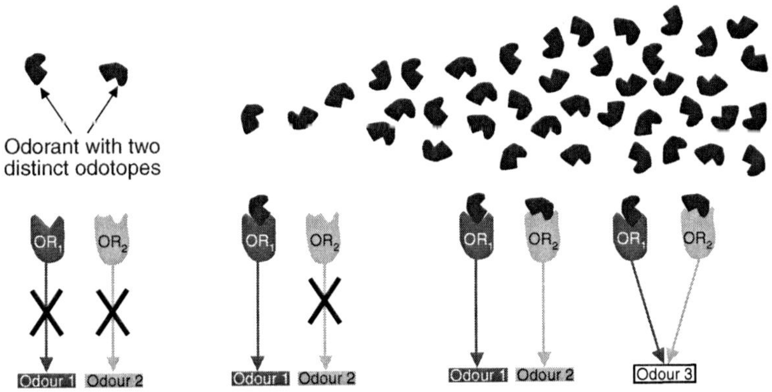


**FIGURE 1.6** Hypothetical explanation of odor change as a function of odorant concentration. (a) A single odorant as two distinct odotopes, which can independently bind two distinct ORs, OR1 and OR2 (coding for Odor 1 and Odor 2, respectively). (b) Supposing that the binding affinity is greater for OR1 than for OR2, then, at low odorant concentration, only OR1 is activated, generating the smell Odor 1; at higher odorant concentration, OR2 also binds the odorant so that either (c) a second distinct smell is perceived or, alternatively, (d) the combinatorial signal generated in the olfactory bulb gives rise to another smell Odor 3. Crosses on arrows indicate neuronal pathway deactivation. (From Pernollet, J.-C. L. Briand, *Flavor Perception*, A.J. Taylor, D.D. Roberts, Eds., Blackwell Publ., Ames, 2004, p. 86. With permission.)

activated on smelling of the odorant mixture. It appears that mixtures of odorants result in a unique pattern of GL stimulation that does not include all odorants present in the mixture. Thus, mixtures are not the sum of its parts but a new entity. This hypothesis is strongly supported by the work of Jinks and Laing [56] demonstrating that humans are very poor at identifying individual components of a mixture. This is particularly fascinating in the sense that flavorists must smell an odorant mixture (e.g., a strawberry) and be able to select components that can be used to recreate that odor (create a strawberry flavor). This ability is counterintuitive and must be a learned ability as opposed to an innate ability. Flavorists typically must work several years learning the odors of individual aroma chemicals and how they might be used in mixtures to become accomplished at their task (see Chapter 12).

In this text, we will spend substantially more time on the aroma of foods than taste or chemesthetic responses. This is largely due to historical reasons. In the past, the flavor industry and those academicians traditionally considered to be flavor chemists have focused their efforts on the aroma sensation. The flavor industry has traditionally sold the food industry mixtures of volatile constituents that characterize this component of flavor. In some instances, elements of bitterness and occasionally umami (savory flavorings) have also been supplied by the flavor industry. The food industry has added the other components of flavor, e.g., sweeteners, acidulants, and salts. Thus, discussions of the flavor industry have largely ignored much of taste. Only recently has the flavor industry become involved in selling more complete flavor profiles due to opportunities in the marketplace.

Similarly to their industrial counterparts, flavor chemists in academic settings have also largely focused their efforts on volatile compounds in foods. Numerous international meetings have been, and continue to be, held that focus on the identification and characterization of volatiles in foods. With few exceptions, research on taste components has been left to sensory scientists or less focused groups. Thus, a rather well-organized group of flavor chemists has provided substantial literature on the volatiles in foods but not expended similar effort on the taste (generally non-volatile) components of foods. Unfortunately, there is much less literature to draw upon for taste and chemesthetic responses than aroma. However, both the academic and industrial communities are beginning to recognize the importance of taste and chemesthetic contributions to overall flavor perception, and activities in these areas are increasing.

## 1.5 SUMMARY

In recent years, we have learned a great deal about how chemicals in foods are detected and translated into nerve signals, and ultimately produce a perceptual response of flavor during eating. At the present, much of this learning is of academic interest with little immediate application to the food and flavor industries. Yet this knowledge may hold the future to truly understanding the process of flavor perception as well as differences in perception (liking) amongst the population (genetic control of perception mechanisms).

At one time, it was thought that flavor could be reproduced by simply determining the volatile components of a food and then reconstituting these volatiles. This lead to long laundry lists of volatile components in foods but has contributed little to solving some of the most vexing flavor problems of the industry. Flavor research has progressively moved from one milestone to another (laundry lists → key volatile components → aroma release/interactions → taste/odor interactions → perception → ?) without still being able to overcome some issues that plague the field such as producing good quality reduced calorie (or fat) foods. It is likely possible that our ability to recreate natural foods/flavors, or make ingredient/processing substitutions without flavor change may be addressed by our understanding of perception as a whole. Creating a perception will only be possible by understanding the stimuli required to form a perception and ultimately delivering these stimuli appropriately.

This book is devoted to an overview of academic and industrial aspects of flavor chemistry and technology as we know it today. It cannot be comprehensive due to the breadth of the field, but it has been thoroughly referenced to permit the reader to find greater detail in each topic area.

## REFERENCES

1. Keast, R.S.J., P.H. Dalton, P.A.S. Breslin, Flavor interactions at the sensory level, in *Flavor Perception*, A.J. Taylor, D. Roberts, Eds., Blackwell Publ., Ames, 2004, p. 228.
2. Taylor, A., Measuring proximal stimuli in flavour perception, in *Flavor Perception*, A.J. Taylor, D.D. Roberts, Eds., Blackwell Publ., Ames, 2004, p. 1.

3. Taylor, A.J., D.D. Roberts, Eds. *Flavor Perception.* Blackwell Publ., Ames, 2004, p. 283.
4. Lindermann, B., Receptors and transduction in taste, *Nature*, 2001(413): p. 219.
5. Rawson, N.E., X. Li, The cellular basis of flavour perception: taste and aroma, in *Flavor Perception*, A.J. Taylor, D.D. Roberts, Eds., Blackwell Publ., Ames, 2004, p. 57.
6. Mann, N.M., Management of smell and taste problems, *Cleveland Clinic J. Med.*, 2002(69): p. 329.
7. Smith, D.V., R.F. Margolskee, Making sense of taste, *Sci. Amer.*, 2001(284): p.32.
8. McGarrick, J., *Taste & Smell,* http://www.umds.ac.uk/physiology/jim. Accessed April 25, 2005.
9. Linden, R.W.A., Taste, in *Clinical Oral Science*, Harris M, M. Edgar, M.S. Wright, Eds., Oxford Publ., Oxford, 1998, p. 167.
10. Hoon, M.A., E. Adler, J. Lindemeier, J.F. Battey, N.J. Ryba, C.S. Zuker, Putative mammalian taste receptors: a class of taste-specific GPCRs with distinct topographic selectivity, *Cell*, 1999(96): p. 541.
11. Hänig, D.P., ZurPsychophysik des Geschmackssinnes, *Phil. Stud.*, 1901(17): p. 576.
12. Erickson, R.P., The evolution of neural coding ideas in the chemical senses, *Physiol Behav.*, 2000, 69(1–2): p. 3.
13. Smith, D.V., S.J. St. John, Neural coding of gustatory information, *Curr. Opinion Neurobiol.* (1999), 9(4), p. 427.
14. Gilbertson, T.A., S. Damak, R.F. Margolskee, The molecular physiology of taste transduction, *Curr. Opinion Neurobiol.*, 2000(10): p. 519.
15. Adler, E., M.A. Hoon, K.L. Mueller, J. Chandrashekar, N.J.P. Ryba, C.S. Zuker, A Novel Family of Mammalian Taste Receptors, *Cell*, 2000(100): p. 693.
16. Heck, G.L., S. Mierson, J.A. Desimone, Salt taste transduction occurs through an amiloride-sensitive sodium transport pathway, *Science*, 1984, 223(403).
17. Avenet, P., F. Hofmann, B. Lindemann, Transduction in taste receptor cells requires cAMP-dependent protein kinase, *Nature*, 1988(331): p. 351.
18. Lin, W., T.E. Finger, B.C. Rossier, S.C. Kinnamon, Epithelial Na+ channel subunits in rat taste cells: localization and regulation by aldosterone, *J. Comp. Neurol.,* 1999, 405/406.
19. Smith, D.V., C.A. Ossebaard, Amiloride suppression of the taste intensity of sodium chloride: evidence from direct magnitude scaling, *Physiol Behav.*, 1995(57): p. 773.
20. Kinnamon, S.C., V.E. Dionne, K.G. Beam, Apical localization of K channels in taste cells provides the basis for sour taste transduction, *Proc. Natl. Acad. Sci.*(USA), 1988(85): p. 7023.
21. Gilbertson, T.A., P. Avenet, S.C. Kinnamon, S.D. Roper, Proton currents through amiloride-sensitive Na channels in hamster taste cells: role in acid transduction, *J. Gen. Physiol.*, 1992(100): p. 803.
22. Nelson, G., M.A. Hoon, J. Chandrashekar, Y. Zhang, N.J.P. Ryba, C.S. Zuker, Mammalian sweet taste receptors, *Cell*, 2001(106): p. 381.
23. Kawai, K., K. Sugimoto, K. Nakashima, H. Miura,Y. Ninomiya, Leptin as a modulator of sweet taste sensitivities in mice, *Proc. Natl. Acad. Sci.*, 2000(97): p. 11044.
24. Ninomiya, Y., N. Shigemura, R. Ohta, K. Yasumatsu, K. Sugimoto, K. Nakashima, B. Lindemann, Leptin and sweet taste, *Vitam. Horm.*, 2002(64): p. 221.
25. Matsunami, H., J.P. Montmayeur, L. Buck, A family of candidate taste receptors in human and mouse, *Nature*, 2000(404): p. 601.
26. Spielman, A.L., T. Huque, H. Nagai, G. Whitney, J.G. Brand, Generation of inositol phosphates in bitter taste transduction, *Physiol Behav.*, 1994(56): p. 1149.

27. Spielman, A.I., H. Nagai, G. Sunavala, M. Dasso, H. Breer, I. Boekhoff, T. Huque, G. Whithey, J.G. Brand, Rapid kinetics of second messenger production in bitter taste, *Amer. J. Cell Physiol.*, 1996(270): p. 926.
28. Naim, M., R. Seifert, B. Nürnberg, L. Grünbaum, G. Schultz, Some taste substances are direct activators of G-proteins, *Biochem. J.*, 1994(927): p. 451.
29. Ikeda, K., On a new seasoning, *J. Tokyo Chem. Soc.* 30(820): p. 1909.
30. Lin, W., S.C. Kinnamon, Physiological evidence for ionotropic and metabotropic glutamate receptors in rat taste cells, *J. Neurophysiol.*, 1999(82): p. 2061.
31. Chaudhari, N., A.M. Landin, S.D. Roper, A metabotropic glutamate receptor variant functions as a taste receptor, *Nat. Neurosci.*, 2000(3): p. 113.
32. Lawless, H.T. *Trigeminal Response Page*, http://zingerone.foodsci.cornell.edu/trigeminal/1996. Accessed October 30, 2004.
33. Decker, J.D., *Packing punch with pungency.* Food Product Design http://www.foodproductdesign.com/archive/1998/0998AP.html, September 1998. Accessed September 15, 2004.
34. Caterina, M.J., M.A. Schumacher, M. Tominaga, T.A. Rosen, J.D. Levine, D. Julius, The capsaicin receptor: a heat-activated ion channel in the pain pathway, *Nature*, 1997(389): p. 816.
35. McKemy, D.D., W.M. Neuhausser, D. Julius, Identification of a cold receptor reveals a general role for TRP channels in thermosensation, *Nature*, 2002(416): p. 52.
36. Peier, A.M., A. Moqrich, A.C. Hergarden, A.J. Reeve, D.A. Andersson, G.M. Story, T.J. Earley, I. Dragoni, P. Mcintyre, S. Bevan, A. Patapoutian, A TRP channel that senses cold stimuli and menthol, *Cell*, 2002, 108(5): p. 705.
37. Nagodawithana, T. *Savory flavors.* Esteekay Associates, Milwaukee, 1995, p. 468.
38. Clapham, D.E., Some like it hot: spicing up ion channels. *Nature*, 1997(389): p. 783.
39. Leffingwell, J.C., *Cool without Menthol & Cooler than Menthol.* Leffingwell & Associates, http://www. leffingwell.com/cooler_than_menthol.htm, 2001. Accessed March 18, 2004.
40. Okazawa, M., T. Terauchi, T. Shiraki, K. Matsumura, S. Kobayashi, l-menthol-induced [$Ca^{2+}$] increase and impulses in cultured sensory neurons, *Neuro Report*, 2000(11): p. 2151.
41 Reid, G., M.L. Flonta, Cold current in thermoreceptive neurons, *Nature*, 2001(413): p. 480.
42. Wei, E.T., D.A. Seid, AG-3-5: a chemical producing sensations of cold, *J. Pharm. Pharmacol.*, 1983(35): p. 110.
43. Ottinger, H., T. Soldo, T. Hoffmann, Systematic studies on structure and physiological activity of cyclic alpha-keto enamines, a novel class of cooling compounds, *J. Agric. Food Chem.*, 2001(49): p. 5383.
44. Kenmochi, H., T. Akiyama, Y. Yuasa, T. Kobayashi, A. Tachikawa, Method for producing para-menthane-3,8-diol, U.S. Patent 5959161, 1999. Takasago International Corporation.
45. Grub, H., R. Pelzer, R. Hopp, R. Emberger, H.-J. Bertram, Compositions which have a physiological cooling effect, and active compounds suitable for these compositions, U.S. Patent 5266592, 1993, Haarmann & Reimer GmbH.
46. Maarse, H., C.A. Visscher, L.C. Willemsens, L.M. Nijssen, M.H. Boelens, *Volatile Compounds in Food: Qualitative and Quantitative Data,* TNO Nutrition and Food Research, Zeist, The Netherlands, 1994.
47. Nijssen, L.M., C.A. Visscher, H. Maarse, L.C. Willemsens, M.H. Boelens, *Volatile Compounds in Food-Qualitative and Quantitative Data,* 7th ed., TNO Nutrition and Food Research Institute, Zeist, The Netherlands, 1996.

48. M.H. Boelens, *Volatile Compounds in Food-Database (from Volatile Compounds in Food, Qualitative and Quantitative Data,* 7th ed., (1996), Supplement 1 (1997) and Supplement 2 (1999), TNO Nutrition and Food Research Institute, Zeist, The Netherlands, 2000.
49. Leffingwell & Associates, 1998., *Flavor-Base,* http://www.leffingwell.com/odorthre.htm, Canton, GA. 1998. Accessed March 24, 2004.
50. Roujou De Boubee, D., C.V. Leeuwen, D. Dubourdieu, Organoleptic impact of 2-methoxy-3-isobutylpyrazine on red Bordeaux and Loire wines: effect of environmental conditions on concentrations in grapes during ripening, *J. Agric. Food Chem.*, 2000(48): p. 4830.
51. Leffingwell J.D., *Olfaction — A Review*, http://www.leffingwell.com/olfaction.htm. 2004. Accessed April 25, 2005.
52. Pernollet, J.-C. L. Briand, Structural recognition between odorants, olfactory-binding proteins and olfactory receptors — first events in odour coding, in *Flavor Perception*, A.J. Taylor, D.D. Roberts, Eds., Blackwell Publ., Ames, 2004, p. 86.
53. Malnic, B., J. Hirono, T. Sato, L.B. Buck, Combinatorial receptor codes for odours, *Cell*, 1999(96): p. 713.
54. Schaefer, M.L., D.A. Young, D. Restrepo, Olfactory fingerprints for major histocompatibility complex-determined body odors, *J. Neurosci.*, 2001(21): p. 2481.
55. Schaefer, M.L., K. Yamazaki, K. Osada, D. Restrepo, G.K. Beauchamp, Olfactory fingerprints for major histocompatibility complex-determined body odors II: relationship among odor maps, genetics, odor composition, and behavior, *J. Neurosci.*, 2002(22): p. 9513.
56. Jinks, A., D.G. Laing, A limit in the processing of components in odour mixtures, *Perception*, 1999(28): p. 395.

# 2 Flavor and the Information Age

## 2.1 INTRODUCTION

As was mentioned in the introduction, this text cannot possibly cover the field of flavor in detail. Thus, it was deemed useful to provide a section of this text detailing how to find information on flavors. Certainly, the information age is revolutionizing this field as all others.

## 2.2 HISTORY OF FLAVOR LITERATURE

Historically, there was little literature in the public domain on food flavors until the mid-1970s. While flavor research has existed for well over 100 years, the vast majority of early flavor research was done within companies and held secret. The industry generally did not even disclose findings via patents but chose to hold findings as trade secrets, which is still common today. Ernest Guenther was one of the first individuals to publish a significant work in the public domain. He published a six-volume collection on essential oils [1]. At the time of the publication, most flavors were based on essential oils, and this collection was regarded as "the" source of information on flavorings. It is still highly regarded and available in print today [2], sold as a hardcover, six-volume set.

Arctander [3,4] published a classic work on perfume and flavoring materials that is also still available today [5]. Merory [6,7] published a book that largely presented formulae for the creation of flavorings. While these flavorings are primitive by today's standards, this book provided a view of how flavor companies created flavorings. The next significant work was that of Heath [8,9]. Heath provided a reference book on the activities of a flavor company overall, and this was the first comprehensive book on the functioning of the flavor industry.

The final historical books to be singled out are those of Fenaroli [10,11]. This offering has been updated by Burdock [12,13] and is a valued source. The current three-volume set [13] provides information on: sensory thresholds, molecular structure, empirical formula/MW, specifications, natural occurrence, synthesis, consumption, uses in foods, and regulations/guidelines related to the use of aroma chemicals.

The above-mentioned texts have been singled out because they were written by single authors and notably, individuals in the industry. As we look at later publications in this field (next section), books became edited as opposed to authored (this has implications in coherency and breadth), and content is dominated by academics as opposed to the industry researchers. To a large extent, the industry has continued to work in privacy and has offered limited, significant literature to the public domain (there are some notable exceptions).

As mentioned earlier, the flavor literature began to grown quickly in the early 1970s. Over a ten-year period (mid-1960s to mid-1970s), academic and government research laboratories became fascinated with this field, and numerous research groups came into existence. These groups were spread across the world with the largest number in Europe, the U.S., and Japan. The groups organized meetings and published the proceedings of these meetings in book form. The Weurman Symposium (1975) was one of the first such symposia to be held and is widely respected today as the premier flavor meeting (held in Europe every three to four years). Other national and international meetings focusing on flavor came into existence (e.g., Greek [Charlambous] and German [Wartburg symposia]). Today, the American Chemical Society (Ag and Food Division) is the U.S. home for the dissemination of flavor research (as it is defined in this book) in meeting (ACS national meetings), journal (*J. Agric. Food Chem.*) and book (ACS books) forms. A key point however, is that the literature changed from primarily single authored to edited volumes, and the work became that of the academic community as opposed to industry. It is this author's opinion that meetings and subsequent symposium proceedings are good mechanisms to disseminate current, original research, but they are less suitable to provide a basis for teaching or obtaining comprehensive coverage of a topic. Thus, in this author's view, the field suffers from an overabundance of academic publications in the flavor field and too few contributions by individuals with experience in the industry. We acknowledge the works of Ashurst [14–16], Saxby [17,18], Ziegler and Ziegler [19] and DeRovia [20] as good quality, recent, industry authored contributions.

There have been few textbooks offered in this area. Heath and Reineccius [21] authored a text (the precursor of this one) combining a view of both academic and industry activities and Taylor [22] has recently edited a textbook that is also a cross section of both segments of the field. (See Table 2.1 for books.)

## 2.3 JOURNALS

A second source of information in the flavor field is journals: scientific as well as trade journals. While flavor research appears in many scientific journals, journals that are particularly well respected and focus on flavor (by this text's definition) include the *J. of Agricultural and Food Chemistry* (American Chemical Society), *Flavour and Fragrance J.* (John Wiley and Sons) and *Perfumer and Flavorist* (Allured Pub. Corp.). As noted, many other journals publish flavor research, but these journals are typically general in scope (e.g., *Food Chemistry, Z. Lebens. Unters. Forschung, J. Science Food Agriculture* and *J. Food Science*) or focus on a commodity/discipline (e.g., *Food Engineering, Cereal Chemistry, J. Dairy Science*, and *Lipid Chemistry*) where flavor may also be relevant. The majority of these journals are now online, so retrieval of articles is simple if one's library/company has a subscription to this service.

Today trade journals abound, and flavor is a common topic (e.g., *Snack Food & Wholesale Bakery, New Products, Prepared Foods, Food Processing, Food Technology,*

**TABLE 2.1**
**Classical References in the Flavor Field**

| Author | Year | Title | Publisher(s) |
|---|---|---|---|
| Guenther, E. | 1948–52 | *The Essential Oils* (6 vols.) | D. Van Nostrand Co., New York, 1948–1952. Reprinted by Allured Pub. Corp, Carol Stream, IL, 1998. |
| Arctander, S. | 1969 | *Perfume and Flavor Materials of Natural Origin* (1960) and *Perfume and Flavor Chemicals* (1969) | Published by the author, Elizabeth, N.J., 1960. Steffen Arctander's Publications, Las Vegas, NV, 1969. Reprinted and in CD-ROM by Allured Pub., Carol Stream, IL, 2000. |
| Merory, J. | 1960, 1968 | *Food Flavorings: Composition, Manufacture and Use* | AVI Pub. Co., Westport, CT. Currently out of print. |
| Fenaroli, G. | 1971, 1975, 1994, 2001 | Fenaroli, G. *Fenaroli's Handbook of Flavor Ingredients*; Furia, T. E. and Bellanca, N., Eds. | Chemical Rubber Company, Cleveland, OH, 1971, 2 vols. Revised and edited by Burdock, G.A. *Fenaroli's Handbook of Flavor Ingredients,* 4th ed., 3 vols., CRC Press, Boca Raton, FL, 2001. |
| Heath, H.B. | 1978 | *Flavor Technology: Profiles Products Applications* | AVI Pub. Co., Westport, CT. |
| | 1981 | *Source Book of Flavors* | AVI Pub. Co., Westport, CT. Revised and edited by G.A. Reineccius *Source Book of Flavors*, 2nd ed., Chapman & Hall, New York, 1994. |

and *Cereal Foods World*). Based on numbers available, it appears that trade journals are a profitable business. Some of these journals are free to professionals in the field being heavily supported by advertising. There is little published in these journals that one would consider to be research in nature but often interesting information is presented on new products or market trends. A special category of trade journals is that produced by a flavor company to show off expertise or inform potential customers of new products. Examples of flavor companies that produce a publication for the trade include Givaudan Inc. *Inspire*, Sigma-Aldrich *Fine Chemicals Quarterly*, Hasegawa U.S. Inc. *Hasagawa Letter*, and a newsletter from Symrise.

## 2.4 PROFESSIONAL SOCIETIES

The last traditional source of information in the flavor area is various professional societies (synonymous with association or organization). A list of the U.S. based societies that have flavor interests is presented in Table 2.2 (note that organizations with similar charters exist in many other countries of the world but could not be listed in this text due to the number). These societies have wide-ranging ties to the flavor field. They may be involved in training or accrediting members of the profession

**TABLE 2.2**
**Professional Societies with Flavor Interests**

| | | |
|---|---|---|
| American Association of Cereal Chemists (AACC)<br>3340 Pilot Knob Road<br>St. Paul, MN 55121–2097<br>www.aacc.org | American Chemical Society (ACS)<br>1155 16th Street N.W.<br>Washington, D.C. 20036–4800<br>www.acs.org | American Chemical Society (ACS)<br>Flavor Subdivision<br>Cynthia Mussinan<br>c/o IFF<br>1515 State Highway 36<br>Union Beach, NJ 07735–3542 |
| American Oil Chemists' Society (AOCS)<br>2211 W. Bradley Ave.,<br>Champaign, Illinois, 61821–1827<br>www.aocs.org | American Pharmaceutical Association (AphA)<br>2215 Constitution Avenue N.W.<br>Washington, DC 20037–2985<br>www.aphanet.org | American Society of Perfumers<br>P.O. Box 1551<br>West Caldwell, NJ 07004–1551<br>www.perfumers.org |
| American Spice Trade Association (ASTA)<br>2025 M Street, NW Suite 800<br>Washington, DC 20036–3309<br>www.astaspice.org | Association for Chemoreception Sciences (AchemS)<br>c/o Panacea Associates<br>744 Duparc Circle<br>Tallahassee, FL 32312–1409<br>www.achems.org | Diane Davis<br>Chemical Sources Association (CSA)<br>3301 Route 66<br>Suite 205, Bld. C<br>Neptune, NJ 07753<br>diane@afius.org |
| Controlled Release Society (CRS)<br>13355 10th Ave. N, Suite 10B<br>Minneapolis, MN 55441–5553<br>www.controlledrelease.org | Cosmetic, Toiletry, and Fragrance Association (CTFA)<br>1101 17th Street NW Suite 300<br>Washington, D.C. 20036–4702<br>www.ctfa.org | Drug, Chemical & Allied Trades Association, Inc. (DCAT)<br>510 Route 130, Suite B1<br>East Windsor, NJ 08520–2649<br>www.dcat.org |
| Flavor and Extract Manufacturers' Association of the United States<br>(FEMA)<br>1620 I Street NW Suite 925<br>Washington, D.C. 20006–4005<br>www.femaflavor.org | Food and Drug Administration (FDA)<br>5600 Fishers Lane<br>Rockville, MD 20857–0001<br>www.fda.gov | Fragrance Materials Association of the United States, Inc. (FMA)<br>1620 I Street NW, Suite 925<br>Washington, D.C. 20006–4005<br>http://www.fmafragrance.org |
| Institute of Food Technologists (IFT)<br>525 W. Van Buren, Suite 1000<br>Chicago, IL 60607–3814<br>www.ift.org | Mint Industry Research Council (MIRC)<br>P O. Box 971<br>Stevenson, WA 98648<br>mirc@gorge.net | National Association of Flavors and Food Ingredient (NAFFS)<br>3301 Route 66<br>Suite 205, Building C<br>Neptune, NJ 07753–2705<br>www.naffs.org |

**TABLE 2.2 (continued)**
**Professional Societies with Flavor Interests**

| | | |
|---|---|---|
| National Association for Holistic Aromatherapy (NAHA)<br>4509 Interlake Ave N., #233<br>Seattle, WA 98103–6773<br>www.naha.org | Research Institute for Fragrance Materials (RIFM)<br>Two University Plaza Suite 406<br>Hackensack, NJ 07601–6209<br>http://www.fpinva.org/Industry/RIFM.htm | Sense of Smell Institute<br>President: Avery Gilbert<br>145 East 32nd Street<br>New York, NY 10016–6055<br>www.senseofsmell.org |
| Society of Flavors Chemists (SFC)<br>Colgate Palmolive Company<br>909 River Road, PO Box 1343<br>Piscataway, NJ 08855–1343<br>http://flavorchemist.org | Synthetic Organic Chemical Manufacturers Association (SOCMA)<br>1850 M Street NW, Suite 700<br>Washington, D.C. 20036–5810<br>www.socma.com | Women in Flavor and Fragrance Commerce, Inc. (WFFC)<br>PO Box 1732<br>Englewood Cliffs, NJ 07632–1132<br>www.wffc.org |

(e.g., SFC), active in establishing regulatory policies consistent with industry wishes (e.g., FEMA), establish buying groups of sufficient size to be influential in the marketplace (e.g., CMA, NAFFS, FMA, ASTA, and SOCMA), or provide a meeting environment to present research or product exhibits (e.g., AOCS, ACS, IFT, AACC). (Note: All acronyms are defined in Table 2.2.)

## 2.5 INTERNET

The last information resource to be discussed is the Internet. The Internet has brought us both convenience (much to our desks) and a greatly expanded world of information. In this final section, we will discuss how the Internet is used to gather information about flavors.

### 2.5.1 Internal Communication

Companies use the Internet (or an intranet) to communicate within a company. One of the major problems a company faces is how to facilitate communication between locations, groups and even within the same group at the same location. All sizes of companies profit from this ability although the larger multinational companies probably benefit the most. There is great benefit in sending communications, reports, questions, and notices to individuals in the company around the world any time of day, to any location, nearly instantly. Most individuals can access their e-mail accounts from home or on the road: people are more accessible and thus able to respond to needs of colleagues in a more timely manner.

An intranet provides the infrastructure to solve another long-term internal problem, that of easily searching and retrieving archived reports within a company. A frequent occurrence in industry is the "reinventing the wheel" because no one is aware that a certain project had already been done. Easy access and searching abilities minimizes this problem.

### 2.5.2 External Communication

This same convenience of communication applies to individuals outside the company. A scientist may wish to find an individual with expertise in a given problem area. This can easily be done through Internet searches. One can access publications (in progress) or reports written normally for internal use instantly, which traditionally have not been readily accessible. It has become easy to work with people on a global basis. The Internet removes boundaries in forming collaborative working relationships. It is nearly as easy to work with a person in another country or continent as in an adjacent building.

### 2.5.3 General Information about the Industry

In preparing to write this chapter, colleagues in both industry and academia were asked how they used the Internet and what sites they found most useful. The most commonly listed site that was that of Leffingwell & Associates (http://www.leffingwell.com/). This site was used in many ways, including staying up on current events in the industry, finding what new chemicals have come on the market or are approved (the current FEMA GRAS list is posted), buying chemicals and software, checking international legislation, and finding contacts and consultants. The site also has links to many flavor companies. The wealth of information on this site will become evident via the frequent reference to it throughout this text.

### 2.5.4 Societies/Organizations

Most organizations now have readily accessible websites, and many have been listed in Table 2.2. Obviously, this list could be greatly expanded upon. The sites noted provide information on short courses and workshops, featured articles or topics in their journals, provide direct access to their journals, information on their national or local meetings, employment information, chat rooms (in some cases) and much more. While the sites provided in Table 2.2 list national organizations, local organizations also have sites. These sites tell us what is happening in our local community.

### 2.5.5 Internet Searches

As mentioned, the entire world awaits one at his/her desktop via the Internet. There is merit in doing very general searches to get a perspective on what is happening across all disciplines. Perhaps one is interested in encapsulation. A search on this word provided 456,000 hits! That is clearly unmanageable, but it crossed all fields. A cursory browsing of 50 to 100 is manageable and may introduce the reader to something he/she had never thought of. This fits into the topic mentioned below (2.5.8 Idea Generators/Promotions). This search would not only be an idea generator but also bring up individuals, companies (service and development), meetings and associations active in this topic area.

One may choose to limit the topic area and focus on a specific aspect of encapsulation, e.g., cyclodextrins. A search on cyclodextrins brings the number of hits down to 89,000 — still a bit unmanageable. But again one can browse the top 50 to 100 hits and get an overview of the broad activity in cyclodextrins or narrow the topic more by putting in qualifying keywords. Simply tying cyclodextrins to flavor brought the number down to 367. That is very manageable. It is worth noting that there are many search engines available and each offers unique search mechanisms/criteria. While we prefer Google as our choice, another engine will give very different hits using the same keywords.

### 2.5.6 Literature Retrieval

One of the most useful services of the Internet is the broad access to the literature. Keeping up on literature is a problem to all scientists — there is more to read and less time to do so today than ever before. Simply making the literature available at one's desktop, not having to go to a library, facilitates reading. Journals may not be available at the library either because the library does not subscribe, the journal is only published electronically, or the journal you want is missing. While one can generally get an interlibrary loan, this may take longer than a person has to work on a project.

There are several ways one can access the literature through the Internet. The National Agriculture Library (http://agricola.nal.usda.gov/) provides free search and retrieval accesses and "for fee" articles can be ordered and received by Internet or by mail. Most libraries have sophisticated searching routines that become available at the desktop. If the library has a subscription to an online subscription, very often entire journal articles can be downloaded. If the library does not have an online subscription, an abstract can likely be downloaded. A second approach to getting literature by Internet is that most organizations permit members to access their journals *if they are subscribers*. For example, the *J. of Agricultural and Food Chemistry* is available through ACS at http://pubs.acs.org/journals/jafcau/index.html. If one has a subscription to the journal online, the articles can be downloaded in HTML or pdf formats.

A final approach is to contact the author directly by e-mail and request the article be returned as an e-mail attachment. Sending an article by e-mail is relatively easy. One simply clicks on Reply and attaches the file for the article. This is much easier than finding the article, putting it in an envelope, addressing the envelope, adding postage and finally mailing it. This ease of providing an article greatly enhances the probability of an author honoring a request for a reprint of his/her article and it is much quicker.

Retrieving patents, either domestic (http://www.uspto.gov/patft/index.html) or international (http://ep.espacenet.com/espacenet/ep/en/e_net.htm), has been made easy via the Internet. It is possible to search and download complete patents at no cost. In the past, obtaining domestic patents was a problem, and international patents were nearly impossible to get in a timely manner. Leffingwell & Associates have provided a compilation of patents (downloadable) at their site for the major flavor companies that have been issued patents since 1976. The patents listed are directly

**TABLE 2.3**
**Publishers of Flavor Books**

| | | |
|---|---|---|
| CRC Press LLC Headquarters<br>2000 NW Corporate Blvd<br>Boca Raton, FL 33431<br>http://www.crcpress.com | Allured Publishing Corporation<br>362 S. Schmale Rd.<br>Carol Stream, IL 60188–2787<br>http://www.allured.com | Marcel Dekker, Inc.<br>270 Madison Ave.<br>New York, NY 10016<br>http://www.dekker.com |
| Kluwer Academic/Plenum Publishers<br>233 Spring St. Fl 7<br>New York, NY 10013–1522<br>http://www.wkap.nl | Oxford University Press<br>2001 Evans Road<br>Cary, NC 27513<br>http://www.oup-usa.org | |
| Wiley Publishers<br>111 River Street<br>Hoboken, NJ 07030<br>http://www.wiley.com | Elsevier Science<br>Customer Support Department<br>P.O. Box 211<br>1000 AE Amsterdam,<br>The Netherlands<br>http://www.elsevier.com | |

linked to the selected patent at the United States Patent and Trademark Office (USPTO) website at http://www.leffingwell.com/patents/patents1976_2001.htm.

It should go without saying that finding books, publishers, and ordering online is simple. Virtually all book publishers have websites to display their products and facilitate ordering. Selected major publishers are listed in Table 2.3.

### 2.5.7 Locating Materials and Equipment Suppliers

In the flavor industry, one of the challenges is to find suppliers of the numerous aroma chemicals needed as inventory. This is simplified again by the power of search engines and corporate websites. Some websites useful in locating flavoring materials include:

| | |
|---|---|
| Allured's FFM | (http://64.78.48.186/ffm) |
| SensoryNet.com | (http://www.fks.com/flavors/home.asp) |
| Chemfinder.com | (http://www.chemfinder.cambridgesoft.com) |
| Perfume Portal | (http://www.perfume2000.com/default.asp) |

The Internet has made it possible to buy either new or used equipment and instrumentation on a national and global basis. Websites such as Labx (http://www.labx.com/) and Boulder Recycled Scientific Instruments (http://www.labused.com/) offer used instrumentation typically at large savings over new products. Equivalent sites are available to find used processing equipment.

### 2.5.8 Idea Generators/Promotions

This use of the Internet is fascinating. One of my colleagues said that he devoted some time each week to surfing universities, organizations and the patent literature to simply generate ideas, crossing into other fields such as pharmaceutics, medicine, or agricultural science to see what is happening. Again, we have the entire world of knowledge at our fingertips via the Internet, not just the world we live and work in.

### 2.5.9 Competition

Companies and academics use the Internet for promotion. This is where they tell their existing and potential customers what they are doing and where they are going. The Internet is a great means of monitoring the competition and getting ideas of how to position and sell your company (or university department).

### 2.5.10 How to Use the Internet Effectively

It is probably most effective for an individual to use the Internet himself/herself as opposed to having this done as a service. The Internet is like a labyrinth where one hit leads to another — it is a logical, albeit, meditative route with the idea or answer at the end. No one can choose the path or follow an interest as well as the person needing the information. While a librarian can do many general searches, there are many tasks best done by the individual. This means an investment of time and money into learning (subscribe to training programs) and trying.

One of my better sources for this chapter added the following sites for those who do not learn to use the Internet effectively: http://www.monster.com/, http://www.flipdog.com, and http://www.hotjobs.com. Let's hope that is not necessary.

## 2.6 SUMMARY

Hopefully, the discussion offered above on sources of flavor information will help readers use the literature and Internet more effectively in their job whether it is in industry, government, or academia. The Internet is rapidly replacing reference books and radically transforming how industry and academia relate internally, and with the rest of the world.

## REFERENCES

1. Guenther, E., *The Essential Oils*, Vol. 1–6, D. Van Nostrand Co., New York, 1948–1952.
2. Guenther, E., *The Essential Oils*, Vol. 1–6, reprinted by Allured Pub. Corp., Carol Stream, IL, 1972–1998, http://www.allured.com/pf/peoeg.html.
3. Arctander S., *Perfume and Flavor Materials of Natural Origin*, published by the author, Elizabeth, 1960.

4. Arctander S., *Perfume and Flavor Chemicals (Aroma Chemicals)*, Steffen Arctander's Publications, Las Vegas, 1969.
5. Arctander S., *Perfume and Flavor Chemicals and Perfume and Flavor Materials of Natural Origin*, 3 vols., reprinted and available in CD-ROM by Allured Pub. Corp., Carol Stream, IL, 2000, http://www.allured.com/pf/ppfcandppfm.html
6. Merory, J., *Food Flavorings: Composition, Manufacture, and Use*, AVI Pub. Co., Westport, 1960.
7. Merory, J., *Food Flavorings: Composition, Manufacture, and Use,* 2nd ed., AVI Pub. Co., Westport, 1968.
8. Heath, H.B., *Flavor Technology: Profiles, Products, Appications*, AVI Pub. Co., Westport, 1978.
9. Heath, H.B., *Source Book of Flavors*, AVI Pub. Co., Westport, 1981.
10. Fenaroli, G., *Fenaroli's Handbook of Flavor Ingredients,* 1st ed., Furia, T. E. and Bellanca, N., Eds., (edited, revised, and translated from the Italian language works of Giovanni Fenaroli), CRC Press, Boca Raton, 1971.
11. Fenaroli, G., *Fenaroli's Handbook of Flavor Ingredients,* 2nd ed., Furia, T. E. and Bellanca, N., Eds., (edited, revised and translated from the Italian language works of Giovanni Fenaroli), CRC Press, Boca Raton, 1975.
12. Fenaroli, G., *Fenaroli's Handbook of Flavor Ingredients,* 3rd ed., Burdock, G.A., Ed., CRC Press, Boca Raton, 1994.
13. Fenaroli, G., *Fenaroli's Handbook of Flavor Ingredients,* 4th ed., Burdock, G.A., Ed., CRC Press, Boca Raton, 2001.
14. Ashurst, P.R., *Food Flavourings*, 1st ed., Blackie Academ. Prof., New York, 1991.
15. Ashurst, P.R., *Food Flavorings*, 3rd ed., Aspen Publishers, Gaithersburg, 1999.
16. Ashurst, P.R., *Food Flavorings*, 2nd ed., Blackie Academ. Prof., New York, 1995.
17. Saxby, M.J., *Food Taints and Off-Flavours*, 1st ed., Blackie Academ. Prof., New York, 1993.
18. Saxby, M.J., *Food Taints and Off-Flavours*, 2nd ed., Blackie Academ. Prof., New York, 1996.
19. Ziegler, E., Ziegler, H., *Flavourings: Production, Composition, Applications, Regulations*, Wiley, New York., 1998.
20. DeRovia, D.A., *The Dictionary of Flavors and General Guide for Those Training in the Art of Flavor Chemistry*, Food Nutrition Press, Trumbull, 1999.
21. Heath, H.B., G.A. Reineccius, *Flavor Chemistry and Technology*, AVI Pub. Co., Westport, 1986.
22. Taylor, A.J., Ed. *Food Flavour Technology,* Sheffield Acad. Press, Sheffield, U.K., 2002.

# 3 Flavor Analysis

## 3.1 INTRODUCTION

Historically, flavor research has largely meant a study of the *volatile* substances in a food or flavoring. The flavor industry sold this component of flavor, and researchers in the academic community have been focused on it. It is obvious even to the layman that aroma is very key to flavor perception. A cold deprives the brain of aroma stimuli and leaves flavor perception to the basic tastes and chemesthetic response. Without aroma, it is very difficult to identify the flavor of a food product. Thus, it is understandable that there has been so much focus on this single component of flavor. Yet, as noted in Chapter 1 of this text, flavor perception is slowly being recognized as being a multifaceted sense, depending heavily upon other sensory input (e.g., taste, texture, appearance, etc.). Thus, the industry and academicians are beginning to broaden their definitions of flavor to include other sensory input, most notably taste, chemesthetics, texture, and appearance.

With this said, the following discussion will be weighted towards the analysis of volatile substances for two reasons. The first is that aroma is unquestionably important to flavor perception. The second is that there is less work published on the analysis of taste (nonvolatile substances). We have a poor understanding how taste/texture stimuli support flavor perception and limited methodology or data to discuss.

## 3.2 AROMA COMPOUNDS

### 3.2.1 Introduction

The task of identifying volatile flavor components (aroma compounds) particularly in a food matrix is one of the most formidable tasks faced by an analytical chemist. A primary obstacle is that laboratory instrumentation is not as sensitive to many odors as is the human olfactory system. Stuiver [2] calculated that as few as 8 molecules of a potent odorant can trigger 1 olfactory neuron and that only 40 molecules may provide an identifiable sensation. Making a few assumptions about air concentration vs. absorption on the olfactory membrane, it is postulated that the nose has a theoretical odor detection limit of about $10^{-19}$ moles, which surpasses even the most sensitive analytical instrumentation. The low concentrations at which these analytes may be present in a food and have sensory significance requires that they be isolated from the food system and concentrated to permit instrumental analysis.

The fact that trace quantities of aroma components are distributed throughout a food matrix further complicates the aroma isolation/concentration process. The isolation of exceedingly low concentrations of flavor compounds from food systems containing sugars, complex carbohydrates, lipids, proteins, and water is problematic. Aroma isolation methods based on volatility are complicated by the fact that water is the most abundant volatile in a food. Thus, any procedure that draws a vacuum, or involves distillation, will also extract/isolate the water from the sample. Isolation methods based on solubility (most aroma compounds are lipophilic), e.g., solvent extraction, will not only extract aroma compounds but lipids. Lipids in the aroma extract preclude gas chromatographic analysis. Also, proteins are great emulsifiers and foam stabilizers, and either one complicates a simple flavor extraction process using organic solvents. Carbohydrates often add viscosity, foaming, or emulsification properties to a product thereby again complicating aroma isolation.

Flavor isolation and analysis are made difficult also by the fact that flavors comprise a large number of chemical classes. If they were comprised of one or just a few classes of compounds, isolation methods could focus on molecular properties characteristic of a given class of compounds. Rather, the chemist must attempt to effectively extract and concentrate alcohols, aldehydes, acids, ketones, amines, carbonyls, heterocyclics, aromatics, gases, nonvolatiles (or nearly so), etc.

The absolute number of flavor compounds in a food further complicates flavor analysis. It is a rather simple, natural flavor that has less than 200 identified constituents. In fact those with less than 200 identified constituents probably have not been adequately researched. It is not uncommon for the browning flavors (e.g., meats, coffee, or chocolate) to be comprised of nearly a 1,000 volatile constituents. To date, over 7,000 volatile substances have been found in foods [3].

A final problem complicating the instrumental study of flavor is instability. The food product being examined is a dynamic system, readily undergoing flavor changes while being stored awaiting analysis to begin. The flavor isolation process may initiate chemical reactions (e.g., thermally induced degradation or oxidations), which alter the flavor profile and introduce artifacts. Thus, we have to be very cautious that the volatile components we find in a food product are truly native to that product.

Unfortunately, once we have considered each of the points above and obtained some instrumental profile of the aroma compounds in a food, we are left with the huge question of attempting to determine the importance of each volatile to the perceived flavor. This has been the topic of countless research articles over the past 30 years. Unfortunately, analytical instrumentation has no sense of taste or smell. Instrument response for the flame ionization detector (most commonly used detector in gas chromatography) is related to the number of carbon-carbon bonds, whereas the human olfactory system varies greatly in response to different odors. For example, 2-methoxy-3-hexyl pyrazine has an odor threshold of 1 part in $10^{12}$ parts water, while pyrazine has an odor threshold of 175,000 parts/$10^{12}$ parts water [4]. On pyrazines alone, the human threshold varies by nearly $2 \times 10^8$. It could be that the smallest peak in a gas chromatogram may be more important to aroma than the largest peak. It must also be recognized that the instrument is providing no appreciation for aroma character of each component. It is not apparent, for example, that peak 3 is buttery while peak 48 contributes oxidized notes. There is no question that flavor analysis offers a most challenging analytical problem.

This chapter will discuss the basis of the methods used in the isolation and analysis of food aroma components. It will be pointed out repeatedly that there is no single method of isolation or analysis that provides a complete view of the aroma compounds found in a food. The goal is to find an analytical method that can measure those components that are of interest to the analyst. They may be, for example, the compounds that give an off-flavor or those that give a fresh note to a food. Unfortunately, any aroma profile will be a partial view of the overall picture. The reader is encouraged to obtain a more complete discussion of this topic than can be provided in this text and suggests references [5–8] as sources.

### 3.2.2 Sample Selection/Preparation

The first step is to select samples of the food most typical of the flavor or off-flavor to be studied. If one is studying an off-flavor problem, the strongest yet characteristically off-flavored samples need to be selected. Recalling the extreme sensitivity of the human olfactory system, one must select the most intense samples in order to improve the probability that our relatively insensitive instruments can detect the volatiles of interest. While this may seem to be an easy task for an off-flavor, often it is not. There may be disagreement amongst individuals as to which samples have the most intense off notes.

If one is considering a study of desirable flavor, the task can become very difficult. For example, what if one was asked to obtain an aroma profile of aged Cheddar cheese? Cheddar cheese flavor varies around the world, and there is absolutely no consensus as to what is a typical Cheddar cheese. Sample selection becomes arbitrary and is left to the prejudices of the researcher.

Assuming one gets by the first hurdle of sample selection, then sample preparation becomes the next issue. One cannot simply put an apple or a piece of Cheddar cheese into an instrument and expect a response (at least a desirable response!), thus one must somehow extract the aroma from the food and concentrate it for analysis. This generally requires that the food be crushed, homogenized, blended, gas stripped, or extracted in some manner. Most fresh plant and animal tissues contain active enzyme systems, which may quickly alter the flavor profile once cellular disruption has occurred [9,10]. Singleton et al. [11] demonstrated how sample handling during aroma isolation may influence the flavor profile of peanuts. Peanuts ground in water following immersion in liquid $N_2$ contained 62% less pentane and 8% less carbonyls than the control, which was ground in water at room temperature. Peanuts ground in liquid $N_2$ showed an 81% decrease in pentane and an increase in total carbonyls compared to dry grinding at room temperature. Blending time, temperature, and pH all were shown to have a pronounced effect upon the flavor profile of peanuts.

The inactivation of enzymes of fresh plant and animal tissue when the isolation procedure exceeds only a few minutes is essential. A common method used to inactivate enzymes is homogenization in methanol [12,13]. This does, however, dilute the sample, decrease the polarity of an aqueous food slurry, and may interfere with later isolation methods. Thermal processes may also be employed if the product is a juice and may be rapidly passed through a high temperature short-time heat exchanger. One must be aware of the artifacts or interferences contributed by the means of enzyme inactivation.

One must be extremely careful of water quality if the sample is mixed with any water (or steam). Organic solvents are seldom sufficiently pure to be used in aroma isolation without additional cleanup (typically distillation). Any polymer-based materials (containers or tubing) are common sources of contamination. Antifoam additives may contribute as many components to an aroma isolate as the food itself. Stopcock or vacuum greases are known sources of contamination. Bottle closures must be Teflon-coated rather than rubber to prevent the closure from both absorbing some aroma components and contributing others.

Long isolation procedures may even permit fermentation to occur. Ribereau-Gayon et al. [14] added sodium fluoride to crushed grapes to inhibit microbial growth. In addition to enzyme or microbially induced changes in flavor profile during the isolation procedure, one must also be aware of chemical changes. Long isolation times may permit oxidative changes to occur. Thought must be given to extracting under $CO_2$ or $N_2$. Some researchers have chosen to add antioxidants such as BHA, BHT, or ascorbic acid to their sample [14]. High temperatures (greater than 60°C) for extended periods can promote nonenzymatic browning reactions. Reduced temperatures (e.g., with vacuum distillations) should be used whenever possible. Inadequate consideration for artifact formation during sample preparation can lead to erroneous results.

### 3.2.3 Principles of Aroma Isolation

#### 3.2.3.1 Introduction

Most of the techniques used in aroma isolation take advantage of either solubility or volatility of the aroma compounds. Inherently, aroma compounds must be volatile to be sensed and thus, it is logical that volatility is a common basis for separation from a food matrix. Likewise, aroma compounds tend to be more soluble in an organic solvent than in an aqueous solution (e.g., a food matrix) and thus, aroma isolates may be prepared by solvent extraction processes.

#### 3.2.3.2 Solubility

As was noted above, most aroma compounds tend to be lipophilic. Log P values for a range of aroma compounds found in foods are presented in Table 3.1. (Log P as defined here is the logarithm of the oil:water partition coefficient and relates to an octanol:water system. The values presented are calculated as opposed to experimentally determined.) If one considers the Log P values presented in Table 3.1, one notices that the vast majority of these aroma compounds will partition into the oil phase (i.e., the organic solvent in an extraction). Thus, solvent extraction of a food can be a very effective means of preparing an aroma isolate for study. The weaknesses of this method include the obvious factor that different aroma compounds have different partition coefficients and thus, will be extracted to varying extents in the extraction process. Certainly, the more water-soluble components such as diacetyl

**TABLE 3.1**
**Partitioning Data for a Range of Typical Aroma Compounds**

| Compound | Log P[a] | $k_{ow}$[b] | Compound | Log P[a] | $k_{ow}$ |
|---|---|---|---|---|---|
| 2,3-butanedione | −1.34 | 0.046 | 1-methyl pyrrole | 1.43 | 26.9 |
| Ethanol | −0.14 | 0.72 | Phenol | 1.51 | 32.4 |
| Furfuryl alcohol | 0.45 | 2.81 | Hexanal | 1.80 | 63.1 |
| 2-acetyl pyridine | 0.49 | 3.09 | Dimethyl trisulfide | 1.87 | 74.1 |
| Propionic acid | 0.58 | 3.80 | Benzothiazole | 2.17 | 148 |
| 2-pentanone | 0.75 | 5.62 | 1-octen-3-one | 2.37 | 234 |
| 2-acetyl furan | 0.80 | 6.30 | 4-ethyl guaicol | 2.38 | 240 |
| Furfural | 0.83 | 6.76 | 1-pentanethiol | 2.67 | 468 |
| Ethyl acetate | 0.86 | 7.24 | Eugenol | 2.73 | 537 |
| Methyl sulfide | 0.92 | 8.31 | 2,4-decadienal | 3.33 | 2138 |
| 2,6- dimethylpyrazine | 1.03 | 10.7 | Anethole | 3.39 | 2455 |
| Butyric acid (butanoic) | 1.07 | 11.7 | Ethyl decanoate | 4.79 | 6465 |
| Ethanethiol | 1.27 | 18.6 | Limonene | 4.83 | 6761 |

[a] From http://esc.syrres.com/interkow/kowdemo.htm.
[b] Oil:water partition coefficient ($k_{ow}$ = [conc. in octanol]/[conc. in water] at equilibrium) based on calculated Log P.

will be poorly extracted while compounds such as eugenol will be very effectively extracted. For example, if one calculates the fractions of diacetyl and eugenol remaining in the aqueous phase (i.e., is not extracted) following extraction of 1 L of aqueous solution containing each volatile with 100 ml of an organic solvent (using equation below), one finds that 95.5% of the diacetyl and 4.1% of the eugenol remain in the aqueous phase (are not extracted). Thus, the solvent extract would very poorly represent the composition of the original aqueous solution for these two volatiles. A second weakness is that solvent extraction will also extract lipids from the food. This will include trigylcerides, phospholipids, waxes, emulsifiers, fat-soluble vitamins, etc. Thus, the extraction of a lipid-containing food will provide an aroma isolate in a lipid base. The lipids preclude effective concentration of the aroma isolate as well as subsequent chromatography. However, if the food is very low in lipids (fruits, vegetables, no fat breads, and most beverages), this can be an excellent means of sample preparation.

The simplest method of solvent extraction is batch extraction using a separatory funnel. The need to use multiple extractions using limited solvent quantities rather than a single extraction using a large solvent volume is obvious from the concern for artifacts introduced by the extracting solvent (minimize solvent use) and extraction efficiency as evident from the following equation:

$$\text{Fraction not extracted} = \frac{\left[V_w\right]^n}{\left[k_{ow}V_o + Vw\right]}$$

where $V_w$, volume of aqueous phase; $V_o$, volume of extracting solvent; $K_{ow}$, partition coefficient; and *n*, number of extractions.

Since the term in parentheses is always a fraction, raising this term to a power decreases its value quickly. Using the example above for the extraction of diacetyl and eugenol from aqueous solution but extracting the solution five times with 20 ml (5 × 20) of organic solvent instead of a single extraction with 100 ml of solvent, would leave about the same amount of diacetyl in the aqueous phase (it prefers the aqueous phase) but only 0.017% of the eugenol. If proper care is given to provide adequate solvent-sample shaking and multiple extractions are used, excellent recovery of most flavor compounds may be achieved via simple solvent extraction.

### 3.2.3.3 Sorptive Extraction

Sorptive extraction (SPME and Stir Bar extraction methods, see 3.2.4.5 Sorptive Extraction for a description of the methods) is an equilibrium method, which like solvent extraction, is dependent upon the extraction phase:water partition coefficient. The theoretical recovery of a given analyte ($m_{extracted}/m_{total}$) from an aqueous solution by a PDMS (polydiemthylsiloxane) SPME fiber is given by David et al. [15] as:

$$\frac{m_{PDMS}}{m_o} = \frac{k_{PDMS/w}/\beta}{1 + k_{PDMS/w}/\beta}$$

where $m_{PDMS}$ is the mass of an analyte in the extraction phase (e.g., the PDMS fiber), $k_{PDMS/w}$ is the partition coefficient of the analyte between the PDMS fiber and water, $m_o$ is the total mass of the analyte in the system, and $\beta$ is phase ratio ($V_{\text{aqueous phase}}/V_{\text{extraction phase e.g., PDMS}}$). Extraction efficiency of a given analyte using this methodology can be predicted by the octanol:water partition coefficient (as is solvent extraction discussed above).

This equation illustrates that not only Log P but also phase ratio ($\beta$) determines extraction efficiency for a given analyte. The higher the partition coefficient, or the $\beta$, the greater the recovery of an analyte. A low phase ratio is problematic for SPME methodologies. A typical SPME fiber, e.g., a 100 μm PDMS fiber, has only about 0.5 μl of phase [15]. This results in the method giving poor recoveries of analytes having $k_{ow}$ <10,000. Also, a low phase ratio gives variable extraction for nonpolar volatiles due to competitive reactions between the SPME fiber, the aqueous phase, the glass walls of the sample vessel, and the stir bar used to stir samples. These difficulties can be minimized by increasing the phase ratio. A Stir Bar method may have a 25–250 μm phase coating over a larger surface area (i.e., a stirring bar) to yield much better extraction efficiencies than SPME. The theoretical extraction efficiency approaches 100% for analytes having $k_{ow}$ of >500. A look at the $k_{ow}$ values in Table 3.1 shows that many aroma compounds have low partition coefficients (<500) and would not be well suited to SPME but are quite well suited to extraction by the Stir Bar method (Figure 3.1).

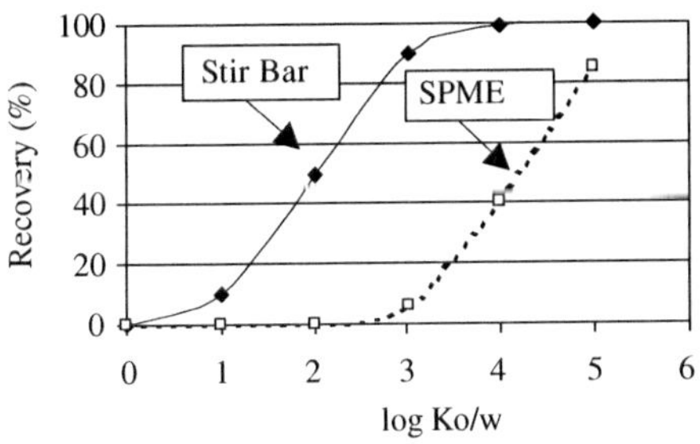


**FIGURE 3.1** Recovery of solutes as influenced by $k_{ow}$ by SPME and Stir Bar methodologies (From David, F.T., B. Tienpont, P. Sandra, *LC/GC*, 21(2), p. 109, 2003. With permission.)

#### 3.2.3.4 Volatility

A consideration of volatility as a means of aroma isolation requires an appreciation for the factors that influence the amount (or proportion) of an aroma compound in the gaseous phase vs. in the food at equilibrium (static headspace isolation methods) and nonequilibrium conditions (dynamic headspace isolation methods). In both cases, our methodology requires that the aroma compound partitions into the gas phase for isolation. Considering equilibrium conditions first, the amount of an aroma compound in the gaseous phase is defined by the gas:food partition coefficient ($k_{gf}$). This can be most simply expressed as:

$$k_{gf} = c_g/c_f \text{ or } [\text{conc. gas phase}]/[\text{conc. food phase}]$$

If our food was a dietetic beverage (almost a pure aqueous solution), we could readily predict $k_{gf}$ for an aroma compound (and therefore recovery during isolation). Unfortunately, the presence of various solutes in the food (simple sugars, complex carbohydrates, lipids, proteins, water, minerals, etc.) enormously complicates this equation and our task.. Some factors can be predicted and thus understood, for example, lipid effects. Others, for example binding by proteins or "salting out" by simple sugars, are not predictable and thus cannot currently be accounted for in a theoretical treatment.

While we have difficulty with complex food systems, we can also estimate the concentration of an aroma component in the gas phase at equilibrium over simple water and oil systems, i.e., food emulsions. In these food systems, the aroma compounds will partition into the oil and water phases based on their $k_{ow}$ and ultimately into the gas phase based primarily on the final concentration in the continuous phase. The concentration in the gas phase ($c_g$) over an emulsion can be expressed as:

$$c_g = k_{gc}c_c\ [(1 - \phi_d) + \phi_d k_{dc}]$$

where $k_{gc}$ is the equilibrium gas phase/continuous phase partition coefficient, $c_c$ is the concentration of an aroma compound in the continuous phase, $\phi_d$ is the volume fraction of the dispersed phase, and $k_{dc}$ is the equilibrium dispersed:continuous phase partition coefficient [16].

This equation illustrates that aroma compounds that have low gas phase partition coefficients (e.g., those that have high solubility in the continuous phase or very low vapor pressures, i.e., large molecular weight molecules) will not be well suited for an equilibrium vapor phase isolation method. One can see the relationship between phase volumes and solubility in the dispersed phase as well. Partitioning into the dispersed phase ($k_{dc}$) becomes very important when there is a large amount of dispersed phase. This equation is manageable but reasonably complex with $c_g$ depending upon the gas phase partition coefficient in the continuous phase, partitioning into the dispersed phase, and phase volumes.

As noted earlier, we have major problems when there are unquantifiable interactions/binding of a given aroma compound with a food constituent. Unfortunately, this is generally the case, and thus we often have little idea of how to estimate $c_g$. The implications are that every aroma compound will have a different $k_{gc}$ and $k_{dc}$, which means that the relationship between what is in the food and the gas phase above the food is complex and largely unpredictable in anything but the simplest of food systems (aqueous or aqueous/lipid systems).

A nonequilibrium isolation method (e.g., dynamic headspace) depends upon many of the same factors as the equilibrium method discussed above, and thus also poorly represents what is found in the food. However, additional considerations must be made for the nonequilibrium nature of the method. In nonequilibrium situations, the *rate* of release of aroma compounds from a food into a gas phase has to be considered. In this method, one may pass a gas stream through a food with the gas stream picking up aroma compounds under nonequilibrium conditions. The aroma compounds would be subsequently stripped from the gas stream to produce an aroma isolate.

We can generally assume that the limiting factor influencing mass transfer into the gas phase is the ability to transfer an odorant across the gas/food interface. Making this assumption, the equation presented below [17] is useful to us in considering how aroma compounds can be transferred from a food to a flowing gas for isolation.

$$dm/dt = 2\,(D_e/\pi t_e)^{1/2}\,A_{gc}\,[c_e^{\,i}\,(t) - c_e(t)] \text{ or } = h_D\,A_{gc}\,[c_e^{\,i}\,(t) - c_e(t)]$$

($h_D$ is the overall mass transfer coefficient and is substituted for the $2\,(D_e/\pi t_e)^{1/2}$ term)

where dm/dt is the rate of mass transfer into the gas phase, $D_e$ is the average diffusion coefficient of free aroma molecules in an emulsion, $A_{gc}$ is the surface area of gas/food interface, $t_e$ is the time that each surface element is exposed to the surface, and $c_e^{\,i}$ (t) and $c_e(t)$ are the concentrations of an aroma compound at the interface and in the emulsion, respectively [16].

While this equation can be expanded to consider partition coefficients, phase volumes, etc. as done above for the equilibrium conditions, we are interested in maximizing dm/dt, and this equation gives us an understanding of some method.

First, it is clear that we want a large $A_{gc}$ (stripping gas:food interface surface area). This is most readily accomplished by bubbling a gas through a food sample (or having very small particles if the food is a particulate) as opposed to simply purging the surface of a food sample. Small bubbles give us large surface areas for mass transfer. If foaming is a problem on bubbling the gas through the food and we are forced to simply purge the surface of a food sample, expect poor recovery of aroma compounds. The $t_e$ term is maximized by mixing of the food. Thus, mixing is important in the method.

The primary interference with this basis of isolation is the water present in a sample. In most foods, the aroma components seldom make up more than 300 ppm (0.03%) of the product. Yet, the moisture content of a food, even a dry food, is generally above 2% and thus isolation methods based solely on volatility will produce a dilute solution of aroma compounds in water. The high boiling point of water precludes a simple concentration and analysis, i.e., the aroma compounds would be lost during concentration since they are present in low concentrations and are often more volatile than water. Thus, most dynamic headspace methods of aroma isolation involve some additional method to remove water from the isolate.

### 3.2.4 Methods of Aroma Isolation

As noted earlier, no method will accurately reflect the aroma compounds actually present in a food or their proportions (Figure 3.2. All bars should be of equal height if the method accurately reflected the composition of the sample). Thus, one has to have the analytical objective clearly in mind when choosing a method of aroma isolation. The objective (and the food) will define the method chosen. For example, attempting to find a given off-aroma in a food is very different from wanting to have an overall aroma profile of that food. The first task only requires that the aroma isolation method isolates the volatile component with the objectionable character. It is not important if any other volatile is present in the isolate other than the one causing the off note. However, if one wants an overall view of the aroma profile, then one will likely combine several methods of aroma isolation, each isolation technique providing a part of the overall profile. With this said, a brief description of the common methods of aroma isolation will be presented including the primary strengths and weaknesses of the method.

#### 3.2.4.1 Static Headspace

Direct analysis of the equilibrium headspace above a food product would appear to be an ideal method for aroma studies. It is very simple, gentle, and easily automated. In this method, one places a food sample into a vessel, closes the vessel with an inert septum (Teflon-lined), allows equilibration (between the food and the sample headspace — 30 to 60 min) and then draws a few ml of headspace above the food into a gas-tight syringe and makes a direct injection into a gas chromatography (GC).

The primary limitation of this method is its lack of sensitivity (Figure 3.2). Since direct headspace injections into a GC are generally limited to 10 mL or less, one can see from Table 3.2 that only volatiles present at concentrations exceeding

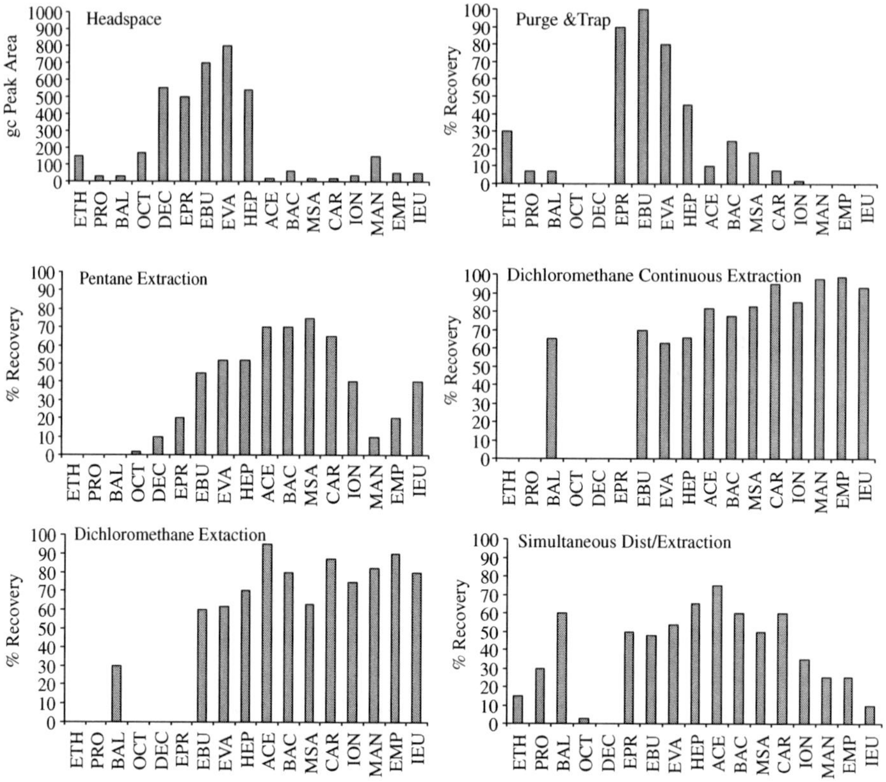


**FIGURE 3.2** The recovery of a range of volatiles by different aroma isolation methods. (ETH–ethanol, PRO–propanol, BAL–butyl alcohol, OCT–octane, DEC–decane, EPR-ethyl propionate, EBU–ethyl butyrate, EVA–ethyl valerate, HEP-heptanone, ACE–acetophenone, BAC-benzylacetate, MSA-methyl salicylate, CAR-carvone, β-ionone, MAN-methyl anthranilate, EMP-ethylmethylphenylglycidate, IEU-isoeugenol.) (Adapted from Leahy, M.M., G.A. Reineccius, *Analysis Volatiles: Methods, Applications, Procedures*, P. Schreier, Ed., de Gruyter, New York, 1984, p. 19. With permission.)

$10^{-7}$ g/L (headspace) will be detected by GC, and only those at concentrations exceeding $10^{-5}$g/L will be adequate for mass spectrometry. Since the concentration of volatiles above a food product generally ranges from about $10^{-4}$ to $10^{-10}$ g/L (or less) [19], only the most abundant volatiles will be detected by direct headspace sampling.

Trace component analysis will require some method of headspace concentration (dynamic headspace analysis), which permits sampling of large volumes of headspace (100–1000 L) thereby compensating for low headspace concentrations.

Headspace methods have a second disadvantage in that it is difficult to do quantitative studies using them. The analytical data one obtains reflects the amount of an aroma constituent in the headspace. As was discussed earlier, the relationship between concentration in the headspace (equilibrium or nonequilibrium) vs. the food can be very complex and must be determined experimentally.

**TABLE 3.2**
**Minimum Concentrations of a Substance in a Given Volume of Air Required for GC Analysis or Identification by Mass Spectrometry**

| Air Volume[a] | GC (g/L) | MS (g/L) |
|---|---|---|
| 1 mL | $10^{-5}$–$10^{-6}$ | $10^{-3}$–$10^{-4}$ |
| 10 mL | $10^{-6}$–$10^{-7}$ | $10^{-4}$–$10^{-5}$ |
| 100 mL | $10^{-7}$–$10^{-8}$ | $10^{-5}$–$10^{-6}$ |
| 1 L | $10^{-8}$–$10^{-9}$ | $10^{-6}$–$10^{-7}$ |
| 10 L | $10^{-9}$–$10^{-10}$ | $10^{-7}$–$10^{-8}$ |
| 100 L | $10^{-10}$–$10^{-11}$ | $10^{-8}$–$10^{-9}$ |
| 1 $m^3$ | $10^{-11}$–$10^{-12}$ | $10^{-9}$–$10^{-10}$ |

[a] Sample volume put into a GC or MS.

*Source:* From Schaefer, J., *Handbuch der Aroma Frischung*, H.B. Maarse, Ed., Academie-Verlag, Berlin, 1981, p. 44. With permission.

Sensitivity of the method may be enhanced to some extent via headspace enrichment. This may be accomplished by preparing a distillate of a food product and analyzing the distillate by headspace methods. Enrichment of the headspace may also be accomplished through the addition of soluble salts to the aqueous food product. The salts tend to drive the organic volatiles from solution into the vapor phase. It is of interest that the enhancing effect is not similar for all volatiles [21]. Thus the use of NaCl to enrich headspace volatiles may further quantitatively distort the headspace profile.

The advantages and shortcomings of static headspace sampling dictate its applications [22]. It is often used in quality control situations where only major components need to be measured. Whereas the components measured may not actually be responsible for the flavor attributes being monitored, if there is a good correlation between flavor quality and the component(s) measured, the goal has been accomplished. For example, Buttery and Teranishi [23] used headspace analysis of 2-methyl propanal, and 2- and 3-methyl butanal as indicators of nonenzymatic browning in potato granules. Sullivan et al. [24] have used this technique to do additional work on the flavor quality of dehydrated potatoes. It is commonly used to indicate the oxidative quality of edible oils (monitor hexanal formation).

#### 3.2.4.2 Headspace Concentration Methods (Dynamic Headspace)

Headspace methods employing some type of gas stripping and concentration are commonly called dynamic headspace methods. In these methods, the sample is purged with an inert gas, such as nitrogen or helium, which strips aroma constituents from the sample (Figure 3.3). The volatiles in the purge gas must then be trapped

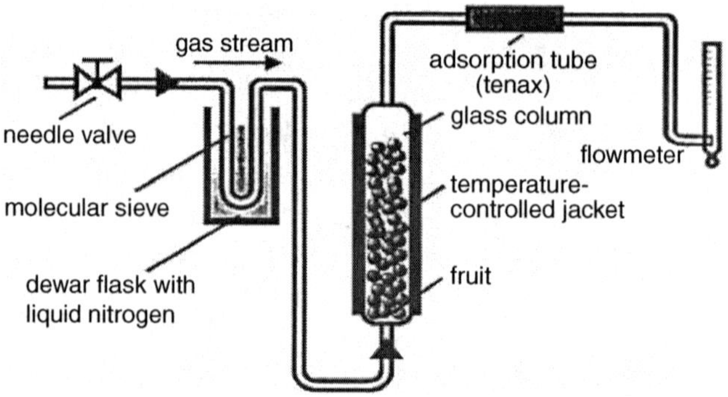


**FIGURE 3.3** Example of a set up for the isolation of volatiles via dynamic headspace methodology (Tenax trap with ambient pressure operation). (Reproduced from Guntert, M., *Techniques for Analyzing Food Aroma*, C. Mussinan, M. Morello, Eds., Amer. Chem. Soc., Washington, D.C., 1998, p. 38. With permission.)

(somehow removed) from the gas stream. The aroma constituents may be trapped via a cryogenic, Tenax (or alternative adsorptive polymer), charcoal, or other suitable trapping system. This approach favors the isolation of constituents with the highest vapor pressure above the food, as has been discussed. The aroma isolate is further distorted due to the aroma-trapping technique. A cryogenic trap is the least selective of the traps. It will remove and contain virtually any aroma constituent if properly designed and operated. The primary problem with a cryogenic trap is that it will also trap water — the most abundant volatile in nearly all foods. Thus, one obtains an aqueous distillate of the product, which must then be somehow treated to remove the water (again more biases enter the profile).

A Tenax trap (vinyl benzene-based polymer) is very widely used for aroma trapping. Despite its wide usage, it has a low surface area and, therefore, a low adsorption capacity. Additionally, Tenax has a low affinity for polar compounds (hence it does not retain much water) and a high affinity for nonpolar compounds. An aroma compound such as hydrogen sulfide would not be retained on this trapping material. Buckholz et al. [26] demonstrated the biases associated with Tenax traps during a study on peanut aroma. They found a breakthrough of peanut aroma (through two traps in series) after only 15 min of purging at 40 mL/min. In an evaluation of the sensory properties of the material collected on the Tenax trap, they found a representative peanut aroma had been collected by the trap after 4 h of purging. Shorter or longer purge times did not produce an aroma isolate smelling like peanuts. In fact, the majority of purging conditions did not yield an aroma isolate characteristic of the sample. The isolate was biased both by the volatiles preferentially going into the purge gas and by the Tenax trapping method (Figure 3.2).

The work of Guntert et al. [27] found that the Tenax trapping method (ambient pressure) did not yield as true of an aroma profile as vacuum distillation (cryotrapping). This difference in performance may have been partially due to the operating conditions used for the Tenax system. As Buckholz et al. [28] noted, operating conditions have a strong influence on the composition of the aroma isolate. Sucan et al. [29] used a response surface methodology to optimize their purge and trap method for a study of dry dog food aroma. This approach, once optimized, was claimed to yield a very good quality aroma isolate.

While Tenax is a common tapping material, activated carbon traps have a strong affinity and large capacity for most aroma constituents and thus, have been used in this application. As little as 1–10 mg of carbon will trap the volatiles from 10–100 L of purge gas [20]. The primary concern with charcoal traps is that they may not give up their aromatic components without artifact production. It has been suggested that this problem can be minimized through the use of a good quality coconut charcoal.

Despite the concerns noted about this approach to aroma analysis, it is commonly used in the field today. Several manufacturers offer completely automated systems, which simplify the task and add considerable precision to the data. Wampler [22] has provided a good review of this technique.

#### 3.2.4.3 Distillation Methods

Distillation can be defined broadly to include high vacuum molecular distillation (Figure 3.4, top of figure), steam distillation (Figure 3.4, bottom of figure), or simple heating of the food and sweeping the distilled aroma constituents into a GC (Figure 3.5, thermal desorption). High vacuum distillation may be applied to pure fats or oils (use a thin film and high vacuum), solvent extracts of fat-containing foods (e.g., cheese), or aqueous-based foods (e.g., fruit). Since fats and oils are essentially anhydrous, additional extractions (or sample manipulations) would not be required to remove any codistilled water. The use of high vacuum distillation for the isolation of volatiles from solvent extracts has frequently been used to provide good quality extracts for Aroma Extraction Dilution Assays ([30] to be discussed later). This distillation process would likely require an additional solvent removal step since diethyl ether is commonly used as the extracting solvent and it will extract some water as well.

The high vacuum distillation of fresh food products uses product moisture to co-distill volatiles (Figure 3.4 top). Water is always the major part of the distillate and a secondary extraction is mandatory. Thus, an extraction is commonly a part of this type of aroma isolation process. The primary sources of aroma profile distortion come from the distillation process and subsequent solvent extraction.

Steam distillation may be accomplished in several ways. The product may simply be put in a rotary evaporator (if liquid, or initially slurried in water if solid) and a distillate collected. This distillate would be solvent extracted to yield an aroma isolate suitable for GC analysis. The most common steam distillation method employs Simultaneous Distillation/Extraction (SDE may be called a Likens-Nickerson method). This is one of the oldest and most popular methods for obtaining aroma isolates. Chaintreau [31] has provided a very good review of this method and its evolution. An atmospheric pressure system is shown in Figure 3.4 (bottom). Vacuum systems must have joints that are air-tight, and all parts of the apparatus must be under rigid temperature control.

In either approach, the aroma profile ultimately obtained is influenced by volatility of the aroma compounds (initial isolation), solubility during solvent extraction

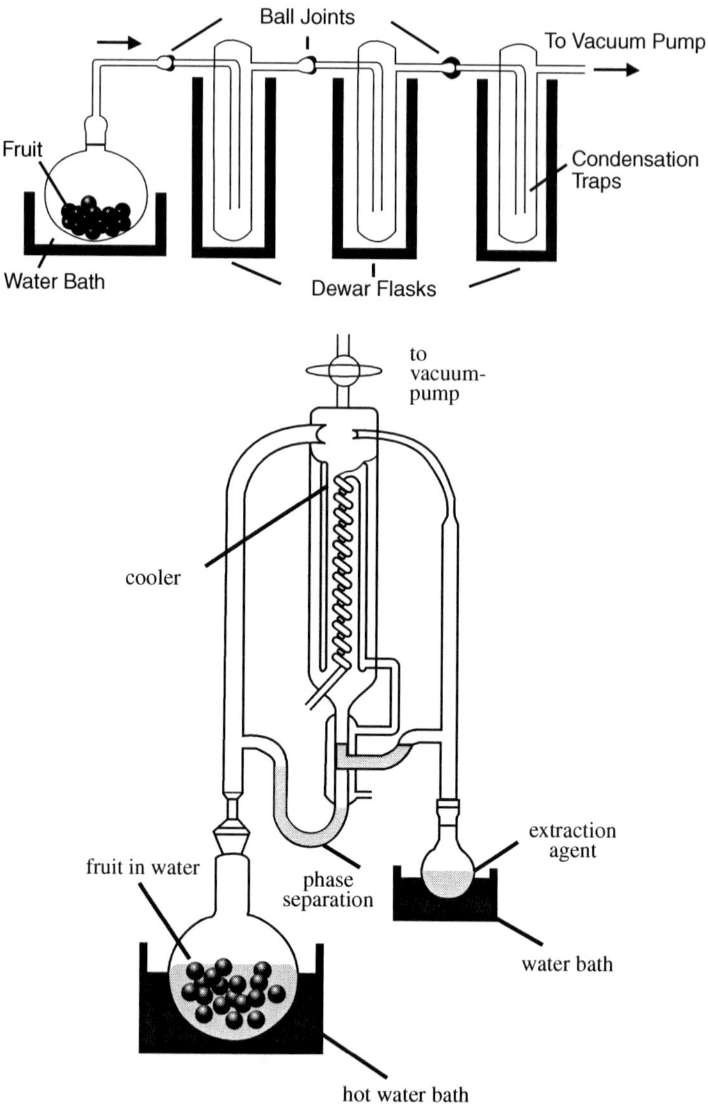


**FIGURE 3.4** Distillation equipment used to isolate volatiles from foods (left: high vacuum distillation of volatiles from an aqueous food; right: simultaneous distillation/extraction apparatus). (Reproduced from Guntert, M., G. Krammer, H. Sommer, P. Werkhoff, *Flavor Analysis*, C.J. Mussinan, M.J. Morello, Eds., Amer. Chem. Soc., Washington, D.C., 1998, p. 38. With permission.)

of the distillate, and, finally, volatility again during the concentration of the solvent extract. The aroma isolate prepared by SDE (atmospheric or reduced-pressure operation) contains nearly all the volatiles in a food, but their proportions may only poorly represent the true profile in a food (Figure 3.2). The popularity of this method

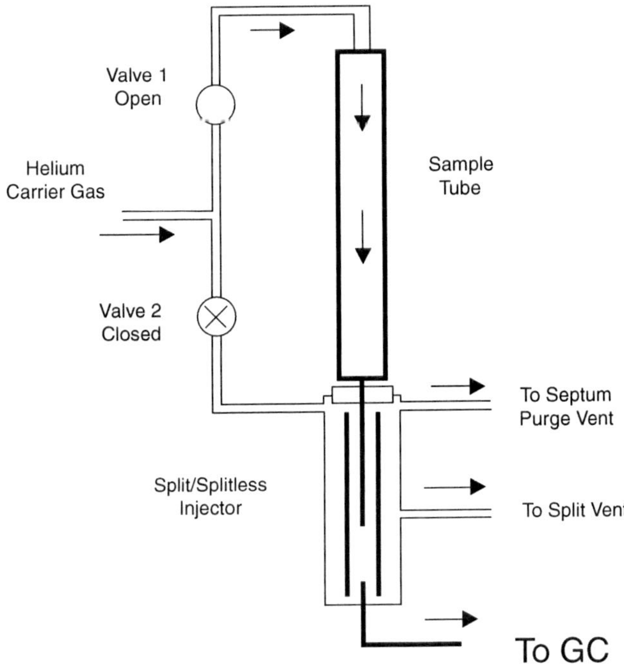


**FIGURE 3.5** Short path thermal desorption apparatus. (From Grimm, C.C., S.W. Lloyd, J.A. Miller, A.M. Spanier, *Flavor, Fragrance, and Odor Analysis*, R. Marsili, Ed., Marcel Dekker, New York, 2002, p. 55. With permission.)

comes from the fact that volatiles with medium to high boiling points are recovered well, and a liquid isolate is obtained. This liquid isolate is quite concentrated, which facilitates mass spectrometry (MS) work or repeated injections for further studies.

Distillation, as defined in this chapter, also includes direct thermal analysis techniques. These techniques involve the heating of a food sample in an in-line (i.e., in the carrier gas flow of the GC) desorber. Generally, aroma compounds are thermally desorbed from the food and then cryofocused to enhance chromatographic resolution. This technique has been used for a number of years for the analysis of lipids and was later modified to include aqueous samples [32,33]. Aqueous samples were accommodated by including a water trap after the desorption cell. This general approach has been incorporated into the short path thermal desorption apparatus discussed by Hartman et al. [34] and Grimm et al. [35]. A schematic of this apparatus is shown in Figure 3.5. In the schematic shown, a sample of food is placed in the desorption tube and quickly heated. The volatiles are distilled into the gas flow that carries them into the cooled injection system where they are cryofocused prior to injection into the analytical column.

The issue of water in the sample often limits sample size even when it is a minor component of the food. Virtually all of these methods/instruments require cryofocusing prior to GC, and small amounts of water will tend to freeze the cryotrap blocking the carrier gas flow. Thus, the sample size (and therefore, sensitivity) is often limited by the moisture content of the sample (method is applied to samples <5% moisture) [36].

The primary bias inherent in this approach is again the relative volatility of the aroma constituents. Additional concerns involve the technique used to remove water from the sample (if aqueous-based sample) and the potential for artifact production due to heating of the sample. There is a substantial body of information demonstrating aroma formation (in this case artifact formation) due to heating. Some reactions proceed rapidly at temperatures as low as 60°C. Therefore, the aroma profile can be greatly altered via the formation of artifacts due to heating the sample during isolation.

#### 3.2.4.4 Solvent Extraction

One of the simplest and most efficient approaches for aroma isolation is direct solvent extraction. The major limitation of this method is that it is most useful on foods that do not contain any lipids. If the food contains lipids, the lipids will also be extracted along with the aroma constituents, and they must be separated from each other prior to further analysis. Aroma constituents can be separated from fat-containing solvent extracts via techniques such as molecular distillation, steam distillation, and dynamic headspace.

A second consideration in solvent extraction is for solvent purity. Solvents must be of the highest quality, which often necessitates in-house distillation prior to use. One must be mindful that various qualities of solvents can be purchased, and GC grade is highly recommended (not high pressure liquid chromatography (HPLC) or other quality). Furthermore, a reagent (solvent) blank must always be run to monitor solvent artifacts irrespective of the quality of the solvent.

Solvent extraction can be as simple as putting a food sample into a separatory funnel (e.g., apple juice), adding a solvent (e.g., dichloromethane), and shaking. The dichloromethane is collected from the separatory funnel, dried with an anhydrous salt, and then concentrated for GC analysis. Alternatively, the process may be much more costly and complicated, involving, for example, a pressure chamber and supercritical $CO_2$ [37]. Supercritical $CO_2$ has the advantage of being very low boiling (efficiently separated from extracted volatiles), leaves no residue to interfere with any subsequent sensory analysis, penetrates food matrices, and its solvent properties can be altered through temperature and pressure changes or chemical modifiers (e.g., methanol). Negative aspects of this solvent include its high cost due to pressure requirements, small sample sizes (most commercial extractors) and its highly nonpolar nature (without modifiers). The use of modifiers such as methanol negate some of the advantages noted earlier. Morello [38] has provided a good example of its use and thoughtful discussion of the technique.

The biases imposed on the aroma profile by solvent extraction relate to the relative solubility of various aroma constituents in the organic/aqueous phases (Figure 3.2, see pentane and dichloromethane extraction). This is discussed in the earlier section (3.2.3.2 Solubility) on principles of aroma isolation. Cobb and Bursey [39] have made a comparison of solvent effect on recovery of a model aroma system

**TABLE 3.3**
**Recovery of Model Compounds from an Alcohol-Water (12% v/v) System**

| Compounds Extracted | Recovery in %[a] Using Different Solvents | | | |
|---|---|---|---|---|
| | Freon II | Dichloromethane | Ether | Isopentane |
| Ethyl butanoate | 66 | 43 | — | 16 |
| 2-Me-l-propanol | 34 | 55 | 22 | 32 |
| 3-Me-l-butanol | 63 | 66 | 50 | 48 |
| 1-Hexanol | 85 | 67 | 23 | 38 |
| Benzaldehyde | 83 | 54 | 18 | 20 |
| Acetophenone | 53 | 41 | 34 | 20 |
| Benzyl formate | 75 | 56 | 21 | 25 |
| 2-Phenethyl butanoate | 46 | 48 | 25 | 17 |
| Me anthranilate | 62 | 59 | 57 | 27 |

[a] Batch separatory funnel extraction, 757 mL of model system extracted 6 × 50 mL solvent.

*Source:* From Cobb, C.S., M.W. Bursey, *J. Agric. Food Chem.*, 26, p. 197, 1978. With permission.

from 12% (v/v) ethanol in water (Table 3.3). Recoveries of aroma constituents were low and variable, depending on the solvent chosen and aromatic component being extracted.

While it is obvious that even a simple solvent extraction introduces substantial bias into an aroma profile, combining solvent extraction with another technique (e.g., to separate aroma components from extracts containing lipids) adds more bias. For example, applying a distillation technique to a solvent extract that contains lipids selects for the most volatile components (now from an oil phase). Despite these considerations, this combination (solvent extraction followed by distillation) has been widely applied due to its efficiency at isolating a broad range of volatiles. The distillation was originally done using a simple high vacuum system (Figure 3.6) but this is fairly tedious due to the requirement that the solvent be slowly added to the high vacuum flask. Engel et al. [40] have developed a much more rapid and yet highly efficient Solvent Assisted Flavor Evaporation (SAFE) distillation head that is now widely used (Figure 3.7).

### 3.2.4.5 Sorptive Extraction

Sorptive extraction (Solid Phase Micro Extraction [SPME] and Stir Bar extraction) are relatively new techniques for the isolation of food aromas. Pawliszyn's group [41] was the first to develop the SPME method, and they applied it in environmental analysis. Since then, it has become a widely used technique for the analysis of volatiles in foods. Harmon [42] and Marsili [43] have provided a comprehensive review and a critical review, respectively, of this technique.

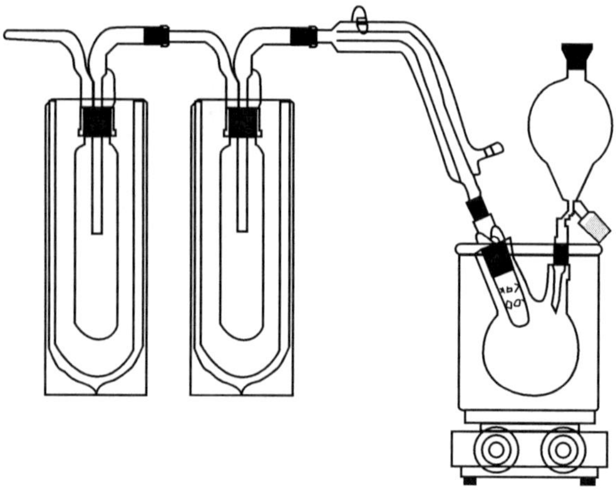

**FIGURE 3.6** High vacuum distillation apparatus for the isolation of volatiles from solvent extracts. (From Engel, W., W. Bahr, P. Schieberle, *Zeitschrift Lebensmit. Untersuch. Forsch.*, 209(3–4), 1999, p. 237. With permission.)

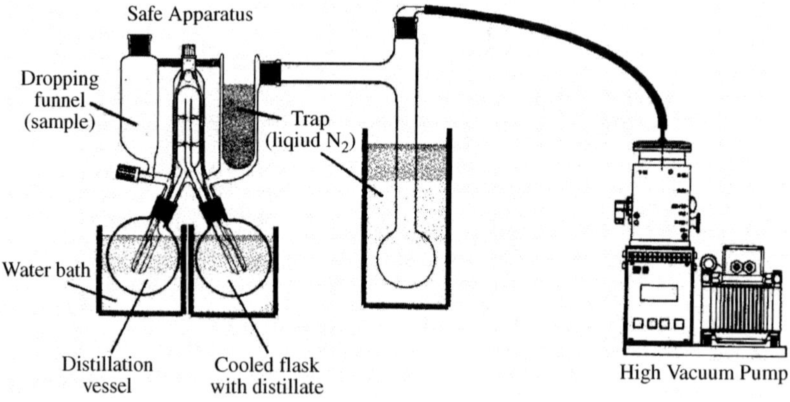


**FIGURE 3.7** Solvent assisted flavor evaporation system for the isolation of aroma compounds from solvent extracts. (From Werkhoff, P., S. Brennecke, W. Bretschneider, H.J. Bertram, *Flavor, Fragrance, and Odor Analysis*, R. Marsili, Ed., Marcel Dekker, New York, 2002, p. 139. With permission.)

In this technique, an inert fiber is coated with an adsorbent (several choices). The adsorbent-coated fiber is placed in the headspace of a sample, or the sample itself if liquid, and allowed to adsorb volatiles. The loaded fiber is then thermally desorbed into a GC carrier gas flow, and the released volatiles are analyzed. A schematic of the device is presented in Figure 3.8. The coated fiber is a modified syringe where the needle is retractable into an outer sheath. The retractable feature affords protection to the fiber against physical damage and contamination. SPME is an equilibrium technique, and therefore the volatile profile one obtains is strongly

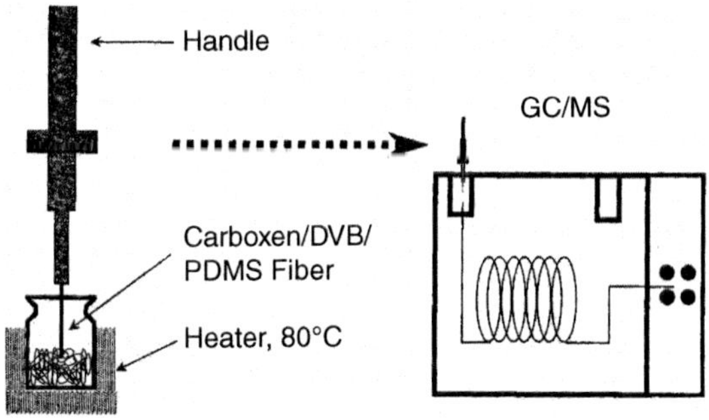


**FIGURE 3.8** Schematic of a SPME device. (From Grimm, C., E.T. Champagne, K. Ohtsubo, *Flavor, Fragrance and Odor Analysis*, R. Marsili, Ed., Marcel Dekker, New York, 2002, p. 229. With permission.)

dependent upon sample composition and careful control of all sampling parameters is required. While Harmon [42] notes that the method can give excellent results, Coleman [45] and Marsili [43] caution that the fibers have a definite linear range, and competition between volatiles for binding sites can introduce errors. Roberts et al. [46], recognizing the sensitivity of the method to sampling errors, proposed a stable isotope method for the accurate quantification of volatiles by SPME.

Similar to all of the other methods described thus far, SPME affords a certain view of the volatile composition of a food. This view is determined by factors common to headspace techniques as well as the unique affects contributed by the adsorption process. To use the method effectively, one has to be very familiar with the factors that influence volatile recovery. These factors have been discussed in detail in the literature (references cited above). If the method provides an isolate that has the component(s) one wishes to measure and it is adequately reproducible, the method is quite attractive. There are no solvents for contamination; it is simple, automated, moderately sensitive and rapid.

As was briefly described in an earlier section, this basic SPME method has been modified to include a Stir Bar sorption version [15]. In this method, a coating of adsorbent phase is placed on an inert stir bar (glass). The coating acts as an extracting solvent as opposed to an adsorbent, thus phase volume as opposed to surface area is important. The stir bar is immersed in the product to be analyzed, allowed to come to equilibrium with the liquid being analyzed (30–240 min), rinsed with water, dried, and then either thermally desorbed into a GC or solvent extracted. It has been found that food samples containing fat levels below 2–3%, or alcohol levels below 10% can be efficiently extracted with this technique. The availability of larger extraction phase volumes results in better quantitative data as well as greatly improved sensitivities. The speed and simplicity of this method, and refinements over SPME make it quite attractive.

### 3.2.4.6 Concentration for Analysis

Some of the aroma isolation procedures discussed above produce dilute solutions of aroma compounds in an organic solvent (e.g., distillation and solvent extraction).

The solvent then needs to be partially evaporated to facilitate GC analysis. Evaporative techniques take advantage of the difference in boiling point between the flavor compounds and solvent. Thus, low boiling point solvents are commonly used in the isolation process to facilitate concentration (e.g., pentane, dichloromethane, diethyl ether, and isopentane).

A disadvantage of evaporative techniques is that volatiles of interest may be lost by codistillation. Unfortunately, the loss of different components will not be uniform or predictable. Therefore, quantitative results may be in error even when multiple internal standards are employed. Since the aroma isolate typically contains a small amount of water (from the food product or distillate), care must be taken to remove water prior to concentration. While this is typically done via the addition of dessicants (e.g., anhydrous magnesium sulfate or sodium sulfate), an alternative would be to freeze out the water [48]. Failure to remove the water will result in steam distillation of the volatile components, and thus cause substantial losses during the concentration step.

Concern also has to be given to oxidation of the flavor components. Times and temperatures used in evaporative methods may be adequate to oxidize the more labile aroma compounds. This problem is often minimized by evaporating under vacuum or in an inert atmosphere.

Equipment used for evaporative concentration may be very simple or quite sophisticated. The simplest approach is to evaporate the solvent from an Erlenmeyer flask using gentle heating under a stream of nitrogen. Assuming one is using a low boiling solvent and the most volatile aroma constituents are not of interest, this is a very suitable method for concentration. However, if one is interested in the low boiling volatiles, a reflux system should be employed. A very efficient reflux setup is the Kuderna-Danish. A high reflux ratio provides effective solvent removal with minimal loss of volatile aromas. A more efficient system is to use a spinning band fractionating column [49]. Proper care in operation will result in nearly complete recovery of volatiles that differ in boiling point from the solvent by as little as 1°C.

Vacuum systems are occasionally used for sample concentration [50,51]. While vacuum systems may be used routinely, they are particularly suited to aroma isolates containing relatively higher boiling and heat sensitive components. More recently, large volume GC injection systems have minimized the need for concentration [52].

#### 3.2.4.7 Aroma Isolation Summary

Since every method preferentially selects those aroma constituents that meet certain physical or chemical criteria (e.g., solubility or volatility), one must make do and compensate for having a very biased analytical view of the aroma constituents in a food product. This biased view does not mean that it is useless or even of lesser value than a truly accurate picture. We need to choose our methodology wisely so that we measure the aroma components we need to monitor to solve our problem, i.e., they are contained in the aroma view we *consciously* select. Furthermore, one must recognize that the most commonly used approaches in the literature may not be the best or even suitable for a given task. A particular task requires a unique method. The frequency of a method appearing in the literature is more often linked to the size of the research group publishing papers using the method than anything else. A given research group may be large and doing similar work and thus, their particular methodology appears frequently. Also, individuals have certain biases — no two researchers will approach the same problem in the same manner. Thus, every aroma isolation task should be approached as a unique analysis.

### 3.2.5 Analysis of Aroma Isolates

The analytical method used to analyze an aroma isolate depends on the task at hand. If one wishes to determine the amount of an aroma compound(s) in a food, gas chromatography (GC) may suffice. If one is looking for odorous compounds in a food (desirable or undesirable), then one will use GC/Olfactometry. If one wishes to identify the aroma compounds in a food, this would require GC and mass spectrometry (or GC/Olfactometry/MS). While other instrumental methods may also be applied (e.g., IR or nmr), the bulk of aroma research is done by these three methods.

#### 3.2.5.1 Prefractionation

Despite the excellent resolving powers of modern capillary chromatography, there are situations where the analyst may chose to prefractionate the aroma isolate. Some of the more common methods to prefractionate flavor isolates prior to GC analysis include acid/base separations, high pressure liquid chromatography (HPLC), silicic acid column chromatography and preparative GC.

Acid/base/neutral fractionation of aqueous or organic solvent-based aroma isolates are relatively simple to accomplish taking advantage of the effect of pH on the solubility of ionizable analytes. One can, by changing pH, selectively partition an aroma isolate into acid, basic, and neutral fractions.

An HPLC method is attractive for flavor fractionation since it uses a different set of physical properties for separation than GC does. A flavor isolate may be separated by adsorption or reverse/normal phase chromatography. Adsorption chromatography is a good initial choice since it has the greatest column capacity and can handle the widest range of types of compounds [53]. Fractions based on adsorption affinity could then be further fractionated on a normal or reverse phase column [54].

A simple inexpensive method for flavor fractionation is via silicic acid [55]. Basically, a flavor concentrate is passed through a column of silicic acid and then eluted with a solvent gradient. This effectively fractionates the flavor isolate by compound polarity. The major problem with this method is potential artifact formation. However, careful control of the activity of the silicic acid and limiting contact time minimizes artifact formation.

Preparative GC offers the greatest resolving power of the fractionation methods. Traditionally, a flavor concentrate was chromatographed on a 0.125-inch or 0.25-inch o.d. packed column and the effluent collected via cold trap in numerous fractions. The solvent-free fractions could be subjected to sensory evaluation at this point to focus on a specific sensory property (e.g., an off-flavor or desirable note) or subjected to additional chromatographic separations. Today it is more common to use large bore, thick film capillary columns with automated heart cutting and rechromatography, or fraction collection. Proper selection of chromatographic cuts can make minor components major components of the fraction. The primary criticisms of preparative GC as a fractionation method are that the sample may be subjected to high temperatures causing chemical changes, and some compounds may be irreversibly bound to the GC column.

#### 3.2.5.2 Gas Chromatography

Aroma research has benefited tremendously from the development of GC. In 1963, only about 500 aroma compounds had been identified in foods. The development of GC in the mid-1960s and subsequent application to flavor research has resulted in over 7,000 compounds being identified to date. Gas chromatography is ideally suited to aroma studies since it has excellent separation powers and extreme sensitivity (picogram detection levels). Resolution and sensitivity are essential for analysis of the complex aroma isolates encountered routinely in flavor work. The primary disadvantage that capillary columns have brought is their low sample capacities. Sample capacity is needed if the analyst wishes to collect a component for further work, e.g., nmr or IR. While capacity is enhanced through the use of thick phase coatings (can work with up to 500 ng/component), capacity is still problematic.

Some of the most difficult flavor studies need to use two-dimensional GC. Two-dimensional GC involves collecting part of a GC run and rechromatographing it on a different chromatographic phase. These systems typically permit the collection of a selected part of several GC runs, which improves on sensitivity. GC as a technique will not be discussed in this book since there are many excellent offerings on this subject [56–58].

#### 3.2.5.3 GC/Olfactometry (GC/O) or GC-MS/Olfactometry (GC-MS/O)

GC/O and GC-MS/O are techniques uniquely applied to aroma studies [59]. In olfactometric techniques, the nose is used as a GC detector. The GC system may be set up such that the column effluent is split so that a portion of the effluent goes to a sniffing port and the remainder goes to a GC detector (flame ionization (FID) or an MS detector), or the GC run may be made by passing *all* of the GC column effluent to the nose at one time, the column is then connected to the instrument detector, and a second run made. This latter alternative provides the maximum amount of sample to the nose one time, and the FID or MS detector the second time enhancing the performance of each detector. The primary disadvantages of making two separate GC runs is analysis time and the difficulty of determining which GC peak is responsible for what odor. In busy chromatograms, there may be several GC peaks in the vicinity of the retention time of a given odor, thereby making it difficult to assign an odor to a specific GC or MS peak.

It is generally most desirable to use a GC-MS/O method (as opposed to a GC/O method). Most commonly, the analyst is interested in the identification of components that have odor (may be a desirable odor or an off-odor). Unfortunately, if one does GC/O work and is able to assign a given odor to a given GC peak, it is problematic to determine which GC peak is which MS peak. While one may think it is simple to make this assignment, the MS works under a vacuum and this changes GC column elution time even when using the same GC column in the GC and MS. A second complication is that a FID detector gives different responses to compounds than an MS detector. This means that retention times may shift and the analytical profiles (peak heights) change between a GC run and a GC-MS run. If one can get an odor profile and a MS profile at one time, the identification of odorants is greatly simplified.

In addition to determining which GC peaks have an odor, GC/O work can aid in compound identifications. An experienced chemist can provide a great deal of information about a chemical by sniffing it as it exits the GC column. The chemist can typically give molecular weight and functional group or compound class estimates. The chemist may be able to identify a chemical simply based on its aroma and GC retention properties. For example, it is relatively easy to distinguish between an alkyl pyrazine and a methoxy pyrazine by odor.

GC/O produces what is called an aromagram: a listing of the odor character of each peak in a GC run. Since more than one sniffer is used in this analysis (minimize data error due to any specific anosmia), data are more often presented in table form (Table 3.4). These data are extremely valuable since they indicate which GC peaks have odors. This information is useful in directing where an analyst should focus his/her efforts. For example, if one is studying a particular off-flavor, once the offending odorants are located in the aromagram, efforts can then be directed towards the identification of the compound, or compounds, of interest. There is no need to identify all of the GC peaks, only selected GC peaks. The aromagram may show that the most important area of the chromatogram is where there are *no* GC peaks. This would suggest going back to the "drawing-board" and trying to do a better job in isolation.

GC/O may be criticized as being a subjective method yielding inconsistent results. If one looks at the descriptors presented in Table 3.4, it is clear that while there is substantial agreement, there also is frequent disagreement amongst assessors. These types of results are common despite the use of well-trained subjects (consistent with sensory panel results). One makes an effort to minimize sensory fatigue by limiting the time a subject is asked to do sniffing (may limit to 20 min) and uses only experienced panelists.

Occasionally GC/O data may be misleading due to concentration effects. Odor characteristics of some flavor compounds can vary as a function of concentration. For example, skatole (3-methyl indole) has a characteristic fecal odor at high levels but becomes pleasant, sweet, and warm at very low levels. Fortunately, there are not many aroma compounds exhibiting such a large concentration-dependent odor character. Additional errors in the perceived intensity of a GC peak can be due to masking in mixtures (i.e., situations where components are not resolved on the GC column). Finally, in-line condensation of some compounds can result in persistent background odors. The GC column effluent splitter and the transfer lines to the sniffer should be well conditioned and adequately heated to render them odor free. Despite these

**TABLE 3.4**
**Sensory Descriptions (and Compound Identifications) of GC Peaks Noted by Three Different Sniffers. Aroma Isolate was Obtained by Dynamic Headspace Analysis from Parmesan Cheese**

| GC Peak Number | Compound | Male Subject | Female Subject 1 | Female Subject 2 |
|---|---|---|---|---|
| 1 | Acetaldehyde | Strong pungent | Strong pungent | Strong pungent |
| 2 | Propanal | Pungent | n.d. | n.d. |
| 3 | 2-Methylpropanal | Strong pungent | Stale bread crust | Strong pungent |
| 4 | Butanal | Weak pungent | Alcoholic, fruity | Weak Parmesan |
| 5 | 3-Methylbutanal | Strong malty | Strong stale | Parmesan-like |
| 6 | 2-Methylbutanal | Strong cocoa | n.d. | n.d. |
| 7 | Diacetyl | Strong buttery | Strong buttery | Diacetyl |
| 8 | 2-Butenal | Weak malty | n.d. | n.d. |
| 9 | 2,3-Pentadione | Weak buttery | Warm buttery | n.d. |
| 10 | Dimethyl disulfide | Putrid, cabbage | Weak buttery | Weak sulfury |
| 11 | Ethyl butyrate | Strong fruity | Sweet, fruity | Strong esters |
| 12 | Hexanal | Green, grassy | Weak green | Rubbery |
| 13 | Isoamyl alcohol | Fruity, alcoholic | n.d. | n.d. |
| 14 | Ethyl pentanoate | Weak fruity | n.d. | n.d. |
| 15 | 2-Heptanone | Strong fruity | n.d. | Weak fruity |
| 16 | Heptanal | Green | n.d. | n.d. |
| 17 | Ethyl hexanoate | Strong fruity | Sweet candy | Ripe banana |
| 18 | 2,6-Dimethylpyrazine | Baked, fried potato | n.d. | n.d. |
| 19 | Dimethyl trisulfide | Strong cabbage | Strong putrid | Heavy gassy |
| 20 | Methional | Strong baked | Cooked cabbage | Burnt, browned |
| 21 | Butyric acid | Cheesy | Cheesy | Dirty socks |
| 22 | Phenylacetaldehyde | Strong floral | n.d. | n.d. |

n.d. = not detected.

*Source:* From Qian, M., G.A. Reineccius, *J. Dairy Sci.*, 85(6), p. 1362, 2002. With permission.

potential pitfalls, GC/O is an invaluable tool to the flavor chemist and has found frequent application in this field [60].

### 3.2.5.4 Mass Spectrometry

The very high sensitivity inherent to MS (10–100 pg) and compatibility with GC makes a GC-MS combination extremely valuable. Mass spectrometers may be classed as low-resolution (LR) or high-resolution (HR) instruments. The LR instruments provide mass measurements to the closest whole unit mass. Since many combinations of atoms may give the same unit mass, LR MS may provide the molecular weight of a compound but does not provide elemental composition. HR instruments provide sufficiently accurate mass measurements to permit determination of elemental composition. The majority of flavor work in the past has utilized LR instruments. This is primarily because LR instruments are cheaper to purchase and operate than HR instruments.

Mass spectrometry is generally used in the flavor area to either determine the identity of an unknown or to act as a mass-selective GC detector. As mentioned, MS as an identification tool is unequaled by other instruments. The systems have largely become turn-key systems that require little or no operator expertise. If the operator can do GC, he/she can do MS. Comprehensive MS libraries and efficient searching algorithms make identification simple; however, here lies a danger. A MS will provide a best match (suggest identity) for any unknown irrespective of the validity of the match. The neophyte often accepts the proposed identifications without question and obtains incorrect identifications. It is essential that all MS identifications be supported by other data, for example, GC retention data, IR, or nmr.

The use of an MS as a GC detector can facilitate some quantification or chromatographic resolution problems that would be difficult or impossible by other techniques [62]. In these applications, the MS is operated in either the selected ion (SIM), multiple-ion detection (MIM) mode, or full scan mode if sensitivity and scan rate are adequate. In the SIM (or MIM) mode, the MS continuously measures only selected ions at very short time intervals throughout a GC run. Allowing the MS to focus on only a few ions permits greater sensitivity and a larger number of scans to be taken than if full scans were taken (see discussion in following paragraph). The monitoring of selected fragment ions makes it possible to separately quantify components, which are coeluting from the GC. The MS is generally set up to measure two or three ions unique to each component being quantified. It is wisest to use two or three ions to minimize the probability that an individual ion is being contributed by another compound. Computer software can then reconstruct *mass* chromatograms for each of the masses to be quantified (Figure 3.9a and b), the individual mass chromatograms integrated, and quantitative data on unresolved peaks obtained [62].

Obtaining full mass spectra to generate mass chromatograms has been problematic with magnetic sector or quadrupole MS since they require significant time to scan a typical mass range. Since the time is relatively long (up to 1 sec for a magnetic sector instrument), the MS can acquire only a limited number of scans of a given GC peak. During this scanning, the concentration within the MS may change significantly (spectrum is skewed) if the GC peak is very sharp. Ion trap and time of flight (TOF) MS instruments can collect spectra much faster (ion trap about 10–15 spectra/sec and TOF up to 500 spectra/sec) and are more suited to capillary column GC, and more recently, fast GC. The TOF instrument uniquely offers the ability to take a large number of spectra across a GC peak and then sum groups of spectra thereby reducing noise, improving sensitivity and detection limits. The ability to take many spectra per unit time offers another advantage in facilitating the deconvolution of mixed spectra, i.e., resolving the MS data from one compound from a mixture of compounds that coelute. While the analyst has traditionally been nearly required to resolve one compound from another to obtain an MS identification, TOF with the proper software is able to do identifications and quantification frequently without the need for peak resolution.

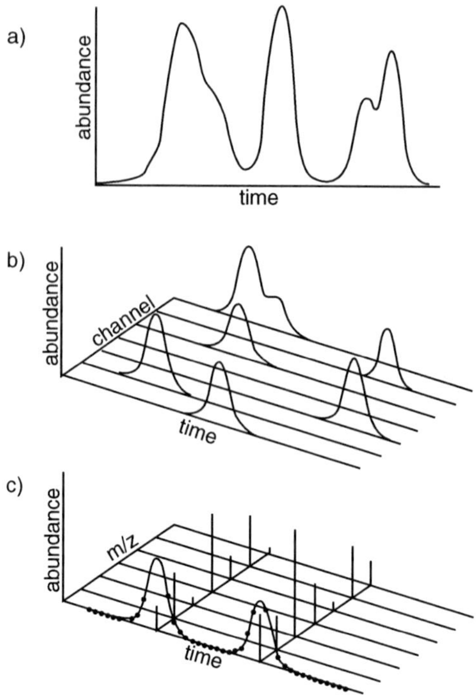


**FIGURE 3.9** Reconstructed mass chromatograms. a) Total ion current plot; b) SIM plot; c) Total ion chromatogram with full mass spectra. (From Holland, J.F., B.D. Gardner, *Flavor, Fragrance, and Odor Analysis*, R. Marsili, Ed., Marcel Dekker, New York, 2002, p. 107. With permission.)

## 3.2.6 Specific Analysis

### 3.2.6.1 Key Components in Foods

Historically, modern aroma research began with the isolation and identification of aroma compounds in foods. It was thought that if we could identify all of the aroma compounds in foods, we would be able to reproduce the aroma of that food by formulating a flavor based on the analytical data. This did not prove to be the case. Researchers found that there were very large numbers of aroma compounds present in foods and not all could possibly be contributors to the aroma of a food. Thus, an era began where researchers attempted to determine which aroma compounds were truly needed to recreate the aroma of a food. It was postulated that somewhere between 20 and 30 compounds should be adequate to reproduce the aroma of a food. The question then was, which compounds were needed? Several approaches were developed to meet this challenge. These methods will be briefly discussed.

The earliest work in this area is now more than 45 years old [63]. Patton and Josephson [63] proposed estimating the importance of an aroma compound to the sensory character of a food by calculating the ratio of the concentration of a compound in a food to its sensory threshold in that food. This ratio is known now as the odor activity value (OAV) (also as: odor value, odor unit, flavor unit, or aroma value). They suggested that compounds present above their sensory threshold concentrations in a food are likely to be significant contributors to its aroma, whereas those occurring below their threshold are not. Patton and Josephson [63] proposed this method as a guide "that may not hold in some instances."

Since the introduction of the OAV concept, various approaches have been extensively used to screen for significant odorants in food. Two major screening procedures for determining the key odorants in food are based on this concept: the Aroma Extract Dilution Analysis (AEDA) developed by Ullrich and Grosch [64] and a variation, the Aroma Extract Concentration Analysis (AECA) by Kerscher and Grosch [65], and CHARM Analysis developed by Acree and Barnard [66]. These two methods evaluate by GC/O a dilution (or concentration) series of an original aroma extract from a food (Figure 3.10). Note is taken of the occurrence of an aroma (its retention time or Kovats index) in each dilution. One then adds the occurrences of an odorant across dilutions. The greater the number of dilutions an odorant is sensed, the higher its CHARM or Dilution Value. One generates a plot of Dilution or CHARM values as a function of linear retention index (Figure 3.10, lower right of figure).

Both AEDA and CHARM methodologies originally proposed that the larger the dilution value (number of dilutions until an odorant cannot be perceived at the sniffing port), the greater the contribution of that compound to the overall aroma. However, with time data interpretation has changed. Researchers now consider AEDA, and CHARM methodologies, even the more thorough OAV methods, to be screening in nature. These methodologies are used to determine which aroma compounds *most likely* make a contribution to the aroma of a food, recognizing that sensory work (e.g., recombination studies) needs to be done to determine which aroma compounds are truly contributory. Interpretation has changed due to a recognition that the methods violate certain sensory rules or psychophysical laws [68–71].

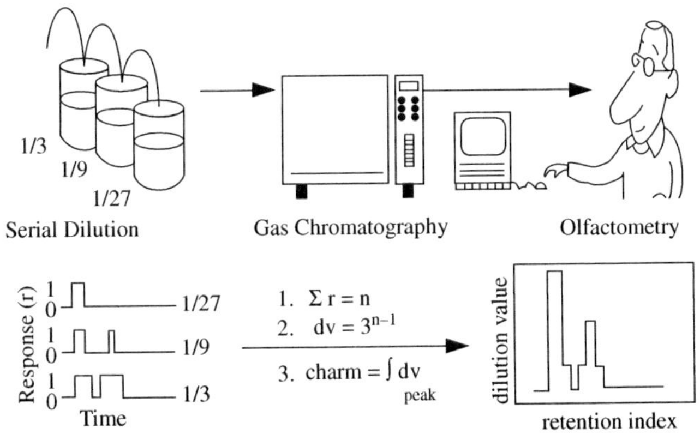


**FIGURE 3.10** Schematic of the GC/O system used in obtaining CHARM data. (From Acree, T.E., *Flavor Measurement*, C.T. Ho, C.H. Manley, Eds., Marcel Dekker, New York, 1993, p. 77. With permission.)

Two other GC/O methods have also found application for this purpose. One is called OSME and the other nasal impact frequency (NIF) or surface of nasal impact frequency (SNIF). OSME was developed by McDaniel et al. [72] and has been applied to wine aroma studies. In her method, a panelist evaluates the aromas eluting from a GC column and responds by moving a variable resister as aroma intensity changes (Figure 3.11). Thus, one is obtaining intensity and duration measurements of each GC peak. There are no dilutions made of the sample that facilitates the use of a larger number of judges as opposed to being limited to two or three judges when using dilution methods (very tedious). This adds further validity to the method. The importance of an odorant to the overall aroma is judged based on relative sensory intensities during sniffing. This is a fundamental difference between the dilution methods (AEDA, OAV, and CHARM), and OSME. The dilution methods are based on the principle that compounds present at the *greatest multiple* of their threshold are most important to aroma. This violates a basic law of sensory science in that there is a power function relationship between concentration and sensory intensity, and that relationship is different from one aroma compound to another. Thus, one cannot rank compound importance based on OAV, CHARM, or AEDA value. As was said earlier, this weakness is recognized and these values are now considered as screening as opposed to hard numbers, i.e., compounds with the highest values are candidates for further study to evaluate their true contribution.

The NIF (or SNIF) method was developed by Pollien et. al. [73] (see also [74]). In this method, eight to ten untrained individuals sniff the GC effluent (one at a time). They simply note when they smell an odor. The aroma isolate used is adjusted

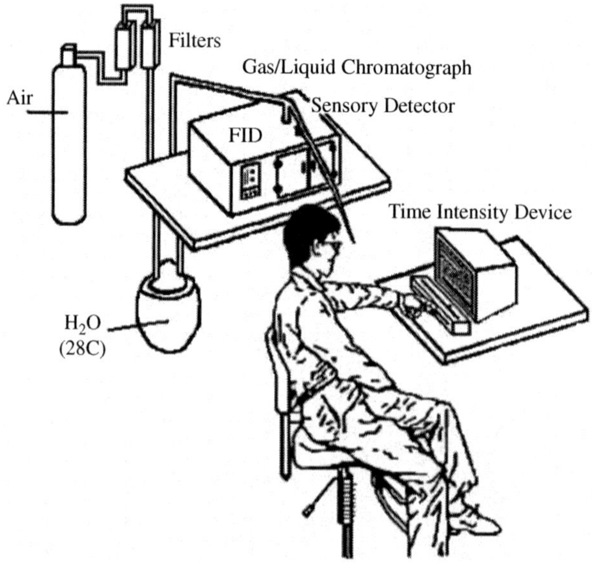


**FIGURE 3.11** GC/O system used by McDaniel for obtaining OSME data. (From McDaniel, M.R., R.Miranda-Lopez, B.T. Watson, N.J. Micheals, L.M. Libbey, *Flavors Off-Flavors '89*, G. Charalambous, Ed., Elsevier Publ., Amsterdam, 1990, p. 23. With permission.)

in strength such that in a single GC run, approximately 30 odorants are perceivable to the sniffers. This adds an element of selection in that only the more intense aroma compounds will be evaluated. The number of sniffers detecting an odorant is tabulated and plotted. Those odorants (GC peaks) being detected by the greatest number of individuals are considered likely to be the most important odorants. This method also suffers from weaknesses. One problem is that for two compounds in an aroma isolate, one may be barely over the sensory threshold of all sniffers while another may be a great distance above its sensory threshold for all sniffers, and yet both of these compounds would be viewed as being equal by this methodology.

There is no clear choice in methodology to use when determining key aroma components of a food. The methods are complicated by biases in preparing aroma isolates for analysis, by anosmia amongst panelists, human variability and bias, as well as problems interpreting the contribution of an aroma compound singly out of the food matrix as opposed to being in a food and in a complex aroma mixture. These weaknesses are acknowledged, but there is no alternative, correct methodology. If one is to do sensory studies to determine what aroma compounds are needed to reproduce the aroma of a food, there must be some preselection of aroma compounds to use in the sensory studies. It is impossible to try all possible combinations of all of the volatiles in a food.

#### 3.2.6.2 Aroma Release During Eating

As was noted earlier, progress in aroma research has been evolutionary, beginning with the long lists of aroma compounds identified in foods and progressing to a determination of key aroma compounds in foods. While this latter work has brought us closer to reproducing the aroma of a food, it has become clear that there are other factors playing into sensory perception of a flavor for one can put exactly the same flavoring into two different food products and find that they have different aromas. It was postulated that this effect is due to how aroma compounds are released from a food. Thus, analytical methods for measuring aroma release during eating have been developed.

One of the earliest methods for measuring aroma release from foods was by MS. One simply passed an inert gas (e.g., helium) over a food in a vial that was shaken with glass beads while held at constant temperature. The gas eluting from the vial was fed directly into an EI (electron impact)-MS [75]. This technique lacked sensitivity and measured total volatiles as opposed to individual aroma components. Oxygen and moisture in the system interfered with the optimal performance of the MS. Soeting and Heidema [76] and Springett et al. [77] addressed these problems by using a membrane MS inlet to remove the interfering components. Unfortunately, this adds elements of compound discrimination, reduces resolution, response time and sensitivity. The best approach to date was developed by Taylor et al. [78] and involves the use of an atmospheric pressure ionization (API) MS. Taylor et al. [78] developed an interface between a human subject and a quadrupole MS. Grab and Gfeller [79] developed an alternative inlet and used an ion trap MS and while Yeretzian et al. [80] used a PTR (proton transfer reaction)-MS. All of these systems are very robust and tolerate the host of nonaroma components, water, and oxygen

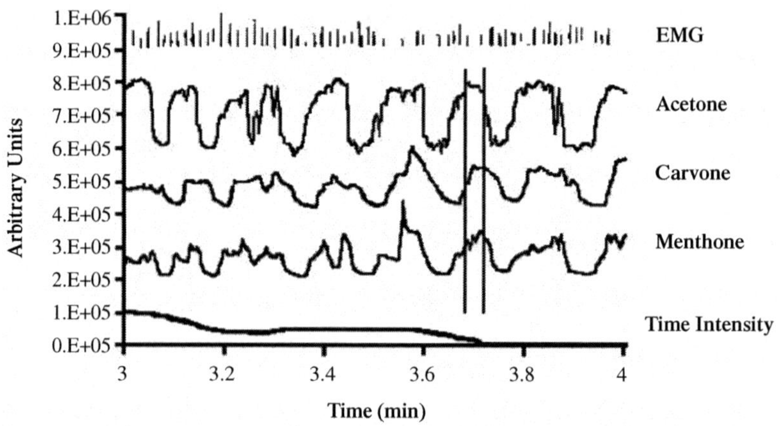


**FIGURE 3.12** Raw data obtained on aroma release during the chewing of a mint gum. (From Taylor, A.J., R.S.T. Linforth, I. Baek, M. Marin, M.J. Davidson, *Frontiers of Flavour Science*, P.E. Schieberle, K.H. Engel, Eds., Deutsche Forschung. Lebensmit., Garching, 2000, p. 255. With permission.)

prevalent in the human breath or above foods since they have been designed to be interfaced with HPLC instruments. Buettner et al. [81] have used other methodologies (not real time MS), including analyzing the amount of an odorant remaining in a food and spit off during eating (Spit Off Odor Measurement — SOOM) and a breath trapping technique (Exhaled Odor Measurement) to quantify the aroma compounds that have been released from food during eating.

The inlet system designed by Taylor et al. [78] draws a breath sample into the API MS ionization source by a venturi effect created by high nitrogen gas flows required by the MS ion source. Volatiles in the breath are ionized in the ion source by the corona discharge pin and drawn into the MS analyzer. The MS then monitors individual ions characteristic of the compounds of interest. One obtains raw data that look like that shown in Figure 3.12 for chewing a mint gum. The top line indicates chewing times, acetone is indigenous to the breath and indicates breathing times, the next two lines (carvone and menthone) reflect their release from the gum during chewing, and the bottom line is time/intensity sensory data from a panelist. These data are tabulated and smoothed to result in the view of aroma release presented in Figure 3.13.

Despite the research area being relatively new, considerable work has been published using this type of methodology [83,84]. Researchers have found that aroma release is influenced by the chemical interactions that occur within a food (thereby altering vapor pressure) and the physical properties of the food. Unfortunately, the interactions that occur are difficult to quantify and mathematically model [83]. Also, the latest results reinforce that sensory perception is multi modal — we must not only consider the aroma portion of a food but the taste, texture, and visual stimuli [1]. Thus, it appears that understanding aroma release from a food is not the final answer in understanding human perception, but we must also understand the cognitive and interactive aspects of these senses.

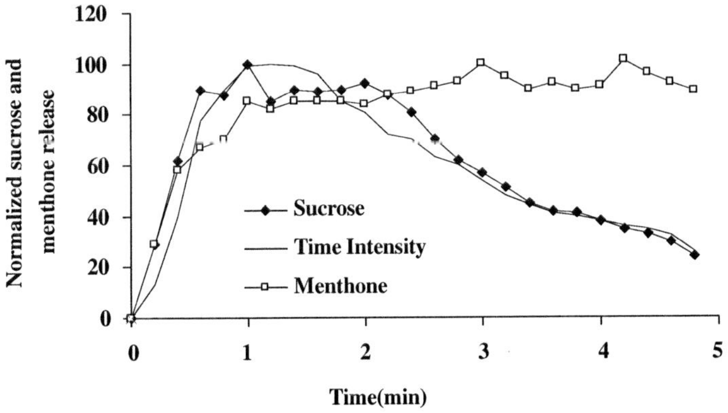


**FIGURE 3.13** The release of menthone, sucrose, and perceived flavor intensity (sensory time/intensity) during the chewing of a mint flavored gum. (From Taylor, A.J., R.S.T. Linforth, I. Baek, M. Marin, M.J. Davidson, *Frontiers of Flavour Science*, P.E. Schieberle, K.H. Engel, Eds., Deutsche Forschung. Lebensmit., Garching, 2000, p. 255. With permission.)

### 3.2.7 Electronic Noses

An electronic nose functions by analyzing a sensor array response to a complete aroma, i.e., there is no separation of aroma components. The sensor array response to any given aroma is correlated (pattern recognition software) to sensory panel data. Using neural network software and many training samples, the system determines a sensor response pattern that is representative of a fresh milk vs. a spoiled milk, for example. This technique is relatively young but has generated a substantial number of publications [5,85]. It is particularly attractive for quality control applications where an acceptable/unacceptable decision is often needed.

The sensors are key components of this system [85]. Currently, there are several types of sensors, including semiconductor gas sensors (metal oxides), surface acoustical wave devices, biosensors, conducting polymers, and mass spectrometry-based sensors [86]. In the current instruments, it is common to combine sensor types to gain a wider range in responses.

At first glance the technique appears to be ideal in that there is no need for separation of volatiles. This can result in very rapid analysis. Also, it seems to be based on a process similar to the human olfactory system in that both the electronic nose and human olfactory systems consist of a host of receptors (sensors) and yield a pattern of response to any given aroma. The brain, in the case of the human, and the computer, in the case of the electronic nose, make judgments based on a pattern recognition process as to the aroma and its quality. Thus, the speed is attractive and theoretical foundation appears to be rational.

The primary weakness of such instrumentation is that one has no clear idea of what the instrument is responding to in making a judgment. One chooses to evaluate some sensory parameter (e.g., staling or rancidity during storage) and then asks the instrument to develop a means of predicting that sensory parameter. In the end, the instrument uses some stimuli/response pattern to make a prediction, but we have no idea *what* the instrument was measuring. For example, roast and ground coffee gives off $CO_2$ during aging. In a storage study where one is determining the sensory quality of coffee (e.g., oxidized flavor) and obtaining electronic nose correlations, the instrument could be responding to changes in $CO_2$ concentration over time as opposed to oxidized flavor. As long as $CO_2$ out-gassing is correlated to lipid oxidation, the relationship is good. If it is not correlated in all situations, then the relationship is likely to be invalid in other systems or studies. The point is that the human brain uses causative input/patterns to make judgments while the electronic nose uses patterns that are not necessarily causative and may be only casually or haphazardly related.

Another potential concern for this type of instrumentation is that some sensors respond to water vapor or $CO_2$ and these responses may dominate or unduly alter sensor patterns. The sensors also deteriorate with time (or can be poisoned), changing response that makes shelf life studies problematic or frequent calibration necessary. These weaknesses largely relegate the technique to quality control situations as opposed to research studies. Despite these words of caution, the technique has substantial potential to be applied in our field. A paper by Marsili [86] illustrates some of these applications.

## 3.3 TASTE COMPOUNDS (NONVOLATILES)

### 3.3.1 Introduction

Taste has generally been thought of as a relatively simple sense being composed of salt, sweet, sour, bitter, and umami sensations (Chapter 1). This simplification is not justified since it is clear that each basic taste sensation has many nuances. Furthermore, it is worthwhile to note that each taste sensation supports a different overall flavor perception. For example, if one uses citric acid in a food system, the citrus notes of the flavor will be enhanced. Phosphoric acid is intimately associated with certain cola flavors. Tartaric acid supports grape flavors. Thus, while each acidulant gives a unique sensory character (taste), it also influences our overall flavor perception (interaction to give an overall flavor perception).

#### 3.3.1.1 Taste Compounds

From an analysis perspective, taste can readily be accounted for through well-established analytical techniques. For example, sour can be readily determined through organic (or inorganic) acid analysis via HPLC. Sweet or salt can likewise be accounted for through the analysis of known mono- and disaccharides or high potency sweeteners (e.g., aspartame, acesulfame k, etc.), or inorganic salts, respectively. Umami is limited primarily to monosodium glutamate or the 5′-nucleotides (granted some peptides are considered to have a umami character). Bitter is more difficult for there are many diverse compounds known to cause a bitter sensation [87]. There is little analytical challenge in analyzing these taste compounds except for bitter. A general overview of methods can be found in basic food analysis texts

[88]. Articles by Engel [89,90] and Preininger et al. [5] and Warmke et al. [91,92] outline analytical protocols for determining *sensory significant* taste components in foods (cheeses). These approaches are similar to those used in determining sensory significant aroma compounds (and thus offer similar strengths and weaknesses).

#### 3.3.1.2 Other Nonvolatile Components of Foods

Nonvolatile components (taste and nontaste) in foods play a greater role in food flavor than just defining the taste sensation. Nonvolatiles in foods are known to interact with some aroma compounds (chemically bind) thereby exerting an additional influence on flavor [93–95]. For example, Hofmann et al. [96,97] reported that melanoidins in coffee reduce the intensity of the roasty-sulfury aroma notes in coffee. They found that melanoidins promote the degradation of 2-furfurylthiol (FFT), 3-methyl-2-buten-thiol, 3-mercapto-3-methylbutyl formate, 2-methyl-3-furanthiol, and methanethiol — all key odorants in coffee. Likewise, Ebeler et al. [98–100] have found an interaction between the polyphenolics in wine and certain aroma compounds. Thus, nonvolatile components in foods that might be considered to have no taste per se may still exert an influence of the flavor of a food. (We might also hypothesize that there may be unrecognized cognitive effects between nonvolatiles and overall flavor perception.) Thus, we may be interested in the analysis of nonvolatiles in foods beyond those that contribute to taste perception.

Our interest in the analysis of nonvolatiles, thus, may involve taste substances or substances that indirectly influence taste or aroma. As mentioned earlier, in the first case, we are interested in the analysis of substances that impart sweetness, tartness, bitterness, saltiness, or unmami sensations. The analysis of these substances is reasonably well defined. In the latter case, the analyses employed are less well defined and are unique to the components one wishes to analyze. For example, we may wish to measure substances (e.g., melanoidins) that interact with sulfur aroma compounds (in coffee). There are no standardized methods for the analysis of melanoidins in foods and thus, the protocols have to be developed. In this chapter, we will only briefly discuss the established methods for the analysis of taste substances. Due to the specificity of methods for the analysis of nonvolatiles that may indirectly influence flavor perception, we will only refer the reader to the literature [93–100].

### 3.3.2 The Analysis of Taste Substances

#### 3.3.2.1 Sweeteners

Sweeteners used in the food industry typically are limited to the bulk sweeteners, sucrose, fructose, glucose, and corn syrups, or the high potency sweeteners, saccharin, aspartame, sucralose, and acesulfame k. While various enzymatic and colorimetric methods may be used, high performance liquid chromatography is the most commonly used technique in this analysis [101]. HPLC offers speed, sensitivity, accuracy, and precision to the analyst. Several types of HPLC columns (anion and cation exchange, normal phase and reversed phases) may be used in conjunction with a suitable detector. Traditionally, refractive index detectors have been used to detect the common bulk sweeteners but they lack sensitivity and cannot be used with gradient elution programs. Thus, electrochemical detectors (e.g., a pulsed amperometric detector) have found recent application.

HPLC has also found substantial application in the analysis of the high potency sweeteners. In one methodology, equilibrium dialysis of a food against water (15 h, 12,000–14,000 dalton cutoff) was used to eliminate interference from lipids, carbohydrates, and proteins. Direct chromatography of the dialysate on a reverse phase column (UV detection @220 ηm) gave very good results for acesulfame-k, saccharin, and aspartame.

### 3.3.2.2 Salt

One may be primarily be interested in the determination of table salt (NaCl) or one of the salt substitutes ($NH_4Cl$ or KCl) depending upon the product application. While there are numerous methods for the determination of minerals (e.g., salt) in foods, table salt is most easily analyzed using a specific ion electrode [102]. The simplicity and sensitivity of this method are very attractive with detection limits being less than 0.1 ppm and response time less than 30 sec.

If one wishes to measure other salts, the choice is generally ion chromatography [103] or atomic absorption/emission spectroscopy [104]. The chromatographic approach uses an ion exchange column coupled with a conductivity detector.

### 3.3.2.3 Acidulants

Both organic and inorganic acids are broadly used in foods. The organic acids used include citric, malic, tartaric, acetic, and lactic while hydrochloric and phosphoric acids constitute the commonly used inorganic acids. These taste components are commonly measured using ion chromatography [103].

### 3.3.2.4 Umami

Monosodium glutamate (MSG) and the 5′-nucleotides are generally recognized as the primary food components that provide the umami sensation [105,106]. MSG is readily measured by ion chromatography or reverse phase HPLC [103]. The 5′-nucleotides are most commonly determined by HPLC as well [107,108] but other methods have found use, e.g., derivative spectrophotometry [108].

### 3.3.2.5 Bitter Substances

As was mentioned earlier, bitter substances are composed of a broad range of chemical structures, thus, there often is little commonality in structure to permit the utilization of a single analytical approach. Methods for this analysis have to be designed for each bitter component (or group of components) to be analyzed. There are numerous methods in the literature, most depending upon HPLC since bitter substances are typically nonvolatile [98,109,110].

## REFERENCES


1. Hollowood, T.A., J.M. Davidson, L. DeGroot, R.S.T. Linforth, A.J. Taylor, Taste release and its effect on overall flavor perception, in *Chemistry of Taste: Mechanisms, Behavior, and Mimics*, P.G.D. Paredes, Ed., Amer. Chem. Soc., Washington, D.C., 2002, p. 166.
2. Stuiver, M., Ph.D. Thesis, Rijks University, Groningen, 1958.
3. TNO, *Volatile Compounds in Foods*, AJ Zeist, The Netherlands, Nutrition and Food Research, Utrechtseweg, 1995.
4. Seifert, R.M.B., R.G. Guadagni, D.G. Black, G. Harris, Synthesis of some 2-methoxy-3-alkylpyrazines with strong bell pepper-like aromas, *J. Agric. Food Chem.*, 18, p. 246, 1970.
5. Marsili, R., Ed., *Flavor, Fragrance, and Odor Analysis*, Marcel Dekker, New York, 2002, p. 425.
6. Ho, C.T., C.H. Manley, Eds., *Flavor Measurement*, Marcel Dekker, NewYork, 1993, p. 379.
7. Reineccius, G.A., Instrumental methods of analysis, in *Food Flavour Technology*, A.J. Taylor, Ed., Sheffield Academ. Press, Sheffield, 2002, p. 210.
8. Mussinan, C., M.J. Morello, Eds., *Flavor Analysis*, Amer. Chem. Soc., Washington, D.C., 1998, p. 389.
9. Dirinck, P.J., H.L. de Pooter, G.A. Willaert, N.M. Schamp, Flavor quality of cultivated strawberries: the role of the sulfur compounds, *J. Agric. Food Chem.*, 29(2), p. 316, 1981.
10. Drawert, F., R. Emberger, R. Tressl, Enzynmatische veranderung des naturichlen apfelaromas bei der aufarbeitung, *Naturwissenschaften*, 52, p. 304, 1965.
11. Singleton, J.A., H.E. Pattee, T.H. Sanders, Production of flavor volatiles in enzyme and substrate enriched peanut homogenates, *J. Food Sci.*, 41(1), p. 148, 1976.
12. Schreier, P., F. Drawert, A. Junker, Identification of volatile constituents from grapes, *J. Agric. Food Chem.*, 24(2), p. 331, 1976.
13. Drawert, F., R. Emberger, R. Tressl, Gas-chromatographische untersuchung pflantzlicher aromen, *Chromatographia*, 2, p. 57, 1969.
14. Ribereau-Gayon, P.B., J.N. Boidron, A. Terrier, Aroma of muscat grape varieties, *J. Agric. Food Chem.*, 23(6), p. 1042, 1975.
15. David, F.T., B. Tienpont, P. Sandra, Stir-bar sorptive extraction of trace organic compounds from aqueous matrices, *LC/GC*, 21(2), p. 109, 2003.
16. Harrison, M., B.P. Hills, J. Bakker, T. Clothier, Mathematical models of flavor release from liquid emulsions, *J. Food Sci.*, 62(4), p. 653, 1997.
17. Coulson, J.M., J.F. Richardson, *Chemical Engineering*, Vol. 1, Pergamon Press, Oxford, 1993, p. 446.
18. Leahy, M.M., G.A. Reineccius, Comparison of methods for the isolation of volatile compounds from aqueous model systems, in *Analysis Volatiles: Methods, Applications, Procedures*, P. Schreier, Ed., de Gruyter, New York, 1984, p. 19.
19. Weurman, C., Sampling in airborne odorant analysis, in *Human Responses to Environmental Odors*, A.J.J. Turk, J.W. Johnston, D.G. Moulton, Eds., Academ. Press, New York, 1974.
20. Schaefer, J., Isolation and concentration from the vapor phase, in *Handbuch der Aroma Frischung*, H.B. Maarse, Ed., Academie-Verlag, Berlin, 1981, p. 44.
21. Roberts, D.D., P. Pollien, Dependence of "salting-out" phenomenon on the physical chemical properties of the compounds, in *Frontiers of Flavour Science*, P. Schieberle, K.H. Engel, Eds., Deutsche Forschung. Lebensmit., Garching, 2000, p. 311.

22. Wampler, T.P., Analysis of food volatiles using headspace-gas chromatographic techniques, in *Flavor, Fragrance and Odor Analysis*, R. Marsili, Ed., Marcel Dekker, New York, 2002, p. 25.
23. Buttery, R.G., R. Teranishi, Measurement of fat oxidation and browning aldehydes in food vapors by direct injection gas chromatography, *J. Agric. Food Chem.*, 11, p. 58, 1963.
24. Sullivan, J.F., R.P. Konstance, M.J. Calhoun, F.B. Talley, J. Cording Jr., O. Panasiuk, Flavor and storage stability of explosion-puffed potatoes: nonenzymatic browning, *J. Food Sci.*, 39(1), p. 58, 1974.
25. Guntert, M., The importance of vacuum headspace methods for the analysis of fruit flavors, in *Techniques for Analyzing Food Aroma*, C. Mussinan, M. Morello, Eds., Amer. Chem. Soc., Washington, D.C., 1998, p. 38.
26. Buckholz Jr., L.L., D.A. Withycombe, H. Daun, Application and characteristics of polymer adsorption method used to analyze flavor volatiles from peanuts, *J. Agric. Food Chem.*, 28(4), p. 760, 1980.
27. Guntert, M., G. Krammer, H. Sommer, P. Werkhoff, The importance of vacuum headspace methods for the analysis of fruit flavors, in *Flavor Analysis*, C.J. Mussinan, M.J. Morello, Eds., Amer. Chem. Soc., Washington, D.C., 1998, p. 38.
28. Buckholz, L.L.J., D.A. Withycombe, H. Daun, Application and characteristics of polymer adsorption method used to analyze flavor volatiles from peanuts, *J. Agric. Food Chem.*, 28(4), p. 760, 1980.
29. Sucan, M.K., C. Fritz-Jung, J. Ballam, Evaluation of purge and trap parameters: optimization using a statistical design, in *Flavor Analysis*, C.J. Mussinan, M.J. Morello, Eds., Amer. Chem. Soc., Washington, D.C., 1998, p. 22.
30. Engel, W., W. Bahr, P. Schieberle, Solvent assisted flavor evaporation: a new and versatile technique for the careful and direct isolation of aroma compounds from complex food matrixes, *Zeitschrift Lebensmit. Untersuch. Forsch.*, 209(3–4), 1999, p. 237.
31. Chaintreau, A., Simultaneous distillation-extraction: from birth to maturity. Review, *Flavour Fragrance J.*, 16(2), p. 136, 2001.
32. Dupuy, H.P., E.T. Rayner, J.I. Wadsworth, M.G. Legendre, Analysis of vegetable oils for flavor quality by direct gas chromatography. *J. Amer. Oil Chem. Soc.*, 54(10), p. 445, 1977.
33. Legendre, M.G., H.P. Dupuy, Flavor volatiles as measured by rapid instrumental techniques, in *Protein Funct. Foods*, Amer. Chem. Soc., Washington, D.C., 1981, p. 41.
34. Hartman, T., J. Lech, K. Karmas, J. Salinas, R.T. Rosen, C.T. Ho, Flavor characterization using adsorbent trapping-thermal desorption or direct thermal desorption-gas chromatography and gas chromatography-mass spectrometry, in *Flavor Measurement*, C.T. Ho, C. Manley, Eds., Marcel Dekker, New York, 1993, p. 37.
35. Grimm, C.C., S.W. Lloyd, J.A. Miller, A.M. Spanier, The analysis of food volatiles using direct thermal desorption, in *Flavor, Fragrance, and Odor Analysis*, R. Marsili, Ed., Marcel Dekker, New York, 2002, p. 55.
36. Rothaupt, M., Thermo desorption as sample preparation technique for food and flavor analysis by gas chromatography, in *Flavor Analysis*, C.J. Mussinan, M.J. Morello, Eds., Amer. Chem Soc., Washington, D.C., 1998, p. 116.
37. Jennings, W.G., M. Filsoof, Comparison of sample preparation techniques for gas chromatography, *J. Agric. Food Chem.*, 25, p. 440, 1977.
38. Morello, M.J., Isolation of aroma volatiles from an extruded oat ready-to-eat-cereal: comparison of distillation-extraction and super critical fluid extraction, in *Thermally Generated Flavors*, T.H. Parliment, M.J. Morello, R.J. McGorrin, Eds., Amer. Chem. Soc., Washington, D.C., 1994, p. 95.

39. Cobb, C.S., M.W. Bursey, Comparison of extracting solvents for typical volatile components of eastern wines in model aqueous-alcoholic systems, *J. Agric. Food Chem.*, 26, p. 197, 1978.
40. Engel, W., W. Bahr, P. Schieberle, Solvent assisted flavor evaporation — a new and versatile technique for the careful and direct isolation of aroma compounds from complex food matrices, *Zeitschrift Lebensmi. Untersuch. Forsch.*, 209(3–4), p. 237, 1999.
41. Pawliszyn, J., Solid phase microextraction, *Advan. Exper. Med. Biology*, 488, p. 73, 2001.
42. Harmon, A.D., Solid-phase microextraction for the analysis of flavors, in *Techniques for analyzing food aroma*, R. Marsili, Ed., Marcel Dekker, New York, 1997, p. 81.
43. Marsili, R., SPME comparison studies and what they reveal, in *Flavor, Fragrance, and Odor Analysis*, R. Marsili, Ed., Marcel Dekker, New York, 2002, p. 205.
44. Werkhoff, P., S. Brennecke, W. Bretschneider, H.J. Bertram, Modern methods for isolating and quantifying volatile flavor and fragrance compounds, in *Flavor, Fragrance, and Odor Analysis*, R. Marsili, Ed., Marcel Dekker, New York, 2002, p. 139.
45. Coleman, W.M., A study of the behavior of Maillard reaction products analyzed by solid-phase microextraction gas chromatography-mass selective detection, *J. Chromatograph. Sci.*, 34, p. 213, 1996.
46. Roberts, D.D., P. Pollien, SPME method development for headspace analysis of volatile flavor compounds, *Book of Abstracts*, 216th ACS National Meeting, Boston, August 23–27, p. AGFD-146, 1998.
47. Grimm, C., E.T. Champagne, K. Ohtsubo, Analysis of volatile compounds in the headspace of rice using SPME/GC/MS, in *Flavor, Fragrance and Odor Analysis*, R. Marsili, Ed., Marcel Dekker, New York, 2002, p. 229.
48. Buttery, R.G., J.A. Kamm, Volatile components in alfalfa: possible insect host attractants, *J. Agric. Food Chem.*, 28, p. 978, 1980.
49. Coleman, E.C., C.T. Ho, Chemistry of baked potato flavor: I. Pyrazines and thiazoles identified in the volatile flavor of baked potato, *J. Agric. Food Chem.*, 28(1), p. 66, 1980.
50. MacLeod, A.J., N.M. Pieris, Volatile flavor components of wood apple (Feronia limonia) and a processed product, *J. Agric. Food Chem.*, 29(1), p. 49, 1981.
51. Peacock, V.E., M.L. Deinzer, L.A. McGill, R.E. Wrolstad, Hop aroma in American beer, *J. Agric. Food Chem.*, 28(4), p. 774, 1980.
52. Teske, J.E., W, Methods for and applications of large-volume injection in capillary gas chromatography, *Trends Analyt. Chem.*, 21(9/10), p. 584, 2002.
53. Teitelbaum, C.L., A new strategy for the analysis of complex flavors, *J. Agric. Food Chem.*, 25, p. 466, 1977.
54. Parliment, T.H., Concentration and fractionation of aromas on reverse-phase adsorbents, *J. Agric. Food Chem.*, 29(4), p. 836, 1981.
55. Schreier, P., The use of HRGC-FTIR in tropical fruit flavour analysis, in *Analysis of Volatiles: Methods and Applications*, P. Schreier, Ed., de Gruyter, Berlin, 1984, p. 269.
56. Jennings, W.M., E. Mittlefehldt, P. Stremple, *Analytical Gas Chromatography*, 2nd ed., Academ. Press, San Diego, 1997, p. 389.
57. McNair, H.M., J.M. Miller, *Basic Gas Chromatography: Techniques in Analytical Chemistry*, Wiley Publ., New York, 1998, p. 200.
58. Scott, R.P.W., *Introduction to Analytical Gas Chromatography*, 2nd ed., Marcel Dekker, New York, 1997, p. 397.
59. Blank, I., Gas chromatography-olfactometry in food aroma analysis, in *Flavor, Fragrance, and Odor Analysis*, R. Marsili., Ed., Marcel Dekker, New York, 2002, p. 297.

60. Leland, J.V., P. Schieberle, A. Buettner, T.E. Acree, *Gas Chromatography-olfactometry,* Amer. Chem. Soc., Washington, D.C., p. 219, 2001.
61. Qian, M., G.A. Reineccius, Identification of aroma compounds in parmigiano-reggiano cheese by gas chromatography/olfactometry, *J. Dairy Sci.*, 85(6), p. 1362, 2002.
62. Holland, J.F., B.D. Gardner, The advantages of GC-TOFMS for flavor and fragrance analysis, in *Flavor, Fragrance, and Odor Analysis*, R. Marsili, Ed., Marcel Dekker, New York, 2002, p. 107.
63. Patton, S., D.V. Josephson, A method for determining significance of volatile flavor compounds in foods, *Food Res.*, 22, p. 316, 1957.
64. Ullrich, F., W. Grosch, Identification of the most intense volatile flavor compounds formed during autoxidation of linoleic acid, *Zeitschrift Lebensmit. Untersuch. Forsch.*, 184(4), p. 277, 1987.
65. Kerscher, R., W. Grosch, Comparative evaluation of potent odorants of boiled beef by aroma extract dilution and concentration analysis, *Zeitschrift Lebensmit. Untersuch. Forsch.*, 204(1), p. 3, 1997.
66. Acree, T.E., J. Barnard, D.G. Cunningham, A procedure for the sensory analysis of gas chromatographic effluents, *Food Chem.*, 14(4), p. 273, 1984.
67. Acree, T.E., Gas-chromatography-olfactometry in flavor analysis, in *Flavor Measurement*, C.T. Ho, C.H. Manley, Eds., Marcel Dekker, New York, 1993, p. 77.
68. Abbott, N., P. Etievant, S. Issanchou, D. Langlois, Critical evaluation of two commonly used techniques for the treatment of data from extract dilution sniffing analysis, *J. Agric. Food Chem.*, 41(10), p. 1698, 1993.
69. Piggot, J.R., Relating the sensory and chemical data to understand flavor, *J. Sensory Studies*, 4, p. 261, 1990.
70. Frijters, J.E., A critical analysis of the odor unit number and its use, *Chem. Senses Flavour*, 3(2), p. 227, 1978.
71. Mistry, B.S., T.A. Reineccius, L.K. Olson, Gas chromatography-olfactometry for the determination of key odorants in foods, in *Techniques for Analyzing Food Aroma,* R. Marsili, Ed., Marcel Dekker, New York, 1997, p. 265.
72. McDaniel, M.R., R.Miranda-Lopez, B.T. Watson, N.J. Micheals, L.M. Libbey, Pinot noir aroma: a sensory/gas chromatographic approach, in *Flavors Off-Flavors '89*, G. Charalambous, Ed., Elsevier Publ., Amsterdam, 1990, p. 23.
73. Pollien, P., A. Ott, F. Baumgartner, R. Munoz-Box, A. Chaintreau, Hyphenated headspace-gas chromatography-sniffing technique: screening of impact odorants and quantitative aromagram comparisons, *J. Agric. Food Chem.*, 45(7), p. 2630, 1997.
74. Chaintreau, A., Quantitative use of gas chromatography-olfactometry: the GC-"SNIF" method, in *Flavor, Fragrance, and Odor Analysis*, R. Marsili, Ed., Marcel Dekker, New York, 2002, p. 333.
75. Lee, W.E., A suggested instrumental technique for studying dynamic flavor release from food products, *J. Food Sci.*, 51, p. 51, 1986.
76. Soeting, W.J., J. Heidema, A mass spectrometric method for measuring flavor concentration/time profiles in human breath, *Chem. Senses Flavor*, 13, p. 607, 1988.
77. Springett, M.B., V. Rozier, J. Bakker, Use of fiber interface direct mass spectrometry for the determination of volatile flavor release from model systems, *J. Agric. Food Chem.*, 47, p. 1125, 1999.
78. Taylor, A.J., R.S.T. Linforth, B.A. Harvey, A. Blake, Atmospheric pressure chemical ionization mass spectrometry for in vivo analysis of volatile flavour release, *Food Chem.*, 71, p. 327, 2000.

79. Grab, W., H. Gfeller, Flavorspace — a new technology for the measurement of fast dynamic changes of the flavour release during eating, in *Frontiers of Flavour Science,* P.E. Schieberle, K.H. Engel, Eds., Deutsche Forschung. Lebensmit., Garching, 2000, p. 261.
80. Yeretzian, C., A. Jordan, H. Brevard, W. Lindinger, Time resolved headspace analysis by proton-transfer-reaction mass-spectrometry, in *Flavor Release,* A.J. Taylor, D.D. Roberts, Eds., Amer. Chem Soc., Washington, D.C., 2000, p. 58.
81. Buettner, A., A. Beer, C. Hannig, M. Settles, P. Schieberie, Quantitation of the in-mouth release of heteroatomic odorants, in *Heteroatomic Aroma Compounds,* T.A. Reineccius, G.A. Reineccius, Eds., Amer. Chem. Soc., Washington, D.C., 2002, p. 296.
82. Taylor, A.J., R.S.T. Linforth, I. Baek, M. Marin, M.J. Davidson, Flavor analysis under dynamic conditions: measuring the true profile sensed by consumers, in *Frontiers of Flavour Science*, P.E. Schieberle, K.H. Engel, Eds., Deutsche Forschung. Lebensmit., Garching, 2000, p. 255.
83. Roberts, D.D., A.J. Taylor, Eds., *Flavor Release,* Amer. Chem Soc., Washington, D.C., 2000, p. 484.
84. Schieberle, P., K.H. Engel, Eds., *Frontiers of Flavour Science,* Deutsche Forschung. Lebenmit., Garching, 2000, p. 602.
85. Hodgins, D., D. Simmonds, Sensory technology for flavor analysis, *Cereal Foods World*, 40(4), p. 186, 1995.
86. Marsili, R., Combining mass spectrometry and multivariate analysis to make a reliable and versatile electronic nose, in *Flavor, Fragrance, and Odor Analysis*, R. Marsili, Ed., Marcel Dekker, New York, 2002, p. 349.
87. Rouseff, R.L. Bitterness in foods and beverages, *Develop. Food Sci.,* 25, p. 1, 1990.
88. Nielsen, S.S., Ed., *Food Analysis,* 2nd Ed., Aspen Publ., Gaithersburg, 1998, p. 630.
89. Engel, E., J.B. Lombardot, A. Garem, N. Leconte, C. Septier, J. Le Quere, C. Salles, Fractionation of the water-soluble extract of a cheese made from goats' milk by filtration methods: behaviour of fat and volatile compounds, *Internat. Dairy J.*, 12(7), p. 609, 2002.
90. Engel, E., C. Septier, N. Leconte, C. Salles, J.-L. Le Quere, Determination of taste-active compounds of a bitter Camembert cheese by omission tests, *J. Dairy Research*, 68(4), p. 675, 2001.
91. Preininger, M., R. Warmke, W. Grosch, Identification of the character impact flavor compounds of Swiss cheese by sensory studies of models, *Zeitschrift Lebensmit. Untersuch. Forsch.*, 202(1), p. 30, 1996.
92. Warmke, R., H.D. Belitz, W. Grosch, Evaluation of taste compounds of Swiss cheese (Emmentaler), *Zeitschrift fur Lebensmit. Untersuch. Forsch.*, 203(3), 1996, p. 230.
93. Pickenhagen, W., C-T Ho, A.M. Spanier, Eds., *The Contribution of Low- and Non-volatile Materials to the Flavor of Foods,* Allured Publ., Carol Stream, 1996, p. 242.
94. Taylor, A.J., R.S.T. Linforth, I. Baek, J. Davidson, M. Brauss, D.A. Gray, Flavor release and flavor perception, in *Flavor Chemistry*, S.J. Risch, C.-T. Ho, Eds., Amer. Chem. Soc., Washington, D.C., 2000, p. 151.
95. Ganga, A., F. Pinaga, A. Querol, S. Valles, D. Ramon, Cell-wall degrading enzymes in the release of grape aroma precursors, *Food Sci. Technol. Internat.,* 7(1), p. 83, 2001.
96. Hofmann, T., P. Schieberle, Influence of Melanoidins on the Aroma Staling of Coffee Beverage. Abstracts of Papers, 222nd ACS National Meeting, Chicago, August 26–30, 2001, p. AGFD-097.

97. Hofmann, T., M. Czerry, S. Calligaris, P. Schieberle, Model studies on the influence of coffee melanoidins on flavor volatiles of coffee beverages, *J. Agric. Food Chem.*, 49(5), p. 2382, 2001.
98. Ebeler, S.E., Phytochemicals and wine flavor, *Recent Advances Phytochem.*, 31, p. 155, 1997.
99. Ebeler, S.E., J.S. Aronson, and D.-M. Jung, Sensory Analysis and Analytical Flavor Chemistry: Some Missing Links. Abstracts of Papers, 224th ACS National Meeting, Boston, MA, United States, August 18–22, 2002, p. AGFD-015.
100. Jung, D.-M., J.S. de Ropp, S.E. Ebeler, Application of pulsed field gradient NMR techniques for investigating binding of flavor compounds to macromolecules, *J. Agric. Food Chem.*, 50(15), p. 4262, 2002.
101. BeMiller, J.N., N.H. Low, Carbohydrate analysis, in *Food Analysis,* 2nd Ed., S.S. Nielsen, Ed., Aspen Publ., Gaithersburg, 1998, p. 167.
102. Hendricks, D.G., Mineral analysis, in *Food Analysis*, 2nd Ed., S.S. Nielsen, Ed., Aspen Publ., Gaithersburg, 1998, p. 151.
103. Rounds, M.A., J.F. Gregory, High performance liquid chromatography, in *Food Analysis,* 2nd Ed., S.S. Nielsen, Ed., Aspen Publ., Gaithersburg, 1998, p. 509.
104. Miller, D.D., Atomic absorption and emission, in *Food Analysis*, 2nd Ed., S.S. Nielsen, Ed., Aspen Publ., Gaithersburg, 1998, p. 425.
105. Maga, J.A., Flavour potentiators, *CRC Crit. Reviews Food Sci Nutrit.*, 18(3), p. 231, 1983.
106. Sugita, Y.H., Flavor enhancers, *Food Sci. Technol.*, 116, p. 409, 2002.
107. Charalambous, G., K.J. Buckner, W.A. Hardwick, T.J. Weatherby, Determination of beer flavor compounds by high pressure liquid chromatography: IV. Nucleotides, *Tech. Quart.-Master Brew. Assn. Am.*, 12(4), p. 203, 1975.
108. Duran Meras, I., F. Salinas, A.M. de la Pena, M. Lopez Rosas, Simultaneous determination of flavor enhancers inosine 5′-monophosphate and guanosine 5′-monophosphate in food preparations by derivative spectrophotometry, *J. Amer. Oil Chem. Soc. Intern.*, 76(4), p. 754, 1993.
109. Hibi, K., M. Bounoshita, Use of high-performance liquid chromatography in the analysis of bitter compounds in food products, *Shokuhin Kagaku*, 31(7), p. 120, 1989.
110. Adler Nissen, J., Bitterness intensity of protein hydrolyzates — chemical and organoleptic characterization, in *Frontiers of Flavor,* G. Charalambous, Ed., Elsevier Science Publ., Amsterdam, 1988, 17, p. 63.

# 4 Flavor Formation in Fruits and Vegetables

Knowledge of precursors and pathways leading to the formation of flavor in fruits and vegetables has progressed slowly over the years. Historically, emphasis has been placed on determining what constitutes flavor rather than on the mechanisms of flavor formation. Also, the task of elucidating metabolic pathways in very complex biological systems is difficult at best. The continued demand for natural flavorings has renewed interest in this research area since a knowledge of biological pathways facilitates their production (or enhancement) under conditions that permit their labeling as natural. This most often involves biotechnology (enzymology or fermentation).

This chapter will provide an overview of how flavor is formed in plants. Discussion will primarily focus on aroma compounds as opposed to taste compounds. The reader is encouraged to read other more comprehensive sources such as Reineccius and Reineccius [1], Taylor and Mottram [2], Teranishi et al. [3], Schieberle and Engel [4] or Rouseff and Leahy [5] for detail in this subject area.

## 4.1 INTRODUCTION

In the biological sciences, when we find some chemical component in a plant, we often search for an understanding of why this component is in the plant, i.e., determine its function. Considering that flavor components are reasonably low molecular weight materials, we might expect that they are a part of the normal biosynthesis process. While this is often true for the nonvolatile components of flavor (i.e., taste stimuli such as sugars or acids), this is generally not true for aroma components. It is interesting that few if any aroma compounds are known to serve a biological function in the plant (fruit or vegetable). In fact, most aroma compounds arise as a result of degradation reactions. In fruits, the plant cell walls soften and internal organization is lost during ripening. This loss of organization permits enzymes normally associated with growth to attack various substrates normally not available to the enzymes. Enzymes typically involved in synthetic processes are involved in degradation reactions. This attack results in the formation of a host of low molecular weight products (volatile aroma compounds) many of which have significant sensory properties.

In vegetables, there is no ripening period — aroma compounds are formed via enzymatic reactions upon *cellular damage*. An onion, for example, has no characteristic aroma until cellular damage occurs. This damage may be the result of preparation ( e.g., slicing, dicing, or cooking), or simply eating (biting and chewing). The enzymatic processes are rapid resulting in full aroma in seconds.

If flavor is the result of an evolutionary process, it may be in how it influences animal behavior or microbial degradation. Animals may be attracted to a pleasant flavor, promoting the consumption of the plant and the spread of its seeds. An unpleasant flavor (unripe) may deter an animal from eating the plant until it has had time to come to reproductive maturity (or at all). Some flavor components have antimicrobial properties that may protect the plant's reproductive elements to enhance survival. While much of this rationale for existence is conjecture, it is unlikely that human pleasure figured into this process. We are likely simple benefactors of a long-term, perhaps evolutionary, process.

## 4.2 BIOGENESIS OF FRUIT AROMA

As described earlier, the typical flavor of fruits (e.g., bananas, peaches, pears, and cherries) is not present during early fruit formation but develops entirely during a rather brief ripening period. This flavor development period, or ripening, occurs during the climacteric rise in respiration. During this period, metabolism of the fruit changes to catabolism, and flavor formation begins. Minute quantities of lipids, carbohydrates, proteins, and amino acids are enzymatically converted to simple sugars or acids and volatile compounds. The rate of flavor formation reaches a maximum during the postclimacteric ripening phase.

An overall view of flavor formation during ripening of fruit is shown in Figure 4.1 [6]. Flavors are formed from major plant constituents under genetic control. Each metabolic pathway is connected to other metabolic pathways. As direct products of a metabolic pathway, or as a result of interactions between pathways or end products, a host of volatile compounds are produced which contribute to the flavor (aroma) of a ripe fruit. The precursors of some aroma compounds are presented in Table 4.1 [7].

### 4.2.1 Aroma Compounds from Fatty Acid Metabolism

Volatile flavor compounds may be formed from lipids via several different pathways. The primary pathways include α- and β-oxidation, and oxidation via lipoxygenase enzymes [8,9]. Some of the earliest work in this area was on the development of banana aroma by Tressl and Drawert [10]. Recent work showing pathways for the formation of several key volatiles derived from apple lipids are illustrated in Figure 4.2. They demonstrated the conversion of labeled acetate to acetate esters and labeled butanoate to butanoate esters by postclimacteric banana slices. They have further shown the conversion of hexanoic acid to hexanol by these tissues.

The formation of flavors via β-oxidation is exemplified by considering flavor formation in pears. The decadienoate esters are generally considered carriers of the flavor of pear [11]. These esters are formed via β-oxidation of linoleic acid (Figure 4.3). Linoleic acid is metabolized, two carbons at a time, to shorter chain:CoA derivatives that react with alcohols to yield esters. During this process, isomerizations may occur to yield the trans-cis isomers.

The widest variety of flavor compounds formed from lipids arises via lipoxygenase activity. Many of the aliphatic esters, alcohols, acids, and carbonyls found

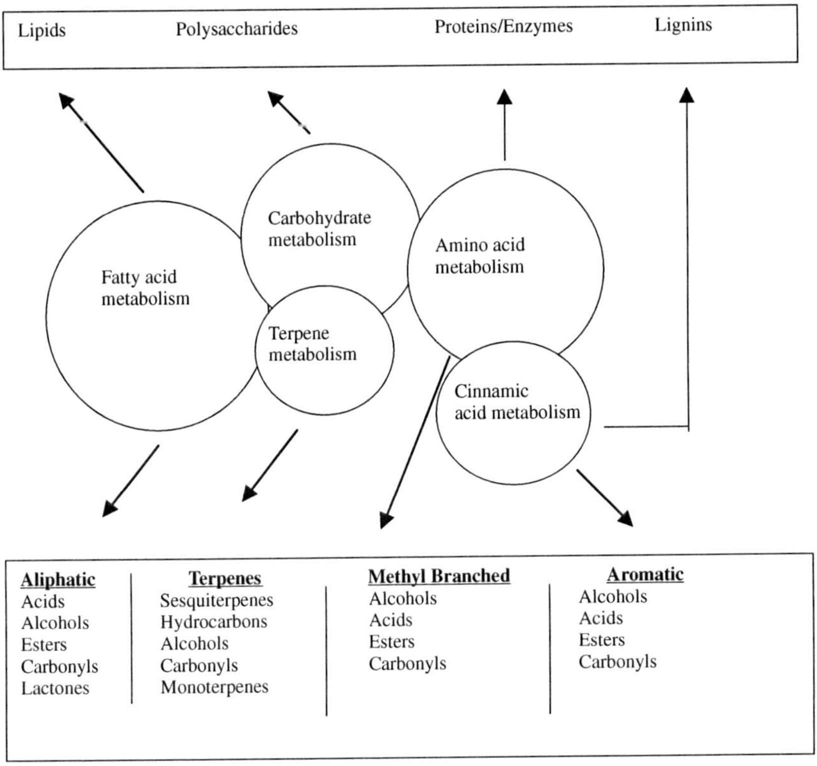


**FIGURE 4.1** The formation of fruit aroma from major food components. (From Tressl, R., M. Holzer, M. Apetz, *Aroma Research*, H. Maarse, P.J. Groenen, Eds., Wageningen, Netherlands, 1975, p. 41. With permission.)

in fruit are derived from the oxidative degradation of linoleic and linolenic acids. Studies on the generation of flavor in banana [10], tomato [12], apple [13], tea and olive oil [14] elucidate and point out the importance of lipoxygenase activity to flavor development in fruit.

The mechanism of lipid oxidation via lipoxygenase in tomato has been studied by several workers. Jadha et al. [15] originally proposed that 9-, 12-, 13- and 16-hydroperoxides were formed via enzymatic action on linolenic acid. Later, Galliard and Matthew [16] reported that only the 9- and 13-hydroperoxides are formed (95:5 ratio, respectively). Further, a hydroperoxide cleavage system was found by Galliard et al. [17] only for the 13-hydroperoxide which makes the minor hydroperoxide the greater contributor to flavor (Figure 4.4). Schreier and Lorenz [18] presented a detailed study on lipoxygenase and the hydroperoxide cleavage-enzyme in tomatoes. The 9-hydroperoxide is not further degraded enzymatically but only slowly decomposes to volatile compounds. The pathway presented in Figure 4.4 outlines only the formation of hexanal and cis-3-hexenal, two of the more abundant volatiles in tomato. In fact, the oxidation of linoleic and linolenic produces a large number of volatiles. Acids and ketones as well as other intermediates in the oxidation

**TABLE 4.1**
**Precursors of Aroma Compounds in Foods**

| Nutrient | Aroma Component |
|---|---|
| **Carbohydrates** | **Organic acids:** |
| Glucose | Pyruvic acid, acetic acid, propionic acid, acetoacetic acid, butyric acid, hexanoic |
| Fructose | acid, octanoic acid |
| Sucrose | **Esters:** |
| | Pyruvates, acetates, propionates, butyrates, acetonacetates, hexanoates, octanoates |
| | **Alcohols:** |
| | Ethanol, propanol, butanol, hexanol, octanol |
| | **Aldehydes:** |
| | Acetaldehyde, propanal, butanal, hexanal, octanal |
| | **Terpenes:** |
| | Monoterpene, linalool, limonene, α-pinene, citronellal, citral, geranial |
| **Amino Acids** | Pyruvic acid, acetaldehyde, ethanol |
| Alanine | Isopropanal, isopropanol, α-keto-isobutyric acid |
| Valine | 3-Methylbutanal, 3-methylbutanol, α-keto-isocaproic acid |
| Leucine | 2-Methylbutanal, 2-methylbutanol |
| Isoleucine | Benzaldehyde, phenylacetaldehyde, cinnamaldehyde |
| Phenylalanine | Hydrocinnamaldehyde, p-hydroxybenzaldehyde |
| | p-Hydroxy phenylacetaldehyde, p-hydroxy cinnamaldehyde, p-hydroxy cinnamaldehyde |
| Serine | |
| Theronine | Pyruvic acid |
| Glycine | Thiazoles |
| Cystine/cysteine | Glyoxal |
| Serine | |
| **Fatty acids** | trans-2-trans-4-Decadienal, hexanal, trans-2-octenal |
| Linoleic acid | trans-2-Pentanal, trans 2-hexenol, hexanal |
| | cis-3-Hexenal, cis-3-hexenol |
| | trans-2-trans-4-Heptadienal, propanal |
| **Vitamins** | |
| Carotene | |
| β-Carotene | β-Ionone |

*Source:* Salunkhe, D.K., J.Y. Do, *CRC Crit. Reviews Food Sci. Nutrit.*, 8(2), p. 161, 1976. With permission.

process are readily converted to alcohols, aldehydes, and esters by other enzyme systems in the plant (Figure 4.5). Both α- and β-oxidation pathways have been demonstrated to exist thereby providing a wide range of volatiles for further conversion to flavor compounds [19].

The discussion above outlines how many volatile compounds are formed from lipid degradation. Yet, one may be inclined to question the importance of these pathways since there is very little lipid in fruits. First, one must recognize that many aroma compounds have low sensory thresholds (ppm). It does not take much precursor to yield ppm quantities of aroma compounds. Second, there is a significant

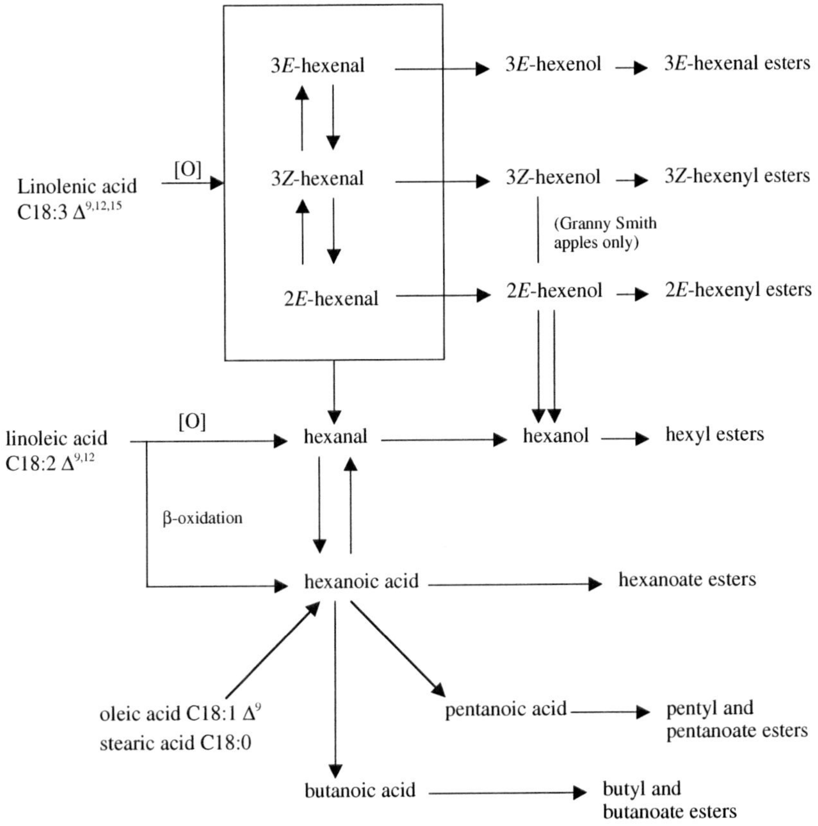


**FIGURE 4.2** Biosynthetic pathways leading to straight-chain ester volatiles in Granny Smith and Red Delicious apples. ([O] is lipoxygenase catalyzed oxidation.) (From Rowan, D., J.M. Allen, S. Fielder, M. Hunt, *J. Agric. Food Chem.*, 47, p. 2553, 1999. With permission.)

quantity of linoleic and linolenic acid in plant chloroplasts. As fruits ripen, they loose their green color due to chloroplast degradation, which then releases membrane lipids which are rich in these key aroma precursors. These precursors then enter the pathways described above to form the host of esters and carbonyls that characterize the aroma of many fruits [13].

### 4.2.2 Aroma Compounds from Amino Acid Metabolism

Amino acid metabolism generates aromatic, aliphatic, and branched chain alcohols, acids, carbonyls, and esters that are important to the flavor of fruit. Yu et al. [20–22] were amongst the earliest researchers to demonstrate that valine, leucine, alanine, and aspartic acid can be converted to short chain carbonyls by tomato extracts. The enzymes involved in these transformations were found to be located in different sites in the tomato. The soluble tomato fraction isolated via centrifugation acted on leucine while the mitochondrial fraction metabolized both alanine and aspartic acid. The

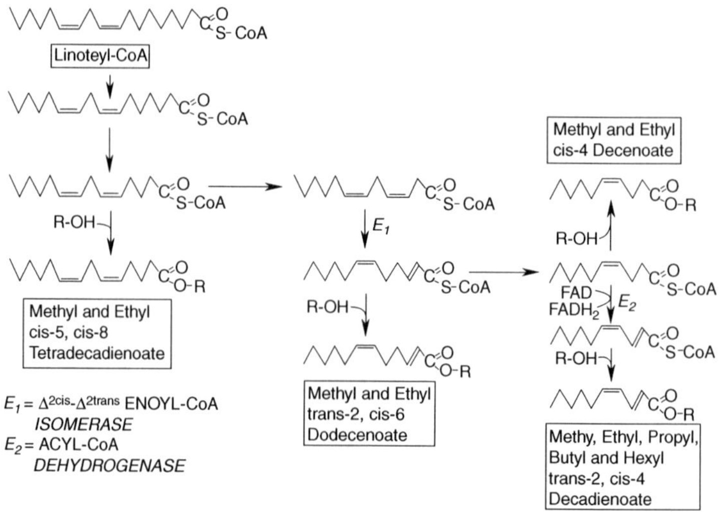


**FIGURE 4.3** Proposed pathway for the production of unsaturated esters in pears, El = Δ3cis-Δ2trans-enoyl-CoA isomerase; E2 = Acyl-CoA dehydrogenase. (From Tressl, R., M. Holzer, M. Apetz, *Aroma Research*, H. Maarse, P.J. Groenen, Eds., Wageningen, Netherlands, 1975, p. 41. With permission.)

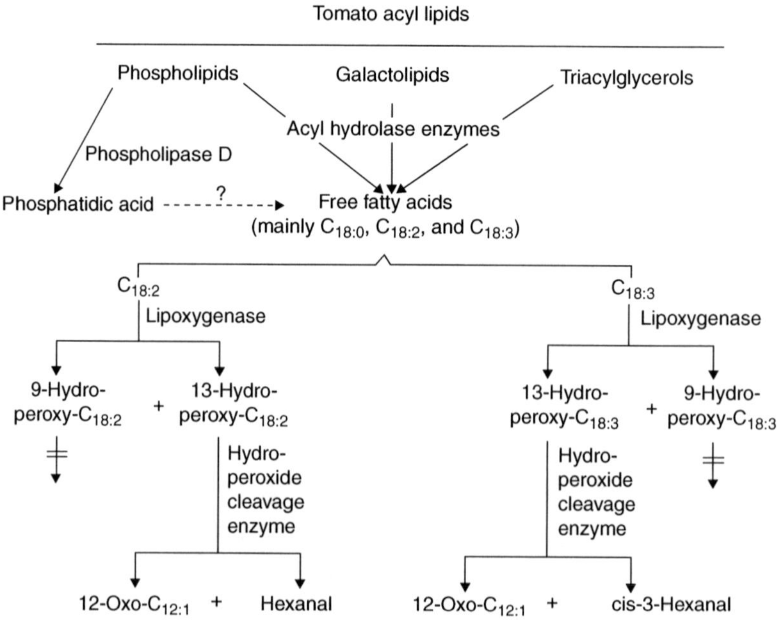


**FIGURE 4.4** Proposed pathway for the formation of short chain carbonyls by enzymatic degradation of acyl lipids in disrupted tomato fruits. (From Galliard, F., J.A. Matthew, A.J. Wright, M.J. Fishwick, *J. Sci. Food Agric.*, 28, p. 863, 1977. With permission.)

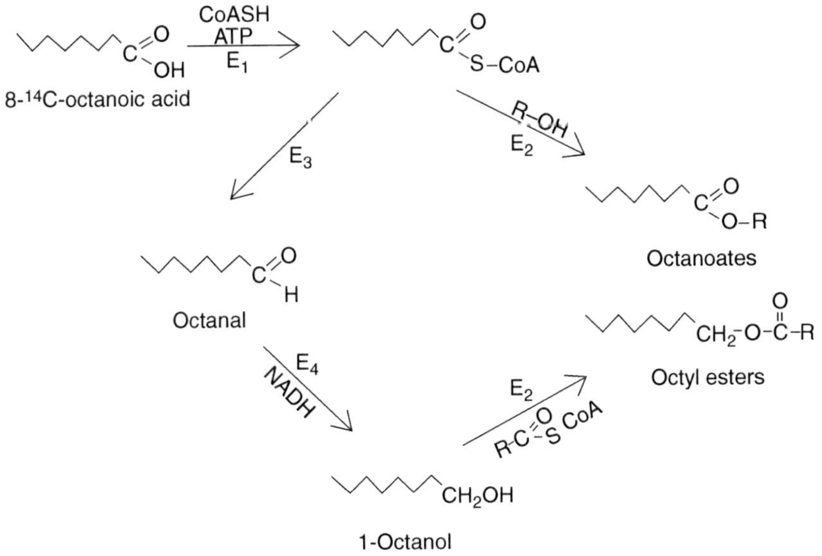


**FIGURE 4.5** Reaction scheme for conversion of octanoic acid into esters, $E_1$, acyl-thiokinase; $E_2$, acyl-CoA-alcohol-transacylase; $E_3$, acyl-CoA reductase; $E_4$, alcohol-NAD oxydoreductase. (From Tressl, R., F. Drawert, *J. Agric. Food Chem.*, 21, p. 560, 1973. With permission.)

activities of these enzymes were substantially higher in field-grown vs. hothouse tomatoes [23]. This undoubtedly accounts for some of the flavor differences between hothouse and field-grown tomatoes.

Studies on banana tissue slices have shown that valine and leucine concentrations increase about threefold following the climacteric rise in respiration [10]. Radioactive labeling studies have shown that valine and leucine are transformed into branched chain flavor compounds that are essential to banana flavor (2-methyl propyl esters and 3-methyl butyl esters, respectively). As can be seen in Figure 4.6, the initial step is deamination of the amino acid followed by decarboxylation. Various reductions and esterifications then lead to a number of volatiles that are significant to fruit flavor (acids, alcohols, and esters). Recent work has shown that amino acids play a role in apple flavor as well. For example, isoleucine is the precursor of 2-methyl butyl and 2-methyl butenyl esters in apples [24,25]. An unusual flavor compound, 2-isobutylthiazole, has been found to be important to the flavor of tomato. It is hypothesized that this compound is formed from the reaction of 3-methyl-1-butanal (from leucine) with cysteamine.

Aromatic amino acids also may serve as important precursors to fruit flavor. Some of the aromatic flavor compounds have been shown to come from tryosine and phenylalanine (Figure 4.7). Odors characterized as phenolic or spicy are formed via this pathway.

It is suggested that the cinnamic acids serve as intermediates for many of the aromatic flavor compounds existing in foods [27]. Simple esterification to methyl and ethyl cinnamate has been reported in strawberry [28] and guava [29]. Ethyl

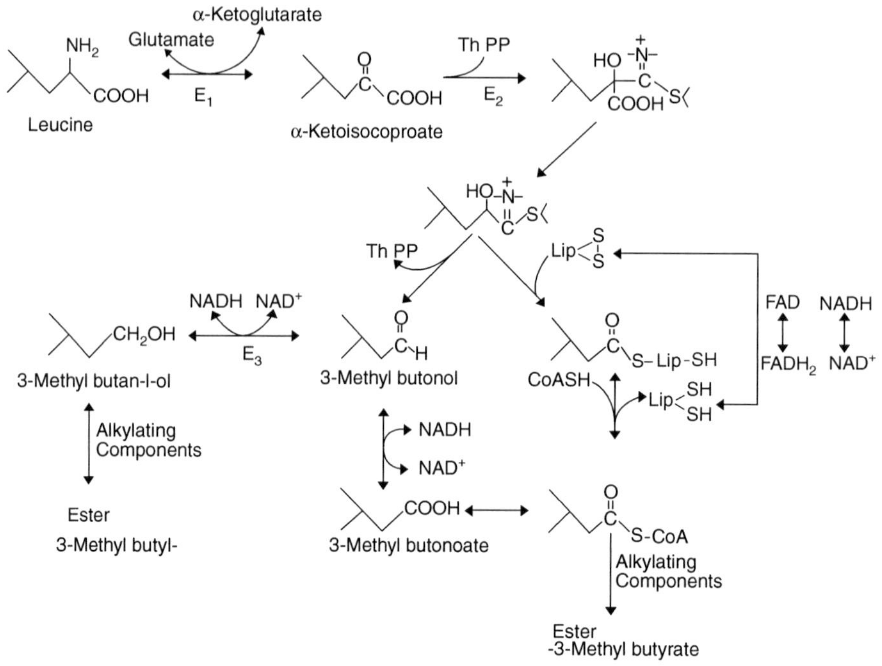


**FIGURE 4.6** Conversion of amino acids into aroma components of banana as illustrated by leucine. $E_1$, L-leucine aminotransferase; $E_2$, pyruvate decarboxylase; $E_3$, aldehyde dehydrogenase; ThPP, thiamin pyrophosphate; oxidized lipoic acid; reduced lipoic acid; FAD flavin-adenine dinucleotide; NAD+, oxidized nicotinamide-adenine dinucleotide; CoA-SH, coenzyme A. (From Drawert, F., *Aroma Research*, H. Maarse, P.J. Groenen, Eds., Pudoc, Wageningen, 1975, p. 245. With permission.)

cinnamate has been found in cranberry [30] and passion fruit [31]. Removal of an acetate group from the cinnamic acids yields benzoic acids. The benzoic acids may be further transformed by esterification to yield benzyl esters and by reduction to yield various benzaldehydes and benzyl alcohols. Decarboxylation would yield phenols.

### 4.2.3 Aroma Compounds Formed from Carbohydrate Metabolism

Returning briefly to Table 4.1, one can see that a large variety of volatile flavors can be traced to carbohydrate metabolism. If one considers plant metabolism, it is clearly established that plants obtain all of their energy directly from photosynthesis. The photosynthetic pathways involve turning $CO_2$ into sugars that are metabolized into other plant needs, e.g., lipids and amino acids. Therefore, one may state that nearly all plant flavors come indirectly from carbohydrate metabolism since all of the other flavor precursors come from carbohydrate metabolism. However, there are few flavor constituents that come directly from carbohydrate metabolism. Terpenes, for example, arise both from carbohydrate and lipid metabolism.

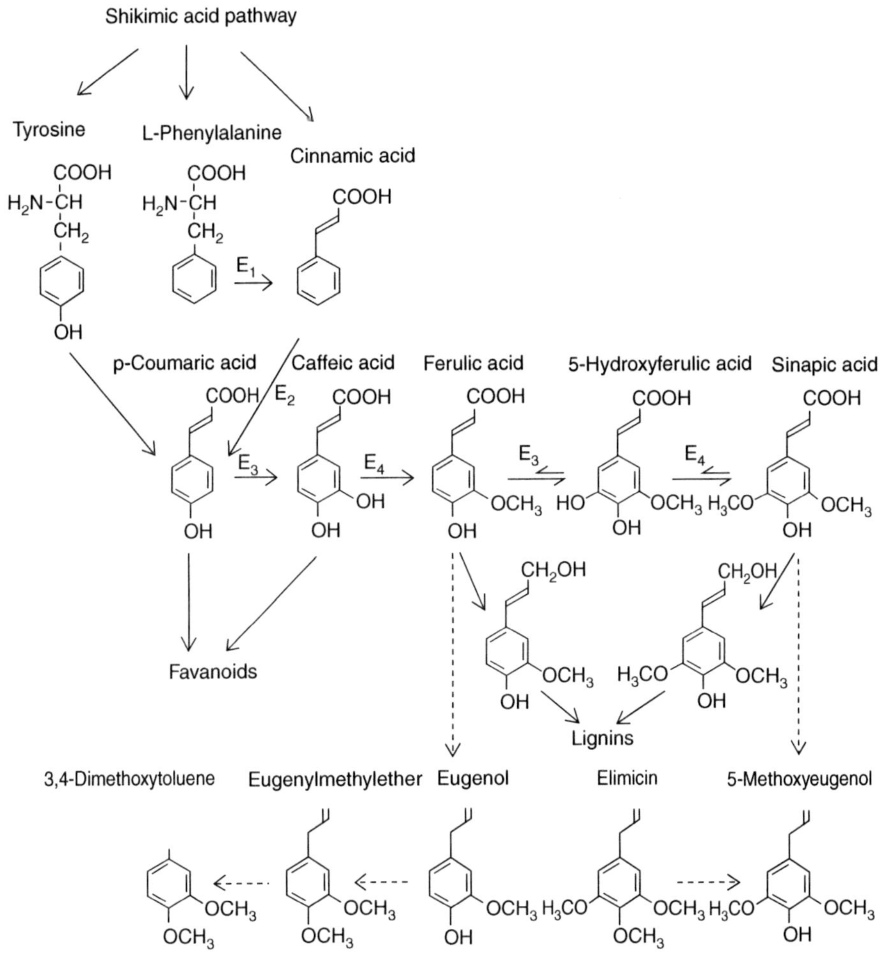


**FIGURE 4.7** Pathways for the formation of phenolic acids and phenol esters in banana. El, L-phenylalanine-ammonia-lyase; E2, cinnamic-acid-hydroxylase; E3, phenolase; E4, methyltransferase. (From Tressl, R., F. Drawert, *J. Agric. Food Chem.*, 21, p. 560, 1973. With permission.)

Terpenes are classified by the number of isoprene units they contain. Monoterpenes contain 2 isoprene units (10 carbons), sesquiterpenes contain 3 isoprene units (15 carbons) and diterpenes contain 4 isoprene units (20 carbons). Of these groups, the monoterpenes, and more specifically, the oxygenated monoterpenes, are considered most important to the aroma of certain fruits, e.g., citrus products. Limonene, a monoterpene hydrocarbon possessing little odor, is the major terpene in most citrus oils accounting for up to 95% of some oils. The oxygenated terpenes, often comprising less than 5% of the oil, generally provide the characteristic flavor of different citrus species. For example, citral is considered the flavor impact component of lemon oil, yet it is unusual to find more than 2% citral in a singlefold oil.

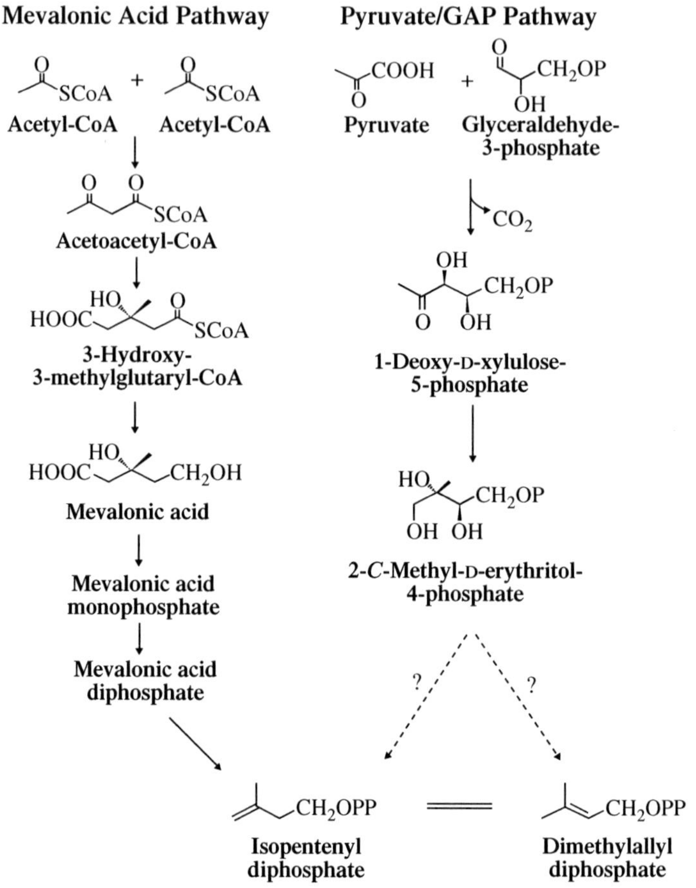


**FIGURE 4.8** Biosynthetic pathway for the formation of isopentenyl diphosphate. (From Little, D.B., R.B. Croteau, *Flavor Chemistry: 30 Years of Progress*, R. Teranishi, E.L. Wick, I. Hornstein, Eds., Kulwer Academ., New York, 1999, p. 239. With permission.)

The biosynthetic pathway proposed for the synthesis of isopentenyl diphosphate (IPP, isoprene building block) is shown in Figure 4.8 [32]. This pathway combines three acetyl-Co-A molecules to form 3-hydroxy-3-methylglutaryl-Co-A (HMG-CoA). The HMG-CoA is reduced to mevalonic acid which is then phosphorylated and decarboxylated to form IPP, the key building block in terpene biosynthesis synthesis. Lichtenthaler [33] has found that IPP can also be formed via a pathway that does not include mevalonic acid. While there is evidence to support the existence of this alternative pathway, it has not been adequately determined.

The next series of steps involves the combination of IPP units to form geranyl diphosphate, farnesyl diphosphate, and geranylgeranyl diphosphate, which serve as the precursors of the mono-, sesqui- and diterpene families, respectively [Figure 4.9]. In effect, all of the other members of these families arise from the cyclization or secondary modification of these three precursors.

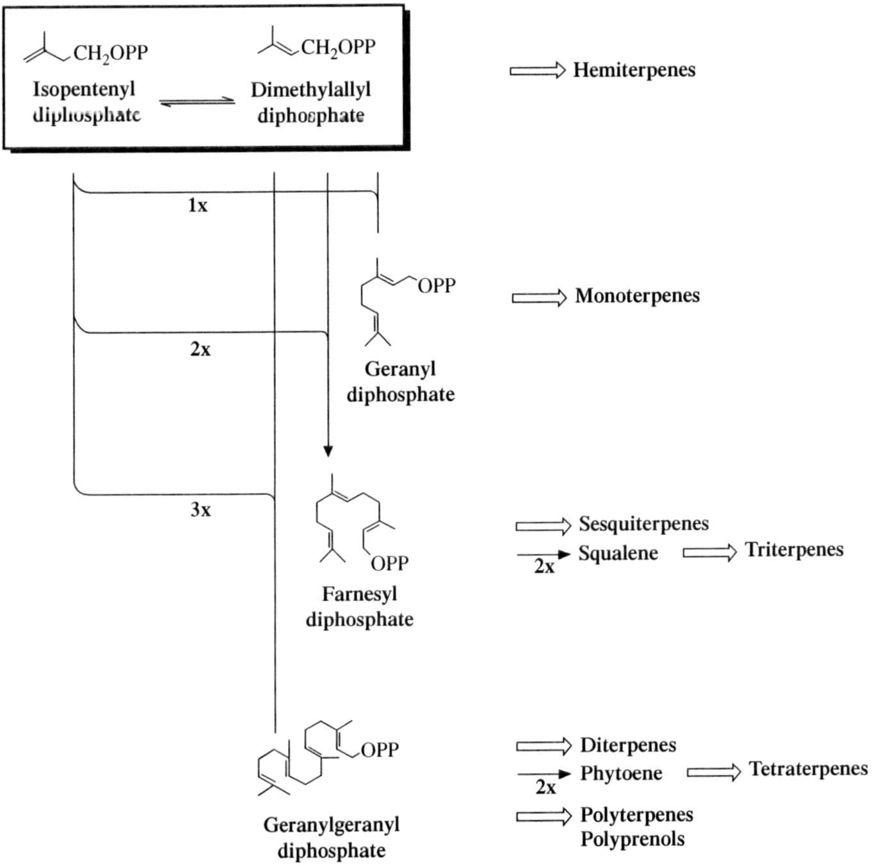


**FIGURE 4.9** Formation of different families of terpenes from the universal precursor isopentylphosphate (IPP). (From Little, D.B., R.B. Croteau, *Flavor Chemistry: 30 Years of Progress*, R. Teranishi, E.L. Wick, I. Hornstein, Eds., Kulwer Academ., New York, 1999, p. 239. With permission.)

In addition to terpenes, carbohydrates serve as a precursor of furanones in plants. The formation of 2,5-dimethyl-4-hydroxy-2H-furan-3-one (DMHF) is an example. This compound, and various forms thereof, are key to the aroma of strawberry. Several researchers have studied the biochemical pathways responsible for the formation of this compound [34,35]. The pathway presented in Figure 4.10 illustrates DMHF formation from its key precursor, 6-deoxy-D-fructose.

## 4.3 BIOGENESIS OF VEGETABLE AROMA

As mentioned earlier, the process leading to aroma formation in vegetables is quite different than that of fruits. Vegetables do not have a ripening period as fruits do. While some vegetables will develop flavor during growth (much of taste and a limited

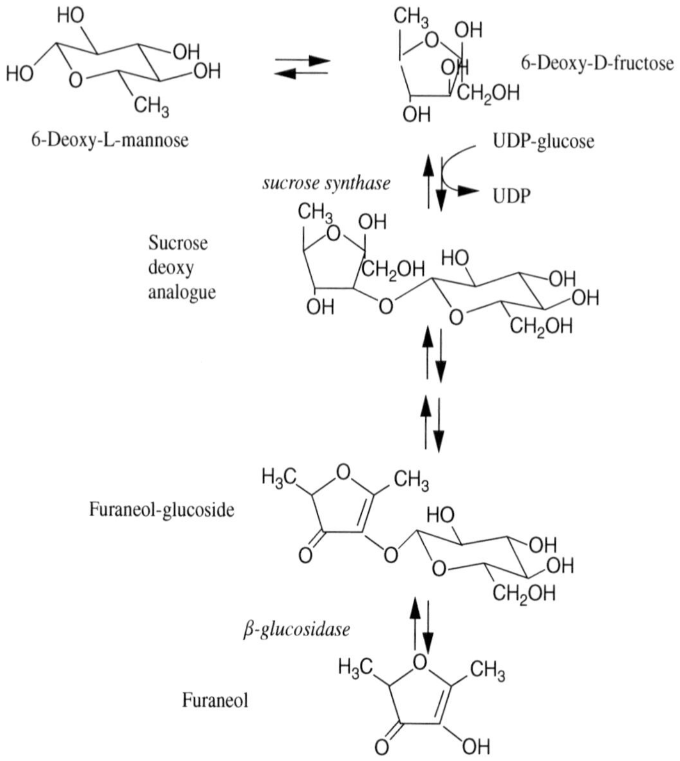


**FIGURE 4.10** The biosynthetic pathway for the formation of furaneol® and its derivatives in strawberry. (From Roscher, R., P. Schreier, W. Schwab, Metabolism of 2,5-dimethyl-4-hydroxy-2H-furan-3-one in detached ripening strawberry fruits, *J. Agric. Food Chem.*, 45(8), p. 3202, 1997. With permission.)

number of aroma components), the remaining flavor (particularly aroma) develops during cellular disruption. Cellular disruption permits the mixing of enzymes and substrates, which had been separated within the cell, thereby resulting in the generation of volatile substances. A few examples of vegetables that contain a typical aroma prior to cellular disruption are celery (contains phthalides and selines), asparagus, (contains 1,2-dithiolane-4-carboxylic acid), and bell pepper (contains 2-methyl-3-isobutyl pyrazine).

An overview of the metabolic pathways leading to vegetable aroma formation is presented in Figure 4.11. Similar to aroma development in fruit, fatty acid, carbohydrate, and amino acid metabolism serve to provide the precursors of vegetable aroma. However, while sulfur-containing volatiles are important to some fruits (e.g., passion fruit, grapefruit, pineapple, and blackcurrant [36]), sulfur-containing volatiles are generally much more important to vegetable flavor. This is due to the type of sulfur-containing flavor precursors present in vegetables vs. fruits. In fresh vegetables, thioglucosinolates and cysteine sulfoxides serve as primary precursors

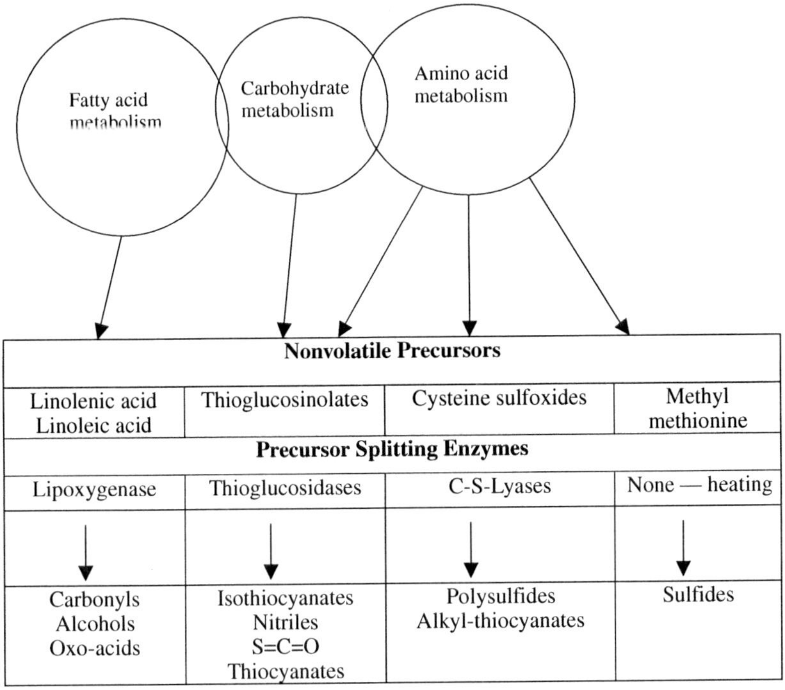


**FIGURE 4.11** The formation of vegetable aroma from major food components. (From Tressl, R., M. Holzer, M. Apetz, *Aroma Research*, H. Maarse, P.J. Groenen, Eds., Wageningen, Netherlands, 1975, p. 41. With permission.)

of volatile sulfur compounds while S-methyl methionine is an important precursor to the flavor of some cooked vegetables, e.g., corn.

### 4.3.1 The Role of Lipids in Vegetable Aroma Formation

Despite the earlier comment about sulfur-containing flavor precursors being key to the flavor of most vegetables, lipid degradation serves as a *contributor* to the aroma of nearly all vegetables, some in a very minor way and others major, e.g., cucumbers and lambs lettuce (Valerianella *locusta*). Lipids contribute to vegetable aroma primarily via lipoxygenase pathways. Similar to lipoxygenase activity in fruits, lipoxygenase attacks linoleic and linolenic acid to yield hydroperoxides. However, the lipoxygenases are unique for each vegetable as are the hydroperoxide lyases that act on the hyroperoxides to yield selected aldehydes and alcohols. As an example of this pathway, a schematic outlining the formation of key aroma components in cucumber is presented in Figure 4.12 [37]. The unsaturation in the fatty acid is carried through to produce very potent unsaturated aldehydes and alcohols that characterize the aroma of cucumber. It is noteworthy that vegetables generally contain very few esters apparently lacking the necessary enzyme systems to form esters.

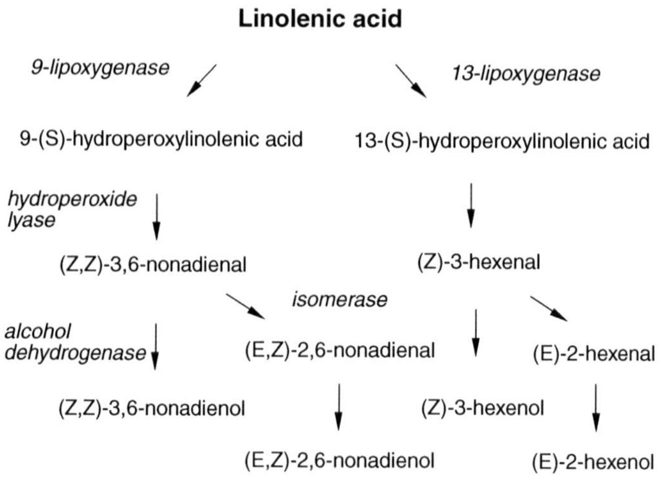


**FIGURE 4.12** Formation of volatiles from linolenic acid via the lipoxygenase pathway in cucumber. (From Takeoka, G., *Flavor Chemistry: 30 Years of Progress*, R. Teranishi, E.L. Wick, I. Hornstein, Eds., Kulwer Academ., New York, 1999, p. 287. With permission.)

### 4.3.2 Aroma Compound Formation from Cysteine Sulfoxide Derivatives

Undoubtedly the best illustration of flavor formation from cysteine sulfoxide precursors comes from studies on onion flavor. Intact onion cells have virtually no characteristic flavor until cellular disruption (cutting, blending, chewing, etc.). Only seconds after cell damage, the onion develops full flavor. This rapid formation of flavor is characteristic of the genus *Allium* as a whole (e.g., onions, garlic, leek, and chives). S-Alk(en)yl-L-cysteine sulfoxides serve as precursors of flavor in this genus. An investigation of 27 *Allium* species found that they could be placed into three groups based on the predominance of certain amino acid precursors [38]. The predominant precursors were S-l-propyl (onion), S-2-propenyl (garlic), and S-methyl cysteine sulfoxides *(A. aflaturense* B. Fedtschenko). A pathway for the formation of S-l-propenyl cysteine sulfoxide was been proposed by Granroth [39]. This pathway starts with the deamination of valine followed by decarboxylation to yield methacrylate. Reaction of methacrylate with L-cysteine followed by decarboxylation yields 1-propenyl-cysteine sulfoxide (Figure 4.13).

The mechanism of flavor formation from the alkyl cysteine sulfoxide (alliin) is outlined in Figure 4.13. While the initial steps of this pathway are enzymatic, reactions beyond sulfenic acid are purely chemical in nature. Sulfenic acid is extremely reactive, readily forming the unstable thiosulfinate intermediate by reaction with a second sulfenic acid molecule. This thiosulfinate decomposes to form a relatively stable thiosulfonate and mono, di- and trisulfides. Considering that several different alkyl precursors are available in each *Allium* species, a host of different mono-, di-, and trisulfides can be formed via different sulfenic acid combinations. It is these mono-, di-, and trisulfides that are most important in determining typical *Allium* flavor.

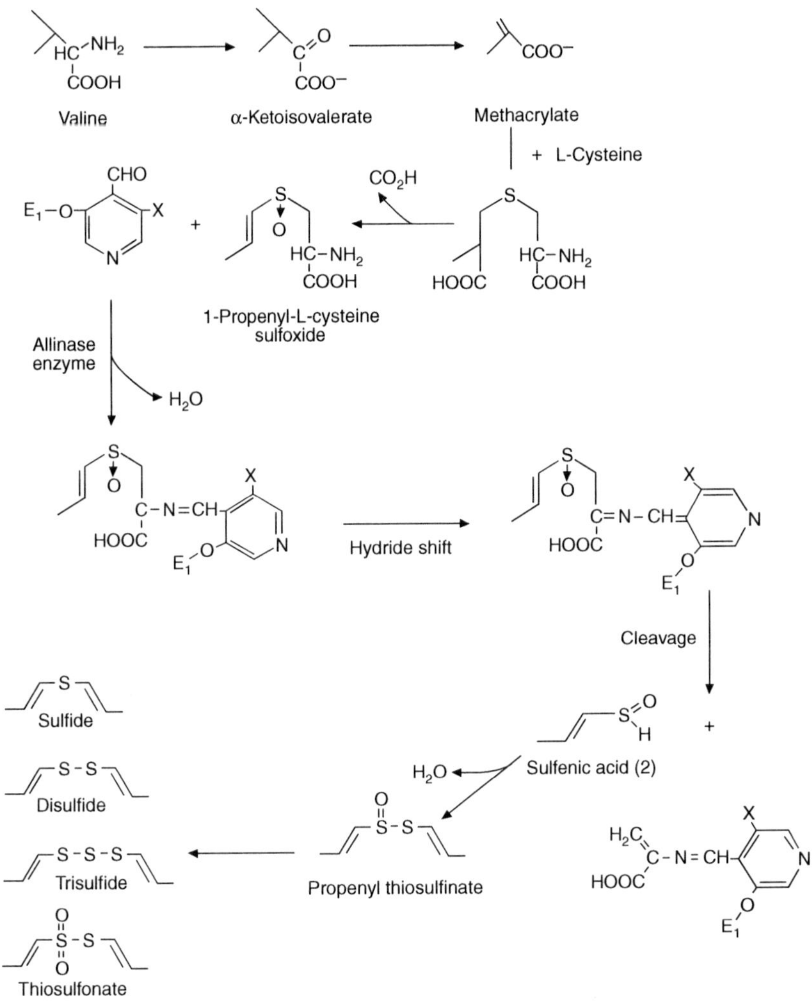


**FIGURE 4.13** Biosynthesis of S-l-propenyl-L-cysteine sulfoxide, and subsequent flavor formation in onions. (From Granroth, B., *Ann. Acad. Sci. Fenn. Ser.*, A2, p. 154, 1970. With permission.)

The lachrymatory factor in onion (thiopropanal-S-oxide) is considered to come from the S-l-propenyl-cysteine sulfoxide precursor [40]. Thiopropanal-S-oxide ($CH_3$-$CH_2$-CH=S=O) is unstable reacting with pyruvate to form propanol, 2-methyl pentanal and 2-methyl pent-2-enal [41]. It is of interest that no lachrymatory properties were found in 25 species of wild onions [42]. This would suggest that the 1-propenyl precursor may be absent from the wild onion.

Matikkala and Virtanen [43] found that nearly half of the cysteine sulfoxide precursors of onion are bound up as a γ-glutamyl peptides. The alliinase enzyme cannot act upon these cysteine sulfoxide flavor precursors until they are freed from the glutamyl peptide. This can be accomplished only via the action of a γ-L-glutamyl transpeptidase enzyme. This enzyme transfers the glutamyl moiety to another amino acid. The glutamyl transpeptidases are present only in sprouted onions, not mature onions [44]. Therefore, nearly one-half of the flavor potential of a mature onion is not available.

The Brassica plants also obtain a flavor contribution from alkyl cysteine sulfoxide precursors. The Brassica group includes broccoli, brussel sprouts, cabbage, cauliflower, and rutabagas. S-Methyl-L-cysteine sulfoxide is the primary cysteine sulfoxide derivative found in this family. An investigation of fresh cabbage flavor found that 6% of the volatiles was dimethyl disulfide, 6% dimethyl trisulfide, 3% dimethyl tetrasulfide, and 1.5% methyl ethyl trisulfide [6].

Rather unusual sulfur-containing volatiles are formed from cysteine sulfoxide precursors in the Shiitake mushroom (Figure 4.14). Similar to the onion, the cysteine sulfoxide flavor precursor is bound to a γ-glutamyl peptide. Therefore, the first step in flavor formation is the removal of glutamic acid by γ-glutamyl transpeptidase. The remaining steps are also similar to onion but yield lenthionine, the major flavor component in this mushroom [45].

### 4.3.3 Glucosinolates as Aroma Precursors in Vegetables

Most of the vegetables belonging to the Cruciferae family depend to varying degrees on glucosinolate precursors for aroma. More than 50 different glucosinolates have been identified [6]. Glucosinolates are nonvolatile flavor precursors, which are enzymatically hydrolyzed to volatile flavors when cellular structure is disrupted. The initial products are isothiocyanates and nitriles. Secondary reactions lead to the formation of several other classes of flavor compounds. Figure 4.15 outlines the pathways proposed to explain the formation of flavor compounds in the radish. The glucose moiety is initially hydrolyzed from the glucosinolate. This results in an unstable molecule that readily splits out $HSO_4^-$. Depending upon molecule rearrangements, either an isothiocyanate or a nitrile result. Further reactions result in the formation of thiols, sulfides, disulfides, and trisulfides. The contribution of isothiocyanates to the flavor of cabbage has been known for many years. Jensen et al. [46] were the first to identify allyl isothiocyanate in cabbage. Bailey et al. [47] added methyl, n-butyl, butenyl, and methyl thiopropyl isothiocyanates to the list of isothiocyanates of cabbage. Allyl isothiocyanate is the most important isothiocyanate to the flavor of cabbage.

### 4.3.4 Additional Pathway for Vegetable Flavor Formation

Terpenes are rather common to vegetables [48]. They arise from mechanisms similar to those active in fruit biosynthesis.

## 4.4 GLYCOSIDICALLY BOUND AROMA COMPOUNDS

Some aroma compounds are bound as glycosides in fruits and vegetables, and thus make no contribution to food aroma unless released [49–51]. They may be released

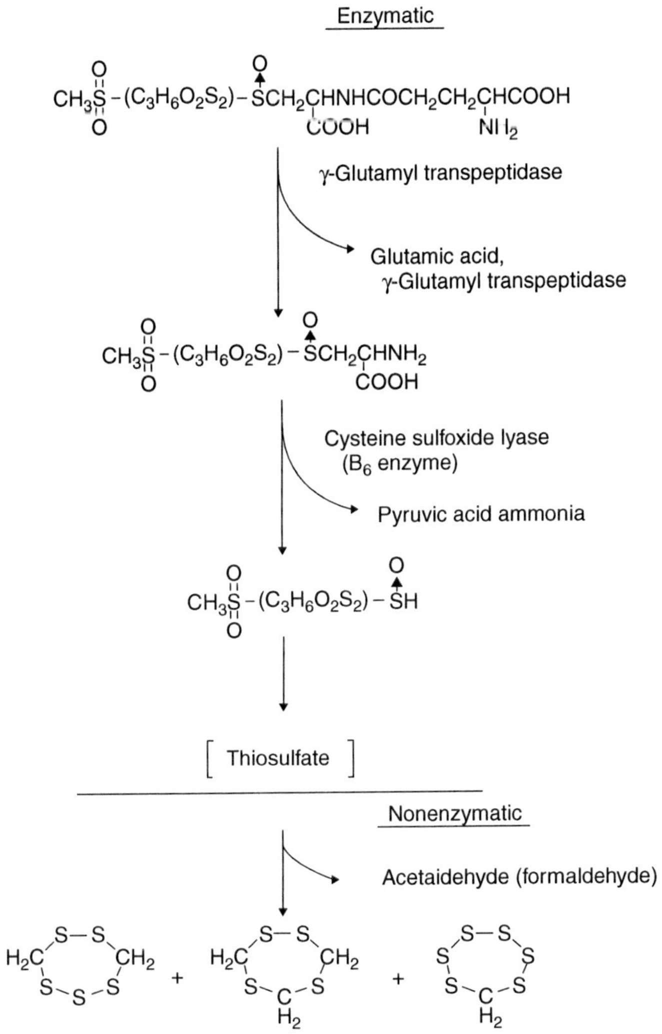


**FIGURE 4.14** Pathway proposed for the enzymatic and nonenzymatic formation of lenthionine from lentinic acid. (From Yasumoto, K., K. Iwami, H. Mitsuda, *Agric. Biolog. Chem.*, 35 (13 Suppl), p. 2070, 1971. With permission.)

from the carbohydrate moiety during maturation, storage, processing, or aging by enzymes, acid, or heat. They were first identified by Bourquelot and Bridel [52] in *Pelargonium odoratissimum* [51]. Since this discovery, this has been a very active area of research. At this time, more than 50 plant families have been found to contain glycosidically bound aroma precursors [53].

Research has been driven by two factors, the first being the desire to enhance flavor strength, or yield, in an application. The pool of glycosidically bound aroma compounds in some plants exceeds the free aroma pool by as much as 10:1, thus

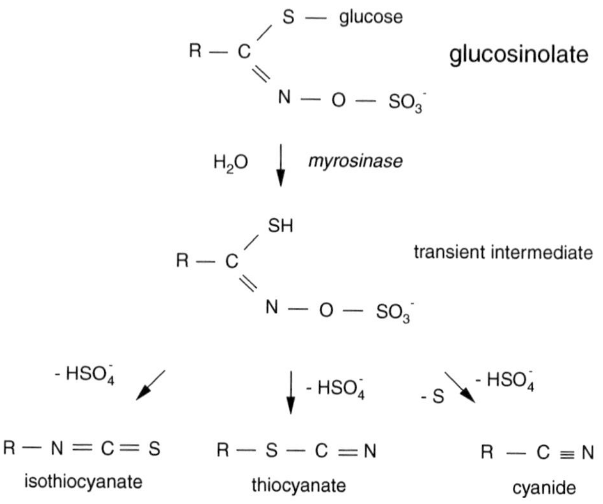


**FIGURE 4.15** The conversion of glucosinolate precursors to aroma compounds by myrosinase. (From Takeoka, G., *Flavor Chemistry: 30 Years of Progress*, R. Teranishi, E.L. Wick, I. Hornstein, Eds., Kulwer Academ., New York, 1999, p. 287. With permission.)

**TABLE 4.2**
**Free and Bound Terpene Alcohols in Grapes, Apricot, Mango, and Passion Fruit**

| Fruit | Free Compounds | Bound Compounds | Bound/Free |
|---|---|---|---|
| Grapes (Muscat) | 1.4 | 6.3 | 5.2 |
| Apricot (Rouge du Roussillon) | 1.0 | 5.2 | 5.2 |
| Mango (African, ungrafted) | 1.4 | 5.3 | 3.8 |
| Purple passion fruit (Zimbabwe) | 1.6 | 4.8 | 3.1 |
| Yellow passion fruit (Cameroun) | 1.5 | 1.5 | 1.0 |
| Maracuja (Columbian) | 0.5 | 1.1 | 2.2 |

*Note:* (mg/kg, expressed as linalool)

*Source:* Crouzet, J., *Functionality of Food Phytochemicals*, T. Romeo, Ed., Plenum Press, New York, 1997, p. 197. With permission.

the ability to free this aroma pool would appear to be desirable (Table 4.2). A second factor driving research is that the aging of some products, most notably wine, was postulated to be related to the freeing of some of the bound aroma constituents over time, thereby maturing or balancing the aroma. If one could determine how to free the bound aroma components, it may be possible to obtain aged wines more quickly. There was not only academic interest behind this research but also considerable financial implications.

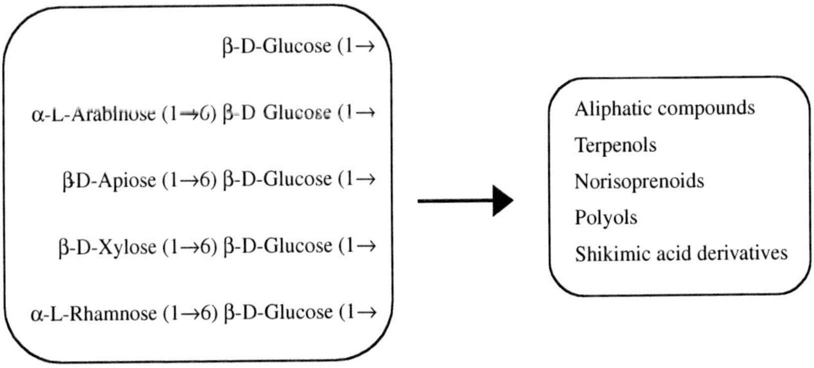


**FIGURE 4.16** Main glycosidically bound aroma compounds found in fruits. (From Crouzet, J., *Functionality of Food Phytochemicals*, T. Romeo, Ed., Plenum Press, New York, 1997, p. 197. With permission.)

### 4.4.1 Glycoside Structure

Glycosidically bound flavor compounds typically exist as glucosides, diglycosides, or triglycosides [51]. The aglycone (flavor compound) is normally bound to glucose and the remainder of the glycoside may be composed of a number of other simple sugars such as rhamnose, fructose, galactose, or xylose. Over 200 aglycones have been found including aliphatic, terpene, and sesquiterpene alcohols, norisoprenoids, acids, hydroxy acids, and phenylpropane derivatives as well as related compounds (Figure 4.16).

### 4.4.2 Freeing of Glycosidically Bound Flavor Compounds

As mentioned earlier, the aglycones may be freed from the glycosides by acid, heat, or enzyme catalysis. In many fruit juices, the pH is low enough to cause slow hydrolysis during storage. This accounts for some flavor changes during aging/storage of fresh plant foods. Hydrolysis may be very rapid when a product is heated. It is impressive that the concentration of α-terpineol in mango pulp increases from 1,800 μg/kg to 55,000 μg/kg during normal thermal processing due to the heat/acid catalyzed release of its glycoside [54]. For a long time it was thought that the characteristic flavor of canned pineapple was due to compounds formed via thermal reaction (e.g., the Maillard reaction). However, it was ultimately found that thermal processing simply caused the release of indigenous glycosidically bound furaneol that gave the tinned flavor character.

The aglycones may also be freed by enzymatic action. An example where this is a major pathway to release is in vanilla. There is very little free vanillin in the green vanilla bean: it is bound as a glucoside. The making of vanilla requires a curing step where glucovanillin is converted to free vanillin by an endogenous β-glucosidase freeing it for extraction [55]. Most often the enzymatic freeing of aglycones is a slow process since the enzymes responsible for this hydrolysis are inhibited by low pH (typical of fruit juices) and the presence of low concentrations of glucose. Also, these enzymes are very substrate specific, which limits the application of microbial enzymes for this application [56].

While research has provided a greatly improved understanding of this whole issue, it has lead to few commercial applications. The Holy Grail would be to be able to shorten the aging of wine through the controlled release of glycosidically bound flavor compounds. Unfortunately, wine flavor is a very delicate balance of numerous flavor components: a balance that is the result of a traditional process (specific acid and enzymatic hydrolysis of indigenous glycosides). This traditional process has not been successfully duplicated though the application of additional enzymes, heating or acidification.

## 4.5 LOCATION OF FLAVOR IN PLANT

As one might expect, flavor is typically nonuniformly distributed in the plant. For example, allyl isothiocyanate is more abundant in the outer leaves than in the inner tissues of cauliflower. Freeman and Whenham [38] dissected onion bulbs separating the bulb into individual leaves and the stem. They examined each section for lachrymator and pyruvate. They found that the dry outer scales have virtually no lachrymator nor pyruvate but levels increase progressively in the leaves moving towards the plant center and stem. As another example, total terpenes are more concentrated in the crown of carrots than in the midsection or tip [57]. Not only is there a difference in total flavor depending upon the plant part being examined, but there may also be a difference in flavor profile [58]. Therefore, if one dissected a plant and ate different individual parts it would not be unusual to find a slightly different flavor character for each part.

## 4.6 INFLUENCE OF GENETICS, NUTRITION, ENVIRONMENT, MATURITY, AND STORAGE ON DEVELOPMENT OF FLAVOR

There is little question that not all apples, oranges, or coffees taste the same. We often select a particular variety of apple (e.g., Delicious) because we prefer the taste. From advertising and experience we know that geographical location influences flavor, e.g., the robust Columbia coffees. We are aware that climate and soil conditions influence flavor, and that often the best fruit comes via handpicking fresh and fully ripened fruit right from the tree. We know that each piece of fruit or vegetable is an individual in appearance and flavor. There is little disagreement that numerous factors influence the flavor in a plant.

There is a great deal of literature reporting on differences in flavor due to genetics, environment, harvesting time, and postharvest treatment [9,12,13,59]. This section of this chapter is devoted to sampling some of this literature to illustrate and explain the flavor differences we observe between foods of the same type. This chapter primarily deals with differences in the flavor of plant foods. While animal products are often affected by similar factors, variations in plant flavor are accepted as natural while variations in animal flavor are generally considered as off-flavors.

One does not accept the varietal differences in beef, for example, as one does in apples.

### 4.6.1 Plant Products

#### 4.6.1.1 Genetics

There is ample evidence both in our daily lives and in the literature to show that plant genetics influences flavor [7,13,57–62]. Genetics determines precursors, enzyme systems, and their activity in flavor formation.

It is most common that varietal differences in flavor are due to quantitative differences in flavor composition rather than qualitative differences. For example, studies have shown that tomatoes may differ in 2-isobutylthiazole content by a factor of two- to threefold.

An additional example of genetic effects on flavor comes from studies on apple flavor. Paillard [63] found that the same volatile constituents were present in all apple varieties studied, but differences in proportions of these volatiles between varieties were observed. The yellow varieties examined tended to produce more acetate esters while the red varieties produced more butyrate esters. Paillard [64] found Golden Delicious parenchyma to rapidly and nearly completely convert butyric acid to acetic acid which resulted in the generation of acetate esters rather than butyrate esters. It was of interest that peels of apples with waxy cuticles (Calville blanc and Starkling) converted fatty acid substrates primarily to short chain alcohols and esters while apples with a corky cuticle (Canada gris) converted the fatty acids primarily to ketones and a small amount of alcohols. These differences in metabolism are often responsible for small but yet significant changes in flavor profile. Varietal differences in flavor have come to be accepted as natural, and in fact, one enjoys the many options provided by nature.

#### 4.6.1.2 Environmental and Cultural Effects on Flavor Development on the Plant

##### *4.6.1.2.1 Soil Nutrition*

Somogyi et al. [65] found that McIntosh apples grown with nitrogen fertilizers produced increased quantities of volatiles. Studies to determine the reason for this increase have been inconclusive [13]. It was hypothesized that increased nitrogen availability would likely result in higher levels of free amino acids during ripening (the precursors of branched chain esters). Yet, there was no significant relationship between N application and free amino acids in the fruit at ripening. Fellman et al. [13] suggested that N application may just enhance the levels of amino acid precursors thereby affecting ester formation.

Very striking changes in the flavor of onion, garlic, *A. vineale,* cabbage, mustard, and watercress have been noted to be dependent on sulfate content of the growth medium [66–69]. This is expected since these plants are very dependent on sulfur-containing precursors for their characteristic flavor. Onions grown under extreme sulfate deficiency lacked typical flavor and lachrymatory potency. Both flavor and lachrymatory activity were found to increase with sulfate availability up to the genetically controlled limits of the plant. Freeman [70] has suggested that the flavor precursors [S-alk(en)yl-L-cysteine sulfoxides in the *Allium* species and glucosinolates in the Cruciferae] are not essential to plant growth but that a certain level of sulfate nutrition is required for plant synthesis of essential amino acids and proteins. The ability to control flavor via plant nutrition could provide the opportunity to produce vegetables of selected flavor strength. It may be desirable to produce a more bland vegetable, e.g., cabbage, to reduce cooking odors, or radishes to reduce harshness.

To the contrary, Wright and Harris [71] found heavy fertilization of tomatoes with nitrogen and potassium reduced sensory scores. They noted an over production of some volatiles may well have resulted in an imbalanced flavor. It appears that flavor depends upon the right balance of nutrients as opposed to excess nutrients.

##### *4.6.1.2.2 Water Availability*

Another factor known to influence flavor is the water content of the soil during plant growth. Plentiful rainfall often results in large, lush vegetables that unfortunately lack flavor [72]. Years with inadequate rainfall often result in fruit and vegetables of smaller size and less attractive appearance but yet with more intense flavor. Freeman [70] suggested that this more intense flavor may be due to increased amounts of flavor precursors caused by stress put on the plant by restricted water availability. There are numerous examples in the literature of plants accumulating low molecular weight metabolites (e.g., sugars, amino acids, and organic acids) when under stress. These metabolites serve as flavor precursors.

##### *4.6.1.2.3 Temperature*

While stress may be due to limited water availability, stress may also be due to temperature. Plants subjected to low temperatures are known to accumulate similar low molecular weight metabolites. This presumably is an evolutionary response directed towards lowering the freezing point of the plant tissue to enhance survival of frost damage.

Additional temperature effects on flavor development may be due to temperature dependency of enzymatic reactions. Generally speaking, enzyme reaction rate will increase with temperature until enzyme denaturation begins. However, enzyme reactions differ in extent of temperature enhancement: one reaction may increase only slightly with a 10°C increase in growing temperature while a different reaction may double with this temperature increase. This may result in differential generation of flavor components.

#### 4.6.1.3 Influence of Maturity and Postharvest Storage on Flavor Development Off the Plant

##### *4.6.1.3.1 Maturity*

Commercial practice in the U.S. is to pick fruit prior to ripening and then control ripening away from the plant. Often fruit must survive substantial handling, be stored for months, and then travel large distances to market. These requirements can only be met by harvesting the fruit before it is ripe. The literature generally indicates that flavor development proceeds differently once the fruit has been picked and ripened away from the plant. An exception would, of course, be those foods that only ripen once removed from the parent plant, e.g., cantaloupe and some varieties of avocado [73]. They would ripen abnormally if left on the parent plant.

Do et al. [74] investigated flavor development in peaches as influenced by harvest maturity and artificial ripening. Peaches obtain most of their characteristic flavor from lactones, and of course, these lactones increase as stage of maturity advances. Artificial ripening of the peaches does not result in similar levels of lactones in the ripe fruit. Artificially ripened peaches contained only about 20% of the total lactone content of tree ripened peaches. Benzaldehyde and total esters in the artificially ripened peaches reached only 20% and 50% of their respective concentrations found in tree-ripened fruit.

Studies on artificially ripened tomatoes show that flavor may not only be reduced in comparison to the vine-ripened product, but also may be abnormal in character. Maul [75] and Maul et al. [76,77] have provided both sensory and instrumental data illustrating poor flavor development in artificially ripened tomatoes. Studies by Yu et al. [20–22,78] and Yu and Spencer [79,80] have shown that immature tomatoes lack equivalent enzyme systems as the vine-ripened tomatoes. Apparently as ripening progresses, more enzyme systems develop that are significant to flavor generation. Therefore, the artificially ripened tomatoes lack the enzyme systems necessary to produce the characteristic tomato flavor.

There is little in the literature on the effect of plant age on the development of volatile flavor components, however Freeman [70] has reported on flavor formation during onion seed germination and growth. It appears that after approximately 20 days, onion flavor potential is completely developed. The seed itself contains no flavor precursors (cysteine sulfoxide derivatives) and only about 3% the alliinase activity (enzyme) of mature onion bulbs. However, the plant quickly develops alliinase activity and a maximum activity is reached after 15–20 days. Therefore, flavor is developed well before the plant is of suitable size for consumption.

While not the subject of this chapter, the effect of plant age on *overall* sensory character is well documented. Young vegetables generally are softer and sweeter than mature or overly mature vegetables. Overall vegetable quality decreases quickly once maturity is achieved.

### *4.6.1.3.2 Postharvest Storage*

Since fruit flavor develops rapidly during the climacteric period and this period commonly occurs away from the plant (early harvest and artificial ripening), storage conditions during this time will influence flavor development [81,12,13.] There is a large body of literature documenting that fruits stored for significant periods under an improper environment (e.g., gas composition [increased $CO_2$ and reduced $O_2$], temperature, humidity, or lighting) will suffer in flavor beyond that associated with ripening away from the plant as discussed above. While the fruit may subsequently appear to ripen, i.e., the color changes and tissue softens, the balance of metabolic processes involved in aroma formation become disrupted and flavor does not develop normally.

An extreme example of the importance of storage conditions on flavor development during ripening comes from studies on banana flavor. Banana flavor production is an exponential function of storage temperature from 5° to 25°C. Prolonged storage at 10–12°C results in substantially reduced (60%) aroma production. Temperatures less than 5°C result in no characteristic aroma development. Chill injury occurs, which changes lipids in subcellular membranes and irreversibly alters normal fruit respiration. Storage temperatures greater than 27°C also result in abnormal flavor development. Characteristic flavor production decreases while products of fermentation, e.g., ethanol and ethyl acetate, predominate.

While we are most aware of temperature sensitivity in banana flavor/ripening, it appears that the flavor quality of other fruits is also influenced by storage temperature. For example, tomato flavor suffers when stored at temperatures <16°C [75]. Therefore, storage temperature must be controlled to permit proper flavor development.

Paillard [59,81], Fellman et al. [13], and Baldwin et al. [12] have summarized additional factors that may influence flavor development during storage. For example, controlled atmosphere storage of apples generally decreases volatile production during ripening. The profound effect of storing apples under refrigeration vs. low oxygen/higher carbon dioxide is illustrated in Figure 4.17. There was very little aroma formation in apples stored in a 1% oxygen environment compared to the refrigerated control.

Low humidity storage of apples was found to increase the emission of hexyl, isopentyl, and butyl acetate and decrease the corresponding alcohols. The low humidity favored ester formation. Illumination of the apples increased volatile production during storage. This increase in volatiles is postulated to be due to the light catalyzing lipid oxidation in the peel. The lipid oxidation products then are metabolized to

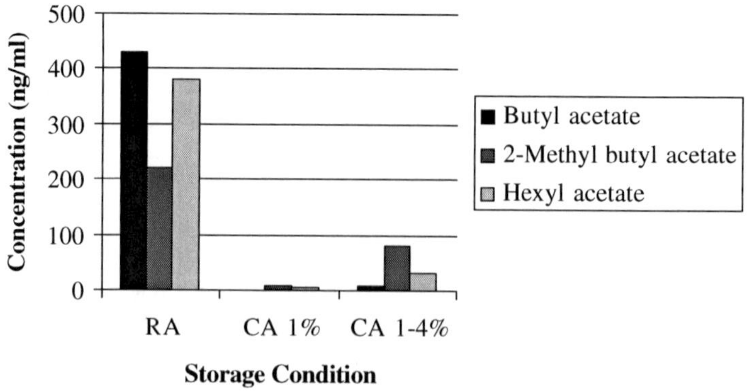


**FIGURE 4.17** The influence of storage conditions on the acetate ester content of Gala apple tissue after 4 months storage at 1°C. RA = refrigerated storage, CA 1% = storage in 1% oxygen, CA-1-4% oxygen levels were raised by 1% each month. (From Fellman, J.K., T.E. Miller, D.S. Mattinson, J.P. Matheis, *Hort Sci.*, 35(6), p. 1026, 2000. With permission.)

short chain esters. Thus, while certain environmental conditions are required for the extended preservation of a fruit during storage, these conditions nearly invariably result in an irreversible loss/change in flavor quality.

## 4.7 ANIMAL PRODUCTS

As was mentioned in the chapter introduction, most variations in flavor of meats are not well accepted by the consumer. For example, a T–bone steak or a hamburger should have a certain flavor with little variation. However, there is ample literature to show that the flavor of meat is particularly dependent upon the animal's diet. These variations (e.g., the rotten egg or mercaptan odor of lambs pastured on oats) are generally considered as off-flavors, and therefore, will be discussed in the off-flavor chapter of this text. One of the few exceptions where flavor differences are accepted between animal species is in fish. The consumer has come to appreciate the wide variety of flavors available from sardines, salmon, sole, catfish, bluefish, etc.

## 4.8 CONCLUSIONS

While flavor development in fruits and vegetables proceeds along many common paths, there are also several contrasts. Fruit flavor develops during a short ripening period, while vegetable flavor develops primarily during cellular disruption. Fruits depend heavily upon esters for their characteristic flavor while sulfur compounds are more important to vegetable flavor. Similarities exist in the formation of alcohols, aldehydes, ketones, and acids by enzymatic pathways. The enzymatically catalyzed oxidation of lipids is a particularly important pathway for flavor formation in both fruits and vegetables. While the pathways leading to flavor formation are known for many compounds, in other cases, many pieces of the synthesis puzzle are missing. It is unfortunate that most of the research on biosynthesis pathways of flavor formation was done in the early 1970s and that this area of study has received limited funding in recent years.

It is obvious from this chapter that many factors influence the flavor of a fresh fruit or vegetable. These factors include plant genetics, soil nutrition, growing environment, stage of maturity, and conditions of storage from harvesting to consumption. Genetics determine the enzyme systems and precursors involved in flavor formation. Soil nutrition provides some of the essentials for flavor development. For example, a radish, which depends upon sulfur compounds for flavor, will not develop characteristic flavor if there is little sulfur in the soil. Growing conditions influence activity of different enzyme systems and can significantly alter flavor development. Certainly the stage of maturity and storage conditions can further influence flavor.

While a number of variables are outside of the control of the plant grower, some are within his control. The grower should consider the influence of plant genetics, soil nutrition, and water regime in determining the flavor of his crop. It seems that appearance and yield per acre are the two major criteria of success in growing produce. Flavor should become a more significant consideration.

## REFERENCES


1. Reineccius, G.A., T.A. Reineccius, Eds., *Heteroatomic Aroma Compounds*, Amer. Chem. Soc., Washington, D.C., 2002, p. 372.
2. Taylor, A.J., D.S. Mottram, Eds., *Flavor Science,* Royal Soc. Chem., London., 1996, p. 476.
3. Teranishi, R., G.R. Takeoka, M. Guentert, *Flavor Precursors: Thermal and Enzymatic Conversions,* Amer. Chem. Soc., Washington, D.C., 1992, p. 258.
4. Schieberle, P., K.H. Engel, Eds., *Frontiers of Flavour Science,* Deutsche Forschung Lebensmittel., Garching, 2000, p. 602.
5. Rouseff, R.L., M.M. Leahy, Eds., *Fruit Flavors: Biogenesis, Characterization, and Authentication,* Amer. Chem. Soc., Washington, D.C., 1995, p. 292.
6. Tressl, R., M. Holzer, M. Apetz, Biogenesis of volatiles in fruit and vegetables, in *Aroma Research*, H. Maarse, P.J. Groenen, Eds., Wageningen, Netherlands, 1975, p. 41.
7. Salunkhe, D.K., J.Y. Do, Biogenesis of aroma constituents of fruit and vegetables, *CRC Crit. Reviews Food Sci. Nutrit.*, 8(2), p. 161, 1976.
8. Rowan, D., J.M. Allen, S. Fielder, M. Hunt, Biosynthesis of straight-chain ester volatiles in Red Delicious and Granny Smith apples using deuterium-labeled precursors, *J. Agric. Food Chem.*, 47, p. 2553, 1999.
9. Paillard, N.M., The flavour of apples, pears and quinces, in *Food Flavours Part C: The Flavour of Fruit*, A.J. McLeod, Ed., Elsevier, Amsterdam, 1990, p. 1.
10. Tressl, R., F. Drawert, Biogenesis of banana volatiles, *J. Agric. Food Chem.*, 21, p. 560, 1973.
11. Jennings, W.G., Peaches and pears, in *Chemistry and Physiology of Flavors*, H.W. Schultz, E.A. Day, L.M. Libbey, Eds., AVI Pub. Co., Westport, 1967, p. 419.
12. Baldwin, E.A., J.W. Scott, C.K. Shewmaker, W. Schuch, Flavor trivia and tomato aroma: biochemistry and possible mechanisms for control of important aroma components, *Hort. Sci.*, 35(6), p. 1013, 2000.
13. Fellman, J.K., T.E. MIller, D.S. Mattinson, J.P. Matheis, Factors that influence biosynthesis of volatile compounds in apple fruits, *Hort Sci.*, 35(6), p. 1026, 2000.
14. Salas, J.J., M. Williams, J.L. Harwood, J. Sanchez, Lipoxygenase activity in olive (*Olea europaea*) fruit, *J. Amer. Oil Chem. Soc.*, 76(10), p. 1163, 1999.
15. Jadha, V., B. Singh, D.K. Salunkhe, Metabolism of unsaturated fatty acids in tomato fruit: linoleic and linolenic acid as presursors of hexanol, *Plant Cell Physiol.*, 13, p. 449, 1972.
16. Galliard, F., J.A. Matthew, Lipoxygenase-mediated cleavage of fatty acids to carbonyl fragments in tomato fruits, *Phytochem.*, 16, p. 339, 1977.
17. Galliard, F., J.A. Matthew, A.J. Wright, M.J. Fishwick, The enzymatic breakdown of lipids to volatile and nonvolatile carbonyl fragments in disrupted tomato fruits, *J. Sci. Food Agric.*, 28, p. 863, 1977.
18. Schreier, P., G. Lorenz, Formation of "green-grassy" notes in disrupted plant tissue: characterization of the tomato enzyme system, in *Flavour '81*, P. Schreier, Ed., Walter deGruyter, Berlin, 1981, p. 495.
19. Engel, K.H., J. Heidlas, R. Tressl, The flavour of tropical fruits (banana, melon, pineapple), in *Food Flavours Part C: The Flavour of Fruit*, A.J. McLeod, Ed., Elsevier, Amsterdam, 1990, p. 195.
20. Yu, M.H., L.E. Olson, D.K. Salunkhe, Precursors of volatile components in tomato fruit: II Enzymatic production of carbonyl compounds, *Photochem.*, 7, p. 555, 1968.
21. Yu, M.H., L.E. Olson, D.K. Salunkhe, Precursors of volatile components in tomato fruits: III. Enzymatic reaction products, *Phytochem.*, 7, p. 561, 1968.

22. Yu, M.H., L.E. Olson, D.K. Salunkhe, Production of 3-methyl butanal from L-leucine by tomato extract, *Plant Cell Physiol.*, 9, p. 633, 1968.
23. Dalal, K.B., D.K. Salunkhe, L.E. Olson, J.Y. Do, Volatile components of developing tomato fruit grown under field and greenhouse conditions, *Plant Cell Physiol.*, 9, p. 389, 1968.
24. Rowan, D.D., H.P. Lane, J.M. Allen, S. Fielder, M.B. Hunt. Biosynthesis of 2-methylbutyl, 2-methyl-2-butenyl, and 2-methylbutanoate esters in Red Delicious and Granny Smith apples using deuterium-labeled substrates, *J. Agric. Food Chem.*, 44(10), p. 3276, 1996.
25. Rettinger, K., K. Volker, H.G. Schmarr, F. Dettmar, U. Hener, A. Mossandl, Chirospecific analysis of 2-alkyl-branched alcohols, acids, and esters: chirality evaluation of 2-methylbutanotes from apples and pineapples, *Photochem. Anal.*, 2(4), p. 184, 1991.
26. Drawert, F., Biochemical formation of aroma components, in *Aroma Research*, H. Maarse, P.J. Groenen, Eds., Pudoc, Wageningen, 1975, p. 245.
27. Schreier, P., *Chromatographic Studies of Biogenesis of Plant Volatiles,* Huethig Verlag., Berlin, 1984, p. 165.
28. Drawert, F., R. Tressl, G. Standt, H. Koppler, Gaschromatographischmassenspektometrische Differenzierung von Erbeerarten, *Z. Naturforsch*, 28C, p. 488, 1973.
29. Torline, P., H.M. Ballschmieter, Volatile constituents from guava. I. A comparison of extraction methods, *Lebensm. Wissen. u-Technol.*, 6(1), p. 32, 1973.
30. Anjou, K., E. Sydow, The aroma of cranberries. II. *Vaccinium macro-carpon, Acta Chem. Scand.*, 21, p. 2076, 1967.
31. Winter, M., R. Kloti, Uber das aroma der gelben Passionfrucht, *Helv. Chim. Acta*, 55, 1972.
32. Little, D.B., R.B. Croteau, Biochemistry of essential oil terpenes, in *Flavor Chemistry: 30 Years of Progress*, R. Teranishi, E.L. Wick, I. Hornstein, Eds., Kulwer Academ., New York, 1999, p. 239.
33. Licthenthaler, H.K., The plant's 1-deoxy-D-xylulose-5-phosphaste pathway for biosynthesis of isoprenoids, *Fett/Lipid*, 100, p. 128, 1998.
34. Roscher, R., P. Schreier, W. Schwab, Metabolism of 2,5-dimethyl-4-hydroxy-2H-furan-3-one in detached ripening strawberry fruits, *J. Agric. Food Chem.*, 45(8), p. 3202, 1997.
35. Perez, A.G., R. Olias, C. Sanz, J.M. Olias, Biosynthesis of 2,5-dimethyl-4-hydroxy-2H-furan-3-one and drivatives in vitro-grown strawberries, *J. Agric. Food Chem.*, 47, p. 655, 1999.
36. Engel, K.H., The importance of sulfur-containing compounds to fruit flavors, in *Flavor Chemistry: 30 Years of Progress*, R. Teranishi, E.L. Wick, I. Hornstein, Eds., Kulwer Academ., New York, 1999, p. 265.
37. Takeoka, G., Flavor chemistry of vegetables, in *Flavor Chemistry: 30 Years of Progress*, R. Teranishi, E.L. Wick, I. Hornstein, Eds., Kulwer Academ., New York, 1999, p. 287.
38. Freeman, G.G., R.J. Whenham, A survey of volatile components of some Allium species in terms of S-alk(en)yl-L-cysteinesulphoxide present as flavor precursors, *J. Sci. Food Agric.*, 26, p. 1869, 1975.
39. Granroth, B., Biosynthesis and decomposition of cysteine derivatives in onion and other Allium species, *Ann. Acad. Sci. Fenn. Ser.*, A2, p. 154, 1970.
40. Brodnitz, M.H., J.H.R. Pascale, Thiopropanol-S-oxide-a lachrymatory factor in onion, *J. Agric. Food Chem.*, 19, p. 269, 1971.
41. Virtanen, A.I., On enzymatic and chemical reactions in crushed plants, *Acta. Biochem. Biophys. Suppl.*, 1, p. 200, 1962.

42. Saghir, A.R., L.K. Mann, The lachrymatory factor in the American wild Alliums, *Suomen Kemistihehti*, B38, p. 78, 1965.
43. Matikkala, E.J., A.I. Virtanen, On the quantitative determination of the amino acids and "γ-glutamyl peptides of onions, *Acta. Chem. Scand.*, 21, p. 2891, 1967.
44. Schwimmer S., S.J. Austin, "γ-Glutamyl transpeptidase of sprouted onion, *J. Food Sci.*, 36, p. 807, 1971.
45. Yasumoto, K., K. Iwami, H. Mitsuda, Enzyme-catalysed evolution of lenthionine from lentinic acid, *Agric. Biolog. Chem.*, 35 (13 Suppl), p. 2070, 1971.
46. Jensen, K.A., J. Conti, A. Kjaer, Isothiocyanates: II. Volatile iso-cyanate in seeds and roots of various Brassicas, *Acta. Chem. Scand.*, 7, p. 1267, 1953.
47. Bailey, S.D., M.L. Bazinet, J.L. Driscoll, A.I. McCarthy, The volatile sulfur components of cabbage, *J. Food Sci.*, 26, p. 163, 1961.
48. Maarse, H., Volatile Compounds in Food, Quantative Data, *Division for Nutrition and Food Research*, TNO, Vols. 1–3, TNO, Zeist., 1984.
49. Stahl-Biskup, E., F. Intert, J. Holthuijzen, M. Stengele, G. Schultz, Glycosidically bound volatiles — a review, *Flav. Frag. J.*, 8, p. 61, 1993.
50. Winterhalter, P., P. Scheier, C13-norisoprenoid glycosides in plant tissues: an overview on their occurrence, composition and role as flavor precursors, *Flav. Frag. J.*, 9, p. 281, 1994.
51. Crouzet, J., Flavor biogeneration, in *Functionality of Food Phytochemicals*, T. Romeo, Ed., Plenum Press, New York, 1997, p. 197.
52. Bourquelot, E., M. Bridel, Synthese du geranylglucoside-B-a l'aide de l'emulsine; sa presence dans les vegetaux, *C.R. Acad. Sci.*, 157, p. 72, 1913.
53. Winterhalter, P., H. Knapp, M. Straubinger, Water soluble aroma precursors: analysis, structure, and reactivity, in *Flavor Chemistry: Thirty Years of Progress*, R. Teranishi, E.L. Wick, I. Hornstein, Eds., Kluwer Academics, New York, 1999, p. 255.
54. Sakho, M., J. Crouzet, S. Seck, Evolution des compeses volatils de la mangue au cours du chauffage, *Lebensm. Wissen. u-Technol.*, 18, p. 89, 1985.
55. Arana, F., Action of B-glucosidase in the curing of vanilla, *Food Res.*, 8, p. 343, 1943.
56. Aryan, A.P., B. Wilson, C.R. Strauss, P.J. Williams, Properties of glycosidase of *Vitis vinifera* and a comparison of β-glucosidase activity with that of exogenous enzymes: assessment of possible applications in enology, *Am. J. Enol. Vitic.*, 38, p. 182, 1987.
57. Simon, P.W., C.E. Peterson, R.C. Lindsay, Genetic and environmental influences on carrot flavor, *J. Amer. Soc. Hort. Sci.*, 105(3), p. 416, 1980.
58. Simon, P.W., R.A. Lindsay, C.E. Peterson, Analysis of carrot volatiles collected on porous polymer traps, *J. Agr. Food Chem.*, 38, p. 549, 1980.
59. Paillard, N., Factors influencing flavour formations in fruits, in *Flavor '81*, P. Schreier, Ed., Walter deGruyter, Berlin, 1981, p. 479.
60. Simon, P.W., C.E. Peterson, Genetics and environmental components of carrot culinary and nutritive value, *Acta. Hort.*, 93, p. 271, 1979.
61. Maga, J.A., The role of sulfur compounds in food flavor. I. Thiazoles, *CRC Crit. Rev. Food Sci. Nutr.*, 6(2), p. 153, 1975.
62. Maga, J.A., The role of sulfur compounds in food flavor. Part III: Thiols, *CRC Crit. Rev. Food Sci. Nutr.*, 7(2), p. 147, 1976.
63. Paillard, N., Analyse des produits volatils emis par quelques varietes de pommes, *Fruit*, 22, p. 141, 1967.
64. Paillard, N., Biosynthesis des produits volatils de la pomme: formation des alcools et des esters a patir des acid gras, *Phytochem.*, 18, p. 1665, 1979.
65. Somogyi, L.P., N.F. Childers, S.S. Chang, Volatile constituents of apple fruits as influenced by fertilizer treatments, *Proc. Amer. Soc. Hort. Sci.*, 84, p. 51, 1964.

66. Freeman, G.G., N. Mossadeghi, Effect of sulfate nutrition of flavour components of onion (*Allium apa*), *J. Sci. Food Agric.*, 21, p. 610, 1970.
67. Freeman, G.G., N. Mossadeghi, Influence of sulfate nutrition on the flavour component of garlic (*Allium sativum*) and wild onion (*Allium vineale*), *J. Sci. Food Agric.*, 22, p. 330, 1971.
68. Freeman, G.G., N. Mossadeghi, Influence of sulphate nutrition on flavour components of three cruciferous plants: Radish (Raphanus sativus), cabbage (Brassica capitata) and white mustard (*Sinapis alba*), *J. Sci. Food Agric.*, 23, p. 387, 1972.
69. Freeman, G.G., N. Mossadeghi, Studies on sulphur nutrition and flavour production in watercress (*Rorippa nasturtium-aquaticum* (L) Hayek), *J. Hort Sci.*, 47, p. 375, 1972.
70. Freeman, G.G., Factors affecting flavour during growth, storage and processing of vegetables, in *Progress in Flavor Research,* H.E. Nurtsen, Ed., Appl. Sci. Publ., London, 1979 p. 225.
71. Wright, D.H. and N.D. Harris, Effects of nitrogen and potassium fertilization on tomato flavor, *J. Agric. Food Chem.*, 33(3), p. 355, 1985.
72. Baldwin, E.A., M. Nisperos-Carriedo, P.E. Shaw, J.K. Burns, Effect of coatings and prolonged storage conditions on fresh orange flavor volatiles, degrees Brix, and ascorbic acid levels, *J. Agric. Food Chem.*, 43(5), p. 1321, 1995.
73. Bliss, M.L., H.K. Pratt, Effect of ethylene, maturity and attachment to the parent plant on production of volatile compounds by muskmelons, *J. Amer. Soc. Hort. Sci.*, 104(2), p. 273, 1979.
74. Do, J.Y., D.K. Salunkhe, L.E. Olson, Isolation, identification and comparison of the volatiles of peach fruit as related to harvest maturity and artificial ripening, *J. Food Sci.*, 34, p. 618, 1969.
75. Maul, F., Harvest maturity and post harvest temperature management can compromise flavor potential of fresh-market tomatoes, in *Horticulture Dept.*, University of Florida, Gainesville, 1999.
76. Maul, F., S.A. Sargent, M.O. Baldwin, D.J. Huber, C.A. Sims, Aroma volatile profiles from vine ripe tomatoes are influenced by physiological maturity at harvest: an application of electronic nose technology, *J. Amer. Hort. Sci.*, 123, p. 1094, 1998.
77. Maul, F., S.A. Sargent, M.O. Baldwin, D.J. Huber, C.A. Sims, Predicting flavor potential for green-harvest tomato fruit, *Proc. Florida State Hort. Soc.*, 111, p. 285, 1998.
78. Yu, M.H., L.E. Olson, D.K. Salunkhe, Precursors of volatile components in tomato fruits: I Compositional changes during development, *Phytochem.*, 6, p. 1457, 1967.
79. Yu, M.H., M. Spencer, Conversion of L-Ieucine to certain keto acids by a tomato enzyme preparation, *Phytochem.*, 8, p. 1173, 1969.
80. Yu, M.H., M. Spencer, a-Alanine aminotransferase from tomato fruit, *Phytochem.*, 9, p. 341, 1970.
81. Paillard, N.M., *The* flavor of apples, pears and quinces, in *Food Flavours Part C. The Flavour of Fruit*, A.J. McLeod, Ed., Elsevier, Amsterdam, 1990, p. 1.

# 5 Changes in Food Flavor Due to Processing

## 5.1 INTRODUCTION

The characteristic flavor of some food products develops during processing or home preparation. Examples of foods where this occurs include sauerkraut, yogurt, meats, chocolate, coffee, baked goods, and deep fat fried foods. The primary routes for flavor formation in these foods are nonenzymatic browning, fermentation, and thermal oxidations of fats (recognizing there are a host of other minor reactions that can significantly influence the flavor of a food, e.g., thiamin degradation in meats). This chapter will present an overview of these key pathways leading to food flavor during processing.

## 5.2 THE MAILLARD REACTION

The Maillard reaction is one pathway falling under the umbrella of nonenzymatic browning (caramelization, Maillard reaction, and ascorbic acid browning). Of these pathways, the Maillard reaction plays the major key role in flavor development. This reaction is responsible for some of the most pleasant flavors enjoyed by man. There is no question that freshly baked bread, a steak, a freshly brewed cup of coffee, or a piece of chocolate is appreciated by the consumer. Yet none of the characterizing flavors existed in the product until the food processor (or cook) heated the product to develop the flavor. While the following will present an overview of this reaction as a generator of flavor, greater detail may be found in several books including those edited/written by Schieberle and Engel [1], Nagodawithana [2], Parliment et al. [3], Tressl and Rewicki [4], and Kerler and Winkel [5]. In the scientific literature, one should search on "C.-T. Ho" who has made major contributions in this field (in the U.S.).

It should be mentioned that today the medical field has become the major discipline researching the Maillard reaction since it plays a role in aging as well as several diseases. The dominance of the medical discipline at the last few international symposia on the Maillard reaction illustrates this occurrence [6–9].

### 5.2.1 General Overview of the Maillard Reaction

Hodge [10] was the first to present an outline of the overall reactions involved in the Maillard reaction. Generally speaking, the Maillard reaction is a reaction between carbonyls and amines. The carbonyls in foods most often are reducing sugars, while the amines come from either amino acids or proteins. In the flavor industry, the

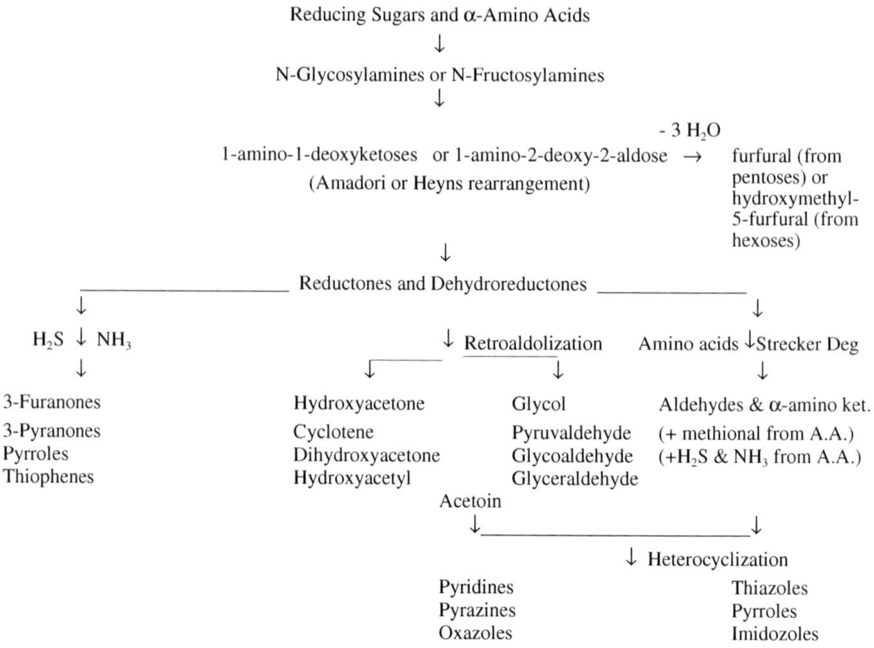


**FIGURE 5.1** Formation of flavor compounds via the Maillard reaction. (From Vernin, G., C. Parkanyi, *Chem. Heterocycl. Comp. Flavours Aromas*, p. 151, 1982. With permission.)

carbonyl may be a pure compound (e.g., diacetyl) and the amine, simply ammonia, or an amine. The major end products of the Maillard reaction are melanoidins and other nonvolatile compounds. However, in excess of 3,500 volatile compounds have been attributed to this reaction with numerous having very low sensory thresholds making them important to food aroma (Figure 5.1). While these volatiles comprise a very minor portion (in mass) of the overall reaction products, they are major contributors to the flavor of these foods. For example, Reineccius et al. [11] have shown that while 1.3 g (per 100 g beans) of reducing sugars and amino acids (total) were lost during the roasting of cocoa beans, only 0.9 mg of pyrazines (characterizing volatiles of chocolate) were formed. Thus, about 0.07% of the reactants were transformed into pyrazines while the remainder of the reactants went into other products. Yet these pyrazines are exceptionally important to the flavor of chocolate. It is typically these minor products that make the largest contribution to flavor.

### 5.2.2 Pathways for Flavor Formation via the Maillard Reaction

As was mentioned earlier, more than 3,500 volatile compounds have been identified as Maillard products. This number documents the large number of pathways potentially active in this reaction. A simplified view of this process begins with the reaction between amino acids and reducing sugars to form imines (Figure 5.2). These imines rearrange to form unstable Amadori (aldose precursor) or Heyns (ketose precursor)

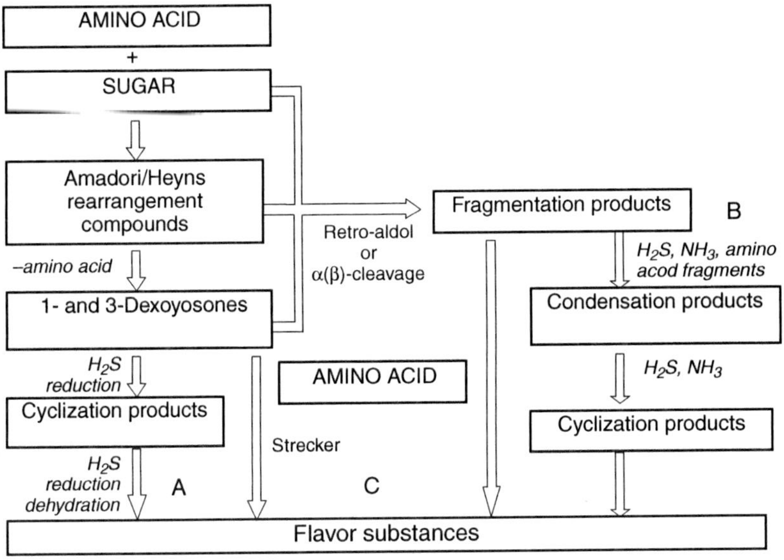


**FIGURE 5.2** Major pathways for the formation of flavor substances during the Maillard reaction. (From Kerler, J., C.Winkel, *Food Flavour Technology*, A.J. Taylor, Ed., Sheffield Publ., Sheffield, U.K., 2002, p. 302. With permission.)

products. Then, they may lose an amino acid to form 1- and 3-deoxyosones, which will undergo numerous other steps to form heterocyclic aroma compounds (Scheme A). Alternatively, the Amadori/Heyns products will fragment through retro-aldol condensation or α-, or β-cleavage. These fragments will readily undergo aldol condensations or fragment/amino acid reaction ultimately resulting in the formation of heterocyclic aroma compounds (Scheme B). The remaining pathway is the Strecker degradation (Scheme C). In this pathway, dicarbonyls or hydroxycarbonyl intermediates deaminate and decarboxylate amino acids to yield the corresponding Strecker aldehydes. Tressl and Rewicki [4] have provided a detailed discussion of these pathways.

A substantial amount of work has been done to determine the pathways for the formation of specific classes or individual aroma compounds. These pathways generally appear in original research articles where a given compound is found since the author will often propose a mechanism of formation.

### 5.2.3 Factors Influencing the Maillard Reaction

Most food science students learn about the Maillard reaction in one of their food chemistry courses. They typically learn about this reaction in the context of either nutritional losses (e.g., the loss of essential amino acids or loss of digestibility) or color formation. It should be recognized that nutritional loss and color formation are generic in the sense that the heating of foods under nearly any conditions will result in the loss of nutritional quality and in color formation. However, the reactions that lead to the generation of any given flavor are extremely specific. Illustrating this point further, heating a food (or model system) with any precursor set under nearly any conditions yields both a loss of reactants and color production, but it will not form a typical roast beef, chocolate, or maple syrup flavor. The pathways leading to flavor are very specific, and thus the reactants, environment, and heating conditions must be chosen carefully to produce the desired flavor. Flavor character can be viewed as a point in space, the space being defined by a complex balance of reactants, environment, and time, and temperature of heating. If any of these components are incorrect, the flavor is incorrect.

Unfortunately, there is a limited amount of published work reporting on the factors that influence flavor formation. In much of the previous work, system reactants (e.g., sugars, amino acids, peptides, proteins, lipids [e.g., phospholipids], and pure chemicals), pH, water activity, and thermal process have commonly been varied. The outcomes of these studies were generally evaluated by sensory and/or instrumental methods. Most of this work is piecemeal and difficult to relate to other literature. Also, in an effort to be able to understand the role of a given compositional factor, environmental or processing variable, simple model systems have been used. Unfortunately, the addition of an additional component (e.g., another amino acid or buffering component) or small change in environment or process can greatly change the reaction rate and direction. Thus, it is extremely difficult to translate model system data to real food data. As a result, our knowledge is incomplete, especially in terms of understanding this reaction in complex food systems.

#### 5.2.3.1 Heating Time/Temperature

One of the most important parameters influencing flavor formation via the Maillard reaction is processing temperature. This effect is obvious if one considers, for example, the sensory quality of roasted vs. stewed meats. Stewed meat lacks flavor notes characteristic of the roasted product. This is primarily because the stewed product has a water activity of approximately 1.0 and never exceeds a temperature of ca. 100°C. The roasted meat, however, dries on the surface so water activity is substantially less than 1.0. Also, since the surface dries, surface temperature may exceed 100°C. The lower water activity and higher surface temperatures favor the production of flavor compounds giving the meat roasted notes from the same basic reactants rather than stewed notes.

The influence of temperature on flavor formation may be understood better by recalling that each particular pathway of flavor formation has its own activation energy. To illustrate this influence, consider the plot shown in Figure 5.3 (note that this plot does not contain valid data but is only for illustrating an idea). In this plot, furanthiol has the highest activation energy, followed by pyrazine and furfural. (Activation energy is equal to the slope of the line divided by R [natural gas constant]). This plot shows that furanthiol would have the highest rate constant (or highest rate of production) at high temperatures followed by pyrazine and then furfural. As temperature is decreased to typical storage temperatures, the furfural would be produced at the highest rate and now furanthiol would have the lowest

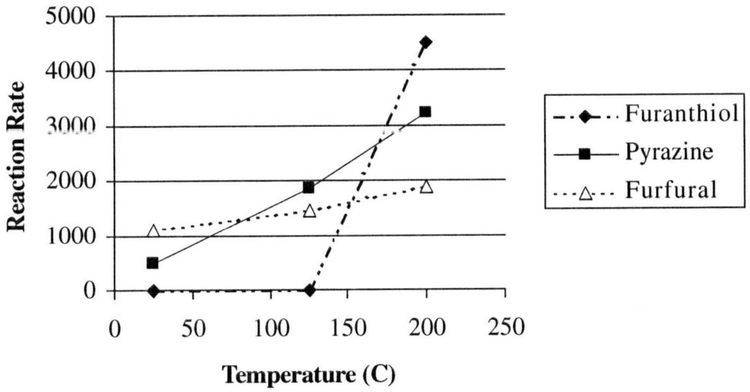


**FIGURE 5.3** Influence of temperature on rate of formation of various aroma compounds.

rate. Since each class of volatiles products, and perhaps even each member of each class, would have its own activation energy, one could put several hundred lines on this figure and realize how important temperature is in influencing the formation of flavor compounds via the Maillard reaction.

There is substantial information on the influence of temperature on pyrazine formation in model systems. While Koehler and Odell [13] reported that essentially no pyrazines were formed at temperatures less than 100°C, Shibamoto and Bernhard [14] and Leahy [15] both reported pyrazine production at temperatures as low as 70°C. In all cases, pyrazine production increased rapidly with temperature. A kinetic study by Leahy (1985) reported an activation energy of approximately 35 kcal/mole for pyrazine and 2-methyl pyrazine, and about 43 kcal/mole for the dimethyl pyrazines. These high activation energies indicate a very strong temperature dependence of pyrazine formation and explain why most pyrazines (roasted, nutty, toasted notes) are not formed during storage but only during high temperature treatment. This statement must be qualified in that some pyrazines can be formed at room temperature if the proper reactants are present. Several investigators [16–19] have found pyrazines to be formed in various cheeses from a reaction of microbial metabolic products. Rizzi [20] reported on the ability to form pyrazines under very mild conditions in model systems if the proper pH and reactants were present. It must be noted that Rizzi's systems were not typical Maillard systems, i.e., reducing sugar and protein/amino acid systems. In these systems select (very reactive) pyrazine precursors were combined to form pyrazines. These precursors typically have to be generated via other pathways of the Maillard reaction.

Processing time is also critical in determining flavor character. While in many situations, one method of getting more of some product is to increase the reaction time. Increasing the reaction time of a Maillard reaction does not necessarily increase flavor intensity but changes the final balance of flavor compounds and thereby changes the flavor character. Work by Yamanishi [21] on the roasting of coffee beans (Figure 5.4) illustrates this point, as one can see that the amount of each component

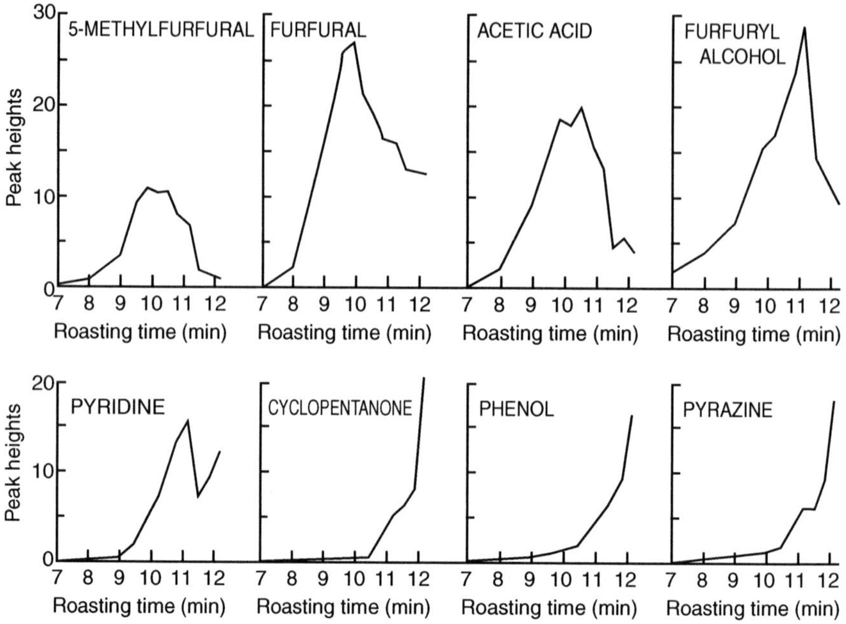


**FIGURE 5.4** The influence of roasting time on the amount of some aroma compounds in coffee beans. (From Yamanishi, T., *Flavor Research: Recent Advances*, R. Teranishi, R.A. Flath, H. Sugisawa, Eds., Marcel Dekker, New York, 1981, p. 231. With permission.)

in the coffee beans changes with roasting time and thus the balance of components, i.e., sensory character, changes with heating time.

### 5.2.3.2 Influence of System Composition

Numerous classical studies have been conducted evaluating the effect of reactant choice (and concentration) on the rate of the Maillard reaction (loss of reactant and/or color formation). One finds that amino acid and sugar type both influence the rate of this reaction. In general, the rate of the reaction is influenced by sugar composition as follows: pentoses (xylose or arabinose) > hexoses (glucose or fructose) > disaccharides (lactose or maltose) > trisaccharides > corn syrup solids > maltodextrins > starches. The reaction rate is also dependent upon the amino acid(s) present with glycine being amongst the most reactive.

If we consider this reaction in terms of flavor formation, the type of carbohydrate tends to play a larger role in determining the rate of the reaction than influencing the flavor character. The rate of reaction largely follows that of the loss of reactants and/or color formation. Sugar type may have some influence on flavor character but the amino acid selection generally plays a much greater role in this respect.

This observation is obvious to those who create process flavors (Chapter 9). When one makes a process flavor, the choice of sugar type is of minimal importance in determining flavor character while the choice of amino acid is all important. As is discussed in this chapter, sulfur-containing amino acids are required to form meat or coffee flavors; valine, leucine, and isoleucine are required for chocolate flavors; and methionine is required to form vegetable (potato) flavors. These flavors demand the presence of specific amino acids and the absence of others in a reaction mixture to produce the desired flavor character.

There has been substantial research on how lipids influence the Maillard reaction. This topic will be discussed later in this chapter (Section 5.3).

#### 5.2.3.3 Influence of Water Activity

Water availability will influence the rate of numerous Maillard pathways thereby influencing the rate of overall flavor formation and possibly flavor character. This is anticipated since some chemical reactions produce water as a byproduct (reaction pathway is inhibited by water) while other chemical reactions consume/require water (reaction pathway is promoted by water).

Leahy and Reineccius [22] found the influence of water activity on the rate of formation of alkyl pyrazines in a model system to parallel that of classical Maillard reaction. Briefly, the accumulation of pyrazines reached a maximum when heated at an Aw of ca. 0.75 and decreased with either increasing or decreasing Aw. This observation suggests that the rate of pyrazine formation is controlled by the initial stages of the Maillard reaction as opposed to an effect on specific pathways of pyrazine formation.

The accumulation of thiazoles during the heating of a cysteine:Dimethylhydroxy-furanone (DMHF) model system decreased with increasing water content while the accumulation of thiophenones and 3,5-dimethyl-1,2,4-trithiolane had an optimum moisture content (ca. 80%) and 2,4-hexanedione and 3-hydroxy-2-pentanone increased with moisture content [23.24]. There is little question that the Aw (or moisture content) influences both the rate of the Maillard reaction and influences the sensory properties of a system on heating.

#### 5.2.3.4 Influence of pH

pH may also influence the rate of specific Maillard pathways thereby changing the balance of volatiles formed. For example, pH has been found to have a very strong influence on pyrazine production [13–15,25]. Leahy [15] reported nearly a 500-fold increase in total pyrazine production between a model system heated at pH 9.0 vs. the same system at pH 5.0 (13 ppm vs. 24 ppb). Mottram and Whitfield [26] demonstrated the effect of pH on the formation of several classes of volatiles during the heating of a model system (Table 5.1). They showed that there may be an optimum pH for the formation of some volatiles, or volatile production may be enhanced or retarded at increasing pH. The situation is analogous to that of water activity effects on the formation of specific volatiles: some chemical pathways are enhanced, others retarded, while others are unaffected by pH of the system. The take home message is that pH may strongly influence the balance of aroma compounds present in a heated food/model system, and thus small changes in pH of a food may significantly change its aroma profile after heating.

**TABLE 5.1**
**The Effect of pH on the Formation of Volatiles During the Heating of a HMF/Cysteine Model System**

| Flavor Compound | pH 2.4 | 4.7 | 7.0 |
|---|---|---|---|
| | — | mg Yield | — |
| 3-Hydroxy-2-pentanone | 16.5 | 27.2 | 8.2 |
| 2,4-Hexanedione | 64.3 | 31.9 | 4.3 |
| 2-Acetylthiazole | 7.5 | 14.1 | 24.7 |
| 3,5-Dimethyl-1,2,4-trithiolanes | 33.7 | 50.7 | 33.9 |
| Thiophenones | 101 | 104 | 23.7 |

*Source:* Mottram, D.S., F.B. Whitfield, *Thermally Generated Flavors*, T.H. Parliment, R.J. McGorrin, C.-T. Ho, Eds., Amer. Chem. Soc., Washington, D.C., 1994, p. 180. With permission.

### 5.2.3.5 Influence of Buffer/Salts

The salt (buffer) type and concentration may also influence reaction rate. While buffers vary in their effect on the Maillard reaction, it is generally accepted that phosphate is the best catalyst [27]. The effect of phosphate on reaction rate is pH dependent with it having the greatest catalytic effect at pHs between 5–7. Potman and van Wijk [27] found the Maillard reaction rate in a phosphate buffered model system increased from 10- to 15-fold compared to a phosphate free reaction system.

In some unpublished work, we chose to thermally process (extrude) low and normal salt (NaCl) lots of a cooked cereal base. We then analyzed the volatile profile of the two extruded products by gas chromatography. We found that the low salt formulation contained substantially less volatiles (quantitatively) than the normal salt product. It appears that the salt levels used in extruded cereal products influenced the rate of the Maillard reaction. This observation is important in efforts to manufacture thermally processed low salt foods. It appears that taking the salt out of a food may influence both *aroma* and taste (saltiness).

### 5.2.3.6 Influence of Oxidation/Reduction State

Oxidation/reduction effects on the Maillard reaction were investigated by Shibamoto and Bernhard [14]. They studied the effects of metal ions, oxygen, antioxidants, and sodium hydroxide on the formation of pyrazines in glucose-ammonia systems. Their results indicated that oxygen does not catalyze pyrazine formation. Propyl gallate significantly inhibited the formation of pyrazines for short reaction times (2 hr) at 100°C, but not for long reaction times (18 hr) as compared to the control. BHT inhibited production slightly with long reaction times. However, reaction under an oxygen blanket gave slightly depressed yields. They concluded that antioxidants may reduce yields for reasons other than reducing the amount of oxygen available to the system. Adding $CuCl_2$ and $ZnC1_2$ also decreased yields of pyrazines while increasing browning, as was noted visually.

### 5.2.4 Kinetics of the Maillard Reaction and Flavor

There is very little work published on the kinetics of aroma formation via the Maillard reaction. However, what work has been done has indicated that volatile formation is very sensitive to both compositional and processing changes. One can generate kinetic data for volatile formation in a simple model system, include another precursor that should have no direct effect on the kinetics of interest, and find major changes in reaction kinetics.

No organized experiments were performed to obtain kinetic data, i.e., studies where the researcher chose to heat a model system at several temperatures over sufficient times to gather reliable kinetic data, until the late 80s. The available literature will be presented.

#### 5.2.4.1 Pyrazines

Perhaps it is not surprising that much of the kinetic data in the literature is on the formation of pyrazines. There has been a long-term interest in pyrazines due to their impact on the flavor of many thermally processed foods [28,29]. In terms of kinetics, Leahy and Reineccius [22,30] found the formation of the pyrazines during the heating of model systems to be zero order. Later work by Huang et al. [31] on aqueous systems (pH 10 at heating temperatures ranging from 120 to 140°C) and Huang et al. [32] in a study of high-pressure effects on pyrazine formation (aqueous and nonaqueous solvents) also noted a linear formation of pyrazines during heating. Thus, they also reported pyrazine formation as being a zero order reaction in both of these studies. The most recent kinetic work [33] carried the reaction further and used a fractional conversion data analysis technique. In this study they found that 2,5-dimethylpyrazine and 2-methylpyrazine formation was initially linear but formation leveled off at later stages of heating. Since zero order kinetics cannot account for the leveling off in formation, they chose to use first order reaction kinetics for these data. They noted that the reactants (glucose and lysine) were consumed well before the pyrazine concentration plateaued, and thus proposed that there is an intermediate rate limiting step in pyrazine formation. Their data treatment yielded $r^2$ values in excess of 0.96, demonstrating the excellent quality of prediction.

In terms of kinetic constants (activation energy [Ea]) for pyrazine formation, some variability exists in the literature. Leahy and Reineccius [22,30] found the Ea for pyrazine formation to range from about 33.4 to 44.8 Kcal/mole. Huang et al. [31] found substantially lower activation energies for similar pyrazines (19.5 to 29.0 Kcal/mole); however, they used arginine as the amino acid source while Leahy [22,30] used lysine. Also, Leahy's work was done at pH's ranging from 5 to 9 while Huang et al. [31] used pH 10. pH is known to greatly influence the rate of pyrazine formation via the Maillard reaction [15]. Huang et al. [32] reported Eas for the formation of pyrazines from intermediate precursors (3-hydroxy-2-butanone/ammonium acetate) in the same range in water (ca. 18.9 Kcal/mole), but Ea decreased to 13.1 Kcal/mole when ethanol was used as the solvent.

The most recent work by Jusino et al. [32] was done at lower moisture contents with glucose/lysine as the reactants. They reported an Ea 13.5 Kcal/mole for 2,5-dimethylpyrazine and 2-methylpyrazine under these conditions (note this is a first order Ea). When they applied pseudo zero order kinetics to the initial linear portion of the formation curves, they calculated Eas of 36.9 and 43.2 Kcal/mole for 2,5-dimethylpyrazine and 2-methylpyrazine, respectively. These values are very close to those calculated by Leahy [15] for her model systems and nonfat dry milk.

Leahy [15] studied the influence of Aw and pH on the formation of pyrazines. She found that pyrazine formation increased linearly with increasing Aw up to an Aw of about 0.75 and then either decreased or was unchanged at an Aw of 0.84, depending upon the pyrazine. A maximum in reaction rate at an Aw around 0.75 is consistent with Maillard browning as measured by the consumption of reactants or the formation of pigments. Leahy [15] reported that pyrazine formation rate increased by an average of 1.37-fold for each increase in Aw of 0.1 units. Similarly, a linear relationship was found between the rate of pyrazine and 2-methylpyrazine formation and pH (within the range of pH 5–9).

### 5.2.4.2 Oxygen-Containing Heterocyclic Compounds

There has been very little kinetic work done on other aroma compounds. Schirle-Keller and Reineccius [34] have published a study on the formation of oxygen-containing heterocyclic compounds during the heating (80–150°C) of aqueous model systems (cysteine and glucose). A summary of their results is presented in Table 5.2 (top). Briefly, they found a linear relationship between time of heating and the formation of volatiles, and thus applied zero order kinetic treatment to the data. The $r^2$ values for first order data treatment were substantially poorer than for the zero order treatment. The Eas are of the same order of magnitude as those for pyrazines reported by Leahy [15].

**TABLE 5.2**
**Kinetic Data for the Formation of Volatiles During the Heating of Model Systems**

| Compound | Ea (Kcal/mole) | Compound | Ea (Kcal/mole) |
|---|---|---|---|
| Furfural | 35.2[a] | 2-Acetylfuran | 36.2[a] |
| 5-Methylfurfural | 37.0[a] | Di(H)di(OH)-6-methylpyranone | 30.7[a] (16)[b] |
| Hydroxymethylfurfural | 28.1[a] | | |
| Isovaleraldehyde | 19[b] | 2-Acetyl-1-pyrroline | 14[b] (12.4 and 16.5)[c] |
| Phenylacetaldehyde | 22[b] | 5-Me-2-phenyl-2-hexenal | |
| 2-Acetylfuran | 18[b] (34.2)[a] | | |

[a] From Schirle-Keller, J.P., G.A. Reineccius, *Flavor Precursors: Thermal and Enzymatic Conversion*, R. Teranishi, G. Takeoka, M. Guntert, Eds., Amer. Chem. Soc., Washington, D.C., 1993, p. 244. With permission.

[b] From Chan, F., G.A. Reineccius, *Maillard Reactions in Chemistry, Food, and Health*, T.P. Labuza, G.A. Reineccius, V. Monnier, J.O. O'Brien, J.W. Baines, Eds., Royal Chem. Soc. London, London, 1994, p. 131. With permission.

[c] Unpublished data.

#### 5.2.4.3 Sulfur-Containing Compounds

Chan and Reineccius [36] have reported on the kinetics of the formation of sulfur-containing aroma compounds during the heating (75 to 115°C) of more complex, buffered (0.1 M phosphate buffer), aqueous model systems (glucose with leucine, methionine, phenylalanine, and proline). They generally found that the amount of sulfur-containing volatiles increased with time or temperature of heating. However, a plateau in concentration was reached for methional and dimethyl disulfide at longer heating times and higher temperatures. This plateau was likely due to the exhaustion of reactants. However, the plateau was not achieved at the same concentration for different pHs, suggesting that the plateau concentration was dependent on secondary degradation reactions (pH dependent).

The activation rates for the sulfur compounds measured ranged from 15 to 33 Kcal/mole. 2-Acetylthiophene had the highest Ea ranging from 22 to 33 Kcal/mole while methional and dimethyl disulfide Eas ranged from 15 to 27 Kcal/mole. The values for the latter compounds are consistent with the Eas for the Maillard reaction in general.

Zheng and Ho [37] reported on the kinetics of hydrogen sulfide formation during the heating (80 to 110°C) of aqueous solutions (pH 3, 5, 7, 9) of glutathione and cysteine. They found that $H_2S$ was released from both precursors following first order reaction kinetics ($r^2$ values ranged from 0.955 to 0.999) and that rate of release increased with pH. Zheng and Ho [37] found activation energies for $H_2S$ to range from 29.4 to 31.8 Kcal/mole for cysteine and 18.8 to 30.8 for glutathione. These data support earlier literature noting that $H_2S$ is more rapidly released from glutathione than cysteine.

#### 5.2.4.4 Miscellaneous Compounds

Chan and Reineccius [35] also have published some kinetic work on other aroma compounds (isovaleraldehyde, phenylacetaldehyde, 2-acetyl-l-pyrroline, 2-acetylfuran, and di(H)di(OH)-6-methyl pyranone). All of these volatiles followed pseudo zero order reaction kinetics in their early stages of reaction (Table 5.2). The fact that the concentrations of isovaleraldehyde, phenylacetaldehyde, and the methylpyranone reached plateaus late in heating suggests that a first order fit, as proposed by Jusino et al. [32], might be more appropriate.

The activation energies of the volatiles presented in Table 5.2 (bottom) are consistent with those found by Ho's group and the Maillard reaction in general. The values found for 2-acetyl-1-pyrroline are very consistent across systems and researchers (values in parentheses). However, the values found for 2-acetylfuran and the methyl pyranone were lower than we have previously observed [34]. This discrepancy in values may be due to the differences in model systems used between the two studies. Chan [38] used a complex amino acid/glucose/phosphate buffered model system while Schirle-Keller and Reineccius [34] used a simple unbuffered glucose/cysteine system. The additional amino acids may serve as good catalysts for the breakdown of glucose into the oxygen-containing volatiles noted, but apparently have little effect on the formation of 2-acetyl-1-pyrroline. This discrepancy in values suggests that the data one obtains from a simple model system may not be valid in complex food systems and care should be exercised in data interpretation based on model systems. This is not a new concern but is reinforced here.

The very low activation energy for 2-acetyl-l-pyrroline may help explain why this volatile is found by sniffing methods in nearly every food studied. It appears that this compound is very readily formed even under mild heating conditions. This fact, coupled with its extremely low sensory threshold, make its detection in foods likely. The Strecker aldehydes, isovaleraldehyde and phenylacetaldehyde, followed similar kinetics during heating. This is due to the similar mechanism of formation and the fact that they are both consumed through reaction to form 5-methyl-2-phenyl-2-hexenal [39].

#### 5.2.4.5 Summary

There has been limited work done studying the kinetics of flavor formation via the Maillard reaction. Of this work, the majority has been on the formation of pyrazines in simple model systems. There is an initial linear period for flavor formation that can be accurately modeled using pseudo zero order reaction kinetics. For longer heating times, flavor concentration in the system will generally plateau (or decrease) and this phenomenon is better modeled using first order kinetics. The activation energies for flavor formation have ranged from about 12 to 45 Kcal/mole. The higher values are for pyrazine formation in simple model systems and represent the linear portion of the formation period. Values for many other volatiles are closer to 20 Kcal/mole, which are more in agreement with the Maillard reaction as characterized by the consumption of reactants or pigment formation.

The work to date has demonstrated that the formation of flavors is very system dependent. For some volatiles, more complex model systems may have quite different reaction kinetics than a simple model system. This suggests caution when one predicts the formation of flavors in real foods based on model system work. Other volatiles, e.g., 2-acetyl-l-pyrroline, appear to be relatively unaffected by the model system complexity and model predictions can be quite accurate.

Reaction rates are generally influenced by pH and Aw of the model system. The effect of pH is variable as one might expect. Some volatiles are formed more slowly at higher pHs while others are formed more quickly. Also, secondary reactions may consume some volatiles, and thus their final concentrations in a system are dependent upon the secondary reaction kinetics as well. Data to date (very limited!) on pH effects on pyrazine formation demonstrates that the rate of formation increases with increasing pH. Data on other volatiles depends upon the volatile compound being considered. At this stage of knowledge, it is impossible to make generalizations in this respect. Overall, we have come to accept that flavor formation/concentration is a dynamic situation. Flavor compounds are formed and reacted to give a blend that we recognize as a given flavor. This balance can be very sensitive to the chemical environment and thermal processing.

### 5.2.5 Flavor Formation via the Maillard Reaction

The most abundant flavor compounds formed via the Maillard reaction are aliphatic aldehydes, ketones, diketones, and lower fatty acids. However, heterocyclic compounds containing oxygen, nitrogen, sulfur, or combinations of these atoms are much more numerous and significant to the flavor of thermally processed foods [40]. The remainder of this section will provide examples of flavor compounds formed via the Maillard reaction, some mechanisms of formation, as well as information on sensory properties. Quite obviously, this section cannot be comprehensive but only a sampling of the diverse number of compounds formed via the Maillard reaction.

### 5.2.5.1 Carbonyl Compounds

The major pathway leading to the formation of carbonyls is the Strecker degradation. This reaction occurs between dicarbonyls and free amino acids. The dicarbonyls involved have vicinal carbonyls (carbonyl groups separated by one double bond) or conjugated double bonds [41]. While these carbonyls typically are intermediates in the Maillard reaction, they may also be normal constituents of the food (e.g., ascorbic acid), be end products of enzymatic browning (e.g., quinones), or be products of lipid oxidation[42].

The end products of the Strecker degradation are $CO_2$, an amine, and the corresponding aldehyde of each deaminated and decarboxylated amino acid. At one time these aldehydes were considered to be quite important to the flavor of heated food products. This was primarily because they are the most abundant volatiles formed via the Maillard reaction (and thus assumed to be important). It is now realized that the heterocyclic volatiles are more important. The Strecker aldehydes are often monitored in foods since they are present in quantity and can serve as indicators of the Maillard reaction [43].

Strecker aldehydes may also be formed via free radical mechanisms [44]. The oxidation of amino acids by hydrogen peroxide or lipid peroxides yields $CO_2$, ammonia, and the corresponding Strecker aldehydes. While this is a viable pathway for the production of Strecker aldehydes, the above-mentioned pathway typically predominates.

The aldehydes formed in this reaction may undergo additional reactions. Pokorny [45] has outlined aldol condensations leading to dimeric unsaturated aldehydes, and eventually, high molecular weight polymers. Takken et al. [46] have demonstrated the reaction of aldehydes with α-dicarbonyls, ammonia, and hydrogen sulfide. A host of thiazoles, dithiolanes, thiazolines, and hydroxythiazolines were found as products of these reactants. An additional fate of the Strecker aldehydes is that they may end up being bound onto various food constituents (e.g., protein). The binding may be sufficiently strong that the aldehydes may not have adequate vapor pressure to contribute to food odor.

### 5.2.5.2 Nitrogen-Containing Heterocyclic Compounds

More than 100 different pyrazines have been identified in various food products. The sensory properties of the pyrazines are quite diverse. The alkyl pyrazines (Figure 5.5a) generally possess roasted, nut-like notes while methoxypyrazines (Figure 5.5b) often possess earthy, vegetable notes [47]. The 2-isobutyl-3-methoxy pyrazine has a freshly cut green pepper flavor with a sensory threshold of 0.002 ppb in water. The acetyl pyrazines typically have a popcorn character, and 2-acetonyl pyrazine has a

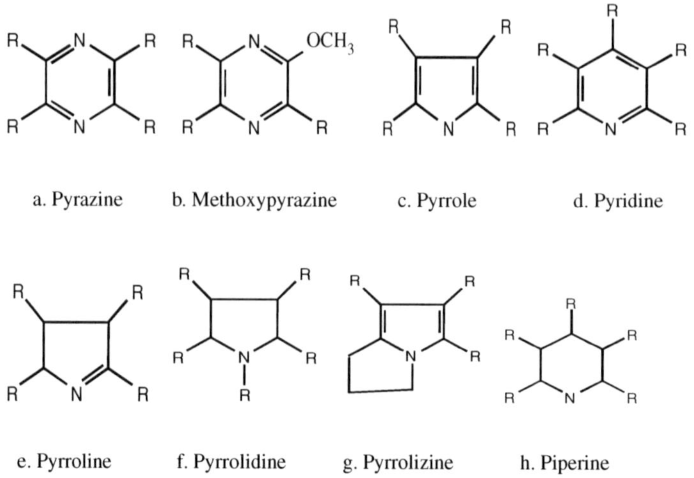


**FIGURE 5.5** Classes of nitrogen-containing heterocyclic volatiles formed via the Maillard reaction (a. pyrazine, b. methoxypyrazine, c. pyrrole, d. pyridine, e. pyrroline, f. pyrrolidine, g. pyrrolizine, and h. piperine).

toasted or burned note. Several bicyclic pyrazines have been identified in foods. The sensory properties of these pyrazines have been described as being burnt, roasted, grilled, and/or animal in character. Several mechanisms have been proposed for the formation of various pyrazines [12]. The alkyl pyrazines are probably formed from the reaction of α-diketones with amino acids to form α-amino ketones (Strecker degradation). These α-amino ketones may condense with other α-amino ketones to yield a heterocyclic compound. This heterocyclic may undergo oxidation to form the tri-unsaturated pyrazines. The alkyl substitution is dictated by the dicarbonyl fragment: the amino acid contributes only the amine to the pyrazine [48].

Pyrroles are nitrogen-containing heterocyclic compounds (Figure 5.5c). 2-formyl pyrrole and 2-acetyl pyrrole are the two most abundant and widely occurring pyrroles in foods [47]. 2-formyl pyrrole has a sweet corn-like odor while the 2-acetyl pyrrole has a caramel-like odor. 1-acetonyl pyrrole has been found to have a cookie or mushroom aroma while the pyrrole lactones have a spicy peppery flavor.

Hodge et al. [49] originally proposed a mechanism for the formation of pyrroles, which is also supported by Tressl et al. [50]. This basically is the participation of proline and hydroxy proline in the Strecker degradation to yield pyrroles. If the food system does not contain either proline or hydroxy proline, then a sugar of at least five or more carbons is required for pyrrole formation [51].

Rizzi [52] proposed an alternate pathway for the formation of acylalkyl pyrroles involving the reaction of an amino acid with the corresponding furan. The amino acid reacts with the number 5 carbon on the furan ring to open the furan ring. The ring closes via dehydration with N in the ring structure. Depending upon the amino acid involved and subsequent rearrangements, a variety of pyrroles may arise.

Pyridine compounds (Figure 5.5d) appear to be less widely distributed in browned foods than the pyrroles and pyrazines [53]. While they possess a wide range in odor properties, green notes are quite common [54]. 3-methyl pyridine has a green odor, as do many of the singly substituted pyrazines. 3-methyl-4-ethyl pyridine has been characterized as being sweet and nutty. Pyridine is judged as being pungent and offensive while 2-acetyl-pyridine is tobacco-like in odor. The contribution of the pyridines to browning flavor is dependent on the individual pyridine formed and its concentration in the food. At low concentrations, the pyridines typically contribute very pleasant notes. However, they generally become harsh and offensive at higher concentrations.

The mechanisms of pyridine formation are poorly characterized. Vernin and Parkanyi [12] have proposed three mechanisms. The first two mechanisms are dependent upon aldol condensations to yield unsaturated aldehydes with side groups. Reaction then with ammonia or an amino acid and ring closure would yield a nitrogen-containing heterocyclic. The oxidation of this heterocyclic would result in the formation of a pyridine. The third pathway involves the reaction of dialdehydes with ammonia followed by dehydration to produce pyridines.

Tressl et al. [50,55,56] have identified an additional 59 nitrogen-containing heterocyclic compounds which, like the pyrroles, are believed to arise from the reaction of proline and hydroxy proline with dicarbonyls via the Strecker degradation. These compounds fall into the classes of pyrrolines, pyrrolidines, pyrrolizines, and piperidines (Figure 5.5e–h, respectively). Odors associated with these compounds are most often cereal or roasted in nature. For example, 2-acetyl-l-pyrroline is characterized as being popcorn-like [50]. l-acetonyl pyrrolidine has a cereal aroma while the furan-substituted pyrrolidines are considered to have sesame-like notes [56]. While the piperidines [e.g., 2-acetyl piperidine] have been characterized as having weak woody odor and very bitter taste, they are readily transformed into the corresponding tetrahydropyridines which have bready, cracker-like aromas [55]. 2-acetyl pyrrolizine has a smoky roasted aroma. Azepine derivatives generally are exceptionally bitter with little odor [50].

### 5.2.5.3 Oxygen-Containing Heterocyclic Compounds

The furanones and pyranones are oxygen-containing heterocyclic compounds associated with both caramelized and Maillard flavors [57]. The odor characters most common to this group of compounds would be caramel-like, sweet, fruity, butterscotch, nutty, or burnt. They predominate both in proportion and absolute amount in condensates of carbohydrates that are subjected to browning reactions [47].

The caramel character is associated with a planar, contiguous C-alkyl-enol carbonyl group (alkyl-C=C(OH)-C:O) in the molecule [58]. Maltol (Figure 5.6a) was one of the first compounds in this class to be identified in foods. Ethyl maltol (2-ethyl-3-hydroxy-4(4H)-pyranone) also has a caramel odor but is approximately four to six times stronger in flavor strength than maltol. Furaneol (4-hydroxy-2,5-dimethyl-3(2H)furanone, Figure 5.6b) is the trade name of a compound used extensively in the flavor industry and, like maltol and ethyl maltol, is a flavor enhancer for sweet products. Furaneol itself has a burnt pineapple odor.

The five carbon analogs of maltol and furaneol possess odor properties similar to their oxygen-containing counterparts. Cyclotene (trade name for compound,

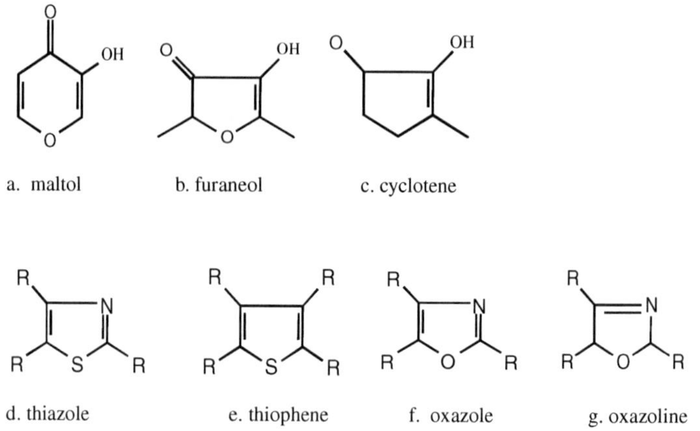


**FIGURE 5.6** Oxygenated and sulfur-containing compounds formed via the Maillard reaction.

Figure 5.6c) has a very characteristic sweet maple character. The related compound, 3-ethyl-2-hydroxy-2-cyclopenten-1-one, also finds extensive use in the flavor industry for similar applications (caramel, nut, maple, and butterscotch flavorings).

Compounds of this class that do not meet the structural requirements for being caramel in character may have quite different sensory properties. For example, 4-methoxy-2,5-dimethyl-3(2H)-furanone has an odor similar to sherry. The n-butyl ether version of this furanone has a jasmine-like odor.

Since this group of compounds does not contain nitrogen, the mechanism of formation generally involves the cyclization of non nitrogen-containing browning intermediates. These intermediates may be products of major browning pathways involving sugar dehydration or the Strecker degradation [59,60].

### 5.2.5.4 Sulfur-Containing Heterocyclic Compounds

Numerous different types of heterocyclic compounds containing sulfur are produced via the Maillard reaction [61,62]. These include thiophenes, dithioles, dithianes, dithiins, trithiolanes, trithanes, tetrathianes, thiazoles, thiazolines, and thiazolidines. The major heterocyclic sulfur-containing compounds produced via the Maillard reaction are thiazoles and thiophenes (Figure 5.6d and Figure 5.6e, respectively).

Thiazoles and pyrazines have somewhat similar sensory properties. Pittet and Hruza [54] and Ho and Jim [63] have reported that the alkylthiazoles give green, nutty, roasted, vegetable, or meaty notes. Trimethyl thiazole is reported to have a cocoa, nutty character. 2-isobutyl thiazole is one of the best known thiazoles and has a strong, green odor of tomato leaf. This compound is considered to be important to tomato flavor. 2,4-dimethyl-5-vinyl thiazole has a nut-like odor. 2-acetyl thiazole is characterized as having a nutty, cereal, and popcorn flavor [47].

With the exception of cranberries, thiophenes have only been found in cooked or roasted food products [47]. Thiophene has a pungent character while 2,4-dimethyl thiophene is well known for its importance to the flavor of fried onions. 2-acetyl-3-methyl thiophene has a honey-like sensory quality at 0.25 g/100 L concentration but is nutty and starchy at 0.11 g/100 L concentration. 5-methyl thiophene-2-carboxaldehyde has a burnt coffee note [51].

These compounds, like the thiazoles and thiophenes, typically are formed via the reaction of sulfur-containing amino acids with intermediates of the Maillard reaction. An alternative mechanism involves the initial formation of $H_2S$ from the sulfur-containing amino acids and then the reaction of the $H_2S$ with browning intermediates [64].

#### 5.2.5.5 Oxygen-Containing Compounds

With the exception of the report by Stoffelsma and Pypker [65], oxazoles and oxazolines (Figure 5.6f and Figure 5.6g, respectively) have been found only in food systems that have undergone Maillard reaction[66,67]. The oxazoles typically have green, sweet, floral, or vegetable-like aromas [68,69].

As an example, 4-methyl-5-propyl oxazole has been characterized as having a green vegetable aroma. However, oxazoles which have a 4 or 5 carbon length chain on the oxazole ring and no alkyl group on carbon-2 or 4 have distinct bacon-fatty notes (e.g., 5-butyl oxazole). When a methyl or ethyl group is substituted on carbon-2 (e.g., 2-ethyl-5-butyl oxazole), the fatty aroma is reduced and sweet-floral aromas become more characteristic. The sweet-floral character is further enhanced by additional methyl or ethyl substitution on carbon-4 [69]. The oxazolines tend to have a wide variety of sensory properties. 2-isopropyl-4,5,5-trimethyl-3-oxazoline has a rum-like note while 2-isopropyl-4,5-diethyl-3-oxazoline has a typical cocoa aroma.

## 5.3 FLAVORS FROM LIPIDS

Lipids may undergo changes during the processing of foods that make a flavor contribution. Later in this text we shall discuss lipid oxidation and lipolysis as undesirable flavors imparted by changes in the lipid fraction of a food (Chapter 7). This chapter discusses lipid changes that impart desirable flavors.

### 5.3.1 Deep Fat Fried Flavor

Deep fat fried foods, e.g., French fried potatoes, doughnuts, and snacks, are generally well liked by consumers. While their popularity may partially lie in the physical properties imparted to the food by fats, e.g., lubricity, richness, and texture, the characteristic fried flavor is very unique and desirable. This flavor comes from both thermally induced changes in the food (Maillard reaction) and flavor developed in the frying oil [70]. It is well known that a deep fat fried product does not obtain its characteristic flavor until the oil has been used for a period of time, i.e., fresh frying oil does not give a good flavored product. The mechanism of flavor development in heated oils is essentially that of lipid oxidation. As Ohnishi and Shibamoto [71] and later, Shibamoto and Yeo [72] have outlined (Figure 5.7), thermally induced oxidation involves hydrogen radical abstraction, the addition of molecular oxygen to form the peroxide radical, formation of the hydroperoxide and then decomposition to form volatile flavor compounds. The products of thermally induced oxidations differ from

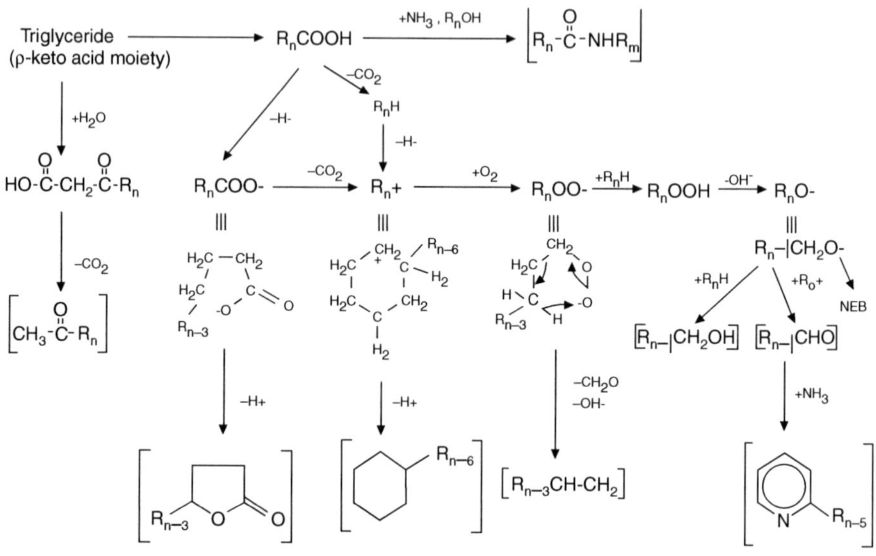


**FIGURE 5.7** Proposed mechanism for the formation of volatiles from heated animal fats. (From Ohnishi, S., T. Shibamoto, *J. Agric. Food Chem.*, 32(5), p. 987, 1984. With permission.)

typical lipid oxidation products formed at room temperature. The reason for these differences goes back to kinetic considerations as was illustrated in the previous section on the Maillard reaction. Each chemical reaction has its own unique activation energy, and therefore, rate at a given process temperature. Thus, the reactions occurring in the frying oil and the volatiles formed are dependent on the temperature of processing.

A second distinguishing factor of thermal oxidations is their more random nature than typical room temperature oxidations. Very high temperatures make more sites available on the fatty acid for oxidation to occur. Thus a wider range in end products (volatile flavor components) will occur. So even though the same chemical mechanisms are involved in flavor formation in deep fat fried foods, the flavor developed is unique to this process.

Chang et al. [73] and Nawar et al. [74] have identified many of the volatiles formed during deep fat frying. They have found numerous acids, alcohols, aldehydes, hydrocarbons, ketones, lactones, esters, aromatics, and a few miscellaneous compounds (e.g., pentylfuran and 1,4-dioxane) as products of deep fat frying. More recently, Wagner and Grosch [75] have studied the key contributors to French fry aroma. The list of key aroma compounds in French fries includes: 2-ethyl-3,5-dimethylpyrazine, 3-ethyl-2,5-dimethylpyrazine, 2,3-diethyl-5-methylpyrazine, 3-isobutyl-2-methoxypyrazine, (E,Z), (E,E)-2,4-decadienal, trans-4,5-epoxy-(E)-2-decenal, 4-hydroxy-2,5-dimethyl-3(2H)-furanone, methylpropanal, 2- and 3-methylbutanal, and methanethiol. If one examines this list, it is obvious that the Maillard reaction (pyrazines, branched chain aldeydes, furanones, and methional), and lipid oxidation (unsaturated aldehydes) are the primary sources of this characteristic aroma.

Chang et al. [76], Nawar et al. [74] and Wagner and Grosch [75] have studied the volatiles produced by different types of oil upon heating. The study reported by Chang et al. [76] reported on qualitative differences while Nawar et al. [74] focused on quantitative differences. Wagner and Grosch [75] considered the influence of oil type on key aroma compounds. The qualitative study noted numerous differences between the volatiles produced by heated corn oil vs. hydrogenated cottonseed oil [76]. This is expected since the oils differ greatly in fatty acid composition. The quantitative study indicated that corn and soybean oils produced similar volatiles but coconut oil was quite different [74]. The coconut oil produced less of the decadienals and unsaturated aldehydes and more saturated aldehydes relative to the corn and soybean oils. Methyl ketones, and γ-lactones were also present in greater quantities in the coconut oil. Wagner and Grosch [75] found that frying potatoes in coconut oil resulted in the presence of γ-octalactone as a key aroma contributor. This finding supports the work of Chang et al. [73] who reported that the γ-lactones with unsaturation at the 2 or 3 position are of particular significance to deep fat fried flavor.

The quantitative effects of heating time and introduction of moisture during deep fat frying were also reported by Nawar et al. [74]. In general, the production of individual volatile components increased with heating time up to 48 h of heating. The introduction of moisture during heating of the oil resulted in a large reduction in volatiles in the oil. This apparent protective effect of water may be due to the steam stripping of volatiles during frying, the displacement of oxygen by the steam, or some combination of both [74].

### 5.3.2 Lactones

Lactones may be formed in foods via microbial action, extensive lipid oxidation (ambient or thermal), or heating [77]. The formation of lactones in heated dairy products is relevant to this chapter. It appears that as a result of inborn errors in lipid metabolism, hydroxy acids may be synthesized and incorporated into the bovine triglycerides [78]. These hydroxy acids are relatively unstable and are readily hydrolyzed from the triglyceride in the presence of water with heat. Hydrolysis of the hydroxy acid results in its cyclization to a lactone. Since the hydroxy acid has no flavor as long as it is bound to the triglyceride, thermal processing frees the acid and a flavorful lactone is produced. This conversion is readily observed if one takes butter and heats it. On heating, the flavor changes to a very pleasant sweet, caramelized, character largely due to the formation of lactones [79,80]. Whitfield [42] has noted that lactones can be formed from other lipids (not hydroxy fatty acids) as well through heat-induced oxidations.

### 5.3.3 Secondary Reactions

Lipids may contribute to food flavor formation through participation in other chemical pathways, most notably, the Maillard reaction. Whitfield [42] has provided a very comprehensive review of how lipids and their degradation products may participate in the Maillard reaction. He lists the primary means of interaction as:

1. The reaction of lipid degradation products with ammonia from the Strecker degradation or the amino groups of cysteine. Thermally induced lipid oxidation provides an abundant source of reactive carbonyls, e.g., aldehydes and ketones. These carbonyls will readily react with free ammonia or amines provided by the Maillard reaction (Figure 5.7). Lipids provide unique carbonyl reactants in that they may have long carbon skeleton chains. The heterocyclic aroma compounds found in heated foods that have R groups of 4 of more carbons are assumed to be derived from lipid sources. This has particular relevance to the snack food industry. One of the issues plaguing this industry is the development of baked snacks with a similar flavor profile as fried snacks. If we look at some of the flavor components identified in fried foods (e.g., 2-pentyl-3,5,6-trimethylpyrazine, 5-heptyl-2-methylpyridine, 5-pentyl-2-ethylthiophene, 2-octyl-4,5-dimethylthiazole and 4-methyl-2-ethyloxazole), we find that these aroma compounds are cyclic and have relatively long alkyl side chains, i.e., they are derived from lipid sources. It then becomes understandably problematic to form these long chain aroma compounds in foods that contain little fat (baked vs. fried foods).
2. The reaction of the $NH_2$ group of phosphatidylethanolamine with sugar derived carbonyls. Lipids may also contribute amine groups for the Maillard reaction (e.g., phosphatidylethanolamine). This would increase the amount of free $NH_2$ for the reaction. An additional comment can be made regarding the participation of phospholipids in the Maillard reaction. There is significant evidence that the presence of phospholipids decreases the amounts of typical heterocyclic compounds formed on heating a food (or model system). It is hypothesized that phospholipid degradation products compete for the available $NH_2$ thereby reducing the heterocyclic compounds normally produced via the Maillard reaction. This would change the distribution of aroma compounds found in the heated food and therefore, its aroma.
3. The participation of free radicals from oxidizing lipids in the Maillard reaction.
4. The reaction of hydroxy or carbonyl lipid degradation products with free hydrogen sulfide from the Maillard reaction. This is essentially analogous to the participation of lipid derived oxidation products with $NH_2$ from the Maillard reaction. One finds many S-containing heterocyclics in heated foods that must have been derived from lipid sources (long R groups).

This interdependency of reactions has been most studied in meats, or model meat reaction systems [42,72,81]. Wasserman [82] was amongst the first to find that the lean portion of the meat supplied the meaty, brothy character and the fat provided the species character much of which is due to lipid/Maillard interactions. This knowledge has long been used in the manufacture of process products (meat flavors). Meat process flavors contain approximately the same sugars and amino acids for the basic meat flavor but contain different fats to give the unique pork, beef, or chicken notes.

## 5.4 FLAVORS FORMED VIA FERMENTATION

Fermentation is a process that provides us with a variety of unique flavors to enjoy. We daily come in contact with a variety of food products produced via fermentation. Some examples include soy sauce, cheese, yogurt, bread, beer, wine, fermented fish products, and sausages. The flavor of these products may be developed from the primary metabolism of the fermentation microorganisms or from residual enzymatic activity once the microbial cell has lysed. Primary metabolism is responsible for much of the flavor of alcoholic beverages while residual enzymatic activity is essential for the development of aged cheese flavor.

Microorganisms can produce a large number of different classes of flavor compounds. Tressl et al. [83] have summarized some of the reactions that microorganisms may accomplish (Table 5.3). This section will present some metabolic pathways leading to the development of flavor in fermented food products. The book by Margalith [84] is dated but still a very good reference for greater detail.

### 5.4.1 Esters

Esters are quite important to the flavor of both natural foods (e.g., fresh fruit) and fermented foods. Of the fermented foods, esters probably are most important to the flavor of some of the alcoholic beverages. The TNO-CIVO [85] compilation of volatiles in foods lists 94 esters that had been identified in beer. Most of the esters found in beer are formed via primary fermentation. They are produced intracellularly in yeast by enzymatic action [86]. Lipid metabolism by the yeast provides a large number of acids and alcohols that may undergo esterification to yield a variety of esters. While pure chemical reactions can lead to ester formation, this reaction is

**TABLE 5.3**
**Reactions Catalyzed by Microbes**

| | |
|---|---|
| **Oxidation** | Acylation |
| Oxidation of CH or C=C | Transglycosidation |
| Dehydrogenation of CHOH | Methylation |
| Dehydrogenation of CH-CH | Condensation |
| | |
| **Reduction** | Cleavage of C-C |
| Hydrogenation C=O | Decarboxylation |
| Hydrogenation C=C | Dehydration |
| | Amination/Deamination |
| **Isomerization** | Halogenation |
| **Esterification/Hydrolysis** | Phosphorylation |

*Source:* From Tressl, R., M.Apetz, R. Arrieta, K.G. Grunewald, *Flavor of Foods and Beverages*, G. Charalambous, G.I. Inglett, Eds., Academ. Press, New York, 1978, p. 145. With permission.

too slow to account for the esters in most foods. Cristiani and Monnet [87] have provided a comprehensive review on the formation of esters in fermented foods.

### 5.4.2 Acids

Acid production via microorganisms is important to the flavor of many fermented foods. The acid of greatest importance to the flavor of fermented dairy products is lactic acid [88]. Lactic acid is an optically active acid existing as the D, L, or optically inactive mixture, depending upon the microorganism involved in its synthesis. There appears to be no difference in flavor between the optical isomers. Lactic acid is characterized as being odorless and having a sour milk taste.

The organisms most commonly associated with lactic acid formation are classified as being either homofermentative (e.g., *Lactobacillus bulgaricus,* which produces primarily lactic acid, >85% of the metabolic end products) or heterofermentative (e.g., *Leuconostoc* sp., which produces lactic acid, acetic acid, ethanol, carbon dioxide, and other metabolites). The homofermentative organisms employ the Embden-Meyerhof pathway for lactose fermentation while the heterofermentation organisms metabolize lactose via the hexose-monophosphate pathway [84]. Lactic acid will reach concentrations of up to 2% or greater in cultures of L. *bulgaricus* or L. *acidophilus* (Figure 5.8).

Lactic acid is also formed in wines during fermentation from L-malic acid. Malic acid is a very tart acid present in grapes, which needs to be converted to lactic acid

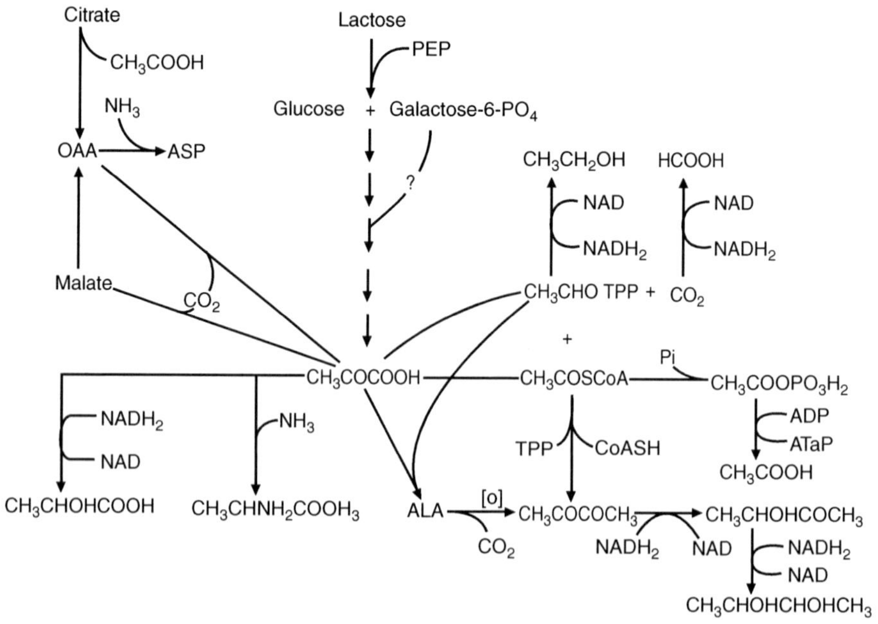


**FIGURE 5.8** Formation of various flavor compounds via fermentation of lactose and citrate. (From Sandine, W.E., P.R. Elliker, *J. Agric. Food Chem.*, 18(4), p. 557, 1970. With permission.)

to smooth out the harsh acidity. This malo-lactic fermentation is the result of bacterial (most commonly *Leuconostoc* sp.) action during the last stages of fermentation.

Numerous other organic and aliphatic acids are produced during fermentation. Most microorganisms have active lipase systems that attack triglycerides to yield glycerol, monoglycerides, diglycerides, and free fatty acids. These free fatty acids can make a substantial contribution to the flavor of various cheeses and sausages. Secondary reactions of the free fatty acids (e.g., oxidation) can again yield a host of new flavor compounds. Acids may also come from the deamination of amino acids. The products are various aliphatic (linear and branched chain) and aromatic acids.

### 5.4.3 Carbonyls

Carbonyl compounds make a particularly significant contribution to the flavor of fermented dairy products. Diacetyl, characterized as having a buttery, nut-like aroma, is one of the most important carbonyls to the flavor of these products. Diacetyl is produced via the fermentation of citrate. The most important citric acid fermenters are *Leuconostoc citrovorum,* L. *creamoris,* L. *dextranicum, Streptococcus lactis* subspecies *diacetylactis,* S. *Thermophilus,* and certain strains of *Proprionibacterium shermani* [90,91]. The metabolic pathway leading to the synthesis of diacetyl involves the degradation of citrate to acetate and oxaloacetate, and the oxaloacetate is then decarboxylated to form pyruvate (Figure 5.9) [92]. Pyruvate plus acetaldehyde forms α-acetolactate, and ultimately, diacetyl. Diacetyl is relatively nontoxic to the bacteria cell so excess pyruvate is channeled into diacetyl.

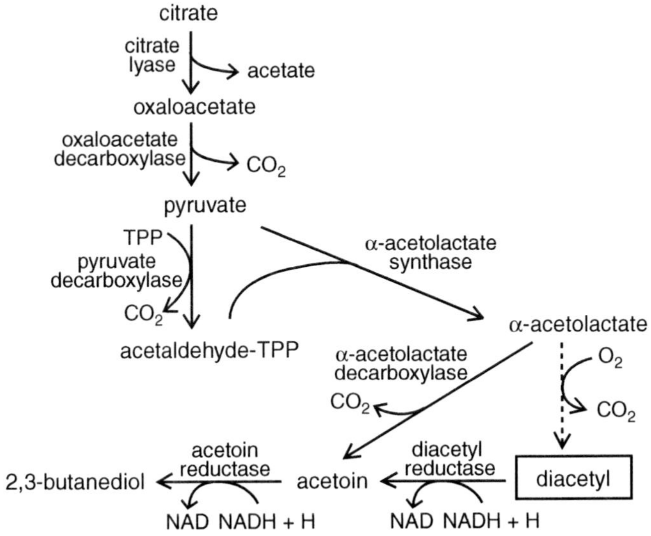


**FIGURE 5.9** Formation of diacetyl during fermentation in dairy cultures. (From Hutkins, R.W., *Applied Dairy Microbiology*, E.L. Marth, J.L. Steele, Eds., Marcel Dekker, New York, 1993, p. 207. With permission.)

Unfortunately diacetyl is not stable in most cultured food products. The microorganisms that synthesize diacetyl also contain diacetyl reductase that reduce diacetyl to acetoin and 2,3-butanediol (Figure 5.9). Thus fermented products such as buttermilk, which depend on diacetyl for flavor have an optimum or peak flavor. As the product is stored, the diacetyl is reduced and flavor strength and quality decrease.

A second dairy product where carbonyls are considered as flavor impact compounds is yogurt. Here acetaldehyde is credited with providing the characteristic pungent, green character. Acetaldehyde is a metabolic end product of L. *bulgaricus* and/or S. *thermophilus* during the fermentation of milk to yogurt [84].

Carbonyls are quite important to the flavor of cheeses. The TNO-CIVO compilation of volatiles in foods lists 29 carbonyls as having been identified in Cheddar cheese at that time. Carbonyls (methyl ketones) may arise in fermented products initially via lipase activity of the starter culture. Dairy products contain a significant quantity of α-keto acids which are readily hydrolyzed from the triglyceride by microbial lipases and then decarboxylated to form odd carbon number methyl ketones.

Methyl ketones and aldehydes may also be formed via microbially induced lipid oxidations [93]. The oxidation may be initiated by microbial lipases, hydrogen peroxide produced by microorganisms and/or lipoxidase-like activity.

Alford et al. [94] have reported on the action of various microorganisms on lard oxidation. They found 10 of the 28 microorganisms studied destroyed the low levels of peroxide in fresh lard. Fourteen of the 28 had no effect on peroxides in the lard, while five strains of streptomyces increased peroxides by threefold. *Pseudomonas ovalis* increased peroxides by 8-fold, and *Micrococcus freudenreichii* increased peroxides 14-fold. The extent of oxidation observed in the lard was found to be very dependent upon the microflora present in the lard.

A final mechanism for the formation of carbonys via microorganisms involves the transamination and decarboxylation of free amino acids. Morgan [95] has demonstrated the production of 3-methyl butanal by S. *lactis* var. *maltigenes* (Figure 5.10). MacLeod and Morgan [96] have demonstrated the production of 2-methyl butanal, methional and phenylacetaldehyde from leucine, methionine and phenylalanine, respectively, by the same microorganism.

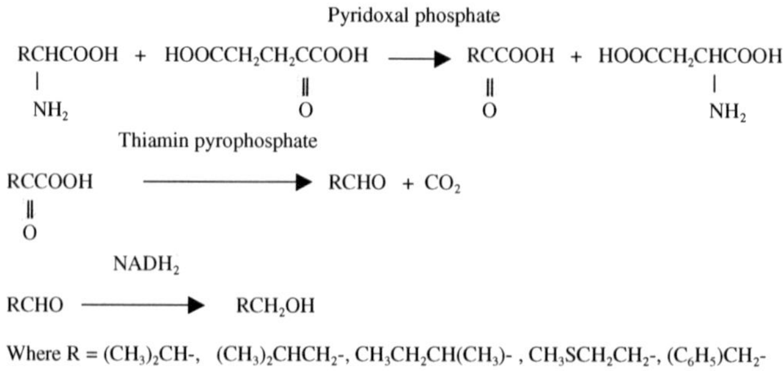


**FIGURE 5.10** Metabolic pathway leading to the formation of aldehydes and alcohols. (From Morgan, M.E., *Biotechnol. Bioeng.*, 18(7), p. 953, 1976. With permission.)

Carbonyl compounds typically are significant to the flavor of most fermented food products. As has been shown, these carbonyls may arise due to carbohydrate or citrate metabolism, lipid oxidation or amino acid degradation.

### 5.4.4 Alcohols

Alcohols typically make a minor contribution to flavor unless present in relatively high concentrations (ppm) or they are unsaturated (e.g., 1-octen-3-ol). Alcohols may arise via primary metabolic activity of a microorganism (e.g., ethanol) or by reduction of a carbonyl to the corresponding alcohol.

Reazin et al. [97] have outlined metabolic pathways leading to the major congeners during grain fermentation (Figure 5.11). This figure illustrates that the fusel alcohols (alcohols larger in size than ethanol) can be formed from either carbohydrate or amino acid metabolism. Alcohol production from amino acids may occur either by transamination, decarboxylation, and reduction, or by oxidative deamination followed by decarboxylation and reduction. In either pathway, the product is always the amino acid minus the amine group and one carbon atom.

The mechanism of fusel oil production from carbohydrates follows the typical Embden-Meyerhof-Parnas pathway to pyruvic acid. Pyruvic acid may be reacted with a second pyruvic acid to yield acetolactic acid. Acetolactic acid enters the valine-isoleucine biosynthesis pathway and is converted to α-keto isovaleric acid. The α-keto isovaleric acid may then be reduced to isobutyl alcohol or isobutyric

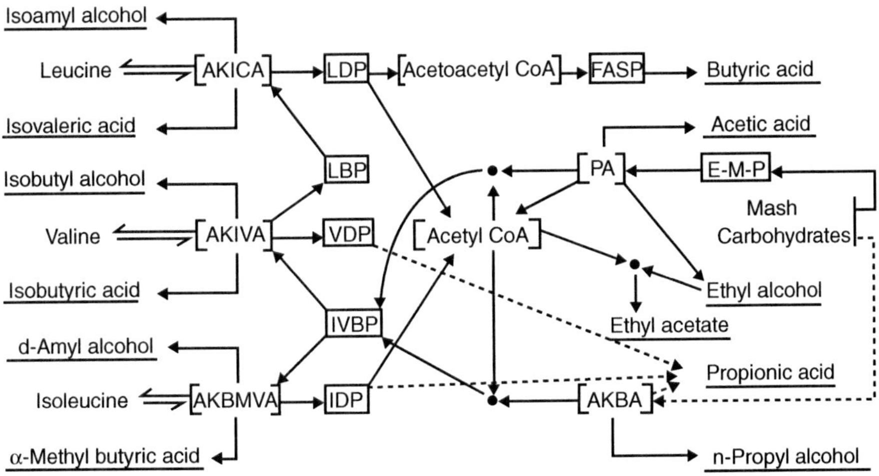


**FIGURE 5.11** Proposed metabolic pathways for the synthesis of whiskey congeners: —, synthetic and degradative pathways; ---, hypothetical pathways; LBP, leucine biosynthesis; IVBP, isoleucine-valine biosynthesis; FASP, fatty acid biosynthesis; LOP, leucine degradation; VDP, valine degradation; IOP, isoleucine degradation; E-M-P, Embden-Meyerhof-Parnas fermentation pathway. (From Reazin, G., H. Scales, A. Andreasen, *J. Agric. Food Chem.*, 18, p. 585, 1970. With permission.)

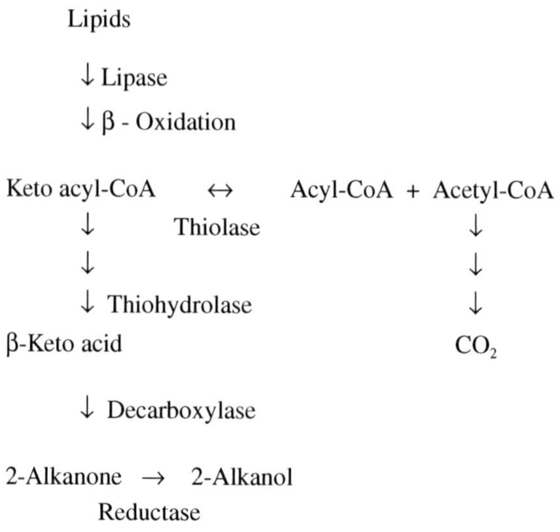


**FIGURE 5.12** Formation of methyl ketones and the corresponding alcohols. (From Lamparsky, D., I. Klimes, *Flavour '81*, P. Schreier, Ed., deGruyter, Berlin, 1981, p. 557. With permission.)

acid. The interconnection of pathways as shown in Figure 5.11 results in the formation of numerous alcohols. Note that Acetyl CoA may also enter into these pathways. Acetyl CoA may come from fatty acid metabolism, which ties fatty acids also into alcohol production.

The alternate metabolic pathway yielding alcohols involves the reduction of the corresponding carbonyl (Figure 5.12). Lamparsky and Klimes [98] demonstrated this mechanism by finding the equivalent alcohols (as aldehydes) in Cheddar cheese. Secondary alcohols corresponding to the methyl ketones found in blue cheeses have also been reported.

### 5.4.5 Terpenes

Mono- and sesquiterpenes are most commonly associated with the flavor of citrus products, spices, and herbs. They are hydrocarbons based upon the five carbon isoprene unit (2-methyl-1,3-butadiene) with structures that may be open chain, closed chain, saturated, or unsaturated, and may contain O, N, or S. While a host of terpenes have been found to be produced by microorganisms, there are no fermented food products commercially available that derive their characteristic flavor from terpenes.

The primary interest in terpene metabolism by microorganisms is due to the potential for the commercial production of specific terpenes via fermentation [99]. In many cases, microorganisms are used to accomplish specific terpene transformations that convert abundant low cost terpenes to high value products. The literature abounds with references to this effort [100–103].

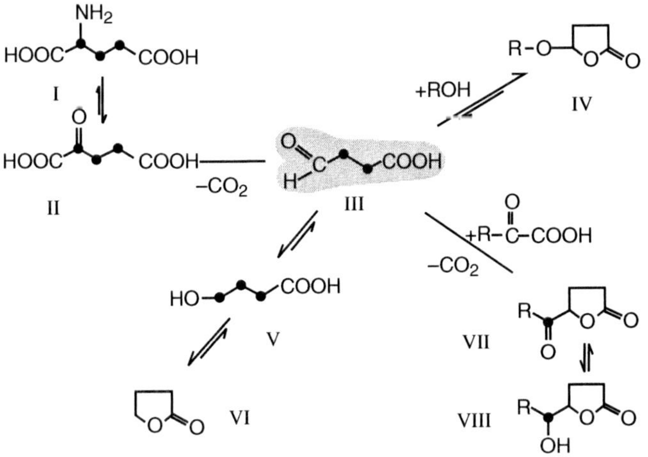


**FIGURE 5.13** Formation of lactones by microorganisms. (From Tressl, R., M.Apetz, R. Arrieta, K.G. Grunewald, *Flavor of Foods and Beverages*, G. Charalambous, G.I. Inglett, Eds., Academ. Press, New York, 1978, p. 145. With permission.)

### 5.4.6 Lactones

Lactones make a significant contribution to the flavor of several fermented food products, particularly dairy products and alcoholic beverages. Tressl et al. [83] have summarized some of the previous literature on lactone synthesis by microorganisms in fermented alcoholic beverages. As can be seen in Figure 5.13, lactones may be formed from amino acids. Pathway A shows the oxidative deamination of glutamic acid to yield 2-oxoglutarate. The 2-oxoglutarate is decarboxylated to produce 4-oxobutyrate, which is then reduced to the 4-hydroxy butyrate. The cyclization of 4-hydroxy butyrate yields $\gamma$-butyrolactone, alkoxy- and acyl-lactones. Pathway B illustrates the conversion of 2-oxobutyric acid into 3-hydroxy-4,5-dimethyl-2(5H)-furanone. This lactone has a burnt aroma and has been identified in aged sake [104].

### 5.4.7 Pyrazines

Pyrazines are generally associated with nonenzymatic browning. However, pyrazines have also been found as a result of microbial action. *Bacillus subtilis* was the first organism found to produce a pyrazine (tetramethyl pyrazine [105]). Since that time, several pyrazines (2-acetyl, dimethyl, 2-methoxy-3-ethyl, 2,5 or 2,6-diethyl-3-methyl, trimethyl, tetramethyl, and trimethyl pyrazine) have been identified in aged cheese [16,106–109]. While some of these pyrazines may be formed via the Maillard reaction during aging of the cheese, it is likely that they were formed via microbial action.

Adachi et al. [110] was the first to postulate a mechanism for the formation of a pyrazine by B. *subtilis* (Figure 5.14). He proposed that this pyrazine could be formed from 2 moles of acetoin and 2 moles of ammonia, both metabolic products

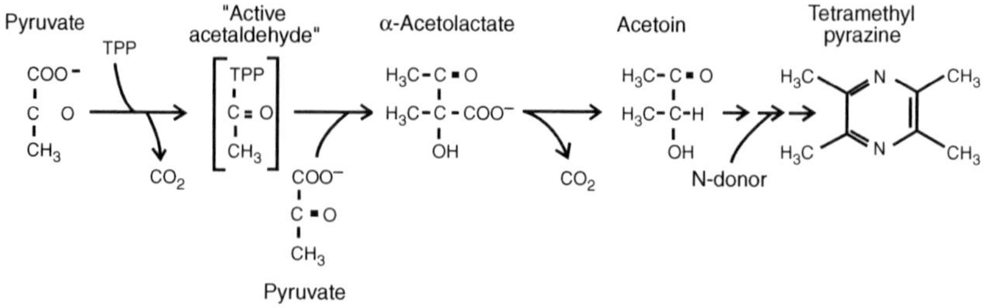


**FIGURE 5.14** Formation of pyrazines by microorganisms. (From Adachi, T., H. Kamiya, T. Kosuge, *J. Pharm. Soc. Japan*, 84, p. 545, 1964. With permission.)

of B. *subtilus.* Acetoin is an intermediate in the biosynthesis of leucine, valine, and pantothenate. Rizzi [20,111] has provided an insight into how pyrazines may be formed under mild conditions and also then their formation in biological systems. The key is the biological production of very reactive pyrazine precursors that can chemically react at low temperatures to form these compounds.

### 5.4.8 Sulfur Compounds

The origin of the majority of sulfur-containing aroma compounds formed by microorganisms is sulfate, which is initially incorporated into the sulfur amino acids (L-methionine and L-cysteine) and the peptide, glutathione [112]. These sulfur-containing precursors are metabolized to a variety of aroma compounds of sensory significance. Spinnler et al. [112] have provided a good discussion of this process.

Cysteine may react with carbonyls to yield flavor compounds (e.g., trithiolanes) or be decarboxylated to give cysteamine, deaminated to provide α-keto-3-thiopropionic acid or degraded to free $H_2S$. While each of these pathways may yield significant aroma compounds, the formation of free $H_2S$ is particularly important. $H_2S$ is a flavor compound in its own right and also is very reactive with carbonyls and free radicals to form very potent aroma compounds (e.g., ethyl sulfide, diethyl disulfide, amyl mercaptan, and 3-methyl-2-butenethiol).

Methionine may be degraded to yield numerous sulfur-containing aroma compounds, the most important of which is methional. Methional has a very low sensory threshold, and thus is found to contribute to the flavor of many foods. Methionine may also be degraded by microbial lyases to yield methanethiol. Methanethiol may make a contribution to flavor itself or through secondary reactions that yield various sulfides, disulfides, tetrasulfides, and thioesters (Figure 5.15).

### 5.4.9 Conclusions

Some of the mechanisms leading to the formation of several classes of flavor compounds via microbial activity are outlined in this chapter. At this point, two additional ideas should be mentioned. First, we have only considered the formation of flavor via metabolic activity of the living microorganism. Microorganisms could

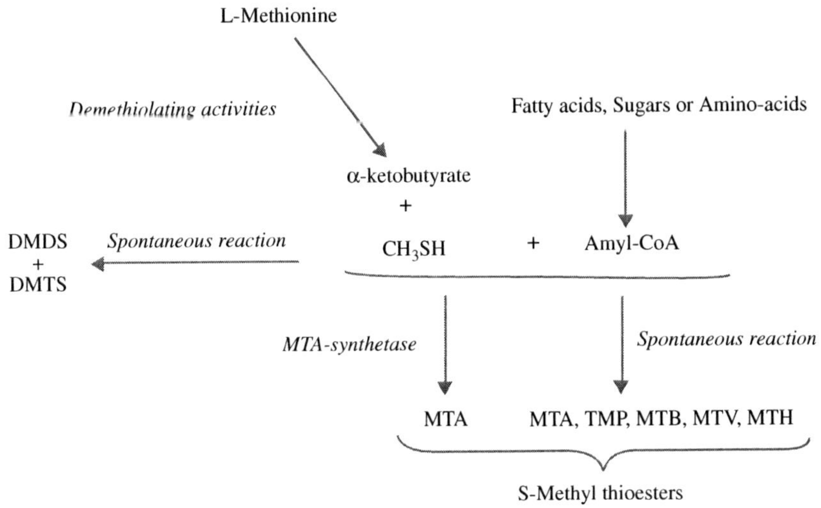


**FIGURE 5.15** Conversion of methionine and Acyl-CoA to thioesters. (From Spinnler, H.E., N. Martin, P. Bonnarme, *Heteroatomic Aroma Compounds*, G.A. Reineccius, T.A. Reineccius, Eds., Amer. Chem. Soc., Washington, D.C., 2002, p. 54. With permission.)

be viewed as a small packet of enzymes, which upon death and lysis of the cell, may continue to function. A good example of the significance of bacterial enzymes to flavor is in aged cheeses. The primary metabolic activity of the microflora in the cheese occurs during the early stages of cheese making and aging. Yet there is little characteristic flavor at this point. Only during aging do we see protein degradation via microbial proteases to yield the soft texture, free amino acids (meaty, brothy flavor background), peptides (bitterness) and free sulfur compounds so important to the flavor of aged cheeses. Changes in lipids during aging of the cheese produce free fatty acids, oxidized fatty acids (e.g., methyl ketones) and a host of other flavor compounds. These changes during aging are occurring when microbial activity is decreasing and enzymes from lysed cells are increasing. One must remember that microbial growth in a food product can have a pronounced effect both on the development of desirable flavor and off-flavor even though no viable bacterial cells exist in the food.

The second idea that should be brought out in closing this section of the text is that the flavor system in a fermenting product is very dynamic [93,113]. A microorganism may produce a metabolite that initially accumulates in the product and may make a significant contribution to flavor. A second organism may then reach significant metabolic activity (population) and start metabolizing this first metabolite. Therefore, the flavor of the product will change. Microorganisms may also alter flavor compounds produced in the product via other mechanisms, e.g., lipid oxidation or nonenzymatic browning. Thus lipid oxidation or the Maillard reaction may result in a different flavor in a fermented product than in a nonfermented product.

## REFERENCES

1. Schieberle, P., K.H. Engel, Eds., *Frontiers of Flavour Science,* Deutsche Forschung. Lebensmittel., Garching, 2000, p. 602.
2. Nagodawithana, T.W., *Savory Flavors*, Esteekay Assoc., Milwaukee, 1995, p. 468.
3. Parliment, T.H., M.J. Morello, R.J. McGorrin, Eds., *Thermally Generated Aromas*, Amer. Chem. Soc., Washington, D.C., 1994, p. 492.
4. Tressl, R., D. Rewicki, Heat generated flavors and precursors, in *Flavor Chemistry: Thirty Years of Progress*, R.G. Teranishi, E.L. Wick, I. Hornstein, Eds., Kluwer Academic/Plenum Publ., New York, 1999, p. 305.
5. Kerler, J., C. Winkel, The basic chemistry and process conditions underpinning reaction flavour production, in *Food Flavour Technology*, A.J. Taylor, Ed., Sheffield Publ., Sheffield, U.K., 2002, p. 302.
6. Labuza, T.P., G.A. Reineccius, V. Monnier, J.O. O'Brien, J.W. Baines, Eds., *Maillard Reactions in Chemistry, Food, and Health*, Royal Chemical Society London, London, 1994, p. 440.
7. Finot, P.A., H.U. Aeschbacher, R.F. Hurrell, R. Liardon, Eds., *The Maillard Reaction in Food Processing, Human Nutrition, and Physiology*, Birkhäuser Verlag., Basel, 1990, p. 516.
8. O'Brien, J., H. Nursten, M.J. Crabbe, J.M. Ames, Eds., *The Maillard Reaction in Foods and Medicine*, Royal Soc. Chem., London, 1988, p. 464.
9. Horiuchi, S., N. Taniguchi, F. Hayase, T. Kurata,T. Osawa, Eds., *The Maillard Reaction in Food Chemistry and Medical Science: Update for Post-Genomic Era*, Elsevier Publ., Amsterdam, 2002, p. 527.
10. Hodge, J.E., Chemistry of browning reactions in model systems, *J. Agric. Food Chem.*, 1, p. 928, 1953.
11. Reineccius, G.A., P.G. Keeney, W. Weissberger, Factors affecting the concentration of pyrazines in cocoa beans, *J. Agric. Food Chem.*, 20(2), p. 202, 1972.
12. Vernin, G., C. Parkanyi, Mechanisms of formation of heterocyclic compounds in Maillard and pyrolysis reactions, *Chem. Heterocycl. Comp. Flavours Aromas*, p. 151, 1982.
13. Koehler, P.E., G.V. Odell, Factors affecting the formation of pyrazine compounds in sugar-amine reactions, *J. Agric. Food Chem.*, 18(5), p. 895, 1970.
14. Shibamoto, T., R.A. Bernhard, Investigation of pyrazine formation pathways in glucose-ammonia model systems, *J. Agric. Food Chem.*, 25(1), p. 143, 1977.
15. Leahy, M.M., The Effects of pH, Types of Sugar and Amino Acid, and Water Activity on the Kinetics of the Formation of Alkyl Pyrazines, Ph.D. Thesis, University of Minnesota, Minneapolis, p. 116, 1985.
16. Liardon, R., J.O. Bosset, B. Blanc, The aroma composition of Gruyere cheese: the alkaline volatile compounds, *Lebensm.-Wiss. Technol.*, 15, p. 143, 1982.
17. Qian, M., G.A. Reineccius, Identification of aroma compounds in Parmigiano-Reggiano cheese by gas chromatography/olfactometry, *J. Dairy Sci.*, 85(6), p. 1362, 2002.
18. Curioni, P.M.G., J.O. Bosset, Key odorants in various cheese types as determined by gas chromatography-olfactometry, *Internat. Dairy J.*, 12(20), p. 959, 2002.
19. Aston, J.W., J.R. Dulley, Cheddar cheese flavour, *Aust. J. Dairy Technol.*, 37(2), p. 59, 1982.
20. Rizzi, G.P., Formation of pyrazines from acyloin precursors under mild conditions, *J. Agric. Food Chem.*, 36(2), p. 349, 1988.

21. Yamanishi, T., Tea, coffee, cocoa and other beverages, in *Flavor Research: Recent Advances*, R. Teranishi, R.A. Flath, H. Sugisawa, Eds., Marcel Dekker, New York, 1981, p. 231.
22. Leahy, M.M., G.A. Reineccius, Kinetics of the formation of alkylpyrazines: effect of pH and water activity, in *Thermal Generation of Aromas*, T.H. Parliment, R.J. McGorrin, C.-T. Ho, Eds., Amer. Chem. Soc., Washington, D.C., 1989, p. 196.
23. Shu, C.-K., C.-T. Ho, Parameter effects on the thermal reaction of cystine and 2,5-dimethyl-4-hydroxy-3(2H)-furanone, in *Thermal Generation of Aromas*, T.H. Parliment, R.J. McGorrin, C.-T. Ho, Eds., Amer. Chem. Soc., Washington, D.C., 1989, p. 229.
24. Shaw, J.J., C.-T. Ho, Effects of temperature, pH, and relative concentration on the reaction of rhamnose and proline, in *Thermal Generation of Aroma*, T.H. Parliment, R.J. McGorrin, C.-T. Ho, Eds., Amer. Chem. Soc., Washington, D.C., 1989, p. 217.
25. Shibamoto, T., R.A., Effect of time, temperature and reactant ratio on pyrazine formation in model systems, *J. Agric. Food Chem.*, 24, p. 847, 1976.
26. Mottram, D.S., F.B. Whitfield, Aroma volatiles from meat-like Maillard systems, in *Thermally Generated Flavors*, T.H. Parliment, R.J. McGorrin, C.-T. Ho, Eds., Amer. Chem. Soc., Washington, D.C., 1994, p. 180.
27. Potman, R.P., T.A. van Wijk, Mechanistic studies of the Maillard reaction with emphasis on phosphate-mediated catalysis, in *Thermal Generation of Aromas,* T.H. Parliment, R.J. McGorrin, C.T. Ho, Eds., Amer. Chem. Soc., Washington, D.C., 1989, p. 182.
28. Maga, J.A., C.E. Sizer., Pyrazines in foods, *CRC Crit. Rev. Food Technol.*, 4(1), p. 39, 1973.
29. Maga, J.A., Pyrazines in flavor, in *Food Flavours, Part A. Introduction*, I.D. Morton, A.J. MacLeod, Eds., Elsevier Scientific Publ., Amsterdam, 1982, p. 283
30. Leahy, M.M., G.A. Reineccius, Kinetics of formation of alkylpyrazines: effect of type of amino acid and type of sugar, in *Flavor Chemistry: Trends and Developments,* R. Teranishi, R.G. Buttery, F. Shahidi, Eds., Amer. Chem. Soc., Washington, D.C., 1989, p. 76.
31. Huang, T.C., L.J. Bruechert, C.-T. Ho, Kinetics of pyrazine formation in amino acid-glucose systems, *J. Food Sci.,* 54(6), p. 1611, 1989.
32. Huang, T.-C., H.-Y. Fu, C.-T. Ho, Kinetics of tetramethylpyrazine formation under high hydrostatic pressure, in *Flavor Technology: Physical Chemistry, Modification and Process*, C.-T. Ho, C.-T. Tan, C.H. Tong, Eds., Amer. Chem. Soc., Washington, D.C., 1995, p. 49.
33. Jusino, M.G., C.-T. Ho, C.H. Tong, Formation kinetics of 2,5-dimethylpyrazine and 2-methylpyrazine in a solid model system consisting of amioca starch, lysine, and glucose, *J. Agric. Food Chem.*, 45(8), p. 3164, 1997.
34. Schirle-Keller, J.P., G.A. Reineccius, Reaction kinetics for the formation of oxygen-containing heterocyclic compounds in model systems, in *Flavor Precursors: Thermal and Enzymatic Conversion*, R. Teranishi, G. Takeoka, M. Guntert, Eds., Amer. Chem. Soc., Washington, D.C., 1993, p. 244.
35. Chan, F., G.A. Reineccius, Reaction kinetics for the formation of isovaleraldehyde, 2-acetyl-1-pyrroline, di(H)di(OH)-6-methylpyranone, phenylacetaldehyde, 5-methyl-2-phenyl-2-hexenal, and 2-acetylfuran in model systems, in *Maillard Reactions in Chemistry, Food, and Health*, T.P. Labuza, G.A. Reineccius, V. Monnier, J.O. O'Brien, J.W. Baines, Eds., Royal Chem. Soc. London, London, 1994, p. 131.

36. Chan, F., G.A. Reineccius, Kinetics of the formation of methional, dimethyl disulfide, and 2-acetylthiophene via the Maillard reaction, in *Sulfur Compounds in Foods*, M.E. Keenen. C.J. Mussinan, Eds., Amer. Chem. Soc., Washington, D.C., 1994, p. 127.
37. Zheng, Y., C.-T. Ho, Kinetics of the release of hydrogen sulfide from cysteine and glutathione during thermal treatment, in *Sulfur Compounds in Foods*, M.E. Keenen. C.J. Mussinan, Eds., Amer. Chem. Soc., Washington, D.C., 1994, p. 138.
38. Chan, F., Reaction Kinetics for the Formation of Strecker Degradation Products in Model Systems, M.S. Thesis, University of Minnesota, Minneapolis, 1994.
39. Lindsay, R., Food additives, in *Food Chemistry*, O. Fennema, Ed., Marcel Dekker Publ., New York, 1985, p. 991.
40. Nursten, H.E., Volatiles produced by the Maillard reaction, in *Frontiers of Flavour Science*, P. Schieberle, K.H. Engel, Eds., Deutsche Forschung. Lebensmittel., Garching, 2000, p. 475.
41. Rizzi, G.P., The Strecker degradation and its contribution to food flavor, in *Flavor Chemistry: Thirty Years of Progress*, R. Teranishi, E.L. Wick, I. Hornstein, Eds., Kluwer Academic/Plenum Publ., New York, 1999, p. 335.
42. Whitfield, F.B., Volatiles from interactions of maillard reactions and lipids, *Crit. Reviews Food Sci. Nutr.*, 31(1–2), p. 1, 1992.
43. Sullivan, J.F., R.P. Konstance, M.J. Calhoun, F.B. Talley, J. Cording Jr., O. Panasiuk, Flavor and storage stability of explosion-puffed potatoes: nonenzymatic browning, *J. Food Sci.*, 39(1), p. 58, 1974.
44. Nilov, V.L., S.J. Ogorodnik, *Kharchove Prom. Nauk.-Virob. ZB No. 1(31), 40,* 1967. Cited by [45].
45. Pokorny, J., Effect of browning reactions on the formation of flavor substances, *Nahrung*, 24(2), p. 115, 1980.
46. Takken, H.J., L.M. van der Linde, P.J. de Valois, H.M. van Dart, M. Boelens, Reaction products of alpha dicarbonyl compounds, aldehydes, hydrogen sulfide and ammonia, in *Phenolic, Sulfur and Nitrogen Compounds in Foods*, G. Charalambous, Ed., Amer. Chem. Soc., Washington, D.C., 1976, p. 114.
47. Ohloff, G., I. Flament, The role of heteroatomic substances in the aroma compounds of foodstuffs, *Forschung. Chem. Org. Naturst.*, 36, p. 231, 1979.
48. Koehler, P.E., M.E. Mason, J.A. Newell, Formation of pyrazine compounds in sugar-amino model systems, *J. Agric. Food Chem.*, 17(2), p. 393, 1969.
49. Hodge, J.E., F.D. Mills, B.E. Fisher, Compounds of browned flavor derived from sugar-amine reactions, *Cereal Sci. Today*, 17(2), p. 34, 1972.
50. Tressl, R., B. Halak, N. Marn, Formation of flavor compounds from L-proline, in *Topics in Flavour Research*, R.G. Berger, S, Nitz, P. Schreier, Eds., H. Eichorn, Marzling-Hangenham, 1985, p. 139.
51. Shibamoto, T., Heterocyclic compounds in browning and browning/nitrite model systems: occurrence, formation mechanisms, flavor characteristics and mutagenic activity, in *Instrumental Analysis of Foods*, G. Charalambous, G. Inglett, Eds., Academ. Press, New York, 1983, p. 229.
52. Rizzi, G.P., Formation of N-alkyl-2-acyl-pyrroles and aliphatic aldimines in model nonenzymatic browning reactions, *J. Agric. Food Chem.*, 22, p. 279, 1974.
53. Maga, J.A., Pyridines in foods, *J. Agric. Food Chem.*, 29(5), p. 895, 1981.
54. Pittet, A.O., D.E. Hruza, Comparative study of flavor properties of thiazole derivatives, *J. Agric. Food Chem.*, 22(2), p. 264, 1974.
55. Tressl, R., D. Rewicki, B. Helak, H. Schroeder, N. Martin, Formation of 2,3-dihydro-1H-pyrrolizines as proline specific Maillard products, *J. Agric. Food Chem.*, 33(5), p. 919, 1985.

56. Tressl, R., D. Rewicki, B. Helak, H. Schroeder, Formation of pyrrolidines and piperidines on heating L-proline with reducing sugars, *J. Agric. Food Chem.*, 33(5), p. 924, 1985.
57. Schieberle, P., T.H. Hofmann, Flavor contribution and formation of heterocyclic oxygen-containing key aroma compounds in thermally processed foods, in *Heteroatomic Aroma Compounds*, G.A. Reineccius, T.A. Reineccius, Eds., Amer. Chem. Soc., Washington, D.C., 2002, p. 207.
58. Pittet, A.O., P. Rittersbacher, R. Muralidhara, Flavor properties of compounds related to maltol and isomaltol, *J. Agric. Food Chem.*, 18, p. 929, 1970.
59. Weenen, H., Reactive intermediates and carbohydrate fragmentation in Maillard chemistry, *Food Chem.*, 2(4), p. 393, 1998.
60. Reynolds, T.M., Flavours from the nonenzymatic browning reaction, *Food Tech. Aust.*, 22(11), p. 610, 1970.
61. Mottram, D.S., J.S. Elmore, Novel sulfur compounds from lipid-Maillard interactions in cooked meat, in *Heteroatomic Aroma Compounds*, G.A. Reineccius, T.A. Reineccius, Eds., Amer. Chem. Soc., Washington, D.C., 2002, p. 93.
62. Mottram, D.S., H.R. Mottram, An overview of the contribution of sulfur-containing compounds to the aroma in heated foods, in *Heteroatomic Aroma Compounds*, G. Reineccius, T.A. Reineccius, Eds., Amer. Chem. Soc., Washington, D.C., 2002, p. 73.
63. Ho, C.T., Q.Z. Jin, Aroma properties of some alkylthiazoles, *Perfum. Flavorist*, 9(6), p. 15, 1985.
64. Sakaguchi, M., T. Shibamoto, Formation of sulfur-containing compounds from the reaction of D-glucose and hydrogen sulfide, *J. Agric. Food Chem.*, 26, p. 1260, 1978.
65. Stoffelsma, J., J. Pypker, *J. Recl. Trav. Chim. Pays-Bas 87, 24,* 1968. Cited by [68].
66. Maga, J., Oxazoles and oxazolines in foods, *J. Agric. Food Chem.*, 29, p. 691, 1978.
67. Maga, J., The chemistry of oxazoles and oxazolines in foods, *Crit. Reviews Food Sci. Nutr.*, 14, p. 295, 1981.
68. Ho, C.T., R.M. Tuorto, Mass spectra and sensory properties of some 4,5-dialkyloxazoles, *J. Agric. Food Chem.*, 29(6), p. 1306, 1981.
69. Jin, Q.Z., G.J. Hartman, C.T. Ho, Aroma properties of some oxazoles. *Perfum. Flavorist*, 9(4), p. 25, 1984.
70. Nawar, W.W., Volatile components of the frying process, *Grasas y Aceites*, 49(3-4), p. 271, 1998.
71. Ohnishi, S., T. Shibamoto, Volatile compounds from heated beef fat and beef fat with glycine, *J. Agric. Food Chem.*, 32(5), p. 987, 1984.
72. Shibamoto, T., H. Yeo, Flavor compounds formed from lipids by heat treatment, in *Flavor Precursors: Thermal and Enzymatic Conversions*, R. Teranishi, G.R. Takeoka, M. Guntert., Eds., Amer. Chem. Soc., Washington, D.C., 1992, p. 175.
73. Chang, S.S., R.J. Peterson, C.-T. Ho, Chemistry of deep fat fried flavor, in *Lipids as a Source of Flavor*, M.K. Supran, Ed., Amer. Chem. Soc., Washington, D.C., 1978, p. 18.
74. Nawar, W.W., S.J. Bradley, S.S. Lomanno, G.C. Richardson, R.C. White-Man, Volatiles from frying fats: a comparative study, in *Lipids as a Source of Flavor*, M.K. Supran, Ed., Amer. Chem. Soc., Washington, D.C., 1978, p. 42.
75. Wagner, R.K., W. Grosch, Key odorants of French fries, *J. Amer. Oil Chem. Soc.*, 75(10), p. 1385, 1998.
76. Chang, S.S., R.J. Peterson, C.-T. Ho, Chemical reactions involved in the deep-fat frying of foods, *J. Amer. Oil Chem. Soc.*, 55(10), p. 718, 1978.
77. Maga, J., Lactones in foods, *Crit. Rev. Food Sci. Nutr.*, 8, p. 1, 1976.

78. Dimick, P.S., N.J. Walker, S. Patton, Occurrence and biochemical origin of aliphatic lactones in milk fat, *J. Agric. Food Chem.*, 17(3), p. 649, 1969.
79. Peterson, D.G., G.A.Reineccius, Determination of the aroma impact compounds in heated sweet cream butter, *Flavour Fragrance J.*, 18(4), p. 320, 2003.
80. Budin, J.T., C. Milo, G.A. Reineccius, Perceivable odorants in fresh and heated sweet cream butters, in *Food Flavors and Chemistry*, Spanier, H., F. Shahidi, T.H. Parliment, C. Mussinan, C.-T. Ho, E. Contis, Eds., Royal Society of Chem., London, 2001, p. 85.
81. Meynier, A., D.S. Mottram, Volatile compounds in meat-related model systems: Investigation on the effect of lipid compounds on the Maillard reaction between cysteine and ribose, in *Progress in Flavour Precursor Studies*, P. Schreier, P. Winterhalter, Eds., Allured Publ., Carol Stream, 1993, p. 383.
82. Wasserman, A.E., Thermally produced flavor components in the aroma of meat and poultry, *J. Agric. Food Chem.*, 20(4), p. 737, 1972.
83. Tressl, R., M.Apetz, R. Arrieta, K.G. Grunewald, Formation of lactones and terpenoids by microorganisms, in *Flavor of Foods and Beverages*, G. Charalambous, G.I. Inglett, Eds., Academ. Press, New York, 1978, p. 145.
84. Margalith, P.Z., *Flavor Microbiology*, Charles Thomas, Springfield, 1981, p. 309.
85. TNO, Volatile Compounds in Foods, Nutrition and Food Research., *Utrechtseweg.*, p. 1122, 1995.
86. Berry, D.R., D.C. Watson, Production of organoleptic compounds, in *Yeast Biotechnology*, D.R. Berry, I. Russell, G.G. Stewart, Eds., Allen and Unwin, London, 1987, p. 345.
87. Cristiani, G., V.M. Monnet, Food micro-organisms and aromatic ester synthesis, *Sciences-des-Aliments*, 21(3), p. 211, 2001.
88. Dumont, J., J. Adda, Flavour formation in dairy products, in *Flavour Research*, D.G. Land, H.E. Nursten, Eds., Appl. Sci. Publ., London, 1979, p. 245.
89. Sandine, W.E., P.R. Elliker, Microbially induced flavors in foods, *J. Agric. Food Chem.*, 18(4), p. 557, 1970.
90. Kempler, G.M., Production of flavor compounds by microorganisms, *Advan. Appl. Microbiol.*, 29, p. 29, 1983.
91. Vedamuthu, E.R., Microbiologically induced desirable flavors in the fermented foods of the west, *Dev. Ind. Microbiol.*, 20, p. 187, 1979.
92. Hutkins, R.W., Metabolism of starter cultures, in *Applied Dairy Microbiology*, E.L. Marth, J.L. Steele, Eds., Marcel Dekker, New York, 1993, p. 207.
93. Haymon, L.W., J.C. Acton, Flavors from lipids by microbial action, in *Lipids as a Source of Flavor*, M.K. Supran, Ed., Amer. Chem. Soc., Washington, D.C., 1978, p. 94.
94. Alford, J.A., J.L. Smith, H.D. Lilly, Relation of microbial activity to changes in lipids of foods, *J. Applied Bacter.*, 34, p. 133, 1971.
95. Morgan, M.E., The chemistry of some microbially induced flavor defects in milk and dairy foods, *Biotechnol. Bioeng.*, 18(7), p. 953, 1976.
96. MacLeod, P., M.E. Morgan, Differences in the ability of lactic streptococci to form aldehydes from certain amino acids, *J. Dairy Sci.*, 41, p. 908, 1958.
97. Reazin, G., H. Scales, A. Andreasen, Mechanism of a major conenger formation in alcoholic grain fermentations, *J. Agric. Food Chem.*, 18, p. 585, 1970.
98. Lamparsky, D., I. Klimes, Cheddar cheese flavor: its formation in light of new analytical results, in *Flavour '81*, P. Schreier, Ed., deGruyter, Berlin, 1981, p. 557.
99. Berger, R.G., *Aroma Biotechnology*, Springer, Berlin, 1995, p. 240.

100. Berger, R.G., E. Latza, F. Neuser, J. Onken, Terpenes and amino acids — progenitors of volatile flavors in microbial transformation reactions, in *Frontiers of Flavour Science*, P. Schieberle, K.H. Engel, Eds., Deutsche Forschung. Lebensmitt, Garching, 2000, p. 394.
101. Sakui, N., S-E Gen, Biotransformation of monoterpenes and several flavored materials by cultured cells of Bupleurum falcatum L., *Foods Food Ingred. J.*, 156, p. 96, 1993.
102. Norouzian, D., A. Akbarzadeh, D. Nouri-Inanlou, B. Farahmand, M. Saleh, F. Sheikh-ul-Eslam, J. Vaez, Biotransformation of alcohols to aldehydes by immobilized cells of Saccharomyces cerevisiae PTCC5080, *Enzyme Microbial Technol.*, 33(2–3), p. 150, 2003.
103. Wache, Y., M. Aguedo,J.M. Nicaud, J.M. Belin, Catabolism of hydroxyacids and biotechnological production of lactones by Yarrowia lipolytica, *Appl. Microbiol. Biotechnol.,* 61(5–6), p. 393, 2003.
104. Takahashi, K., M. Tadenuma, S. Sato, 3-Hydroxy-4,5-dimethyl-2(5H)-furanone, a burnt flavoring compound from aged sake, *Agric. Biol. Chem.*, 40, p. 325, 1976.
105. Kosuge, T., H. Kamiya, Discovery of a pyrazine in a natural product: tetramethylpyrazine from cultures of a strain of *Bacillus subtilis*, *Nature*, 193, p. 776, 1962.
106. Sloot, D., P.D. Harknes, Volatile trace components in Gouda cheese, *J. Agric. Food Chem.*, 23, p. 356, 1975.
107. Lin, S.S., Alkyl pyazines in processed American cheese, *J. Agric. Food Chem.*, 24, p. 1252, 1976.
108. Qian, M., G. Reineccius, Static headspace and aroma extract dilution analysis of Parmigiano Reggiano cheese, *J. Food Sci.*, 68(3), p. 794, 2003.
109. Biede, S.L., E.G. Hammond, Swiss cheese flavor. I. Chemical analysis, *J. Dairy Sci.*, 62(2), p. 227, 1979.
110. Adachi, T., H. Kamiya, T. Kosuge, Studies on the metabolic products of *Bacillus subtilus*, *J. Pharm. Soc. Japan*, 84, p. 545, 1964.
111. Rizzi, G.P., The biogenesis of food-related pyrazines, *Food Rev. Int.,* 4(3), p. 373, 1988.
112. Spinnler, H.E., N. Martin, P. Bonnarme, Generation of sulfur flavor compounds by microbial pathways, in *Heteroatomic Aroma Compounds*, G.A. Reineccius, T.A. Reineccius, Eds., Amer. Chem. Soc., Washington, D.C., 2002, p. 54.
113. Schreier, P., The role of micro-organisms in flavour formation, in *Prog. Flavour Res.*, D.G. Land, H.E. Nursten, Eds., Applied Sci. Publ., London, 1979, p. 175.

# 6 Flavor Release from Foods

## 6.1 INTRODUCTION

The term "flavor release" may broadly encompass the release of flavor from a food or food ingredient during manufacture, storage, preparation, or eating. However, today it most commonly refers to the release of flavor components from a food during eating. Flavor release is influenced by numerous factors, including chemical interactions between the food and the flavoring, physical considerations (physical barriers to release) as well as human factors such as the number of teeth, chewing efficiency, chewing time, breathing process, etc.

Historically there have been numerous studies on how aroma components interact with the major food constituents. One speaks of interactions since any type of interaction between a flavor compound and a food constituent that restricts the movement of a flavor stimulus to a sensory receptor influences perception. This interaction may be chemical (e.g., hydrogen, hydrophobic, ionic, or covalent bonding), e.g., a chemical interaction may reduce the vapor pressure of an aroma substance thereby reducing the driving force for its evaporation in the oral cavity and reducing its movement to the olfactory receptors.

Physical considerations include those where the food matrix physically interferes with a flavor chemical reaching a sensory receptor, i.e., the food matrix serves as a barrier to the movement of a flavor substance. The barrier may be small such as a food having some viscosity that reduces mixing or the surface area of a food in the mouth thereby lowering the possibility of evaporation in the oral cavity or contact with taste receptors in the mouth. The barrier may be very large, for example, a dry food product that does not dissolve very quickly in the mouth. In this case, the flavor would remain locked in the food matrix until the food matrix dissolves thereby freeing the flavor. In either of these situations, the flavor component is hindered in its transmission to the oral cavity and ultimately the olfactory epithelia for smelling or the tongue for tasting.

One can also conceive of interactions that favor the release of flavor substances. The most obvious is the "salting out" of a flavor compound from solution. In this case, the flavorant is pushed out of the food matrix into the oral cavity, enhancing perception. Another example is the evaporation of $CO_2$ or ethanol from a food that may carry aroma compounds thereby enhancing aroma release.

None of the scenarios described above would be particularly problematic to the flavor of a food if the effects described operated across all aroma components equally. If this were the situation, then one would simply add more or less flavoring to a food depending upon the overall balance of promotion vs. inhibition factors. Unfortunately, these factors act upon each flavor constituent individually — some flavor components are driven out of solution while others are tightly held in the food depending upon the net balance of interactions of *each* food constituent upon *each* flavor molecule. The end result is the balance of flavor compounds delivered to one's sensory receptors may be very strongly influenced by the composition/physical state of the food system. This effect is responsible for a given flavor having one flavor profile in one food and a different flavor profile in another food.

This phenomenon affects the inventory of flavor companies. A typical flavor company may have several hundred different strawberry flavors, for example. This is not necessarily because the customer has hundreds of concepts of strawberry flavor to be met, but that a flavor must be designed to undergo a multitude of interactions with the food and ultimately provide a release profile that is liked by the consumer. This phenomenon is also obvious in the recent low-fat food trend. Unquestionably, this concept has been a failure since food companies could not deliver a low-fat product to the customer that had a satisfactory sensory profile. Commonly, people would try a low-fat version of a food product once but then never buy it again. The low-fat and normal-fat products generally used the same flavor system, but they released flavor differently due to formulation variations. These variations may have altered chemical or physical interactions thereby altering the flavor release profile.

In this chapter we will provide a broad overview of the flavor interactions that may occur in foods considering how flavors interaction with nonvolatiles in foods [6]. Initially the interaction of flavorings with the major food constituents (e.g., lipid, carbohydrates, and proteins) will be discussed. The final section will include some discussion of interactions with minor constituents (e.g., melanoidins, polyphenolics, and high potency sweeteners) as literature permits. The reader is encouraged to go to more detailed reviews included in books edited by Taylor [1] or symposia proceedings such as Roberts and Taylor [3], Schieberle and Engel [4], Teranishi et al. [2], or Taylor and Mottram [5].

## 6.2 LIPID/FLAVOR INTERACTIONS

Of the food matrix:flavor interactions, the effect of fat/oil on flavor release is most understandable and thus predictable. In the case of fat/flavor interactions, the overriding effect is the role fat plays as a flavor solvent. Unlike carbohydrates or proteins where numerous undefinable chemical interactions come into play, fat has little true chemical interaction, and thus its effect on flavor release is largely quantifiable. (The discussion that follows will look very much like that provided in Chapter 3 where partition coefficients in solvent extraction [and headspace methods] were discussed.)

The primary determinants of the amount of an aroma compound in the headspace above a liquid food, or released during eating, are: 1) the amount of a given flavorant in the continuous phase of that food, and 2) the air:food partitioning of the aroma compound. The partitioning relationship between the continuous phase and air largely establishes the driving force for volatile release. If the continuous phase is aqueous, the presence of fat as a second phase may have a pronounced effect on the amount of each volatile in the aqueous phase and thus, the amount released. If the continuous phase is lipid, then the concentration of the volatile in that lipid is generally determining. From a taste standpoint, we are largely interested in the amount of flavorant in the aqueous phase (a combination of the saliva and water in a food) for taste compounds must be in the aqueous phase to be tasted.

## 6.2.1 Effect of Fat: Flavor Interactions on Aroma

### 6.2.1.1 Equilibrium Conditions

The amount of an aroma compound in air above a food at *equilibrium* depends upon the equilibrium air:continuous phase partition coefficient. This is given as:

$$k_{a/c} = \text{(conc. in the air in g/L)/(concentration in the continuous phase in g/L)}$$

Most commonly, the continuous phase is water, and so this would be the air:water partition coefficient of an aroma component. If we wish to add oil/fat to this system, we need to include a consideration of odorant partitioning into air, water, and oil. De Roos [7] has provided a good explanation of how oil content of a system influences the amount of an odorant in the air above the system (Figure 6.1).

In the left side diagram of Figure 6.1, the two aroma compounds are partitioned between the aqueous phase and the air. The lipophilic aroma compound tends to be driven into the air since it has little solubility in for the aqueous phase. The hydrophilic compound is soluble in the continuous phase and partitions normally between the two phases as predicted by Raoult's law (under ideal conditions). When oil is added to the food system (right side of Figure 6.1) forming an emulsion, the lipophilic odorant prefers to reside in the particulate oil phase and its concentration in the continuous phase is decreased greatly thereby lowering its concentration in the air phase. Considering the hydrophilic odorant (right side of Figure 6.1), this compound has little solubility in oil, and thus its concentration in the aqueous phase may have increased (less water in system because it was displaced by oil), but this effect is small and the concentration in the air phase is largely unchanged. Since

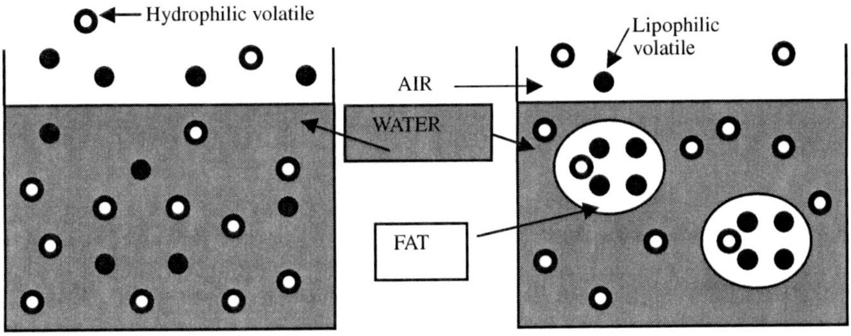


**FIGURE 6.1** Effect of fat on the equilibrium headspace concentrations of hydrophilic and lipophilic aroma compounds above an aqueous solution. (From De Roos, K.B., *Food Technol.*, 51(1), p. 60, 1997. With permission.)

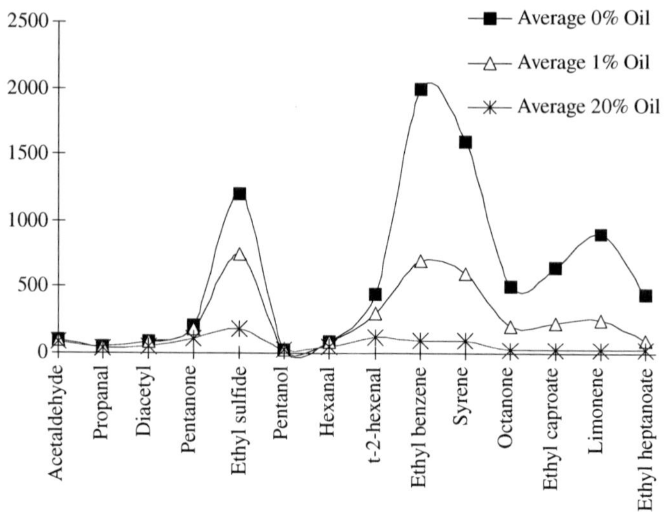


**FIGURE 6.2** The effect of fat content in a model system on the amounts of different volatiles in the equilibrium headspace above the system. (From Schirle-Keller, J.P., Ph.D. thesis, University of Minnesota, Minneapolis, MN, 1995, p. 237. With permission.)

most aroma compounds tend to be lipophilic (but vary greatly in lipophilicity, see Table 3.1), the presence of oil in a food system tends to have a significant but compound dependent effect on aroma release into the air and thus perception.

If we take this simple two odorant system and expand it to include a larger number of odorants, we get a picture such as shown in Figure 6.2 where fourteen odorants were placed in 0, 1, and 20% oil added systems [8]. After equilibration, the headspace concentration profile of this odorant mixture is quite different for the three systems. One would expect these headspace differences to translate into differences in perceived aromas. If we take this situation a bit closer to reality, perhaps include 200 odorants (a simple natural flavor), we can imagine that the resulting air profile becomes much more complicated since each odorant has a different oil:water partition coefficient and thus a different effect of including oil in the system on air:system partitioning. One *expects* the aroma character of the food to change if one changes the oil content.

There are situations where the inclusion of an oil phase does not significantly change the perceived flavor, only the flavor intensity. These situations exist when flavor character is carried by a small number of aroma compounds that have similar oil:water partition coefficients. One may consider vanilla flavor in this category. Vanilla flavor is largely carried by vanillin and ethyl vanillin (artificial vanilla flavors). These two compounds have similar structures and thus partition similarly. If one changes the fat content of an ice cream, for example, the vanilla character may change slightly but is seldom a problem. One simply increases the amount of flavor used as the fat content is increased.

### 6.2.1.2 Dynamic Conditions

The discussion above focused on equilibrium conditions as opposed to dynamic conditions such as one would find during the eating of a food. One never achieves equilibrium conditions in the mouth, so some consideration of how static vs. dynamic conditions influence aroma release is desirable. For static conditions, one considers partitioning phenomena while for dynamic conditions, one considers both partitioning and mass transport phenomena. Mass transport is influenced by factors such as texture (e.g., viscosity, gel strength, brittleness), fat melting point, rate of surface renewal (mixing, the breakdown of solids, and hydration or dissolution of solids) and surface area, none of which have a significant effect upon the equilibrium release of aroma compounds. However, these factors are key to dynamic flavor release [7] and thus will be discussed in greater detail later in this chapter.

De Roos [7] has provided some data to illustrate some of the differences between dynamic vs. static release (Figure 6.3). The top of Figure 6.3 (water:CMC systems) shows that the inclusion of carboxymethyl cellulose (CMC) does not reduce the *equilibrium* release of the model aroma system due to any flavor:CMC interactions. Essentially, there is no effect on the headspace concentration of these volatiles when CMC is included in the formulation (CMC was needed in the system to stabilize the oil added in the Bottom part of the figure). The middle of Figure 6.3 shows the release rate of each volatile from water, vs. water with 1% CMC under dynamic conditions. It is noteworthy that the relative release *profiles* are similar between dynamic release and equilibrium release (water Figure 6.3 top vs. middle). This suggests that the vapor pressure is the *dominant* factor controlling release. However, while similar, there is a difference in amounts of volatiles released under dynamic conditions when CMC is present or not (Figure 6.3 middle) and when oil is present or not (bottom). The reduction in release observed in the middle figure is due to the influence of CMC on dynamic release. The CMC provides some viscosity to the system, which reduces mixing and spreading that would provide reduced surface area for release. The effect of oil on dynamic release is illustrated in bottom of Figure 6.3. One can see that the inclusion of oil in the model system reduced the release of most volatiles: most markedly the most lipophilic compounds. This is similar to the effect noted for equilibrium release.

Much of the literature data on dynamic flavor release comes from the use of artificial "mouths." An artificial mouth is a mechanical device designed to recreate the dynamics of fluid and gas flows that occur in the mouth in a more controllable instrument than the human mouth. Artificial mouths are thermostated to control temperature (37°C), mixed to simulate chewing and tongue movement, have air flow to simulate breathing, and provide dilution to simulate saliva flow [9,10]. In the last decade, there has been substantial data obtained from real-time breath analysis of people during eating. These data more accurately represent the true eating process but contain large variability due to variable human habits and different physiology. This inherent variability in real-time breath analysis often masks subtle effects that are seen in the more controllable artificial mouths.

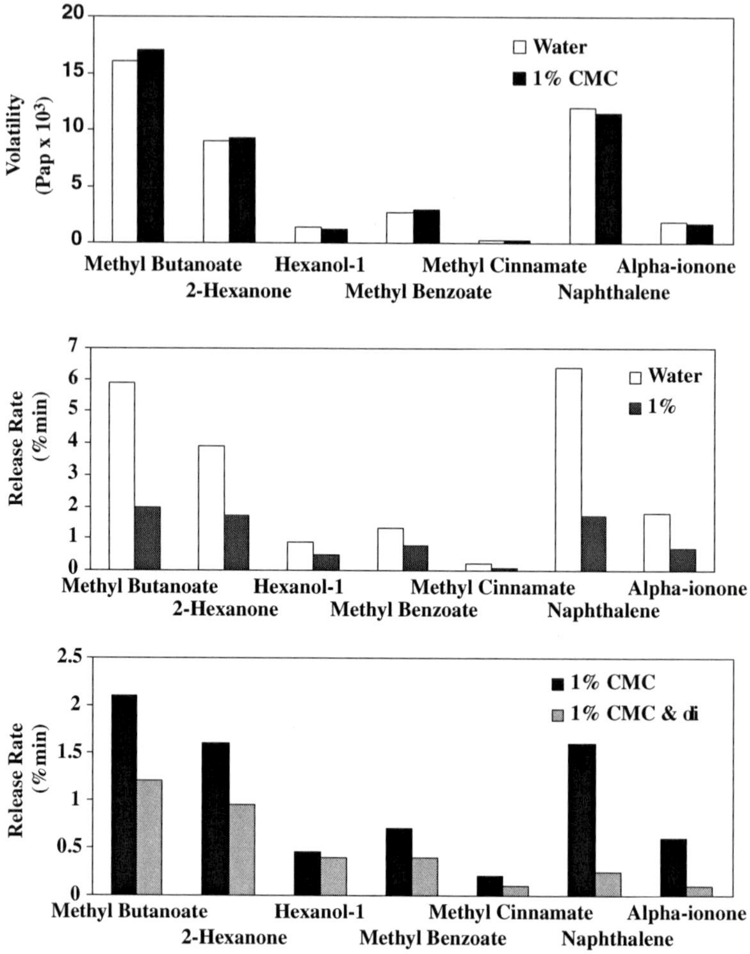


**FIGURE 6.3** The influence of carboxymethylcellulose addition on the equilibrium headspace concentration (top) or dynamic release (middle) of aroma compounds from aqueous solution. Bottom: effect of adding oil on dynamic release of aroma compounds from aqueous solution. (From De Roos, K.B., *Food Technol.*, 51(1), p. 60, 1997. With permission.)

### 6.2.2 Effect of Fat: Flavor Interactions on Taste

The effects of fat on taste are simpler. As noted earlier, taste perception requires that the stimulatory compound be in the aqueous phase. Taste compounds may reside to varying degrees in a lipid phase but only what partitions into the aqueous phase (saliva) may be perceived. If one considers the taste stimuli, the vast majority of these compounds are aqueous soluble. They will not partition into a lipid phase to any significant degree (an exception may be some bitter compounds). Thus one would expect the presence or absence of lipid to have a limited effect on taste perception. This is largely true. However, if fat is present in a formulation, it will replace some other component (perhaps water) potentially increasing the concentration of taste stimuli in the aqueous phase thereby exhibiting an effect. There is also the possibility that fat will coat taste receptors thereby hindering stimuli from moving to the taste buds.

## 6.3 CARBOHYDRATE: FLAVOR INTERACTIONS

Carbohydrates include a very diverse group of food components, ranging from simple sugars to very complicated polysaccharides. Their functions in foods are extremely broad, including sweetening, bulking, viscosity builders (stabilizers), gelling, emulsification, soluble and insoluble fiber, nutraceuticals (e.g., oat fiber), probiotics, encapsulation wall materials, inclusion hosts (cyclodextrins) structural agents, and so on. Carbohydrates are the major components of many foods.

If we consider how carbohydrates will interact with flavors, one has to consider both their chemical and physical properties. Their chemical properties will determine chemical interactions with flavorings and their physical properties will determine their influence on flavor compound mass transport both within a food, and from a food. Chemically, they are built on a hydrated carbon backbone. This backbone generally affords little opportunity for chemical binding other than some very weak hydrogen bonding. However, the side chains on some carbohydrates offer ionizable groups ($SO_3^=$, $COO^-$, $NH_4^+$), peptides or proteins (e.g., gum acacia), or lipophilic sites (e.g., octenyl succinate derivatized starches) for chemical interaction. Furthermore, some carbohydrates that appear to have no opportunity for binding may assume certain conformational forms that provide hydrophobic binding sites (helical structures associated with bread staling, sucrose clusters that bind hydrophobic flavor molecules, or hydrophobic cores in cyclodextrins). Thus, there are opportunities for chemical interactions between flavor compounds and carbohydrates that may influence the volatility of aroma compounds, i.e., reduce their vapor pressure. The degree of interaction (vapor pressure reduction) will depend upon both the chemical makeup of the odorant and the carbohydrate.

The broad diversity in functionality of carbohydrates also offers substantial opportunity for them to influence the mass transport of flavorants. Their ability to form gels, impart viscosity, or promote emulsion formation in food systems are all factors that influence mass transport (dynamic release during eating) of odorants during eating. This section of this chapter will provide an overview of how carbohydrates will influence flavor release from foods. Discussion is organized by carbohydrate type.

### 6.3.1 Simple Sugars: Aroma Interactions

Defining simple sugars as the mono- and disaccharides commonly used in the food industry, this list commonly includes glucose, fructose, sucrose, lactose, sorbitol, and maltose. These simple sugars are generally used for sweetening purposes although they may be bulking materials in some applications. They offer little in chemical structure to suggest a significant chemical interaction to influence aroma release, i.e., no significant functional groups for interaction. However, there is evidence that sucrose in solution tends to reduce the vapor pressure of hydrophobic aroma compounds [11,12]. It is hypothesized that sucrose makes an aqueous solution more hydrophobic therefore affording more solubility to lipophilic odorants. If there is enough sucrose in solution (>40%), there may be a salting out effect, but this is limited to lipophilic aroma components. The net effect of reduced volatility of lipophilic aroma components and an increased volatility of lipophilic components could result in a change in perceived flavor character due to an imbalancing of aroma compounds. None of the other simple sugars appears to solubilize hydrophobic compounds.

From a practical standpoint, it is generally considered that the addition of sweeteners to a food system (that is normally expected to be sweet) enhances the perceived flavor. It is not particularly clear whether this is due to any physical effect (e.g., salting out effect due to interactions [not likely]) or whether it is due to cognitive effects (more likely). If a food is normally sweet, a person anticipates sweetness and the flavor is flat or lacking if the taste stimulus is lacking. Harvey et al. [13] illustrated this effect in their study of flavor perception in chewing gum (Figure 6.4). They monitored the perceived flavor intensity of peppermint chewing gum as well as analytically measured the aroma delivered to the olfactory region (nosespace analysis) and the concentration of sucrose in the saliva as a function of chewing time. It was fascinating to note that the decrease in perceived mint flavor paralleled the decrease in sucrose concentration in the mouth, not the concentration of menthol in the breath. While one should consider that the intensity of flavor may have decreased due to adaptation as opposed to the lack of support by sucrose, the data suggest that taste is exceptionally important to both aroma and overall flavor perception.

### 6.3.2 High Potency Sweeteners: Aroma Interactions

While high potency sweeteners are not carbohydrates, their role in replacing the traditional bulk sweeteners permits their discussion here. There are several high

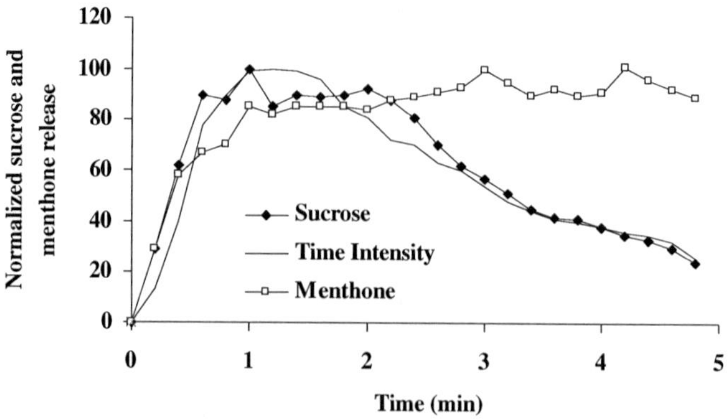


**FIGURE 6.4** Sucrose and menthone release from a market peppermint chewing gum stick compared with sensory flavor intensity. (From Harvey, B.A., J.M. Davidson, R.S.T. Linforth, A.J. Taylor, *Frontiers of Flavour Science*, P. Schieberle, K.H. Engel, Eds., Deutsche Forschung. Lebensmittel, Garching, 2000, p. 271. With permission.)

potency sweeteners on the world market today, including aspartame, saccharine, acesulfame K, Neotame®, sucralose, and sodium cyclamate. There is little question that foods, particularly beverages, sweetened with high potency sweeteners have a different flavor profile than those sweetened with the traditional bulk sweeteners. One is inclined to assume that this difference in flavor is due to sweetener:aroma compound interaction (and thus release). While there are a number of papers on sweetener:aroma interactions, the work of Nahon et al. [12,14] and Deibler and Acree [15] will be highlighted.

Deibler and Acree [15] reported on a study of the influence of beverage base (including sweetener) on volatile release using an artificial mouth (Retro nasal Aroma Simulator) [9]. They found that the aroma release profile was dependent upon the high potency sweetener used in the formulation (Table 6.1). No sensory data were presented to determine the sensory significance of their instrumental data.

Nahon et al. [12,14] used both sensory and instrumental methods to study the influence of sweetener type on aroma release from beverages. Their work showed that sweetener type influences both instrumental and sensory flavor profiles. Unfortunately, they did not find that the instrumental data supported the sensory data, i.e., there were no significant differences in the instrumental data for the aroma compounds expected to *cause* the observed change in specific sensory attributes (orange flavor). In addition, they did not find a significant effect of NaCyclamate on volatile release

**TABLE 6.1**
**Influence of High Potency Sweeteners on Volatile Release from Model Beverage Systems**

| Odorant | Aspartame | Sucralose | Acesulfame K |
|---|---|---|---|
| Isoamylacetate | | + | |
| Citronellyl acetate | | − | |
| Cinnamyl acetate | + | + | − |
| Ethyl valerate | | + | |
| Ethyl decanoate | | − | |
| Perilla aldehyde | + | + | |
| Octanol | + | + | |
| Terpenes | + | | − |
| Nonanal | | + | |
| Fenchol | + | + | + |
| Camphor | | + | |

*Note:* + indicates significant ($p > 0.05$) increase in concentration in the volatile phase of the odorant while the factor increased — indicates significant ($p > 0.05$) decrease in concentration in the volatile phase of the odorant while the factor increased

*Source:* From Deibler, K., T. Acree, *Frontiers of Flavour Science*, P. Schieberle, K.H. Engel, Eds., Deutsche Forschung. Lebensmittel, Garching, 2000, p. 283.

but found a sensory effect. Their final conclusions were that the flavor differences noted between beverages were likely due to differences in the flavor of the sweetener rather than the effect of the sweetener on aroma release. This conclusion brings up a relevant question about the importance of differences in instrumental data on aroma release related to the high potency sweeteners, i.e., is the difference in flavor noted for high potency sweetened beverages due to subtle differences in aroma:sweetener interactions or is it due to cognitive effects? The brain may well note that an aspartame sweetened beverage, for example, is not the same as a sucrose sweetened beverage in viscosity, sweet character, or perhaps even other unknown stimuli. Perhaps the brain simply decides that the perceived flavor stimulus pattern (all inputs) in high potency sweetened beverages is not correct (the same as a sucrose sweetened beverage) and the perceived sensory differences may have nothing to do with flavor:sweetener interactions.

Quite apart from the mild chemical or physiological interactions discussed earlier, aspartame (APM) will undergo chemical reactions resulting in its loss and the component it reacts with. APM is unique amongst the high potency sweeteners in that it has a free amino group that can react with carbonyl-containing food ingredients. This reactivity limits its use (or shelf life) in some applications. For example, APM breaks down in carbonated beverages (pH controlled) [16,17], reacts with flavorings in foods (i.e., flavorings containing aldehydes such as cherry or lemon) [8], and is reactive in thermally processed foods (heat catalyzed reactions) [18]. Thus, APM not only offers the previously discussed interactions but chemical reactions that destroy it and its reactant (e.g., aroma).

### 6.3.3 Complex Carbohydrate (Polysaccharide): Aroma Interactions

Complex carbohydrates are a chemically diverse group of food ingredients. While they may have numerous functions in foods, most commonly they are used for physical stabilization through emulsification or increased viscosity or gelation. They may influence aroma release through vapor pressure reduction (chemical binding of some type) or by influencing mass transfer rate (presenting some barrier to mass transfer). Complex carbohydrates offer many more possibilities for chemical interaction than the simple sugars due to the diversity of functional groups available. If we consider the diversity of structures presented in Figure 6.5, we can see that there are opportunities for ionic, hydrophobic, covalent, Van der Waals, and hydrogen binding. Additionally, since these ingredients are often used to impart viscosity or gel structure, they may also offer significant resistance to mass transfer. Viscous solutions provide both reduced surface area as well as decreased internal mixing, which decrease mass transfer from the food to the oral cavity. Gel structures offer the same resistances but they may be even more pronounced since the gels must be broken down in the mouth to release aroma. There is substantial information in the literature on this topic, particularly on flavor:starch interactions, using both static and dynamic testing methods.

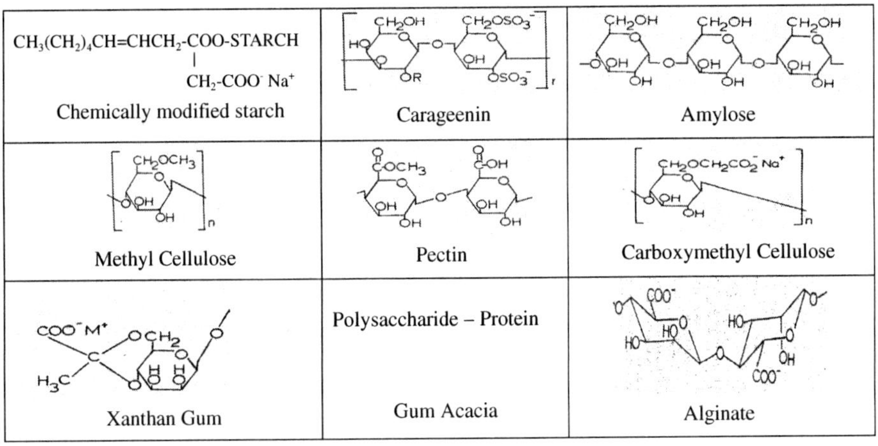


**FIGURE 6.5** Structures of selected polysaccharides.

### 6.3.3.1 Chemical Interactions

Studies on starch demonstrate that starch will include aroma compounds; the extent of inclusion will depend upon the starch (proportion of amylose to amylopectin), its processing history, and the aroma compound. Hau et al. [19] have presented a view of the extent of interactions one observes via static methods (measure of chemical interactions) between selected aroma compounds and starch (Figure 6.6). It is clear that the amount of interaction is quite significant for some compounds (nearly 80% of the hexanol is bound). The interaction is both compound and time dependent (although most compounds reached a plateau in binding after only 1 hr). It is postulated that free amylose forms a helical structure that has hydrophobic areas that will include certain aroma compounds. It should be noted that this research

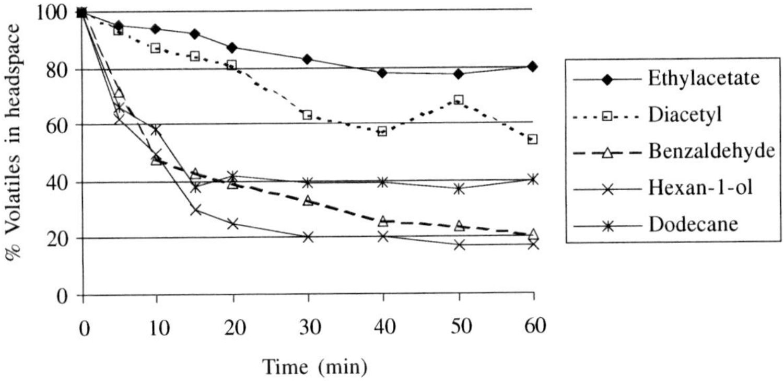


**FIGURE 6.6** The influence of starch on the equilibrium headspace concentration of selected volatiles. (From Hau, M.Y.M., D.A. Gray, A.J. Taylor, *Flavor-Food Interactions*, R.J. McGorrin, J.V. Leland, Eds., Amer. Chem. Soc., Washington, D.C., 1996, p. 109. With permission.)

used a static method and thus would not reflect viscosity effects (resistance to mass transfer) of the starch solution on aroma release.

Langourieux and Crouzet [20] used a dynamic method (exponential dilution) to study the effect of modified and waxy starches, dextrin, dextran, hydroxypropyl celluloses, galactomanans, and β-cyclodextrin on flavor release. Their method involved measuring the dilution of the headspace above a sample over time. In this method, the greater the release of test compounds from a solution, the slower the headspace above the solution could be depleted of these test compounds.

They noted a difference in aroma release from waxy vs. normal starches. For normal starches (25% amylose and 75% amylopectin), aroma release was constant until about 5% added starch, it then decreased rapidly with increasing starch concentration to ca. 20% starch, and then decreased slowly with increasing additions of starch. However, for the waxy starch, decreases with increasing starch were slow and constant: there was no fast rate and then slow. Langourier and Crouzet [20] proposed that the initial rapid rate of loss was due to the formation of inclusion complexes by the amylose. As these binding sites were saturated, the rate of decrease in release slowed. They noted that some of the effects seen may also have been due to residual lipid carried through the refining process.

In terms of interactions with galactomanans (guar and locust bean gums), substantial interaction was observed. They found that there was a 70% decrease in limonene release and 25 to 40% decrease in ethyl hexanoate release for a 1% solution of galactomanan. Interestingly, there was a plateau in the data (release vs. hydrocolloid concentration). They found a very similar relationship for volatile release for hydroxypropyl cellulose, i.e., a rapid decrease in rate of release followed by a plateau.

Langourieux and Crouzet [20] found that the addition of dextrans (60,000–200,000 mwt) to water resulted in an increase in aroma release (limonene and ethyl hexanoate). While they attributed this increase as being due to a salting out effect, this effect was noticed even for low concentrations of dextran (as low as 1%). One would not expect such large molecular weight compounds to exhibit a salting out effect even at high concentrations.

Langourieux and Crouzet [20] published some work on the inclusion (interaction) of β-cyclodextrin and volatiles (β-ionone and limonene). They found strong interactions between the cyclodextrin and both test volatiles. Recent work by Reineccius et al. [21], which included all three cyclodextrins (α, β, and γ) as well as a larger pool of volatiles, supports this work. As is seen in Figure 6.7, while the β-cyclodextrin bound the volatiles to the greatest extent, all of the cyclodextrins had substantial interaction with virtually all model volatiles. This interaction was dependent upon the aroma compound, the cyclodextrin, the concentrations of aroma compound and cyclodextrin, and temperature. Of the carbohydrates studied, the cyclodextrins exhibit the greatest amount of interaction with aroma compounds. This interaction has been shown to have significant effects on the sensory character of foods decreasing flavor intensity and changing character [21]. Goubet et al. [22] have provided a good overview of the interactions that occur between cyclodextrins and aroma compounds.

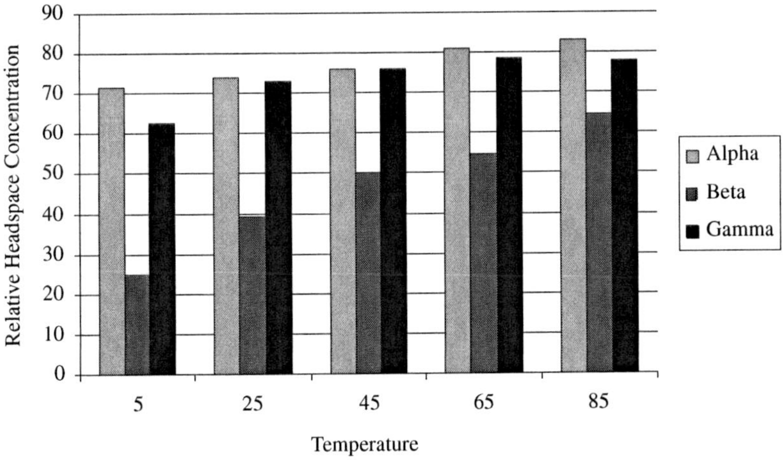


**FIGURE 6.7** The reduction in headspace volatiles (average of six) above an ethanol:water solution (3.6% alcohol) as influenced by temperature and cyclodextrin type (1% cyclodextrin). Y axis is relative to amounts of volatiles in headspace with no cyclodextrin. (From Reineccius, T.A., G.A. Reineccius, T.L. Peppard, *J. Food Sci.*, 68(4), p. 1234, 2003. With permission.)

### 6.3.3.2 Resistance to Mass Transfer

Any food component that increases the viscosity or provides a structure that reduces the generation of surface area during eating will reduce aroma release from a food. One can readily envision that increased viscosity hinders mixing in a food thereby reducing the surface concentration of volatiles for evaporation into the oral cavity. Additionally, increased viscosities will reduce spreading in the mouth, limiting surface area for evaporation. (Roberts et al. [23] have provided a good theoretical discussion of the role of viscosity on reducing aroma release from foods.)

If a food is a solid, the rate that the food breaks down during eating will influence flavor release. The rate of break down of a food during eating will depend upon the chewing process as well as the physical properties of the food. Flavor release will depend upon the rate of dissolution (e.g., a confectionery product), rate of hydration (e.g., a baked good), gel strength (for gels such as jellies and gelled confections), or friability (products that will fragment on eating). The role of food texture on aroma release will be illustrated using the work of Carr et al. [24].

Carr et al. [24] studied the release of model aromas by both analytical and sensory means from gels of different gel strengths made with gelatin, starch, or carrageen. As can be seen in Figure 6.8 (an overall summary of all the experiments), there is a reasonable relationship between gel strength and sensory intensity (this was supported by instrumental data), i.e., as gel strength increased, the aroma intensity decreased. If one considers the effect of gelling agent on aroma release, starch was generally found to limit aroma release to the greatest extent (for a given gel strength) followed by carrageen and then gelatin. The reduction in aroma release

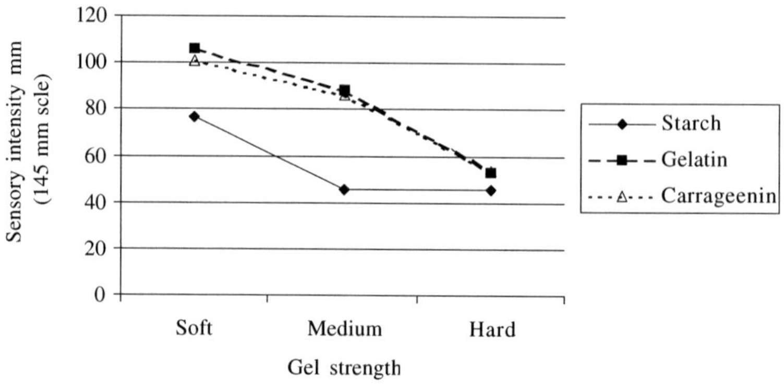


**FIGURE 6.8** The influence of gel strength (starch, gelatin, or carrageenin-based gels) on sensory intensity of model aroma compounds. (From Carr, J., D. Baloga, J.X. Guinard, L. Lawter, C. Marty, C. Squire, *Flavor-Food Interactions*, R.J. McGorrin, J.V. Leland, Eds., Amer. Chem. Soc., Washington, D.C., 1996, p. 98. With permission.)

is likely due to some combination of binding and mass transfer effects, i.e., not all of the effect seen was due to mass transfer considerations. This difference in effect at a given gel strength is likely due to unique binding effects associated with starch, carrageen, and gelatin.

When one considers the effects of chemical interactions (carbohydrate:aroma) on sensory perception, one would expect that aroma character and/or intensity will be altered. This is because the various polysaccharides will interact with different aroma compounds to varying degrees, i.e., change the balance of aroma compounds liberated from a food. However, the effect of carbohydrates on limiting mass transfer would be expected to likely affect flavor intensity and not character. This is because mass transfer effects would be more uniform across aroma compounds.

### 6.3.4 Carbohydrate: Taste Interactions

While carbohydrates may impart a sweet taste, their presence in a food will potentially influence the other taste sensations as well [25]. This influence can be the result of competition for sensory receptors, cognitive inhibition, or a mass transfer effect [26].

In terms of competition for sensory receptors, it is recognized that sensory receptors generally are sensitive to more than one class of stimuli (See Chapter 1). For example, a bitter receptor will also respond to sweetness [27]. Thus, if there is sugar in a sample, the sugar may reach the bitter receptors first (or just overwhelm) and thereby tie up the receptors so they do not respond to the bitter stimuli. Thus, sugars may mask the bitterness of a sample. This type of interaction amongst taste stimuli is fairly common.

Cognitive effects have also been observed [28]. In this case, there is an inhibition of a given stimulant not through competitive action at the receptor but in the brain. It is as though the brain is busy focusing on one stimulus and ignores the other. Interestingly, the two halves of the tongue are independent of each other in sensory channels. Thus, one can put one stimulant on one side of the tongue and another stimulant on the other side and observe an inhibition of one stimulant by the other. In this case, the two different stimuli cannot compete for receptors and yet we may still observe suppression of a given signal.

The final means for carbohydrate:taste interactions to occur is through their effect on mass transfer in the mouth. While hydrocolloids are quite tasteless, they are often used to contribute viscosity to a food and thereby reduce mass transfer from the food to the taste receptors. Pangborn et al. [29] reported that a sample viscosity greater than 12–16 cP results in a significant reduction in sweetness. Work by Vaisey et al. [30] showed that not only is intrinsic viscosity important but the overall rheological properties of the food. For example, hydrocolloids that readily shear thin (e.g., Gelan gum) reduce sweetness less than those hydrocolloids that do not shear thin.

## 6.4 PROTEIN: FLAVOR INTERACTIONS

Some of the earliest work on flavor:food interactions was done on flavor:protein interactions [31]. The driving force for this work was largely the proposal that meat proteins were costly and perhaps could not be a part of the diet in the future, animals being much too inefficient in their conversion of feed stocks to edible protein. However, each of the plant proteins suggested as meat substitutes had indigenous off-flavors and proved problematic upon subsequent flavoring (attempting to get them to taste like meats). While this rationale for study has largely disappeared, there is a resurgence in research targeting protein:flavor interactions since soy has been suggested to have health benefits. Also, the current trend towards low-carb and low-fat diets dictates the greater use of proteins, particularly vegetable proteins. Unfortunately, the issues of flavor binding (bringing in off notes) and being difficult to flavor (binding of desirable flavors) has remained.

### 6.4.1 Protein: Aroma Interactions

The issues of flavor interactions with proteins are similar to those discussed above for carbohydrate:flavor interactions, i.e., we expect to see chemical interactions and mass transfer effects. However, the effects may be more diverse since proteins offer a much wider range of chemical structures for interaction (e.g., amino acid side chains and terminal ends of proteins as well as hydrophobic pockets, Figure 6.9), and mass transfer must consider both the viscosity and gel structures often associated with proteins.

#### 6.4.1.1 Chemical Interactions

In terms of chemical interactions, reversible weak hydrophobic interactions, as well as stronger ionic effects and irreversible covalent bonds, may be formed (e.g., the reaction of aldehydes with the $NH_2$ and SH groups of proteins) between aroma compounds and proteins. The interactions one would expect are influenced by the type and amount of protein (amino acid composition), types of flavoring components,

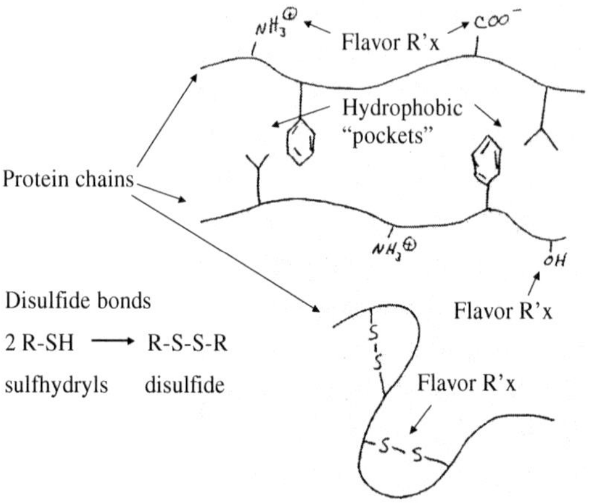


**FIGURE 6.9** Opportunities for flavor to interact with protein molecules.

presence of other food components, processing history (protein ternary structure as influenced by heating), ionic strength (salts), pH (ionic form and conformation), temperature and time [31,32].

In terms of the influence of type of odorant on binding, there is a general observation that for odorants that exhibit only hydrophobic interactions, binding increases with increasing carbon chain length. This would be expected due to the mode of interaction. Losses of odorants that will react with a protein functional group (e.g., –OH, $-NH_2$, or –SH) may be extensive [8,33]. For example, when Mottram and Nobrega [33] added disulfides to an aqueous solution of protein, they recovered as little as 1% of the added disulfide. They attributed the loss to interchange redox reactions between the disulfides (odor compounds) and the disulfide and sulfhydryl groups of the proteins.

The extent of flavor binding depends upon the type and amount of protein. It is generally accepted that the following order of binding is observed: Soy > Gelatin > Ovalbumin > Casein > Corn. Soy appears to be the most problematic of the commonly used food proteins. It is logical that pH also has an effect on binding. pH may affect odorant interactions with proteins in different ways. pH will change the ternary structure of a protein (as would ionic strength), its ionic form (acid and amine groups), and its basic reactivity (e.g., Schiff's base formation). Each of these changes would be expected to influence how a protein interacts with a given aroma compound. For example, Overbosch [34] reported that butanal, hexanal, and diacetyl binding increase with pH while there was no effect of pH on the binding of 2-heptanone.

Denaturation of a protein by either chemical or thermal means tends to open up the protein structure thereby making the hydrophobic pockets of the protein more accessible to binding [31]. Thus, denatured proteins will bind more flavor than native proteins. The means of preparing high protein content ingredients (e.g., acid or salt precipitation or acidification after heating) can significantly change the extent of protein:aroma interactions.

#### 6.4.1.2 Resistance to Mass Transfer

Among the myriad of functions proteins may serve in foods, they are often used to impart viscosity or cause gelation. If they are used for either of these purposes, the same reasoning as provided earlier for carbohydrates will apply to the influence of proteins on aroma release from foods (Section 6.3.3.2), i.e., we would expect increased viscosity or gelation to reduce aroma release. Since each protein (or hydrocolloid) has unique rheological properties, the substitution of one viscosity or gelling agent for another would be expected to have an effect on aroma release that may be adequate to result in a significant effect on sensory properties.

### 6.4.2 Protein Hydrolysate: Aroma Interactions

The hydrolysis of proteins largely removes any opportunity for hydrophobic bonding with aroma components but does not reduce their inherent chemical reactivity. Removing hydrophobic binding opportunities may result in lesser flavor binding effects with a protein hydrolysate than an intact protein. Note that protein hydrolysates also will not build viscosity or form a gel, thereby eliminating any mass transfer aspects. Protein hydrolysates are, however, quite reactive due to the large number of terminal amino groups and the general availability of previously blocked side chains (protein structure may have rendered some groups inaccessible). Thus, one would expect protein hydrolysates to react with carbonyls in a flavoring and there to be more sulfur interchange than in the intact protein. Thus, one may still see some effect of adding a protein hydrolysate to a food on flavor release from that food.

### 6.4.3 Protein: Taste Interactions

The discussion for protein:taste interactions would parallel that of the discussion on carbohydrate:taste interactions. One finds literature on sweet proteins and more recently proteins that make sour substances taste sweet. The addition of proteins or a change in type of protein used in a food will influence the taste of that food [35,36].

## 6.5 MINOR FOOD COMPONENTS: AROMA INTERACTIONS

There is very little published on the interaction of either taste or aroma components with minor components in foods. Earlier in this chapter (Section 6.3.2), some discussion was presented on the interaction of high potency sweeteners with aroma. There is substantial interest in this particular interaction since the industry would like to know why diet products do not taste like the full sugar versions and vice versa. There are few other flavor:food interactions that have such a strong economic link to support this type of research activity. One other area that has received some attention is the interaction of melanoidins with aroma compounds in roasted coffee. This is again driven by economic considerations. A brief discussion of this interaction and pH effects follows.

### 6.5.1 Melanoidin: Flavor Interaction

Work in this area has focused on the interaction of aroma compounds with melanoidins in coffee [37]. (Melanoidins are a chemically undefined class of high molecular weight compounds formed during the Maillard reaction, e.g., coffee roasting). Hofmann et al. [37] added a mixture of volatiles that had been determined to define the aroma of coffee to both water and water plus coffee melanoidins. They found that the model mixtures in water, and water plus melanoidins, were initially similar in aroma but that the aroma of the solution containing melanoidins changed on storage (30 min at 40°C). The mixture lost its characteristic sulfury-roasty notes. Analytical studies showed that approximately 50% of the furfuryl mercaptan was lost after 20 min of storage and it was nearly absent after 30 min of storage. They found similar losses of 3-methyl-3-butene-1-thiol and 3-mercapto-3-methylbutyl formate. These three sulfur volatiles are considered to give the sulfury-roasted aroma of coffee.

They fractionated the melanoidins by gel filtration to determine which molecular weight melanoidins were responsible for this loss. They found that all of the fractions reacted with the sulfur-containing volatiles. Attempts to free the melanoidin-bound volatiles through the addition of other free thiols were unsuccessful, suggesting that the sulfur compounds were not simply involved in disulfide interchange as observed for proteins. Further work strongly supports the hypothesis that the thiols are covalently bound to pyrazinium ions which are oxidation products of 1,4,-bis-(5-amino-5-carboxy-1-pentyl) pyrazinium radical ions.

This loss of aroma compounds through chemical reaction is of interest first since it may partially explain the difference in aroma of freshly brewed coffee vs. aged brewed coffee (or instant coffee), and second because it demonstrates that relatively minor components of a food (melanoidins are present in brewed coffee at only ca. 1.25%) may have an impact on the aroma of a food.

### 6.5.2 Hydrogen Ion Effects (pH effects)

One finds that pH influences both the taste and the aroma of a food. In terms of taste, hydrogen ion concentration is generally linked to the tartness of a food: the lower the pH, the more tart the food tastes. While the effect of pH on taste is well recognized, pH also influences the release of some aroma chemicals, i.e., those that act as acids or bases. For example, one would expect the contribution of volatile acids to aroma to be enhanced in aqueous solution at lower pHs, i.e., those below the pKa of the acid. At low pHs, the acid would be in its protonated form (not ionized) and thus be less soluble in the aqueous phase. This would tend to drive the acid into the sample headspace, increasing its contribution to aroma. Basic odorants (e.g., amines or pyrazines) would behave in an opposite manner. These compounds would become more soluble in the aqueous phase below their pKa since they would be ionized and thus more soluble. This would decrease their contribution to aroma at low pHs. The effect of pH on aroma is obvious when one increases the pH of a traditional acidic food or tries to produce a good chocolate flavor in low pH foods (typically neutral or slightly basic).

## 6.6 SUMMARY

Over the last several years (and most significantly, with the advent of instruments to quantify aroma release during eating) we have learned a lot about what influences aroma release during eating. We have begun to understand the effects of various chemical interactions and food rheology on aroma release (and thus perception). Of the chemical interactions, we have been able to model and thereby predict aroma compound interaction with fats and oils quite well. The effects are quantifiable, and thus theoretically, we may be able to adjust a flavor formulation to balance a flavor in a given food based on its fat/oil content. Unfortunately, this theoretical understanding may not prove helpful in some instances since we do not always have control over the flavoring composition. Often natural flavors are used (e.g., chocolate, essential oils, or plant extracts) in a food where we have little or no control over composition, and thus we cannot change the flavoring composition to function in a given fat content food. It is also possible that fat may influence flavor perception at a cognitive level. We have little understanding of this potential effect.

When we consider the other chemical interactions (e.g., flavor interactions with carbohydrates, proteins, and high potency sweeteners), we understand that interactions take place but they are so complex that we cannot model them to even attempt to make corrections in the flavorings. Thus, we know that a substitution of one protein for another or one hydrocolloid for another will affect the flavor of a product, but we cannot quantify these effects and therefore make changes in flavor formulations. At this time, we can only expect effects and attempt to deal with them in a less than scientific manner, i.e., empirical efforts.

We will likely have better success understanding/predicting the effects of food rheology on aroma release than chemical interactions. Mass transfer is a mature science whose principles can be directly applied to our task. Yet, the rheological properties of a food can be extremely complex and thus be problematic as well. Unfortunately, flavor release from foods is only beginning to be understood — substantial work remains to be done before science will guide this aspect of flavor development.

## REFERENCES

1. Taylor, A.J., Ed., *Food Flavour Technology*, Sheffield Academ. Press, Sheffield, 2002, p. 302.
2. Teranishi, R., E.L. Wick, I. Hornstein, Eds., *Flavor Chemistry: Thirty Years of Progress*, Kluwer Academ./Plenum Publ., New York, 1999, p. 439.
3. Roberts, D.D., A.J. Taylor, Eds., *Flavor Release,* Amer. Chem. Soc., Washington, D.C., 2000, p. 484.
4. P. Schieberle, K.-H. Engel, Eds., *Frontiers of Flavour Science*, Deutsche Forschungs. Lebensmittel., Garching, 2000, p. 602.
5. Taylor, A.J., Mottram, D.S., Eds., *Flavour Science; Recent Developments*, Royal Society Chem., London, 1996, p. 476.
6. Noble, A.C., Taste-aroma interactions, *Trends Food Sci. Technol.*, 7(12), p. 439, 1996.
7. De Roos, K.B., How lipids influence food flavor, *Food Technol.*, 51(1), p. 60, 1997.

8. Schirle-Keller, J.P., Flavor Interactions with Fat Replacers and Aspartame, Ph.D. thesis, University of Minnesota, Minneapolis, 1995, p. 237.
9. Roberts, D.D., T.E. Acree, Simulation of retronasal aroma using a modified headspace technique: investigating the effects of saliva, temperature, shearing, and oil on flavor release, *J. Agric. Food Chem.*, 43(8), p. 2179, 1995.
10. Odake, S., J.P. Roozen, J.J. Burger, Changes in flavour release from cream style dressing in three mouth model system, in *Frontiers of Flavour Science*, P. Schieberle, K.H. Engel, Eds., Deutsche Forschung. Lebensmittel, Garching, 2000, p. 287.
11. Reineccius, G.A., J.P. Schirle-Keller, L.C. Hatchwell, The interaction of aroma compounds with simple sugars, in *Interactions of Food Matrix with Small Ligands Influencing Flavour and Texture*, P. Schieberle, Ed., European Commission, Brussels, Belgium, 1998, p. 27.
12. Nahon, D.F., P.A. Navarro y Koren, J.P. Roozen, M.A. Posthumus, Flavor release from mixtures of sodium cyclamate, sucrose, and an orange aroma, *J. Agric. Food Chem.*, 46(12), p. 4963, 1998.
13. Harvey, B.A., J.M. Davidson, R.S.T. Linforth, A.J. Taylor, Real time flavour release from chewing gum during eating. in *Frontiers of Flavour Science*, P. Schieberle, K.H. Engel, Eds., Deutsche Forschung. Lebensmittel, Garching, 2000, p. 271.
14. Nahon, D.F., J.P. Roozen, G. de Graaf, Sweetness flavour interactions in soft drinks, *Food Chem.*, 56(3), p. 283, 1996.
15. Deibler, K., T. Acree, The effect of soft drink base composition on flavour release, in *Frontiers of Flavour Science*, P. Schieberle, K.H. Engel, Eds., Deutsche Forschung. Lebensmittel, Garching, 2000, p. 283.
16. Tateo, F., L. Triangeli, F. Berte, E. Verderio, The stability and reactivity of aspartime in cols-type drinks, in *Frontiers of Flavor*, G. Charalambous, Ed., Elsevier Publ., Amsterdam, 1988, p. 217.
17. Le Quere, J.L., I. Leschauve, D. Demaizieres, S. Issanchou, R. Delache, Chemical and sensory effects of intense sweeteners on the flavour of diet orange soft drinks, in *Trends in Flavour Research*, H. Maarse, V.D. Haij, Eds., Elsevier Publ. Co., Amsterdam, 1994, p. 387.
18. Hussein, M.M., R.P. D'amelia, A.L. Manz, H. Jacin, W.T. Chen, Determination of reactivity of aspartame with flavor aldehydes be gas chromatography, HPLC and GPC, *J. Food Sci.*, 49, p. 520, 1984.
19. Hau, M.Y.M., D.A. Gray, A.J. Taylor, Binding of Volatiles to Starch, in *Flavor-Food Interactions*, R.J. McGorrin, J.V. Leland, Eds., Amer. Chem. Soc., Washington, D.C., 1996, p. 109.
20. Langourieux, S., J. Crouzet, Interactions between polysaccharides and aroma compounds, in *Recent Development in Food Science and Nutrition*, G. Charalambous, Ed., Elsevier Science, Amsterdam, 1995, p. 1173.
21. Reineccius, T.A., G.A. Reineccius, T.L. Peppard, Comparison of flavor release from alpha- beta- and gamma-cyclodextrins, *J. Food Sci.*, 68(4), p. 1234, 2003.
22. Goubet, I., J.L. LeQuere, E. Semon, A.M. Seuvre, A. Voilley, Competition between aroma compounds for the binding on β-cyclodextrins: study of the nature of interactions, in *Flavor Release*, D.D. Roberts, A.J. Taylor, Eds., Amer. Chem. Soc., Washington, D.C., 2000, p. 98.
23. Roberts, D.D., J.S. Elmore, K.R. Langley, J. Bakker, Effects of sucrose, guar gum, and carboxymethylcellulose on the release of volatile flavor compounds under dynamic conditions, *J. Agric. Food Chem.*, 44(5), p. 1321, 1996.

24. Carr, J., D. Baloga, J.X. Guinard, L. Lawter, C. Marty, C. Squire, The effect of gelling agent type and concentration on flavor release in model system, in *Flavor-Food Interactions*, R.J. McGorrin, J.V. Leland, Eds., Amer. Chem. Soc., Washington, D.C., 1996, p. 98.
25. Stevens, J.C., Detection of tastes in a mixture with other tastes: issues of masking and aging, *Chem. Senses*, 21, p. 211, 1996.
26. Godshall, M.A., How carbohydrates influence food flavor, *Food Technol.*, 51(1), p. 63, 1997.
27. Walters, D.E., How are bitter and sweet tastes related?, *Trends Food Sci. Technol.*, 7, p. 399, 1996.
28. Keast, R.S.J., P.H. Dalton, P.A.S. Breslin, Flavor Interactions, in Abstract of Papers, 26th ACS National Meeting, New, York, NY, Sept 7–11, 2003.
29. Pangborn, R.M., I.M. Trabue, A.S. Szczesniak, Effect of hydrocolloids on oral viscosity and basic taste intensities, *J. Texture Studies*, 4, p. 224, 1973.
30. Vaisey, M., R. Brunon, J. Copper, Some sensory effects of hydrocolloid sols on sweetness, *J. Food Sci.*, 34, p. 397, 1969.
31. O'Neill, T.E., Flavor binding by food proteins: an overview, in *Flavor-Food Interactions*, R.J. McGorrin, J.V. Leland, Eds., Amer. Chem. Soc., Washington, D.C., 1996, p. 59.
32. Fischer, N., S. Widder, How proteins influence food flavor, *Food Technol.*, 51(1), p. 68, 1997.
33. Mottram, D.S., Nobrega, Ian C. C., Interaction between sulfur-containing flavor compounds and proteins in foods, in *Flavor Release*, D.B. Roberts, A.J. Taylor, Eds., Amer. Chem. Soc., Washington, D.C., 2000, p. 274.
34. Overbosch, P., W.G.M Afterof, P.G.M. Haring, Flavor release in the mouth, *Food Reviews Internat.*, 7(2), p. 137, 1991.
35. Temussi, P.A., Why are sweet proteins sweet? Interaction of brazzein, monellin, and thaumatin with the T1R2-T1R3 receptor, *FEBS Lett.*, 526(1–3), p. 1, 2002.
36. Spadaccini, R., F. Trabucco, G. Saviano, D. Picone, O.Crescenzi, T. Tancredi, P.A. Temussi, The mechanism of interaction of sweet proteins with the T1R2-T1R3 receptor: evidence from the solution structure of G16A-MNEI, *J. Molecular Biol.*, 328(3), p. 683, 2003.
37. Hofmann, T., Czerny, M., Schieberle, P., Model studies on the influence of coffee melanoidins on flavor volatiles of coffee beverages, *J. Agric. Food Chem.*, 49, p. 2382, 2001.

# 7 Off-Flavors and Taints in Foods

## 7.1 INTRODUCTION

The major cause of consumer complaints to the food industry is an objectionable flavor [1]. Objectionable (or unacceptable) flavor may be due to incidental contamination of the food from environmental sources (e.g., air, water, or packaging materials) and is termed tainted, or may arise in the food itself as a result of degradation of some food component (e.g., lipid oxidation, nonenzymatic browning, or enzymatic action) and is considered off-flavored [2]. A third cause of unacceptable flavor, which is not generally recognized and is much more poorly defined, is when the flavor is not satisfactory due to the loss of characteristic flavor (evaporation or reaction of characterizing flavor components with the food itself) [3]. The food has neither been tainted nor developed an off-flavor. This source of an unacceptable flavor is difficult to diagnose and correct.

While there are numerous articles in the literature on objectionable flavor in foods, an article by Goldenberg and Matheson [4], despite being dated, and a book edited by Saxby [2] have to be singled out as indispensable references. There has been less published on off-flavors in foods in recent years than in the 70s and 80s. In recent years most of this type of work has been done in industry, and industry is reluctant to publish their problems or solutions. Thus, many of the references will be twenty or more years old. Unfortunately, they are as relevant today as then.

In the real world, one is concerned with determining if a food has an objectionable flavor (sensory methods), and what is causing it (combined sensory/instrumental methods) since this ultimately suggests how to eliminate it in the future. In this chapter we will consider the approaches for determining if a food has an objectionable flavor and its source. The instrumental aspects of this problem have been considered in Chapter 3.

## 7.2 SENSORY ASPECTS OF OFF-FLAVOR TESTING

Kilcast [5] has provided an excellent discussion on this topic and will serve as the primary reference for this chapter. The reader is most strongly advised to read his work for the detail he provides.

Some of the more frustrating aspects of the task of dealing with off-flavors are the difficulties one may have to determine if the food truly has an off-note, how to describe it, and determine the importance to a company. The event is usually initiated by a consumer complaint (but may also originate within the company, e.g., in the quality assurance laboratory). Consumers are notoriously poor at describing an off-note in a food. They know that product flavor does not meet their satisfaction but they generally lack the experience to accurately describe the off-flavor. The descriptions they provide come from their experiences, which vary from person to person. Thus, since the descriptions are so varied and inconsistent, it may be difficult to determine if a series of complaints are the result of the same problem. Thomson [6] illustrated this problem by having 32 untrained subjects describe the character of pork that had a boar taint. The 32 subjects used 22 different terms to describe a note that is generally characterized by the sensory descriptors of urinous, sweaty, or animal. The primary piece of information that one gets from a consumer complaint is that there is a flavor problem that some segment of the population finds unacceptable.

Unfortunately, there often is very poor consensus even among trained individuals. While the consensus improves as individuals become more familiar with an off-note, there still can be very poor agreement in descriptors used, or estimate of intensity of an off-note. This difference can come from experience of the panelists, innate ability to verbalize a sensory stimulus, and/or sensitivity to the stimulus.

Due to differences in sensitivity to a stimulus, some individuals may not detect an off-note while others may be greatly offended by it. Kilcast [5] illustrates this with a typical plot of the distribution of people who can detect a hypothetical aroma chemical as a function of concentration (Figure 7.1). One can see that at the sensory threshold concentration, 50% of the subjects can detect the chemical — this is the standard definition of detection threshold. If we go down the chart, we find that at a concentration ten times lower than the determined threshold, 5% of the population would be expected to still detect the aroma chemical; 95% cannot sense it. Thus, a subset of consumers may be sensitive to a given defect and have a legitimate complaint while the majority of consumers (or even company tasters) may not detect a problem.

Despite the problem in getting good sensory characterization of an off-note, this characterization is very important in determining the source of the problem and its

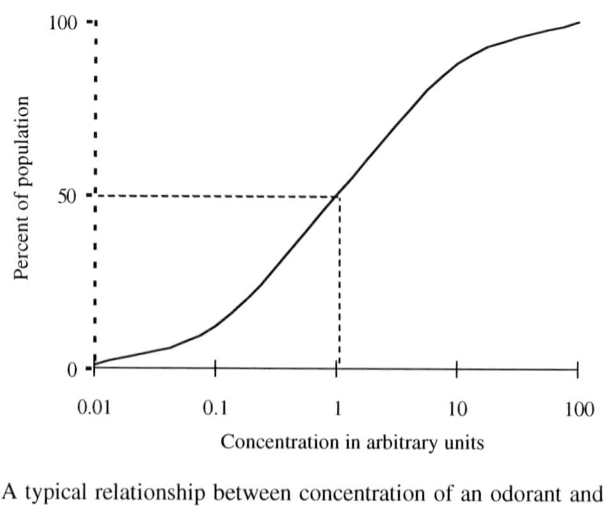


**FIGURE 7.1** A typical relationship between concentration of an odorant and percentage of the population that can perceive it. (From Kilcast, D., *Food Taints and Off-Flavours,* M.J. Saxby, Ed., Blackie Acad. Prof., London, 1996, p. 1. With permission.)

ultimate resolution. If the off-note is commonly a problem in the product (e.g., lipid oxidation), a sensory description by an experienced individual may be adequate to determine the source of the off-note and lead to its resolution. If not, the description is valuable in selecting and identifying the agents responsible for the offending flavor.

Kilcast [5] provides a brief overview of sensory methods employed in the field overall and then focuses on applying sensory testing to taints and off-flavors. He provides some guidelines for sensory testing that illustrate how this application of sensory differs from traditional sensory testing. His points can be summarized as:

1. When the source of the taint or off-flavor is known, use panelists that are sensitive to the objectionable compounds. He points out that it is difficult to have a sensory panel selected for every potential flavor defect, so one may have to select panelists generally known to be sensitive and reliable. (This is not so different from traditional sensory testing.)
2. If a sensory panel cannot be selected that is known to be sensitive to the flavor defect, use a large panel to improve the probability that panelists will be in the group that are sensitive. He suggests a minimum of 15 people (preferably 30) should be used for a triangle test. He cautions that replication is not useful in this type of testing — for if an individual is not sensitive to a flavor defect, testing again does not help.
3. Use a high sensitivity sensory test. He suggests using difference tests as opposed to profile-type tests. Difference tests are more rapid and require less training.
4. Maximize the amount of information gained from a sensory test. Traditional sensory testing has strict rules about getting a limited amount of information from a given test. Yet it is desirable to use a difference test and then ask for preference information (does the panelist prefer a given sample? He or she may like the objectionable sample) and how confident the panelist was in his or her choice.
5. Use statistical methods appropriate to off-flavor testing. This may require that the level of significance may be relaxed to 20% as opposed to the traditional 1 or 5% for accepting there is an off-flavor.
6. Irrespective of the testing outcome, look closely at minority opinions, especially if from typically reliable judges.

From personal experience, most food companies do not apply formal sensory testing to off-flavor problems unless litigation is involved. The purpose is then to establish that there is in fact a problem and establish the magnitude of it. If litigation is not involved, the sensory work is generally done informally by people familiar with the product or off-flavor testing in general.

## 7.3 TAINTS IN FOODS

This section is organized by means of contamination of a food. While it could have been organized around characteristic off-notes or chemical compounds, discussing sources may make the reader more aware of how foods can become contaminated. This could lead to better control of contamination.

### 7.3.1 Airborne Sources

The air can be a very effective means of conveying a taint to a food product. Airborne contamination can enter a food in production (e.g., into milk through odors in the barn, or chicken meat/eggs in chicken coups), in manufacturing, through packaging contamination (will be discussed under packaging related tainting in Section 7.3.4), or during storage and distribution. We will provide brief descriptions of some of these mechanisms of contamination.

Taints can easily enter food products via the lungs of animals and thereby taint eggs, milk, or meat. The dairy industry has long recognized that milk is readily tainted in this manner [7–8]. A milk taint described as “barny” or “cowy” is known to come from cows housed in a poorly ventilated barn. This taint has become relatively rare due to better, cleaner cow environments. A second milk taint related to the environment is a “feed” taint. If cows are fed odorous feeds (e.g., silage, green grass, or oxidized feeds) within four to five hours of milking, the odor of the feed may be transferred to the milk.

An example of how poultry products can become tainted due to air contamination relates to finding 2,3,4,6-tetrachloroanisole in eggs and broiler meat. It was found that the wood shavings used in the hen house came from treated wood (phenol-containing) and microorganisms in chicken litter converted the phenols to the corresponding chloroanisoles [9]. The broilers and eggs assumed a musty taint from breathing the chloroanisole-contaminated air of the hen house.

The finished food products susceptible to airborne tainting are generally limited to goods that are baked and then cooled open to the environment. While the taint may enter the product during cooling, it may also be a problem once packaged. The packaging on these products typically presents little barrier to the transmission of volatile organic compounds. Table 7.1 lists some of the food taints reported in the literature due to airborne contamination. It is of interest that in one case, the source of contamination was 8 km from the food production facility. An examination of the compounds contributed to foods by airborne contamination in Table 7.1 shows that a variety of sources may be involved in the production of airborne taints.

Saxby [2] focused on the role of chlorophenols and chloroanisoles in this context. He cited examples where di-, tri-, tetra- and pentachlorophenols/anisoles were found in wooden flooring, pallets, or shipping containers. Much of the soft timber in the U.S. and Britain (perhaps the world) was at one time treated with chlorophenols (technical grade pentachlorophenol) as antisapstain agents [2]. Therefore, shavings as byproducts or the wood product itself (as noted above) contained significant levels of these chemicals. Chlorophenol contamination has been occasionally traced to accidental contamination through spillage. A third route is that they may arise due to treatment of wood with hypochlorite. In this latter case, naturally occurring phenols in the wood (or paint) may be converted to chlorophenols by hypochlorite treatment [11]. Chloroanisoles may exist as such but generally arise due to microbial conversion of the corresponding chlorophenol [9].

Chloroanisoles and chlorophenols are responsible for musty and medicinal taints, respectively, in foods. If one searches the literature, the majority of reported cases of contamination with these chemicals are in Europe, Australia, and Japan; there are

**TABLE 7.1**
**Examples of Airborne Contamination of Foods**

| Food | Compound(s) or Sensory Character | Mechanism of Entry | Ref. |
|---|---|---|---|
| Biscuits | Herbicides (Chlorophenols) | Carried by wind from a herbicide factory (8 km) | 4 |
| Biscuits | Tar | Tar boiler gave smell due to gaps in the wall | 4 |
| Chocolates | | Anise seed-type odor from adjacent factory | 4 |
| Cakes | Catty off-flavor | Reaction of ketones with sulfur compounds in food | 4 |
| Wrapped bakery goods | Diesel oil fumes | Depots, loading docks, warehouses. Fumes penetrate packaging to contaminate product | 4 |
| Cakes, biscuits, meat pies, chocolate, etc. | Paint | Painted floors, walls, etc. | 4 |
| Biscuits | Oxidized oil | Oil used for spraying biscuits was oxidized. Contaminated product and other products | 4 |
| Cocoa mass | Chlorophenols—musty character | Contaminated shipping container | 10 |

few cases reported in the U.S. This does not mean that tainting by them has not occurred in the U.S., but simply we have few reported cases of such contamination. The U.S. food industry is less likely to make public the occurrence of off-flavor problems or their resolution than European food processors.

An example of chloroanisole/chlorophenol contamination involved the occurrence of a musty taint in dried fruit [1]. It was found that the outer fiberboard cartons were made from recycled waste paper which contained high levels of chlorophenols that were converted (microbially) to the corresponding chloroanisoles during shipping. These chloroanisoles permeated the cartons and contaminated the dried fruit. Interestingly, when the industry changed to virgin fiberboard, the problem decreased but did not disappear. Tindale and Whitfield [12] found an incident where cartons were stored in an area that was cleaned with disinfectants high in available chlorine. The chlorine reacted with the degraded lignin in the board to produce chloroanisoles.

Chloroanisoles are frequently responsible for the tainting of wines. Cork taint is a well-known problem to the wine industry. This topic will be discussed under a subsequent heading titled "Microbial Off-Flavors"(Section 7.6) since this specific tainting is generally associated with the microbial conversion of phenolic substances in cork to chloroanisoles.

A most challenging tainting problem was the occurrence of a "catty" note in some foods. "Catty" is used to describe the odor of tom cat urine. Pearce et al. [13] found that an unsaturated ketone, mesityl oxide, (4-methylpent-3-en-2-one) can react with sulfur compounds in a food, particularly hydrogen sulfide, to form a compound with a catty odor. This compound has been identified as 4-mercapto-methylpentan-2-one [14]. This reaction has been responsible for several tainting events in the industry. In one case, the mesityl oxide was a trace component (0.5%) in a solvent for a polyurethane paint that was applied in a meat chill room [15]. While xylene and pentoxone (major components of the paint solvent) were found in the beef carcasses, 4-mercapto-methylpentan-2-one was found to be responsible for the offensive (catty) taint. Apparently the mesityl oxide reacted with sulfur-containing amino acid degradation products in the meat to form the catty note. The unique difficulty this taint provides is that the offending compound is not present in the environment but is formed in the product. Mesityl oxide has been linked to other occurrences of taints including a variety of canned goods [16]. This taint will be discussed under packaging contamination later in this chapter (Section 7.3.4).

Airborne contamination is generally exceedingly difficult to identify as the source of product contamination. This is partly due to the fact that the taint cannot be reproduced in the product at will for studying. This is contrasted with an off-flavor arising from lipid oxidation, for example, during storage of food. The oxidized off-flavor can be reproduced and subjected to analysis, identification, and eventual management. The airborne taint may occur randomly, depending on the production schedule of the offending source of the taint, wind direction and speed, production schedule of the food processing plant, etc. The offending chemicals also typically have very low sensory thresholds, so they are difficult to isolate and identify by established analytical techniques. An example is the sensory threshold of 6-chloro-*o*-cresol in biscuits ($5 \times 10^{-5}$ ppm). This compound was found to contribute a disinfectant taint to the biscuits [4].

## 7.3.2 Waterborne Sources

Large amounts of water are typically used in most food processing operations. If this water contains significant quantities of tainting components, the potential exists for the taint to be transferred into the food being processed. However, this is generally not a problem unless the water becomes a significant component of the food product or the taint is very intense. Water becomes a significant part of the food if the food is reconstituted or diluted with water prior to processing (e.g., spray drying of cheese or reconstitution of concentrated juices for retail sale), or if direct steam injection is used for product heating.

Taints may also be introduced into a food via water in an indirect manner, e.g., the absorption of taints from habitat water by fish and shellfish. There has been a substantial amount of work on taints in fish and shellfish due to water contamination [17]. Fish are very susceptible to absorbing chemicals from their environment resulting in a taint. The taint may come from water pollution (chemical) or bacterial growth in the water.

In terms of fish tainting from pollution, Vale et al. [18] and Shipton et al. [19] have reported finding a "kerosene" taint in mullet. The tainted mullet were found to contain high levels of hydrocarbons from polluted waters. Bemelmans and den Braber [20] reported an iodine-like taint in herring from the Baltic Sea. They traced this taint to o-bromophenol. While the source of the o-bromophenol was not determined, it is possible that it came from industrial water pollution.

The ease with which chemicals are absorbed into fish (particularly fresh water fish) is a major problem facing aquaculture. While water pollution is not typically

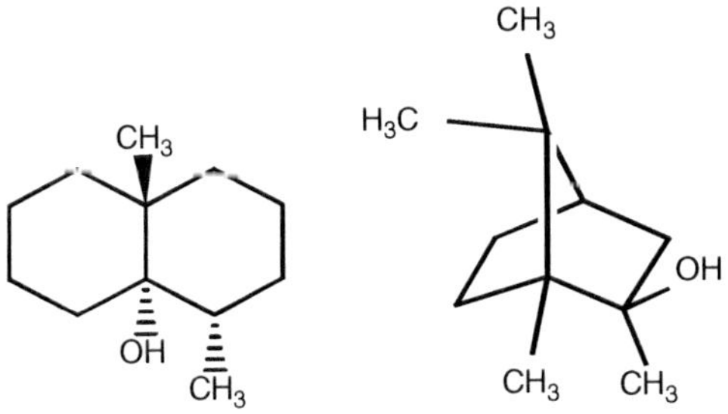


Geosmin Isoborneol

**FIGURE 7.2** Geosmin (*trans*-1,10-dimethyl-*trans*-9-decalol*)* (left) and 2-methylisoborneol (right), causes of musty, earthy taints in water.

a problem, the growth of bacteria or tainting algae in the pond does present taint problems. Lovell and Sackey [21] reported that catfish developed an objectionable earthy-musty taint within two days when holding in tanks containing odor-causing algae. This taint was associated with the presence of *Anabaena*, *Ceratium*, and *Pediastrum* algae [22–23]. The tainting component was likely geosmin (Figure 7.2, trans-1,10-dimethyl-trans-9-decalol) and/or 2-methylisoborneol (Figure 7.2) since these compounds have been identified in muddy tasting fish [24]. Geosmin and 2-methylisoborneol have a very low odor thresholds, 0.05 ppb in water and 6 ppb in fish) and 0.029 ppb in water, respectively, which contribute greatly to their tainting potential. Magligalig et al. [25] demonstrated how quickly water taints can be transferred to fish muscle. They found an exposure of 10 min to 15 ppm 2-pentanone or dimethyl sulfide resulted in significant taint. Recent work by Jirawan and Athapol [26] on the tainting of Nile tilapia by geosmin showed that this earthy taint is acquired in less than two hours when fish are placed in water containing only 5 ng/l geosmin. The fish had to be held in clean water at least 16 days to then eliminate this taint. This very rapid absorption of odorants makes fish very susceptible to picking up taints from their environment.

Some taint problems appearing in the literature from water sources are presented in Table 7.2. The causes of these taints are quite diverse.

### 7.3.3 Disinfectants, Pesticides, and Detergents

Disinfectants, pesticides, and detergents comprise a group of chemicals that occasionally contribute taints to foods. This may be the result of inadequate care in their application by the end user or errors on the part of the cleaning compound manufacturers. The summary of taints observed at Marks and Spencer (Table 7.3) by Goldenberg and Matheson [4] provides a useful illustration of how these chemicals may enter a food and produce a taint. They noted several incidences of taints in food due to detergent contamination, which are not included in this table.

The compounds responsible for these types of taints are often phenols or chlorinated phenols. As noted earlier, phenols can contribute taints to foods but their

**TABLE 7.2**
**Taints Due to Waterborne Contaminants**

| Food | Contaminant | Method of Transmission | Ref. |
|---|---|---|---|
| Canned fruits and canned vegetables | Chlorophenols | Chlorinated water used for cooling lithographed cans subsequently used as boiler feed water. Chlorine reacted with phenol-based lithographic print. Resulting chlorophenols volatized in steam subsequently used for heating brines and syrups by direct steam injection | 4 |
| | | Peas transported to cannery in open steel troughs on newly tarred road. Peas contaminated by tar covered stones flying up from road. Peas contaminated by tar blanched in chlorinated water and chlorophenol | 4 |
| Canned products | Boiler water treatment chemicals | Can cause off-flavors if live steam is used for direct steam-injection for heating | 4 |
| Spray-dried cheese | Disinfectant? | Source of off-flavor water powder borne. Problem solved by use of activated carbon water filter | 4 |
| Beer | Iodinated trihalomethanes | Water in city reservoir was low, resulting in decay of organic matter. Combination of chlorination of water, organic matter, and geological abundance of iodine and bromine | 27 |

relatively high sensory thresholds make them less problematic than their chlorinated counterparts. The chlorinated phenols are soapy, medicinal, or disinfectant in character and have very low sensory thresholds (often < 1 ppb). While the chlorinated phenols may occur as a contaminant in a disinfectant or cleaner, they are also formed from the chlorination of free phenols. Saxby et al. [28] have demonstrated that o-cresol can react with hypochlorite to form 6-chloro-o-cresol (0.05 ppb threshold in biscuits) and the chlorination of mixed phenols typically yields significant quantities of 2-chlorophenol (threshold of about 2 ppb in milk). There are several examples in the literature illustrating how this reaction has caused taints in foods. One example was where a meat processor was washing carcasses with chlorinated water. The water, on occasion, contained trace levels of phenols that were converted to the chlorinated phenols during canning of the meat [2]. Another example involved the transfer of beer through Nylon tubing [30]. The Nylon had been sterilized with a hypochlorite solution. Apparently, the hypochlorite formed a N-chlorinated derivative of the Nylon, which then served as a good chlorinating agent for phenols in the beer. Disinfectant taints in coffee sold in vending machines may have been formed by similar mechanisms [2].

### 7.3.4 Packaging Sources

Food packaging materials are frequent contributors to taints found in food products [31]. Virtually the only food packaging material that will not potentially contribute a taint to foods is glass. Since food is not sold in glass ampoules (i.e., sealed entirely

**TABLE 7.3**
**Taints Due to Disinfectants, Pesticides, and Detergents**

| Food | Contaminant | Method of transmission | Ref. |
|---|---|---|---|
| Cake | Chlorophenol | Due to sultanas, contaminated on board ship by chlorophenol. On a previous voyage, holds had been used for cargo of herbicide containing 0.5% chlorophenol as impurity. Hold washed out but timber still impregnated. Finally timber parts had to be destroyed | 4 |
| Spray dried milk | Chlorophenol | Water contaminated with chlorophenol-cleaner | 4 |
| Canned raspberries | Pesticide | "Catty" taint due to mesityl oxide in solvent | 4 |
| Pork | Disinfectant | Preventive spraying with disinfectant in slaughterhouse against spread of swine vesicular disease | 4 |
| Tomatoes | Pesticide | Appeared to be chlorophenol type contaminant | 4 |
| Smoked bacon | Disinfectant | Hypochlorite used as sterilizing agent; excess used and inadequately washed off. Chlorine combined with phenolic compounds present in smoked bacon to give chlorophenol. Bacon uneatable. Sterilizing agents containing chlorine should not be used where smoked meats or fish are handled, unless special precautions are taken | 4 |
| Beer | Permethrin (pesticide) | High levels of pesticide on malt carried taint into beer | 29 |

in glass) but needs a closure of some type, a small potential exists for food contamination due to this closure even for glass packaging.

Some occurrences of food contamination by packaging materials resulting in tainting are presented in Table 7.4. The contamination seldom comes from the major packaging component. Most often the taint is due to trace amounts of residual solvents, unanticipated contaminants, or the degradation of some packaging component. This can involve the bulk packaging material (e.g., styrene monomer), trace contaminants in the packaging material (e.g., ethyl benzene in plastics or residual inks in recycled paper), solvents used in printing inks, or solvents in adhesives. In some cases the contaminant responsible for the taint is a normal constituent of the packaging system but has not been adequately removed by the processing. This typically occurs due to pushing production too fast. The alternative is that the solvents may, on occasion, be contaminated with higher boiling constituents that are not removed by established process conditions. Tice [31] has presented a good discussion of printing processes and their potential to contribute taints to foods, which will be summarized in the following section.

A variety of printing processes and thus, materials are used to print food packages. One printing system, litho printing, uses a petroleum-based solvent system to carry the inks (resins). While the petroleum-based solvent is chosen to be low in odor, inadequate solvent removal or trace contaminants can be problematic. The

**TABLE 7.4**
**Taints Due in Foods Due to Packaging Materials**

| Food | Packaging Material | Source and Method of Transmission |
|---|---|---|
| | **Printing Systems** | |
| | Printed packaging film[47] | Musty odor in printed film. Found to be due to 4,4,6-trimethyl-1,3-dioxane. Formed by reaction of printing components |
| Yogurt containers[a] and confectionery[4] | | Inadequate curing of inks used in printing |
| Chocolate cakes[4] | Coated paper-print on one side | Inadequate curing of inks used in printing |
| Cook in the bag ham[48] | Plastic wrap | Mesityl oxide formed from the dehydration of diacetone alcohol — a component in red printing ink — catalyzed by ionomer layer in bag |
| | **Paper** | |
| Cakes, biscuits, chocolate[4] | Chipboard | Off-flavor due to either preservatives added to repulped waste paper or compounds in print in waste paper |
| Doughnuts[a] | Coated paper divider plus cellophane overwrap | Inadequate removal of hydrocarbon solvents used to apply divider coating |
| Cold cereals [43,44] | Glassine liner coupons | Residual solvents |
| | **Plastics** | |
| Orange and lemon drinks[4] | PVC bottle | Mercaptide stabilizer used in PVC-split off in processing of beverage |
| Chocolate and lemon cookies[45] | Polystyrene trays wrapped in printed cellophane | Residual styrene monomer |
| TV dinners[a] | Polystyrene trays | Residual styrene monomer |
| | **Laminates** | |
| Fruity soft drinks[45] | Polyester/aluminum foil/polyethylene laminate | Residual toluene in laminate adhesive-faulty drying process |
| | **Metal-Paper** | |
| Refrigerated doughs[a] | Paper tube with metal caps | Residual hydrocarbon solvents in triglycerides used to lubricate metal caps to facilitate packaging operation |
| | **Metal** | |
| Canned pork products[4] | Lacquered can | "Catty" taint due to use of lacquer to cover side seam to prevent blackening. Solvent used contained mesityl oxide which reacted with sulfur compounds in food to cause taint |
| Refrigerated milk beverage[a] | Lacquered can | Trace contaminant (isophorone) in solvents used in lacquer which did not "flash off" |
| Beer[46] | Can | Lubricants used in can manufacture |

[a] Reineccius, unpublished.

resins themselves can also be responsible for tainting. These resins are alkyd types and thus can oxidize in the drying process to form unpleasant aldehydes and ketones.

An alternative printing process, flexographic printing, does not use hydrocarbon solvents but alcohols (ethanol, isopropyl alcohol, and n-propyl alcohol) or esters (ethyl acetate, isopropyl acetate and n-propyl acetate) as solvents. Ethyl acetate and ethanol are the most commonly used solvents. The ink formulation will also contain plasticisors and a slow drying solvent (e.g., propylene glycol monomethyl ether). The resins are polyamides, cellulose nitrates, or acrylics. The tainting potential is primarily from residual solvent (ethyl acetate) since the resins are quite stable. Gravure printing systems use similar solvents and inks as flexographic printing, however, water-based systems are generally used for food packaging applications.

The solvent-less UV printing systems use acrylate monomers with a photo initiator such as benzophenone. The acrylate monomers are polymerized under UV light to form a hard, dry print. Residual odors in this printing system would likely come from unpolymerized acrylate, reaction byproducts, and residual benzophenone. Benzophenone has a geranium-like odor. Potential byproducts that may taint a food include benzaldehyde and alkyl benzoates.

In recent years, there has been a movement to eliminate organic solvents from food manufacturing operations. Various packaging components may be delivered in a water-based or a solvent-less system, the latter employing ultraviolet irradiation to cure and polymerize the resin components (photo initiators are incorporated into the resin). This change has resulted in less food tainting with the usual volatiles but has resulted in a wide range of new taints. Since the new water-based packaging materials (e.g., coatings, inks, overlays, and adhesives) are not *solubilized* in an organic solvent but *dispersed* in an aqueous system, new chemicals are needed to keep these packaging components dispersed. The solvent-less UV printing systems may afford fewer possibilities for contamination, but yet they employ new materials (monomers and photo sensitizers), which again offer possibilities for tainting.

One example of contamination from the newer materials is an off-taste that was found in cookies that were packaged in a clear film. The cookies gave a mouth-tingling, numbing effect where they touched the clear packaging. After substantial research effort, it was found that the packaging was contaminated with spilanthol. Spilanthol is a very unusual compound that is found in a rare essential oil, jambou. This is the first time spilanthol was found as a food contaminant. The implications of this finding are noteworthy.

In the past, a limited number of known solvents (and common contaminants) were likely to be found in food packaging materials. With the advent of water-based delivery systems, the potential list of contaminants has been expanded upon greatly. More significantly, these materials are likely to be nonvolatile and thus not measurable by standard methods used in quality assurance laboratories. Traditionally, food companies have established standards for volatile residues and the testing methods are routine [31]. Unfortunately, there are no routine or all inclusive methods for the testing of nonvolatiles. Thus, the food industry is starting to find new packaging related tainting problems and lacks suitable methods for their study.

There are three examples of food tainting from paper-based packaging materials listed in Table 7.4. Paper or paper-based products come in contact with or form the outer package of a host of food products. For example, liquid dairy products (plastic/board composite), breakfast cereals (variable inner barrier but board exterior), flour (kraft bag), butter and related products (greaseproof paper), chocolates (glassine), cookies (kraft wrap), numerous dry mixes (inner barrier plus outer board), and frozen foods (outer box) have either direct contact or form a secondary package. Even when a contaminated paper product forms a secondary or outer barrier, tainting of the food is possible since the primary, or food contact package, is seldom a good barrier to organic volatiles.

Paper packaging may become tainted in several ways. Paper generally contains various sizing agents, fillers, defoaming agents, and slimicides. The paper may be chemically bleached to provide whiteness. Paper surfaces may also be coated with china clay or calcium carbonate imbedded in a synthetic resin (e.g., a styrene/butadiene copolymer or a styrene/acrylate copolymer) to give the desirable appearance, print surface, or durability [31]. Often we see multilayered packaging where a paper is laminated onto a plastic or foil layer (e.g., milk cartons). Lamination requires the use of adhesives. Each of these packaging materials may be printed upon and further coated with a varnish to give the desired surface properties. Thus, there are many opportunities for the paper-based product to become contaminated and for this contamination to reach the food product. It should be noted that paper-based materials may become more odorous on storage as opposed to less odorous. This is related to the oxidation of resins or lipids in the paper.

The use of plastics in food packaging has become extremely pervasive. Plastics have replaced paper, glass, and metal packaging in a host of food packaging applications. Plastics always begin with the polymerization of monomeric materials. Typically very small quantities of these monomeric materials remain in the polymer, and thus offer little opportunity for food tainting unless the monomer has a very low sensory threshold and is easily recognized as foreign to a food product, e.g., styrene. In addition to the base polymer, additives such as antioxidants, stabilizers, slip agents, antistatics, and colorants may be present in the finished polymer. Each of these components may bring a taint to a food product [31].

Some of the polymers used in food packaging include: PVC (polyvinyl chloride), PET (polyethylene terephthalate), PC (polycarbonate), polystyrene, and PE (polyethylene — either low or high density). Of these polymers, PVC was found in a study of mineral waters to provide the least off-notes. Other than one occurrence (noted in Table 7.4) where a PVC stabilizer was found to produce a taint in beverages, there are *no reports* in the literature of taints by this material. PET is widely used for packaged liquids (mineral water, carbonated beverages, ciders, and beers) and ovenproof food trays. PET is based on the polymerization of either terephthalic acid or dimethylterephthalic acid and ethylene glycol. Both of these monomers have little odor so they cause little problem in terms of tainting. However, a significant amount of acetaldehyde may be formed during the manufacturing or processing of the PET into a final package. This potential is well recognized in the industry, and therefore specifications are in place for its permitted levels. The polyolefins (various low and high density products) are the most common of all food packaging polymers [31]. The odors associated with the polyolefins are seldom from residual monomers since hydrocarbons have high sensory thresholds but arise from the oxidation of the polymers. Koszinowski and Piringer [32] found that two components of heated low density polyethylene, 1-nonene and 2-ethyl-hexanol, are readily oxidized to 2-nonenal and 1-heptene-3-one, respectively. Both of these oxidation products have very low sensory thresholds and have been identified as tainting compounds in LDPE-coated paper board. The packaging most commonly associated with food taints is polystyrene. Styrene can be present in the polymer as a result of incomplete polymerization or thermal degradation. Styrene monomer readily migrates to a food and causes a very characteristic easily recognized plastic taint.

Several other cases of packaging related taints are listed in Table 7.4. These taints involved laminates, metal-paper combinations, and metal cans. Laminates need adhesives to bind the layers together, and the adhesives were contaminated with residual solvents. The paper-metal tube taint involved closing the package before the solvent used to apply a triglyceride lubricant had an opportunity to evaporate from the metal tube lid. The solvent was sealed inside the tube with the product. The final examples in Table 7.4 are tainting from metal cans. In each of these cases, the contamination came from the lacquer lining used on the inner can surface. Foods must be protected from the can and the can from the food. This necessitates the application of a polymer film on the inside of the can surface. Any residual solvent or trace contaminants in this lacquer would potentially taint the can contents.

A final consideration within this topic heading is the loss of flavor either to absorption of the flavor by the packaging (scalping) or its loss through the packaging to the environment (permeation). These processes do not result in a tainted product, but the loss of desirable flavor can expose off-notes to decrease product quality or simply result in a product lacking in flavor.

One has to recognize that permeation can be a two-way process: the loss of flavor from a food product to the environment or the intrusion of odors in the environment into the product. Intrusion of off-odors from the environment may result in tainting of foods and has been discussed earlier in Section 7.3.1 Airborne Sources. One should recognize that food packages range in quality from dust covers, i.e., those packages that present little or no barrier to moisture, oxygen, or organic vapor permeation, to those that are very good barriers to all of these potential migrants. Unfortunately, packaging materials, i.e., polymer-based materials, that offer a good barrier to organic vapors are expensive and seldom used in the industry. Thus, if one is using a polymer-based food package, it is likely to be involved in flavor scalping and offers little to no protection against tainting by permeation.

The amount of scalping or permeation one experiences is related both to the properties of the aroma compounds (polarity and molecular size) and those of the packaging polymer (chemical makeup). Small aroma molecules diffuse better through polymers as do aroma compounds of similar polarity as the packaging polymer. This combination of solubility and molecular dimensions determine the diffusion of a given aroma molecule into and through a package. Since there are numerous references on this topic including symposia proceedings [33–35,29], we will not deal with the details of this process. Our point is that flavor scalping occurs and can result in significant flavor loss and potential flavor imbalances (selective scalping of parts of a flavor). These effects may result in a food product being perceived as being off-flavored when in fact it is not: the food has lost its characterizing flavor.

## 7.4 OFF-FLAVORS DUE TO GENETICS OR DIET

It is obvious to each of us that genetics and diet have a strong influence on the flavor of both plants and animals. We appreciate the differences in flavor found between the numerous varieties of apples, grapes, oranges, etc. We do not, however, accept differences in meat flavor due to either genetic or diet differences between animals. We have come to expect all meats (of the same species) to taste the same. For example, if one beef steak tastes different from another, we consider it to have an off-flavor.

### 7.4.1 Genetics

Two animal species in which genetics plays an important role in producing an off-flavor are swine and ovine. Swine products (pork) occasionally contain an off-flavor characterized as early as 1936 by Lerche as being perspiration or onion-like in character. More recent sensory studies have noted that this off-note will vary depending upon the animal. Sensory descriptors including urine, sweaty, chemical, rancid, turpentine, stable, viscera, pig/animal, manure, and naphthalene-like have been used to describe this off-note [36]. This off-flavor (most commonly referred to incorrectly as a taint) occurs primarily in sexually mature uncastrated male hogs (boars). It has received the most attention of any off-flavor or taint in the literature (194 references in the literature from 1969 to 2002). This is due to its severe impact on the economics of producing pork. Barrows and gilts (no off-flavor) do not gain weight nearly as quickly or efficiently as a boar (uncastrated animal). Thus, substantial research has been directed towards finding means to test for the defect at the time of slaughter to isolate the meat from off-flavored animals or find ways to prevent its occurrence in the male animal [37]. Prevention has typically focused on vaccination or breeding programs.

It is of interest that 46% of the male population and 92% of the female population can smell this off-odor [38], and thus females are most commonly offended by this off-note. It is caused by some combination of 5-α-androst-16-ene-2-one [39], 3-hydroxy-5-α-androst-16-ene [40], indole, skatole [41] and potentially, 4-phenyl-3-buten-2-one [42]. Despite the large amount of research devoted to this sensory defect, the problem persists throughout the industry.

Like swine, the mature ovine animal is also considered to have an off-flavor. (However, this is not a true off-flavor in the sense that the offending odor is a normal characteristic of all mature ovine). This off-flavor is termed "mutton flavor" and is most commonly described as being "sweaty-sour" in nature [49]. The source of this off-flavor has been traced to the unique synthesis of 4-methyl octanoic and 4-methyl nonanoic acids in the mature animal [50]. Jamora et al. [51] have provided a review on this defect and the factors that influence its intensity in meats.

### 7.4.2 Diet

Animal diet can have a pronounced influence on the flavor quality of meats, especially in nonruminant species. Earlier in this chapter, mention is made of musty taints in broilers due to the inhalation of chloroanisoles. While the wood litter was primarily responsible for chloroanisoles in the broiler, Bemelmans and Noever de Brauw [52] also found chloroanisoles in poultry feed. Thus, a tainted diet may also contribute musty taints to the poultry meat. Taints may also enter poultry meat via the diet from the consumption of significant levels (>2%) of highly unsaturated fats [53]. For example, turkeys fed 2% tuna oil develop a fishy defect when cooked [54]. The fishy character was found to be due to oxidation of the highly unsaturated fatty acids during cooking of the turkey. Poultry very readily deposit dietary fat. Goldenberg and Matheson [4] reported similar tainting problems with the feeding of fish oil to hogs. They recommended limiting fish oil to less than 1% of the animal diet.

Tainting of fish via the diet has already been partially discussed in this chapter. Mention was made of muddy taints being the result of the fish eating certain algae *(Anabaena, Ceratium,* and *Pediestrum).* Petroleum-like taints have been found in fish consuming *Limacina helicina* as a major part of their diet [55]. Earthy taints have been associated with the consumption of certain *Actinomyces* species [56]. In addition to off-flavors arising in fish from certain algae in their diet, may also occur if the fish are fed turkey livers or cereal. Maligalig et al. [57] found turkey liver notes in pond-reared catfish after 19 days of feeding livers and cereal notes in the fish following 33 days on a cereal diet. Minimizing taints in fish farming operations is a challenging task.

The contribution of bromophenols to seafood flavor and tainting has been clarified [58–59]. It appears that bromophenols are natural components of seabed algae and other sea creatures (e.g., worms). They are considered to contribute to the characteristic desirable flavor of the seafood when at low levels. However, high levels of these bromophenols in the diet result in tainted product. Boyle et al. [59] found about 3 ppb was a normal bromophenol level in seafoods. However, levels as high as 38 ppb in herring, 30 ppb in salmon, and 100 ppb in shrimp have been found. These levels would be expected to produce taints. Whitfield et al. [58] found 2,6-dibromophenol as the primary source of taint in prawn with 2- and 4-bromophenol, 2,4-dibromophenol, and 2,4,6-tribromophenol as secondary contributors. Whitfield et al. [60] identified various bromophenols as the source of an iodine-like taint in blackfish (Australian origin).

There are numerous studies showing how diet can produce taints in lamb [17]. Park et al. [61] reported on a sickly and nauseating aroma in lamb that had been fed rape. They postulated that the natural glucosinolates of the rape were metabolized to aliphatic sulfides, disulfides, or mercaptans that caused the taints. Park et al. [62] also noted that feeding vetch had an influence on lamb flavor. The vetch-fed lambs had a sweeter and stronger meat flavor. Melton [63] has provided a review of the effect of diet on the flavor of red meat.

Jeon [7] has presented a summary of taints found in milk as a result of pasture weeds. He has noted that land cress (tainting compound, benzyl mercaptan), penny cress (also known as French weed or stinkweed, tainting compound, allylisothiocyanate), penny royal (tainting compound, pulegone), pepper grass (tainting compound, indole), and onion/garlic (tainting compounds, di-n-propyl sulfide, isopropyl mercaptan, and propionaldehyde) have been found to produce tainted milk.

## 7.5 OFF-FLAVORS DUE TO CHEMICAL CHANGES IN THE FOOD

Foods contain a number of constituents that may either react with other constituents or simply degrade into off-flavor components. The major reactions leading to off-flavors include lipid oxidation, nonenzymatic browning, enzymatic changes, and photo-catalyzed reactions. This section will present some examples of off-flavors arising in foods from these reactions.

### 7.5.1 Lipid Oxidation

Lipid oxidation (LO) is with few exceptions (e.g., American milk chocolate and deep fat fried flavor) a detrimental reaction. This reaction is probably the most common source of off-flavors in foods during storage. Off-flavors resulting from LO are described as being: cardboard, beany, green, metallic, oily, fishy, bitter, fruity, soapy, painty, rancid, grassy, buttery, tallowy, or oxidized [64]. The most common descriptors are stale (reaction has not progressed to stage it is recognized yet), cardboard (mild oxidation), tallowy (moderate), metallic, painty, and/or fishy (severe). The sensory character depends upon the degree of oxidation as well as the food product. While dry foods are most susceptible to off-flavors from lipid oxidation, nearly all foods are susceptible to deterioration via this reaction. Grosch [65] and Kochhar [64] have presented excellent reviews of lipid oxidation as it relates to off-flavor production. The reader is recommended to either of these references for detail beyond what is presented here.

Lipid oxidation typically involves the reaction of molecular oxygen with unsaturated fatty acids via a free radical mechanism. The mechanism may be presented as follows:

Initiation

$$R_1H \rightarrow R_1\bullet + H\bullet$$

$$R_1H + O_2 \rightarrow R_1OO\bullet + H\bullet$$

Propagation

$$R\bullet + O_2 \rightarrow R_1OO\bullet$$

$$R_1OO\bullet + R_2H \rightarrow R_1OOH + R_2\bullet$$

Termination

$$ROO\bullet + R\bullet \rightarrow ROOR$$

$$R\bullet + R\bullet \rightarrow RR$$

$$ROO\bullet + ROO\bullet \rightarrow ROOR + O_2$$

where RH is unsaturated lipid; R·, lipid radical; and ROO·, lipid peroxy radical.

Alternatively, lipid oxidation can be initiated by light (photosensitized oxidation). This process is presented below.

$$Sens \rightarrow {}^1Sens^* \rightarrow {}^3Sens^*$$

$${}^3Sens^* + {}^3O_2 \rightarrow {}^1O_2^* + {}^1Sens$$

$${}^1O_2^* + RH \rightarrow ROOH \rightarrow \text{decomposition to form free radicals that continue autoxidation}$$

The Sens is a sensitizer that acts as a means of transferring light energy to oxygen, in essence, activating the oxygen so it is able to directly attack an unsaturated fatty acid. The sensitizer is often chlorophyll, pheophytin, myoglobin, or erythrosine [66–67]. The excited singlet oxygen is extremely reactive, reacting with methyl linoleate, for example, $10^3$ to $10^4$ times faster than normal oxygen. The hydroperoxide formed will decompose, resulting in free radicals thereby initiating free radical autoxidation. Thus, light serves as a good initiator of oxidation in foods.

Foote [68] has proposed a slightly different mechanism for the initiation of photo-sensitized oxidation. In this mechanism, the sensitizer forms an active complex with oxygen, which then attacks the fatty acid. This process is shown below and is believed to function in the riboflavin sensitized oxidation process.

$${}^3Sens + {}^3O_2 \rightarrow {}^1[Sens\text{-}O_2]$$

$${}^1[Sens\text{-}O_2] + RH \rightarrow ROOH + {}^1Sens$$

One can see that this is a self-propagating reaction, i.e., once started it continues itself. It is interesting that an initial radical will produce about 100 hydroperoxides before its termination occurs [69,70].

Off-flavors arise from LO via secondary reaction products. Lipid hydroperoxides are very unstable and break down to produce short chain volatile flavor compounds. Keeney [70] has shown some of the general reactions leading to flavor production:

(A)

$$\underset{\substack{|\\ \bullet O{-}OH}}{R{-}CH{-}R} \rightarrow \underset{\substack{|\\ O\bullet}}{R{-}CH{-}R} + \bullet OH$$

(B)

$$\underset{\substack{|\\ O\bullet}}{R{-}CH{-}R} \rightarrow \underset{\substack{\|\\ O}}{R{-}CH} + R\bullet$$

(C)

$$\underset{\substack{|\\ O\bullet}}{R{-}CH{-}R} + R_1H \rightarrow \underset{\substack{\|\\ O}}{R{-}C{-}R} + R_1\bullet$$

(D)

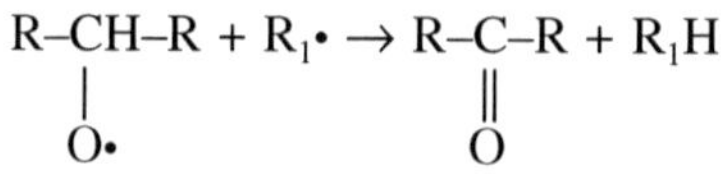

(E)

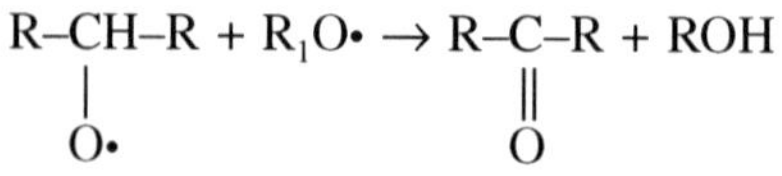

Steps B through E result in the production of short chain saturated or unsaturated aldehydes, ketones, and alcohols. These primary reaction products can undergo further oxidation if unsaturated or secondary reactions to yield a host of off-flavor volatiles (see Figure 7.3). These final products are generally aldehydes, ketones, acids, alcohols, hydrocarbons, lactones, or esters (Table 7.5). The unsaturated aldehydes and ketones have the lowest sensory thresholds and are, therefore, most often credited with being responsible for oxidized flavors. The metallic taint in oxidized butter, for example, has been attributed to oct-l-ene-3-one [71].

An insight into the factors that enhance lipid oxidation will help one understand why some food products develop oxidized off-flavors very rapidly. A primary consideration is the number and type of double bonds in a fatty acid. The relative rates of oxidation for arachidonic, linolenic, linoleic, and oleic acid are 40:20:10:1. In some food applications, the highly unsaturated fatty acids, e.g., linolenic acid, must be selectively hydrogenated in order to permit an acceptable shelf life. For example, the presence of any linolenic acid in a dry coffee whitener would limit shelf life to less than three months.

Trace metals, particularly copper, cobalt, and iron, greatly increase the rate of LO and influence the direction of peroxide decomposition [72]. These metals function both to reduce the induction period and increase reaction rate by decomposing hydroperoxides. Trace levels of these catalysts, e.g., as little as 0.3 ppm iron or 0.01 ppm copper, will result in prooxidant effects [73]. Iron may exist in foods in the free form or as a part of an enzyme (contain organically bound haem, $Fe^{+2}$ or haemin, $Fe^{+3}$). Enzymes containing haematin compounds include catalase and peroxidase (plant tissues) and haemoglobin, myoglobin, and cytochrome C (animal tissues). While heat treatment results in denaturation of the enzymes, it frees the iron to greatly enhance its catalytic properties. This is particularly relevant in the formation of warmed-over off-flavor in cooked meats.

There is substantial literature demonstrating a strong relationship between water activity and rate of LO [66]. This relationship is shown in Figure 7.4. One can see that very dry products or intermediate moisture products are very susceptible to LO. Dry foods such as dry milk, potatoes, potato chips, and dry cereals are particularly susceptible to LO because they have substantial area for oxygen exposure, are permeable to oxygen, and may be at water activities (less than the monolayer) such that oxidation rate is high.

Lipid oxidation rate is temperature dependent. While LO may be initiated by enzymes (e.g., lipoxygenase), it most often is a pure chemical reaction responding predictably to reaction temperature. The higher the temperature, the faster the rate

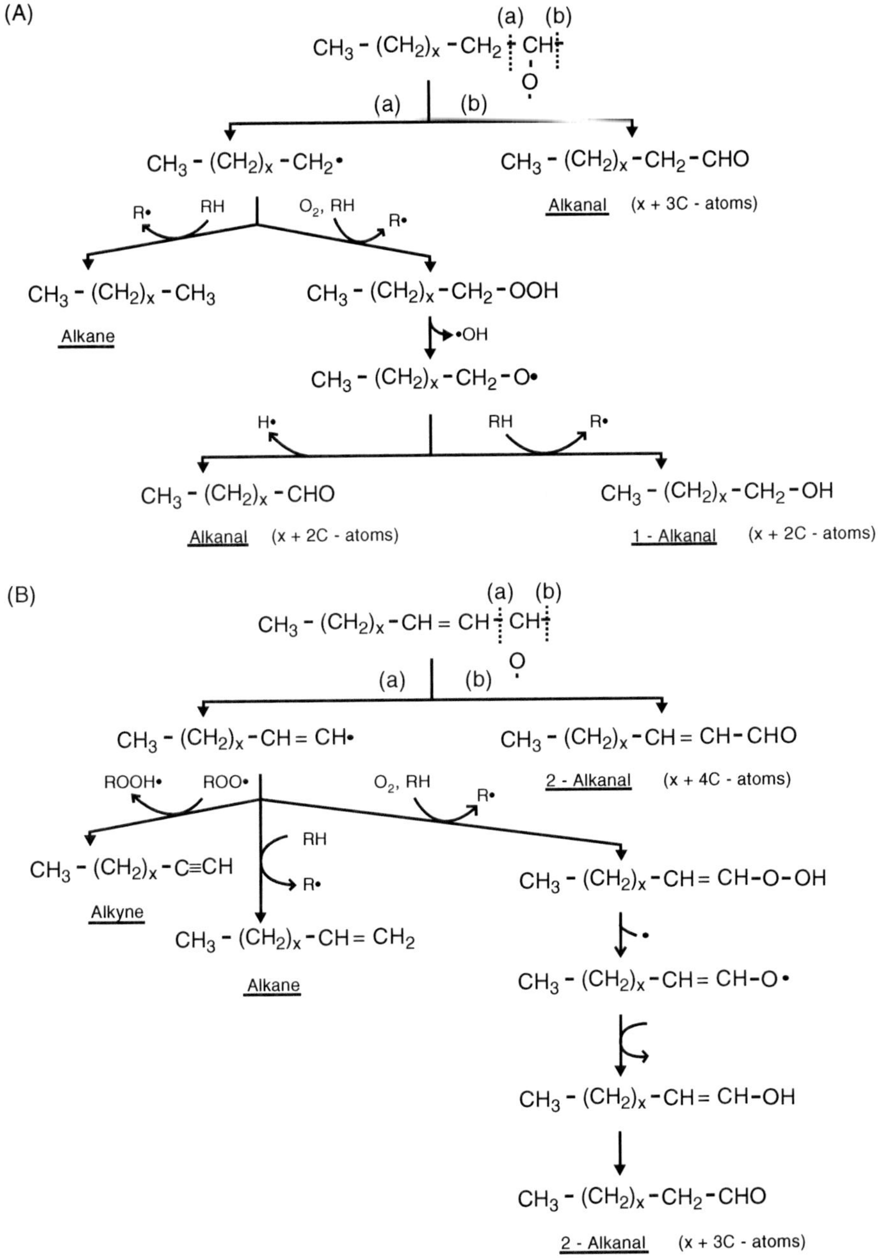


**FIGURE 7.3** Mechanisms for the formation of volatile flavor compounds via lipid oxidation. (A) R' is saturated. (B) R' contains an "ene" system. (From Grosch, W., *Food Flavours*, Part A: Introduction, I.D. Morton, A.J. MacLeod, Eds., Elsevier, New York, 1982, p. 325. With permission.)

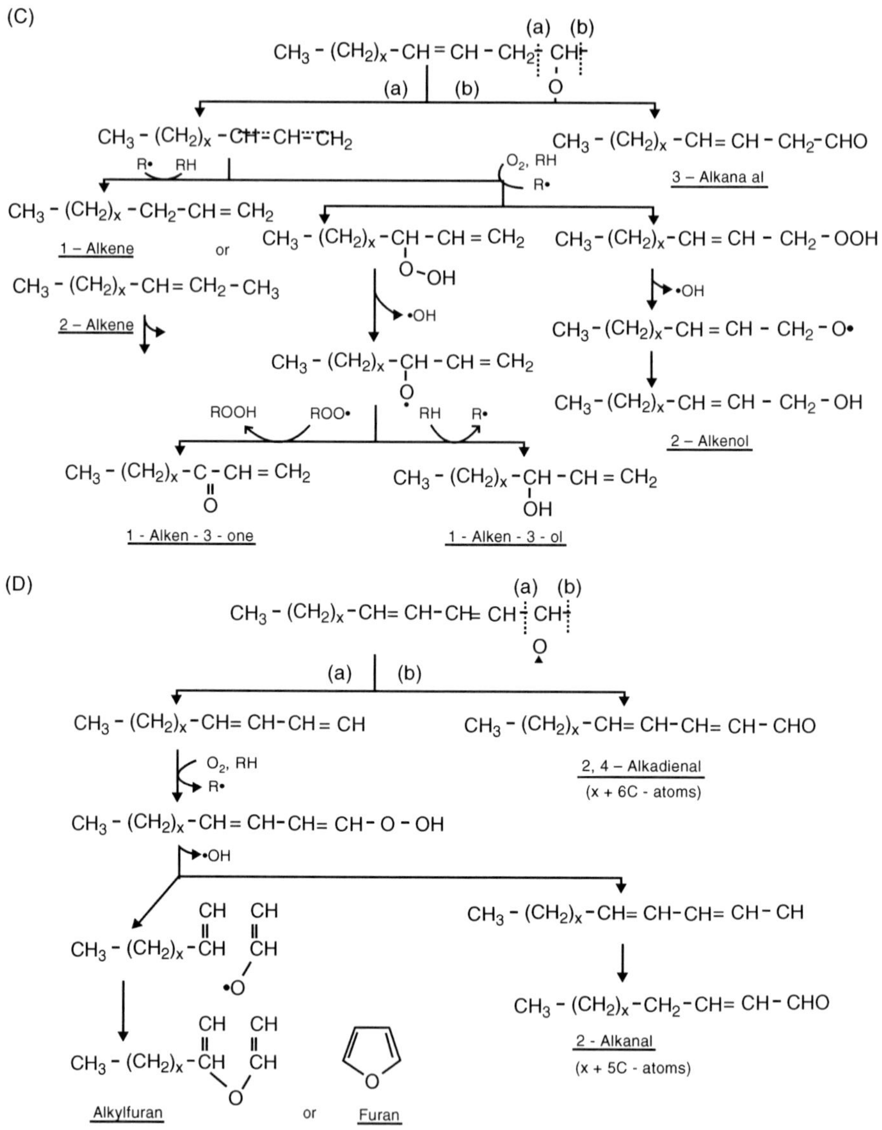


**FIGURE 7.3** (Continued). (C) R' contains an "allylic" system. (D) R' contains a "diene" system. (From Grosch, W., *Food Flavours*, Part A: Introduction, I.D. Morton, A.J. MacLeod, Eds., Elsevier, New York, 1982, p. 325. With permission.)

of LO. When LO is catalyzed by enzymes, the reaction will generally have its highest rate of reaction in the 30 to 45°C range with rate decreasing when the enzyme is denatured [64]. The enzyme catalyzed reaction is pH dependent and fatty acid specific (i.e., methylene interrupted, cis, double bonds).

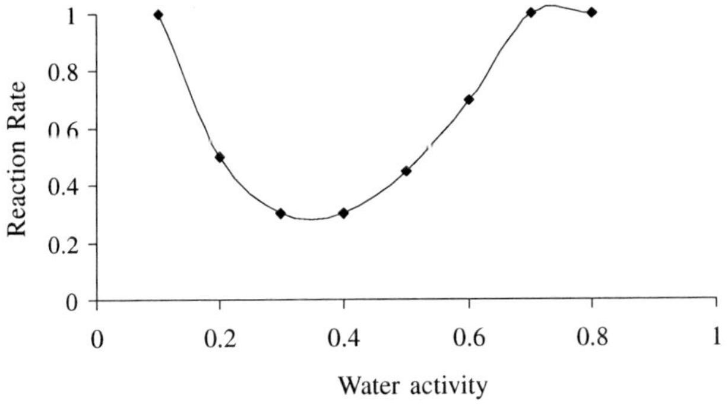


**FIGURE 7.4** Relationship between rate of lipid oxidation and water activity.

As discussed later, light is an effective initiator of LO. The shorter wavelengths are higher in energy and therefore, are more effective in enhancing the rate of LO. A final factor determining rate of LO is the presence of antioxidants. Most of the vegetable oils contain some tocopherols and flavonoids that are quite effective antioxidants. The amount of natural (or added synthetic) antioxidant will strongly influence rate of LO. Kochhar [64] has provided considerable discussion of the role of antioxidants (primary and secondary) in inhibiting off-flavors (OF) in foods.

If one considers the factors that influence rate of LO, one can see that very dry products high in polyunsaturated fatty acids are most likely to develop off-flavors due to LO. However, one must also recognize that virtually all foods contain sufficient lipid to develop off-flavors via LO. It is only a question of whether some other mode of product failure occurs before LO becomes significant. It is a question of time. It was thought that low-fat (or no fat) foods would be less likely to suffer from LO. However, it has been found that these products are actually *more* susceptible to off-flavors from LO than the full fat versions of these products since fat tends to mask off-flavors. Reducing the fat content reduces any masking, making off-flavors more visible to the consumer.

While we usually think of LO in terms of dry milk, breakfast cereals, and snacks, occasionally, we find fish and meats developing off-flavors via oxidation. Fish are very susceptible due to their very highly unsaturated fatty acids. Mendenhall [74] reported significant oxidation in fresh mullet and bluefish after three and five days storage, respectively. However, most fresh fish spoil via microbiological means before oxidation becomes significant. Frozen fish have a sufficiently long shelf-life that oxidation becomes the major mode of deterioration. While other meats are not nearly as susceptible to oxidation (nor bacterial spoilage), one occasionally finds oxidized, frozen meats, particularly from foreign countries where the meat has been stored for several months during shipping, etc.

It should be mentioned that foods containing carotenoids also undergo oxidative deterioration. Carotenoids contain two end groups (often cyclic) joined by a series of conjugated trans, double bonds. The oxidation of carotenoids yields highly unsaturated compounds plus cyclic derivatives of the carotenoids. Ayers et al. [75] were the first to identify α-ionone, β-ionone, and β-ionone-5,6-epoxide as off-flavor

**TABLE 7.5**
**Volatile Compounds Responsible for Off-Notes Resulting from Lipid Oxidation**

| Flavor | Compounds |
|---|---|
| Cardboard tallowy | Octanal; alkanals (C9–C11): alk-2-enal (C8–C9); 2,4-dienals (C7–C10); nona-2-*t*-6-t-dienal |
| Fatty/oily | Alkanals (C5–C7): hex-2-enal; 2,4-dienals (C5–C10); 2-*t*-pentenylfuran |
| Painty | Alkanals (C5–C10); alk-2-enals (C5–C10); hepta-2-t,4-t-dienal; 2-heptanone; pent-2-*t*-enal; but-2-*t*-enal |
| Oxidized | Oct-1-ene-3-one; octanal; hept-2-enal; 2,4-heptadienal; alkanols (C2–C9) |
| Fishy | Alkanals (C5–C10); alk-2-enals (C5–C10); hepta-2-*t*,4-*t*-dienal; 2-alkanones (C3–C11); oct-1-en-3-one; deca-2,4-*c*-7-*t*-trienal; pent-1-en-3-one |
| Grassy | 2-*t*-Hexenal; nona-2,6-dienal; 2-*c*-pentenylfuran; hexanal; heptanal |
| Mild pine-like | 3-*t*-Hexenal |
| Rotten apple | 2-*t*,4-*c*-Heptadienal |
| Rancid (hazelnut) | 2-*t*,4-*t*-Heptadienal |
| Green-beany | 3-c-Hexenal |
| Beany | Alkanals; non-2-enal |
| Deep-fat fried | 2-t,4-*t*-Decadienal |
| Sweet aldehydic | 2-*t*,4-*c*-Decadienal |
| Mushroom moldy | Oct-l-en-3-one: oct-1-en-3-o1 |
| Cucumber-like | Nona-2-*t*,6-*c*-dienal; non-2-*t*-enal |
| Melon-like | Nona-3-*c*,6-*c*-dienals; non-6-*c*-enal |
| Potato-like | Penta-2,4-dienal |
| Lemon | Nonanal |
| Sharp | Octanal; pentanal |
| Brown beans | Oct-2-enal |
| Metallic | Pent-l-en-3-one; oct-l-en-3-one; 2-*t*-pentenylfuran; l-*c*,5-octadien-3-one |
| Rancid | 2-*t*-Nonenal: volatile fatty acids (C4–C10) |
| Nutty | 2-*t*,4-*t*-Octadienal; 2-*t*,2-*c*-octenal |
| Creamy | 4-*c*-Hcptcnal |
| Buttery | 2-*c*-Pentenylfuran; diacetyl; 2,3-pentanedione |
| Fruity | Alkanals (C5, C6, C8, C10); aliphatic esters; isobutyric acid |
| Green putty | 3-*t*-Hexenal; 4-*c*-heptenal |
| Hardened hydrogenation | 6-*t*-Nonenal |
| Liquorice | 2-*t*,1-Pentenylfuran; 2-*c*,1-pentenylfuran; 5-pentenyl-2-furaldehyde |
| Soapy/fruity | Alkanals (C7–C9) |
| Bitter | Pentanal; hexanal; 2-*t*-heptenal |

Data compiled by Kochhar [64] from Smouse and Change [80]; Badings [81]; Forss [82]; Kochhar and Meara [83]; Swoboda and Peers [84]; Chang et al. [85]; Grosch [86]; Hsieh et al. [87].

components of dehydrated carrots. The ionones have a floral character and are typically used in the formulation of raspberry flavorings. If dehydrated yellow or red vegetables (containing carotenoids) are to be used in a food product, an antioxidant must be used to provide adequate shelf life. Products such as dry carrot cake mixes and foods containing dehydrated tomato products commonly suffer from carotenoid oxidation. It is also interesting that frozen tomato products, e.g., pizza, will undergo substantial carotenoid deterioration during the average distribution and storage system. It is in fact carotenoid deterioration that typically limits the shelf life of frozen pizzas.

Warmed-over flavor (WOF) is an off-flavor associated with precooked meats [76]. This off-flavor is generally accepted as being due to the oxidation of intramuscular, phospholipids fatty acids. Oxidation proceeds at a very rapid rate following heating of the meat due to denaturation of heme proteins and destruction of cellular structure thereby permitting intimate mixing of cellular constituents. While the reaction is probably catalyzed by both heme and nonheme iron, the nonheme iron is considered to be of greater importance. WOF can be inhibited by the same methods that are used for lipid oxidation. This includes the use of metal chelators, primary and secondary antioxidants (including natural spice extracts), nitrite, Maillard products, vacuum packaging, etc. [77]. Grosch and coworkers [78,79] have determined that the characteristic sensory defect is due to the loss of desirable aroma components (e.g., two furanones) and the formation of oxidation products, primarily n-hexanal and trans-4,5-epoxy-(E)-2-decenal.

### 7.5.2 Nonenzymatic Browning

Nonenzymatic browning (ascorbic acid browning and the Maillard reaction) is a common source of off-flavor in foods. While the Maillard reaction is very important to the production of desirable flavors (Chapter 5), it is also a source of undesirable flavors in foods. (Ascorbic acid browning has no benefits — it is only negative.) Undesirable flavors may be the result of unwanted Maillard reactions during processing or in subsequent storage. An example of the formation of undesirable flavors during processing is the characteristic flavor of many canned goods. Whether one wishes to say that canned peas, for example, have an off-flavor due to the thermal processing is a matter of semantics. Canned foods do not taste like their fresh counterparts. The primary reason for a flavor change is Maillard reactions, which occur during the canning process. Virtually any food product that undergoes substantial heat treatment develops a flavor from the Maillard reaction.

It is fascinating that the Maillard reaction that occurs during storage is undesirable while that which occurs during heating (e.g., toasting of cereal) is desirable. Why is it that a loaf of bread does not develop a better flavor as it is stored? Browning is occurring in the bread during storage, and browning produces much of the characteristic desirable flavor of the bread during baking. The reason, of course, goes back to reaction kinetics. As explained in Chapter 5, the Maillard reaction is exceedingly complex with a host of possible pathways. The particular pathways active depend on many factors, including water activity, reactants present, temperature, and pH. The primary difference between developing desirable flavors via the Maillard reaction as contrasted to developing off-flavors during storage is temperature. At storage temperatures, the reactions leading to toasted, roasted, nutty, and meaty flavor compounds are not favored, and therefore, these notes ultimately disappear from the product during storage. Unfortunately the Maillard pathways leading to stale, sour, green, sulfury notes can and do progress at typical storage temperatures. Therefore, off-flavors develop in foods during storage due to this reaction. Like lipid oxidation, nearly all foods have sufficient reactants (reducing sugars and amino acids) for the Maillard reaction to occur. Therefore, it will ultimately result in an off-flavor unless some other mode of deterioration renders the product off-flavored first.

The off-flavors due to the Maillard reaction are most typically labeled as being stale in character. Stale is one of the more nebulous terms used in sensory evaluation. Stale is a rather nondescript term denoting a lack of fresh character. Benzothiazole and O-aminoacetophenone are two compounds believed to be formed via Maillard browning, which are responsible for this stale flavor. These compounds were found in stale, dry milk by Parks et al. [88]. O-aminoacetophenone along with some furans were also found to be partially responsible for the "gluey" flavor of old casein [89]. There is also some work in this area related to fruit juices.

Fruit juices generally develop undesirable color changes and a stale flavor during storage [90]. While off-flavors due to the Maillard reaction are quite common, few studies other than those noted have been successful in identifying the components responsible for staleness.

### 7.5.3 Photo-Induced Off-Flavors

Some quite unique off-flavors may be produced via photo-catalyzed reactions in foods. "Skunkiness" in beer and "burnt feathers" in dairy products are two of the more unique examples of off-flavors in foods due to light exposure.

At one time, light-induced (alternatively called light-activated or sunlight) off-flavors in milk were very common since milk was bottled in clear bottles and left on home doorsteps exposed to light. Changes in packaging (plastic-coated paper) and distribution (supermarkets) nearly eliminated this problem. However, several years ago the industry decided to package milk in clear or translucent plastic jugs and display the product in brightly illuminated diary displays. This resulted in the revival of this off-flavor. Fortunately, the industry is now using opaque plastic packaging with good light blockers, and the off-flavor is again on the decline.

Light-induced off-flavors in dairy products can be considered as two very different off-flavors. One is due to typical lipid oxidation and the other to protein (amino acid) degradation [91]. Amino acid degradation produces the light-activated off-flavor while lipid oxidation produces rather typical oxidized notes. The burnt feather or light-activated flavor initially predominates (as little as two hours light exposure); however, the milk would be judged as typical oxidized after longer light exposure.

Amino acid degradation is catalyzed by light and riboflavin as is shown:

$$\text{Amino acids} \xrightarrow[\text{Riboflavin}]{h\eta} \text{Volatile compounds} + NH_3 + CO_2$$

While all amino acids undergo this degradation, the extremely low sensory thresholds of compounds arising from the sulfur-containing amino acids (particularly methionine) are believed to be most responsible for this defect. Patton [92] and Allen and

Parks [93] initially suggested that methional was the key contributor to this defect. However, Samuelsson [94] and Forss [152] and more recently, Jung et al. [95] have suggested that dimethyldisulfide is the key contributor with other sulfur compounds (methanethiol and dimethylsulfide) potentially contributing as well.

Off-flavors due to light exposure have also been reported in champagne [96] and beer [97]. Similar to dairy products, the photo-induced changes in champagne involve the riboflavin catalyzed degradation of methionine. Light-induced changes in beer flavor are the result of more complex reactions involving photo-oxidation of lipids (later stages of the off-flavor), photo-catalyzed decarboxylation and deamination of amino acids, and hydrolysis of the isohumulones. Beer will take on an initial off-flavor often characterized as being "skunky" or "leek-like" [98]. This off-flavor is attributed to the photo-catalyzed degradation of the isohumulones [99]. Tressl et al. [100] have provided a pathway illustrating the photo degradation of isohumulone (Figure 7.5). Note that the 4-methyl-3-pentenoyl side chain of isohumulone is cleaved to yield 2,7-dimethyl-2,6-octadiene and 4-methyl-2-pentenal as major end products. This pathway, in the presence of cysteine, has been shown to yield 3-methyl-2-butene-l-thiol (prenyl mercaptan) via free radical mechanisms. Prenyl mercaptan has a sensory threshold of 7 ng/l in beer and is considered to be the offending aroma compound [101]. Control of this off-flavor is typically accomplished by packaging the beer in light impermeable bottles or cans. The alternative is to use hops, which have had the iso-α-acids (isohumulones) reduced to their saturated counterparts. Reduced α-acids will not form the objectionable thiols. They will, however, still undergo photo-catalyzed cleavage into "green" off-notes and result in a somewhat less objectionable off-flavor [100].

A final example of a unique off-note coming from light-induced reactions involves cottonseed oil. Potato chips that are fried in cottonseed oil and then exposed to light develop a distinctive off-flavor also termed "light struck." Fan et al. [153] found this defect to be due to the formation of 1-decyne that arises from the light-induced oxidation of sterulic acid (a cyclic fatty acid) common to cottonseed oil. They found that chlorophyll was a photo sensitizer for this reaction.

Light is recognized as a catalyst for other chemical reactions, most importantly, lipid oxidation. There are numerous examples in the literature of off-flavors arising in foods due to light-induced lipid oxidation (see review by Borle et al. [154]). The sensory character of this defect varies with the food product.

### 7.5.4 Enzymatic Flavor Changes

One must remember that virtually all foods were once living. These living entities depended on enzymes for metabolism and subsequent growth. Therefore, foods contain a variety of enzymes. While many are inactivated by processing or may be inactive due to inadequate conditions for functioning (e.g., low-water activity, temperature, and pH), some foods do contain active enzymes, which may cause OF to develop in the food product. The three enzymes most commonly associated with OF in foods are lipoxygenase, lipase, and various proteases.

Lipoxygenase is an enzyme that will attack cis-cis double bonded, methylene interrupted fatty acids (most commonly linoleic and linolenic acids). This enzyme

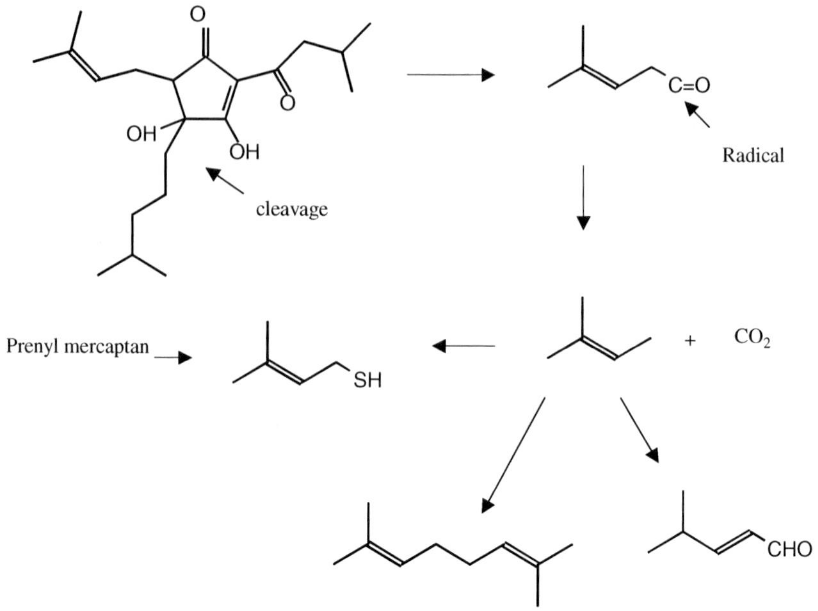


**FIGURE 7.5** Photodegradation of iso-α- humulones in a cysteine model system. (From Saxby, M.J., S. Wragg, *Ind.*, p. 310, 1985. With permission.)

initiates lipid oxidation by abstracting a hydrogen radical from fatty acids. Once lipid oxidation has been initiated by this enzyme, the process of oxidation follows the typical auto-oxidative process leading to OF, which has been previously described in this chapter.

Lipoxygenase enzymes are quite common to plant tissues. The legumes (especially soybeans) contain substantial quantities of lipoxygenase enzymes. The beany flavor of soybeans is believed to be due to lipoxygenase activity in the bean once tissue damage has occurred [102–104]. The reversion flavor of soybean oil is also believed to be due to lipoxygenase activity in the oil. Hsieh et al. [155] have shown that 2-pentylfuran is the primary oxidation product responsible for this OF.

Lipoxygenase also plays a role in the deterioration of frozen vegetables. Many vegetables must be blanched prior to frozen storage in order to inactivate lipoxygenase enzymes, which otherwise result in oxidized OF [105] during storage.

Off-flavors may arise from lipase activity in foods. Lipolyzed flavor is due to the hydrolysis of fatty acids from triglycerides. Since lipase is quite widely occurring in food products, one would expect to find lipolyzed OF more commonly than encountered. The major reason lipolyzed OF is not encountered more frequently is that most lipids in foods contain only mid- to long-chain fatty acids. These fatty acids (>12 carbons) are too large to make a significant contribution to either odor or taste. Therefore, the only foods where lipolysis produces significant OF are those that contain short chain fatty acids (e.g., dairy or coconut oil products, Table 7.6).

As one would expect based on the flavor descriptors of the fatty acids (Table 7.6), the OFs arising from lipase activity in dairy products are characterized as being

**TABLE 7.6**
**Flavor Characteristics and Thresholds of Free Fatty Acids in Selected Media**

| Fatty acid | Flavor Character | Threshold Milk | (ppm) Oil |
|---|---|---|---|
| Butyric acid | Butyric | 25 | 0.6 |
| Caproic acid | Cowy (Goaty) | 14 | 2.5 |
| Caprylic acid | Cowy (Goaty) | — | 350 |
| Caprylic acid | Rancid unclean bitter soapy | 7 | 200 |
| Lauric acid | Rancid unclean bitter soapy | 8 | 700 |

"butyric," "goaty," and/or "bitter" in nature [7,106]. There appear to be two lipases in dairy products: one associated with the fat globule membrane and the other with the proteins. While the point is controversial, it seems that these two lipases are identical but located in different parts of the milk. Milk lipases are highly specific for the primary positions of the triglyceride. Unfortunately, nature has also shown a preference to put the most sensorially significant fatty acids (i.e., butyric and caproic acids) in the primary positions.

The lipases of milk are rather unique in that they are inactive until activated by some treatment (the exception being rare that a cow will produce spontaneously rancid milk). Activation of the lipase enzyme system may be accomplished in several different ways. Occasionally, fluctuations in temperature such as cooling, warming and then cooling, will cause the lipase to become active. Mechanical agitation or foaming can also activate the lipase systems. This was a particular problem when pipeline milking systems were initially introduced. Homogenization of raw milk is especially effective in producing lipolyzed flavor. In a matter of seconds, the homogenization of raw milk results in so many free acids that it will nearly burn the mouth. The milk fat globule membrane is inadequate to cover the new surface generated upon homogenization of the milk. This unprotected fat is particularly susceptible to lipase attack.

It is of interest that lipase activity can be influenced by the presence of certain chemicals [107]. Aureomycin, penicillin, streptomycin, and terramycin were shown to provide 7.6 to 49.8% inhibition of lipase.

While dairy products may become lipolyzed due to indigenous enzymes, the lipases may also be from outside sources. Bacterial activity, spices, and some fruits contain lipases that may attack the triglycerides of milk. It is relevant that different lipases (different sources) will have a different preference for lipolysis of fatty acids. Milk lipase and other ruminant sources of lipases tend to preferentially lyse butyric acid from triglycerides. Lipases from *Pseudomonas* fluorescens and all porcine lipases have little preference for short chain fatty acids. Lipases from molds and *Chromobacterium* viscosum produce less butyric acid but more capric acid [108]. Lipases that preferentially lyse the short chain fatty acids have a greater impact in the production of off-flavors than do those lipases that lyse the longer chain fatty.

This suggests that an analytical method that measures total free fatty acids (e.g., acid degree value) may be a poor measure of the off-flavor potential of these fatty acids.

The other food products (ingredients), which are subject to lipolyzed OF are the tropical oils, most notably coconut oil. Coconut oil contains lauric acid (C12) as a major fatty acid. Lauric acid tastes soapy when hydrolyzed from the triglyceride. While coconut does not contain a significant amount of lipases, coconut is often used in foods with other ingredients that do contribute lipases. It should be mentioned that lipases are reasonably heat stable enzymes. They will often survive a thermal treatment that one would expect to be adequate to denature the lipase.

A beverage where "soapy" flavors are a problem is the pina colada. Pina colada is an alcoholic drink consisting primarily of coconut and pineapple juice. The pineapple is an excellent source of quite heat stable lipases. If one encounters a soapy OF in a food product, the first step is to search the ingredients for coconut oil and then a source of enzyme.

Bitterness in foods is an additional OF that may occur due to enzyme activity. In this case, we are considering bitterness arising from proteolytic activity on food proteins. The organisms that produced the enzymes responsible for bitterness may have been killed by a heat treatment but some proteolytic enzymes may survive rather severe heat treatments to remain active in the food thereby causing off-notes. Murry and Baker [109] are often credited with being the first to determine that this bitterness was due to short chain peptides.

Bitter peptides have been found to produce OF in soy products [110–112], zein [114], casein [115–117], and Cheddar cheese [119]. Arai [113] pointed out that most of these peptides have leucine at the C-terminal end. Otherwise there appears to be little similarity between peptide structures (Table 7.7).

Probably the largest amount of research on bitterness due to peptides has focused on aged cheeses [120–124]. Proteolytic activity in the cheese comes from the rennin, starter organisms, and/or contaminating microorganisms. Rennin initially breaks a

**TABLE 7.7**
**Bitter Peptides Identified in Protein Hydrolysates and Cheese Products**

| |
|---|
| Soy globulin hydrolyzed with pepsin |
| Arg-Leu, Gly-Leu, Leu-Lys, Phe-Leu, Gln-Tyr-Phe-Leu, Ser-Lys-Gly-Leu |
| Zein hydrolyzed with pepsin |
| Ala-Ile-Ala, Gly-Ala-Leu, Leu-Val-Leu, Leu-Pro-Phe-Ser-Gln-Leu |
| Casein hydrolyzed with trypsin |
| Gly-Pro-Phe-Pro-Ile-Ile-Val, Phe-Ala-Leu-Pro-Gln-Tyr-Leu-Lys |
| Cheddar cheese |
| Pro-Phe-Pro-Gly-Ile-Pro |
| Pro-Phe-Pro-Gly-Pro-Ile-Pro-Asn-Ser |
| Leu-Val-Tyr-Pro-Phe-Pro-Gly-Pro-Ile-Pro |

*Source*: From Arai, S., *The Analysis and Control of Less Desirable Flavors in Foods and Beverages*, G. Charalambous, Ed., Acad. Press, New York. p. 1, 1980. With permission.

portion of the protein down into large peptides. These peptides then are further degraded into smaller peptides that may cause bitterness unless they are further metabolized by the starter organisms [125]. Therefore, starter failure, retardation via chemicals or contaminating organisms may end up causing bitterness in a cheese. The proper starter organisms and their growth is essential to minimize the possibility of bitterness during aging of cheese [126].

## 7.6 MICROBIAL OFF-FLAVORS

Microorganisms (MO) are a common source of OF in foods [127]. If the product is not sterile (via heating or irradiation), or preserved via antimicrobial agents (e.g., humectants or chemical preservatives), microorganisms may grow in the product and produce OF.

Off-flavors may arise in food products via microbial sources in many different ways. The OF may be the result of a planned fermentation going wrong. The fermented product may then develop an OF. An example of this would be a starter culture failure in a cheese vat. Failure of the starter culture would result in poor acid production and the opportunity for the growth of other organisms. These contaminating organisms often result in bitterness and unclean OF. This type of OF development is possible in virtually all types of fermented foods. Only through very careful handling of the starter cultures and manufacture of the fermented products can this problem be minimized.

A second way in which microorganisms may produce OF in foods is somewhat indirect. Enzymes from lysed MO can catalyze reactions in foods leading to OF. Microbial enzymes are frequently quite stable to heat denaturation surviving pasteurization and sterilization processes [7,128]. Therefore, a thermal process may kill the MO but leave active enzymes (e.g., lipases or proteases), which can produce OF on storage. One can appreciate that a food may develop an OF in this manner but yet show very low *viable* cell counts. The low viable cell count could mislead the investigator into believing that the OF was not due to MO.

The final means by which MO can produce OF in foods is the most common. That is from growth in the food and contamination with metabolic products. Fresh fish [17] and dairy products [129] are two foods that are very susceptible to OF from microbial sources. Fresh fish is generally considered to have very little odor. However, if fish is stored above freezing temperatures, the fish develops a "fishy" OF initially that later becomes putrid and foul in character. Fish is particularly susceptible to microbial growth since the fish muscle has substantial soluble material and cells tend to lyse once the fish has been killed. The fish muscle provides a rather ideal growth medium for MO.

The fishy OF is primarily due to the generation of trimethylarnine via bacterial action [130,131]. Trimethylamine is formed from trimethylamine oxide, which is a natural constituent of fish muscle. This reduction is accomplished through bacterial enzymes and involves a coupled oxidation of lactic acid to acetic acid and $CO_2$ [132]. The latter stages of fish spoilage involve the production of various nitrogen- and sulfur-containing compounds. These compounds produce putrid, sulfury notes in the fish.

It should be noted that different MO vary in their ability to produce OF in foods. The ability to produce an OF is dependent upon the metabolic compounds which are produced and excreted into the growth medium (food) by the MO. Work on OF in fish demonstrated this very nicely. Herbert and Shewan [133] have shown that Gram-negative bacteria of the genera Pseudomonas, Achromobacter, and Vibrio are most often responsible for OF in fresh fish. Within these *genera*, only a few species can cause spoilage odors. Adams et al. [134] found that typically less than 10% of the bacterial population present on fish can cause OF. This should make one cautious in concluding that high plate counts (or the converse) necessarily mean an OF came from microbial sources. It is the specific organisms and populations that are significant in determining the contribution to OF.

Shipe [129] and more recently Joen [7] have reviewed some of the OFs that may occur in milk due to MO. Prior to wide usage of pasteurization and refrigeration, milk would most commonly spoil due to *Streptococcus* lactis growth. This would result in an acid, sour OF. Pasteurization quite effectively kills *S.* Lactis, and its growth is not favored by storage at refrigeration temperatures, so this organism (and OF character) is no longer a common means of spoilage.

Today the psychrotrophic bacteria are typically responsible for OF in milk. These MOs can multiply at or below 7°C so they will grow at typical refrigeration temperatures. Even though pasteurization destroys these MOs, they produce substantial quantities of heat stable lipases and proteases that may produce OF later during storage. They also are common post pasteurization contaminants that will grow in the milk and produce OF due to metabolic activity. The OFs arising from psychrotrophic bacteria are often characterized as being unclean, foreign, fermented, or bitter.

Two quite characteristic OFs found occasionally in milk are fruity and malty OFs. Fruity defects have been often traced to the presence of *Pseudomonas* fragi. *Pseudomonas* fragi produces a lipase, which initially hydrolyzes short chain fatty acids from milk lipids and then esterifies these acids to form their corresponding ethyl esters. Ethyl butyrate and ethyl caproate have been found to be the OF compounds responsible for this defect [91].

Malty OF occurs occasionally in milk, which has had some temperature abuse (storage at 10°C or above). The malty OF is caused by *Streptococcus* lactic subsp. Maltigenes [135] or *Lactobacillus* maltaromicus [136]. The malty character itself is due primarily to 3-methyl butanal production from leucine.

While geosmin has been mentioned earlier as the compound responsible for earthy taints in fish, it has also been linked to taints in bread [137], wheat flour 138], navy beans [139], and clams [140]. Molds and Actinomycetes growing in these products were found to produce this taint. These are similar organisms to those that produce earthy taints in soils that often end up in root crops such as beets or sugar beets (and sugar made there from).

It is somewhat ironic that molds have been found to grow on foods that have been treated with sorbic acid (a mold inhibitor), degrading the mold inhibitor to 1,3-pentadiene, which has a petroleum-like odor [141]. While sorbic acid is effective in inhibiting molds, if a mold can grow on parts of the food where the inhibitor may

$$CH_2CH{=}CHCH{=}CHCOOH \rightarrow CH_3CH{=}CHCH{=}CH_2 \text{ (kerosene taint)}$$

**FIGURE 7.6** Degradation of sorbic acid to 1,3-pentadiene.

be too low to be effective, the mold will potentially produce this taint (Figure 7.6). While the majority of such taints have been in cheese (Feta — a plastic, paint, or kerosene defect [142]; cheese spread — kerosene defect [143] ), other foods that are treated with sorbic acid and are high in moisture have also been found to have "kerosene" taints due to this compound [2].

Sorbate has also been implicated as the precursor of a flavor defect in wine that has a geranium character. Sorbate may be added in the winemaking process as a secondary yeast inhibitor to reduce the amount of sulfur dioxide used. While sorbate is a good yeast inhibitor, it does not inhibit the growth of bacteria. Crowell and Guymon [144] have found that certain bacteria in wines can convert sorbate to 2-ethoxyhexa-3,5-diene, which has a geranium character.

It is interesting that styrene in food products is generally assumed to arise from packaging contamination (polystyrene polymers), but it has been found to also arise from microbial action. Saxby [145] described an off-flavor problem where a yeast, *Hypopichia burtonii,* growing in the presence of a cinnamon flavoring produced styrene. Cinnamon flavor is based on cinnamic aldehyde: decarboxylation of cinnamic aldehyde yields styrene. This defect was found in a spiced, baked good.

A microbial conversion of sensory significance already discussed to some extent is the conversion of phenols to chloroanisoles (airborne taints). A defect in wine recognized as cork taint is considered to be primarily due to this conversion. It was estimated that cork taint results in the rejection of 2 to 6% of commercial wines [146]. While this reference is somewhat dated, it is of relevance that this is the second most researched (based on number of publications) sensory defect in foods (after boar taint), which suggests that the defect is still of importance to the industry. Chloroanisoles can be formed from the microbial conversion of pentachlorophenol (a component of some insecticides and wood preservatives) in the cork or from hypochlorite treated corks (forms chlorophenols) with subsequent microbial methylation to the anisoles. In either event, the chloroanisoles (primarily 2,4,6-trichloroanisole) are formed on/in the cork and leach into the wine [127]. This has lead to a movement within the wine industry to use synthetic bottle stoppers in place of cork.

One must recognize that microorganisms can accomplish a wide range of chemical transformations thereby potentially producing a host of different off-flavors. Earlier in this section, several off-notes have been described as being the result of microbes. One can add other sensory notes such as: faceal (indole and skatole in rotting potatoes [147]), musty potato (2-methoxy-3-isopropylpyrazine [148]), mousy (wine, 2-acetyl-1,4,5,6-tetrahydropyridine [149]), and musty, "dirty dishcloth" (machine oils 2,6-dimethyl-3-methoxypyrazine [150]).

Additional examples of OF produced in foods from MO are provided by Goldenberg and Matheson [4]. A summary of OF they have encountered at Marks and Spencer is presented in Table 7.8. It is quite obvious from their tabulation that OF due to MO can occur in a wide variety of products.

**TABLE 7.8**
**Off-Flavors Due to Microbial Spoilage**

| Food | Type of Spoilage | Comments |
|---|---|---|
| Cakes | Bacteria | Off-flavor due to bad egg; either use of poor quality egg or careless handling during or after thawing of frozen egg |
| Cakes; angel cakes; apple tarts; marzipan | Mold | Mold due to contamination by flour dust in bakery |
| Cakes | Lipolytic bacteria | Soapy off-taste in coconut |
| Apricot jam | Bacteria-producing butyric acid from sugar | Inadequate processing or inadequate cleaning or washing of equipment |
| Fresh cream | Bacteria | Contamination by yogurt |
| Cream | Lipolytic bacteria | Cheesy off-flavor due to growth of lipolytic bacteria either in milk or during processing before pasteurization |
| Cream | Bacillus *cereus* | Causes bitter flavors |
| Chocolate tea cakes | Osmophilic yeasts | Caused fermentation and beery off-flavor |
| Biscuits | Molds | Musty flour used and not examined before used |
| Chicken | Molds | Wood shavings used as litter in chicken broiler-house made from wood preserved with chlorophenols. This metabolized to chloroanisoles and absorbed by chicken flesh |
| Canned ham | Fecal streptococci | Cheesy off-flavors caused by fecal streptococci that survive heat process, especially in large 15-lb packs |
| Wilshire bacon | Bacteria | Development of sulfide off-flavor caused by contamination of bacon brines by foreign bacteria |
| Minced beef pies | Lactic acid bacteria | Contamination of cooked beef mix by lactic acid bacteria |
| Sausages | Lactic acid bacteria | Souring of sausages caused by selective growth of lactic acid bacteria in presence of sulfite |
| Ice Cream | Bacteria | Smoky defect noted in rework. Conversion of vanillin to guaiacol and 2-ethoxyphenol |

*Source*: From Saxby, M.J., *Food Taints and Off-Flavors*, 2nd ed., Blackie Acad. Prof., London, 1996, p. 326, and Goldenberg, N., H.R. Matheson, *Chem. Ind.* (London), (13), p. 551, 1975. With permission.

## 7.7 SUMMARY

This chapter briefly discusses some of the OF that have been found to occur in foods. Unfortunately, the occurrence of OF is quite common and the causes are very diverse. It is impossible to present a detailed discussion of all of the OF reported in the literature. In some cases, the source of the OF has never been found. Despite the sophisticated instrumentation available to the flavor chemist, identification of OF in foods is literally like looking for a needle (one or two OF compounds) in a haystack (the normal flavor profile of the food). When one appreciates the very low sensory thresholds one may encounter in OF flavor compounds (e.g., bis-(2-methyl-3-furyl)-disulfide, 2 parts in $10^{12}$ parts water [151]), the task of finding and then identifying the OF becomes very formidable. Hopefully, this chapter provides a better appreciation for the diversity and the host of sources of OF in foods.

## REFERENCES


1. Whitfield, F.B., K.J. Shaw, Analysis of off-flavors, in *Progress in Flavour Research*, J. Adda, Ed., Elsevier, Amsterdam, 1985, p. 221.
2. Saxby, M.J., *Food Taints and Off-Flavors*, 2nd ed., Blackie Acad. Prof., London, 1996, p. 326.
3. Reineccius, G.A., Flavor issues in maintaining freshness, in *Freshness and Shelf Life of Foods*, K.R. Cadwallader, H. Weenen, Eds., Amer. Chem. Soc., Washington, D.C., 2003, p. 42.
4. Goldenberg, N., H.R. Matheson, Off-flavors in foods: summary of experience, 1948–74, *Chem. Ind.* (London), (13), p. 551, 1975.
5. Kilcast, D., Sensory evaluation of taints and off-flavours, in *Food Taints and Off-Flavours*, M.J. Saxby, Ed., Blackie Acad. Prof., London, 1996, p. 1.
6. Thomson, D.M.H., The sensory characteristics of three compounds which may contribute to boar taint, in *Progress in Flavour Research*, J. Adda, Ed., Elsevier, Amsterdam, 1984, p. 97.
7. Jeon, J., Undesirable flavors in dairy products, in *Food Taints and Off-Flavours*, M.J. Saxby, Ed., Blackie Acad. Prof., London, 1996, p. 139.
8. Bassette, R., D.Y.C. Fung, V.R. Mantha, Off-flavors in milk, *CRC Crit. Reviews Food Sci. Nutr.*, 24, p. 1, 1986.
9. Curtis, R.F., C. Dennis, J.M. Gee, M.G. Gee, N.M. Griffiths, D.G. Land, J.L. Peel, D. Robertson, Chloroanisoles as a cause of musty taint in chickens and their microbiological formation from chlorophenols in broiler house litter, *J. Sci. Food Agric.*, 25, p. 811, 1974.
10. Saxby, M.J., S. Wragg, Contamination of cocoa liquor by chlorophenols, *Chem. Ind.*, p. 310, 1985.
11. Whitfield, F.B., T.H. Ly Nguyen, C.R. Tindale, Shipping container floors as sources of chlorophenol contamination in non-hermetically sealed foods, *Chem. Ind.*, p. 458, 1989.
12. Tindale, C.R., F.B. Whitfield, Production of chlorophenols by the reaction of fireboard and timber components with chlorine-based cleaning agents, *Chem. Ind.*, p. 835, 1989.
13. Pearce, T.J.P., J.M. Peacock, F. Aylward, D.R. Haisman, Catty odours in food: reactions between hydrogen sulohide and unsaturated ketones, *Chem. Ind.*, p. 1562, 1967.
14. Patterson, R.L.S., Catty odours in food: confirmation of the identity of 4-mercapo-4-methylpentan-2-one by gas chromatography and mass spectrometry, *Chem. Ind.*, p. 48, 1969.
15. Patterson, R.L.S., Catty odours in food: their production in meat stores from mesityl oxide in paint solvents, *Chem. Ind.* (London), p. 548, 1968.
16. Anderton, J.I., J.B. Underwood, *Catty Odours in food*, BFMIRA Tech. Circular No. 407, 1968.
17. Reineccius, G.A., Symposium on meat flavor: off-flavors in meat and fish — a review, *J. Food Sci.*, 44 (1), p. 12, 1979.

18. Vale, G.L., G.S. Sidhu, W.A. Montgomery, A.R. Johnson, Studies on kerosine-like taint in mullet (*Mugil cephalus*): 1. General nature of the taint, *J. Sci. Food Agric.*, 21, p. 429, 1970.
19. Shipton, J., J.H. Last, K.E. Murrai, G.L. Vale, Studies on kerosine-like taint in mullet: 2. Chemical nature of the volatile constituents, *J. Sci. Food Agric.*, 21, p. 433, 1970.
20. Bemelmans, J.M.H., H.J.A. den Braber, Investigation of an iodine-like taste in herring from the Baltic Sea, *Wat. Sci. Tech.*, 15, p. 105, 1983.
21. Lovell, R.T., L.A. Sackey, Absorption by channel catfish of earthy-musty flavor compounds synthesized by cultures of blue-green algae, *Trans. Am. Fisheries Soc.*, 102, p. 774, 1973.
22. Nichols, J.D., R.D. Lacewell, A marketing system — the step beyond production, *Am. Fish Farm. World Aquaculture News*, 2, p. 18, 1971.
23. Brewer, F., Catfish experiments in Kansas, *Kansas Fish and Game*, 29, p. 5, 1972.
24. Yurkowski, M., J.A.L. Tabachek, Identification, analysis and removal of geosmin from muddy flavored trout, *J. Fish Res. Board Can.*, 31, p. 1851, 1974.
25. Maligalig, L.L., J.F. Caul, R. Bassette, O.W. Tiemeir, Flavoring live channel catfish (*Icalurus punctatus*) experimentally: effects of concentration and exposure time, *J. Food Sci.*, 40, p. 1242, 1975.
26. Jirawan, Y., Athapol, N., Geosmin and off-flavor in Nile tilapia (*Oreochromis niloticus*), *J. Aquatic Food Product Tech.*, 9(2), p. 29, 2000.
27. Jackson, A., B. Hodgson, P. Torline, A. de Kock, L. van der Lunde, M. Stewart, Beer taints associated with unusual water supply conditions, *Tech. Q. Master Brew. Am. Assoc.*, 31(4), p. 117, 1994.
28. Saxby, M.J., M.A. Stephens, J.P. Chaytor, Detection and prevention of chlorophenol taints in liquid milk, *Leatherhead Food RA Food Ind. J.*, p. 416, 1983.
29. Gough, A.G.E., Pesticides and brewing raw materials, *Fermentnaya i Spirtovaya Promyshlennost'*, 3(6), p. 373, 1990.
30. Wenn, R.V., G.E. Macdonell, R.E. Wheeler, A Detailed Mechanism for the Formation of a Chlorophenolic Taint in Draught Beer Dispense System: The Rationale for Taint Persistence and Ultimate Removal, Proc. 21st Congress European Brew. Convention, Madrid, 1987.
31. Tice, P., Packaging material as a source of taints, in *Food Taints and Off-Flavours*, M.J. Saxby, Ed., Blackie Acad. Prof., London, 1996, p. 226.
32. Koszinowski, J., O. Piringer, Evaluation of off-odors in food packaging: the role of conjugated unsaturated carbonyl compounds, *J. Plastic Film Sheeting*, 2, p. 40, 1986.
33. Risch, S.J., C.T. Ho, Eds., *Flavor Chemistry*, Amer. Chem. Soc., Washington, D.C., 2000, p.180.
34. Nielsen, T., M. Jaegerstad, Flavour scalping by food packaging, *Trends Food Sci. Tech.*, 5(11), p. 353, 1994.
35. Askar, A., Flavor changes during processing and storage of fruit juices. II. Interaction with packaging materials, *Fruit Processing*, 9(11), p. 432, 1999.
36. Font-i-Furnols, M.G., L. Serra, X. Angels-Rius, M. Angels-Oliver, Sensory characterization of boar taint in entire male pigs, *J. Sensory Studies*, 15(4), p. 393, 2000.
37. Oliver, M., M. Font-i-Furnols, M. Gispert, The acceptability of pork according to consumer sensitivity to androstenone, *Eurocarne*, 10(84), p. 56, 2000.
38. Griffiths, N.M., R.L.S. Patterson, Human olfactory responses to 5 alpha-androst-16-ene-3-one, principal component of boar taint, *J. Sci. Food Agric.*, 21, p. 4, 1970.
39. Patterson, R.L.S., 5-Alpha-androst-16-ene-3-one, compound responsible for taint in boar fat, *J. Sci. Food Agric.*, 19, p. 31, 1968.

40. Beery, K.E., J.D. Sink, S. Patton, J.H. Ziegler, Characterization of the sex odor components in porcine adipose tissue, *J. Amer. Oil Chem. Soc.*, 46, p. 439A, 1969.
41. Hansson, K.E., K. Lundsorm, S. Fjelkner-Modig, J. Persson, The importance of androstenone and skatole for boar taint, *Swed. J. Agr. Ras.*, 10, p. 167, 1980.
42. Rius, M.A., J.A. Garcia-Regueiro, Skatole and indole concentrations in longissimus dorsi and fat samples of pigs, *Meat-Sci.*, 59(3), p. 285, 2001.
43. Heydanek, M.G., Tracing the origin of off-flavors in breakfast cereal, *Anal. Chem.*, 49, p. 901a, 1977.
44. Heydanek, M.G., G. Woolford, L.C. Baugh, Premiums and coupons as a potential source of objectionable flavor in cereal products, *J. Food Sci.*, 44(3), p. 850, 1979.
45. Passey, N., Off-flavors from packaging materials in food products, some case studies, in *Instrumental Analysis of Foods, Recent Progress*, G. Charalambous, G. Inglett, Eds., Academ. Press, New York, 1983, p. 413.
46. Hardwick, W.A., Ultimate flavor influence on canned beer caused by lubricants employed in two-piece can manufacture, *Brew Dig.*, 60(6), p. 22, 1985.
47. Gorrin, R., T.R. Pofahl, W.R. Croasmun, Identification of the musty component from an off-odor packaging film, *Anal. Chim.*, 59(18), p. 1109A, 1987.
48. Franz, R., S. Kluge, A. Lindner, O. Piringer, Cause of catty odour formation in packaged food, *Packaging Tech. Sci.*, 3(2), p. 89, 1990.
49. Wong, E., L.N. Nixon, C.B. Johnson, Volatile medium chain fatty acids and mutton flavor, *J. Agric. Food Chem.*, 23(3), p. 495, 1975.
50. Wong, J.W., S.E. Ebeler, R. Rivkah-Isseroff, T. Shibamoto, Analysis of malondialdehyde in biological samples by capillary gas chromatography, *Anal. Biochem.*, Jul. 220(1), p. 73, 1994.
51. Jamora, J.J., K.S. Rhee, Flavor of lamb and mutton, in *Quality Attributes of Muscle Foods*, Y.L. Xiong, C.T. Ho, F. Shahidi, Eds., Kluwer Academ. Publ., New York, 1999, pp. 135–145.
52. Bemelmans, J.M.H., M. Noever de Brauw, Chloroanisoles as off-flavor components in eggs and broilers, *J. Agric. Food Chem.*, 22(6), p. 1137, 1974.
53. Crawford, L., M.J. Kretsch, D.W. Peterson, A.L. Lilyblade, The remedial and preventive effect of dietary alpha-tocopherol on the development of fishy flavor in turkey meat, *J. Food Sci.*, 40(4), p. 751, 1975.
54. Crawford, L., M.J. Kretsch, GC-MS identification of the volatile compounds extracted from roasted turkeys fed a basal diet supplemented with tuna oil: some comments on fishy flavor, *J. Food Sci.*, 41(6), p. 1470, 1976.
55. Motohiro, T., Studies on petroleum odor in canned chum salmon, *Memoirs Faculty Fish.*, Hokkaido Univ., 10, p. 1, 1962.
56. Thaysen, A.C., The origin of an earthy muddy taint in fish: I. The nature and isolation of the taint, *Ann. Appl. Biol.*, 23, p. 99, 1936.
57. Maligalig, L.L., J.F. Caul, O.W. Tiemeier, Aroma and flavor of farm-raised channel catfish: effects of pond condition, storage and diet, *Food Prod. Devel.*, 7(4), p. 86, 1973.
58. Whitfield, F., J.H. Last, K.J. Shaw, C.R. Tindale, 2,6-Dibromophenol: the cause of an iodoform-like off-flavour in some Australian crustacea, *J. Sci. Food Agric.*, 46, p. 29, 1988.
59. Boyle, J., R.C. Lindsay, D.A. Stuiber, Bromophenol distribution in salmon and selected sea foods of fresh and saltwater origin, *J. Food Sci.*, 57, p. 918, 1992.
60. Whitfield, F., K.J. Shaw, D. Svoronos, Effect of natural environment on the flavour of sea foods: the flavour of *Girella tricuspidata*, *Devel. Food Sci.*, p. 417, 1994.

61. Park, R.J., J.L. Corbett, E.P. Furnival, Flavour differences in meat from lambs grazed on lucerne (Medicago sativa) or phalaris (*Phalaris tuberosa*) pastures, *J. Agric. Sci.*, 78(1), p. 47, 1972.
62. Park, R.J., R.A. Spurway, J.L. Wheeler, Flavour differences in meat from sheep grazed on pasture or winter forage crops, *J. Agric. Sci.*, 78(1), p. 53, 1972.
63. Melton, S.L., Effects of feeds on flavor of red meat: a review, *J. Animal Sci.*, 68 (12), p. 4421, 1990.
64. Kochhar, S., Oxidative pathways to the formation of off-flavours, in *Food Taints and Off-Flavours*, M. Saxby, Ed., Blackie Acad. Prof., London, 1996, p. 168.
65. Grosch, W., Lipid oxidation products and flavour, in *Food Flavours, Part A: Introduction,* I.D. Morton, A.J. MacLeod, Eds., Elsevier, New York, 1982, p. 325.
66. Labuza, T.P., Kinetics of lipid oxidation in foods, Crit. *Rev. Food Technol.*, 2, p. 355, 1971.
67. Chan, H., Photosensitized oxidation of unsaturated fatty acid methyl esters: the identification of different pathways, *J. Amer. Oil Chem. Soc.*, 54, p. 100, 1977.
68. Foote, C., Photosensitized oxidation and singlet oxygen: consequences in biological systems, in *Free Radicals in Biology*, W.A. Pryor, Ed., Academ. Press, New York, 1976, p. 85.
69. Bolland, J., G. Gee, Kinetic studies in the chemistry of rubber and related materials: III. Thermo chemistry and mechanisms of olefin oxidation, *Trans. Faraday Soc.*, 42, p. 244, 1946.
70. Keeney, M., Secondary degradation products, in *Lipids and their Oxidation*, R.O. Sinnhuber, Ed., AVI Pub. Co., Westport, 1962, p. 79.
71. Stark, W., D.A. Forss, A compound responsible for metallic flavour in dairy products: I. Isolation and identification, *J. Dairy Res.*, 29, p. 173, 1962.
72. Labuza, T.P., J.F. Maloney, M. Karel, Autoxidation of methyl linoleate in freeze-dried model system: II. Effect of water on cobalt catalyzed oxidation, *J. Food Sci.*, 31, p. 885, 1966.
73. Dutton, H.J., A.W. Schwab, H.A. Moser, J.C. Cowan, The flavour problem of soybean oil: IV. Structure of compounds counteracting the effect of prooxidant metals, *J. Amer. Oil Chem. Soc.*, 25, p. 385, 1984.
74. Mendenhall, V.T., Oxidative rancidity in raw fish fillets harvested from the Gulf of Mexico, *J. Food Sci.*, 37, p. 547, 1972.
75. Ayers, J.E., M.J. Fishwick, D.G. Land, T. Swain, Off-flavor in dehydrated carrot stored in oxygen, *Nature*, 203, p. 81, 1964.
76. Tims, M.J., B.M. Watts, Protection of cooked meats with phosphate, *Food Tech.*, 12, p. 240, 1958.
77. Mielche, M., G. Bertelsen, Approaches to the prevention of warmed-over flavor, *Trends Food Sci. Tech.*, 5(10), p. 322, 1994.
78. Grosch, W., C. Konopka, H. Guth, Characterization of off-flavors by aroma extract dilution analysis, in *Lipid Oxidation in Food*, A.J. St. Angelo, Ed., Amer. Chem. Soc., Washington, D.C., 1992, p. 266.
79. Kerler, J., W. Grosch, Odorants contributing to warmed-over flavor (WOF) of refrigerated cooked beef, *J. Food Sci.*, 61(6), p. 1271, 1996.
80. Smouse, T., S.S. Chang, A systematic characterization of the reversion flavour of soya bean oil, *J. Amer. Oil Chem. Soc.*, 44, pp. 509–514, 1967.
81. Badings, H., Cold Storage defects in butter and their relation to the autoxidation of unsaturated fatty acids, *Neth. Milk Dairy*, 24, p. 147, 1970.

82. Forss, D.A., Odor and flavor compounds from lipids, in *Progress in Chemistry of Fats and other Lipids*, R.T. Holman, Ed., Pergamon Press, Oxford, 13, 177–258, 1973.
83. Kochhar, S., M.L. Meara, A survey of the literature on oxidative reactions in edible oils as it applies to the problem of off-flavours in foodstuffs, *Leatherhead Food RA Scient. Technol. Survey*, p. 87, 1975.
84. Swoboda, P.A.T., K.E. Peers, Metallic odour caused by vinyl ketones formed in the oxidation of butterfat: the identification of octa-1-cis-5-dien-3-one, *J. Sci. Food Agric.*, 28(11), p. 1019, 1977.
85. Chang, S.S., G. Shen, J. Tang, Q. Jin, H. Shi, J.T. Carlin, C. Ho, Isolation and identification of 2-pentenylfurans in the reversion flavor of soybean oil, *J. Am Oil Chem. Soc.*, 60(3), p. 553, 1983.
86. Grosch, W., Enzymic formation of aroma compounds from lipids, *Lebensmittel. Gerichtliche Chemie*, 41(2), p. 40, 1987.
87. Hsieh, R.J., J.E. Kinsella, Lipoxygenase generation of specific volatile flavor carbonyl compounds in fish tissues, *J. Agric. Food Chem.*, 37 (2), p. 279, 1989.
88. Parks, O.W., D.P. Schwartz, Identification of O-aminoacetophenone as a flavour compound in stale dry milk, *Nature*, 202, p. 185, 1964.
89. Ramshaw, E.H., E.A. Dunstone, Volatile compounds associated with the off-flavour in stored casein, *J. Dairy Research*, 36, p. 215, 1969.
90. Askar, A., Flavor changes during processing and storage of fruit juices. I. Markers for processed and stored fruit juices, *Fruit Processing,* 9 (7), p. 236, 1999.
91. Shipe, W.F., R. Bassette, D.D. Deane, W.L. Dunkly, E.G. Hammond, W.J. Harper, DA. Kleyn, M.E. Morgan, J.H. Nelson, R.A. Scanlan, Off-flavors of milk: nomenclature, standards, and bibliography, *J. Dairy Sci.*, 61, p. 855, 1978.
92. Patton, S., The mechanism of sunlight flavor formation in milk with special reference to methionine and riboflavin, *J. Dairy Sci.*, 37, p. 446, 1954.
93. Allen, C., O.W. Parks, Evidence for methional in skim milk exposed to sunlight, *J. Dairy Sci.*, 58, p. 1609, 1975.
94. Samuelsson, E.G., Experiments on sunlight flavour in milk — S35-labeled milk, *Milchwissenchaft*, 17, p. 401, 1962.
95. Jung, M.Y., S.H. Yoon, H.O. Lee, D.B. Min, Singlet oxygen and ascorbic acid effects on dimethyl disulfide and off-flavor in skim milk exposed to light. *J. Food Sci.,* 63(3), pp. 408–412, 1998.
96. Charpentier, N., A. Maujean, Sunlight flavours in champagne wines, in *Flavour,* P. Schreier, Ed., de Gruyter, New York. p. 609, 1981.
97. Kuroiwa, Y., N. Hashimoto, Composition of sun-struck flavor substance and mechanism of its evolution, *Amer. Soc. Brew. Chem.*, 19, p. 28, 1961.
98. Bennett, S., J. Murray, Beyond the gate — a study of in-trade quality, *Tech. Quart.*, Master Brewers Assoc. Amer., 37(4), p. 465, 2000.
99. Hashimoto, N., Stale flavor derived from higher alcohols and isohumulones in beer, *Report Res. Lab. Kirin Brew*, 19, p.1, 1976.
100. Tressl, R., D. Bahri, M. Kossa, Formation of off-flavor components in beer, in *The Analysis and Control of Less Desirable Flavors in Foods and Beverages*, G. Charalambous, Ed., Acad. Press, New York, 1980, p. 293.
101. Irwin, A.J., L. Bordeleau, R.L. Barker, Model studies and flavor threshold determination of 3-methyl-2-butene-1-thiol in beer, *J. Amer. Soc. Brew. Chem.*, 51 (1), p. 1, 1993.
102. Rackis, J.J., D.H. Honig, D.J. Sessa, H.A. Moser, Lipoxygenase and peroxidase activities of soybeans as related to the flavor profile during maturation, *Cereal Chem.*, 49, p. 586, 1972.

103. Sessa, D.J., J.J. Rackis, Lipo-derived flavors of legume protein products, *J. Amer. Oil Chem. Soc.,* 54, p. 468, 1977.
104. Kalbrener, J.E., K. Warner, A.C. Eldridge, Flavors derived from linoleic and linolenicacid hydroperoxide, *Cereal Chem.*, 51, p. 406, 1974.
105. Whitfield, F.B., J. Shipton, Volatile carbonyls in stored unbalanced frozen peas. *J. Food Sci.*, 31, p. 328, 1966.
106. Shipe, W.F., G.F. Senyk, R.A. Ledford, D.K. Bandler, E.T. Wolff, Flavor and chemical evaluations of fresh and aged market milk, *J. Dairy Sci.*, 63(Suppl. 1), p. 43, 1980.
107. Chandran, R.C., K.M. Shahani, Milk lipases: a review, *J. Dairy Sci.*, 47, p. 471, 1964.
108. Kwak, H., I.J. Jeon, S.K. Perng, Statistical patterns of lipase activities on the release of short-chain fatty acids in Cheddar cheese slurries, *J. Food Sci.*, 54, p. 1559, 1989.
109. Murry, T.K., B.E. Baker, Studies on protein hydrolysis I — preliminary observations on the taste of enzymic protein hydrolysates, *J. Sci. Food Agric.*, 3, p. 470, 1952.
110. Fujimaki, M., M. Yamashita, Y. Okazawa, S. Arai, Diffusible bitter peptides in peptic hydrolyzate of soybean protein, *Agric. Biol. Chem.*, 32, p. 794, 1968.
111. Noguchi, M., S. Arai, H. Kato, M. Fujimaki, Applying proteolytic enzymes on soybean: 2. Effect of Aspergillopeptidase A preparation on removal of flavor from soybean products, *J. Food Sci.*, 35, p. 211, 1970.
112. Arai, S., M. Yamashita, H. Kato, M. Fujimaki, Applying proteolytic enzymes on soybean: Part V. A nondialzable bitter peptide hydrolyzate of soybean protein and its bitterness in relation to the chemical structure, *Agric. Biol. Chem.*, 34, p. 729, 1970.
113. Arai, S., The bitter flavor due to peptides or protein hydrolysates and its control by bitterness — masking with acidic oligopeptides, in *The Analysis and Control of Less Desirable Flavors in Foods and Beverages*, G. Charalambous, Ed., Acad. Press, New York. p. 1, 1980.
114. Wieser, H., H.D. Belitz, Bitter peptides isolated from corn protein zein by hydrolysis with pepsin, *Zeitschrift fuer Lebensmittel-Untersuchung und -Forschung,* 159(6), pp. 329–336, 1975.
115. Matoba, T., C. Nagayasu, R. Hayashi, T. Hata, Bitter peptides from tryptic hydrolysate of casein, *Agric. Biol. Chem.*, 33, p. 1662, 1969.
116. Matoba, T., R. Hayashi, T. Hata, isolation of bitter peptides from tryptic hydrolysate of casein and their chemical structure, *Agric. Biol. Chem.*, 34, p. 1235, 1970.
117. Clegg, K.M., C.L. Lim, W. Manson, The structure of a bitter peptide derived from casein by digestion with papain, *J. Dairy Res.*, 41, p. 283, 1974.
118. Huber, L., H. Klostermeyer, Isolierung und Identifizierung eines Bitterstoffes aus Butterkase, *Milchwiss*, 29, p. 449, 1974.
119. Hamilton, J.S., R.D. Hill, A bitter peptide from cheddar cheese, *Agric. Biol. Chem.*, 38, p. 375, 1974.
120. Dulley, J.R., B.J. Kitchen, Phosphopeptides and bitter peptides produced by rennet, *Aust. J. Dairy Technol.*, 3, p. 10, 1972.
121. Harwalkar, V.R., J.A. Elliot, Isolation of bitter and astringent fractions from cheddar cheese, *J. Dairy Sci.*, 54, p. 8, 1971.
122. Emmons, D.B., W.A. McGugan, J.A. Elliot, Effects of strain of starter culture and of manufacturing procedure on bitterness and protein breakdown in cheddar cheese, *J. Dairy Sci.,* 45, p. 332, 1962.
123. Harwalkar, V.R., Characterization of an astringent fraction from cheddar cheese, *J. Dairy Sci.*, 55, p. 735, 1972.
124. Richardson, B.C., L.K. Creamer, Casein proteolysis and bitter peptides in cheddar cheese, *New Zealand J. Dairy Sci. Technol.*, 8, p. 46, 1973.

125. Lawrence, R.C., H.A. Heap, G. Limsowtin, A.W. Jarvis, Symposium: Research and development trends in natural cheese manufacturing and ripening cheddar cheeses starters: current knowledge and practices of phage characteristics and strain selection, *J. Dairy Sci.*, 61, p. 1181, 1978.
126. Sullivan, J.J., M. Lynette, J.J. Rood, G.R. Jago, The enzymatic degration of bitter peptides by starter streptococci, *Aust. J. Dairy Technol.*, 3, p. 20, 1973.
127. Springett, M., Formation of off-flavours due to microbiological and enzymic action, in *Food Taints and Off-Flavours*, M. Saxby, Ed., Blackie Acad. Prof., London, 1996, p. 274.
128. Weihrauch, J., Lipids of milk: deterioration, in *Fundamentals of Dairy Chemistry*, N. Wong, Ed., Van Nostrand Reinhold, New York, 1988, p. 215.
129. Shipe, W.F., Analysis and control of milk flavor, in *The Analysis and Control of Less Desirable Flavors in Foods and Beverages*, G. Charalambous, Ed., Academ. Press, New York, 1980, p. 201.
130. Davies, W.L., E. Gill, Investigation on fishy flavors, *Chem. Ind.* (London), 55, p. 1415, 1936.
131. Stansby, M.E., Speculations on fishy odors and flavors, *Food Technol.*, 16, p. 28, 1962.
132. Watson, D.W., Studies on fish spoilage: the bacterial reduction of trimethylamine oxide, *J. Fish Res. Bd.* (Canada), 4, p. 252, 1939.
133. Herbert, R.A., J.M. Shewan, Roles played by bacteria and autolytic enzymes in the production of volatile sulphides in spoiling North Sea cod (Gadus morhua), *J. Sci. Food Agric.*, 27, p. 89, 1976.
134. Adams, R.E., I. Farber, D. Lerke, Bacteriology of spoilage of fish muscle: 2. Incidence of spoilers during spoilage, *Appl. Microbiol.*, 12, p. 277, 1964.
135. Bradfield, A. A.H. Duthie, *Vt. Agric. Expt. Sta. Bull.*, p. 645. Cited by #129.
136. Morgan, M.E., The chemistry of some microbially induced flavor defects in milk and dairy foods. *Biotechnol. Bioengin.*, 18 (7), p. 953, 1976.
137. Harris, N.D., C. Karahadian, R.C. Lindsay, Musty aroma compounds produced by selected molds and actinomycetes on agar and whole wheat bread, *J. Food Protection*, 49 (12), p. 964, 1986.
138. Whitfield, F., K.J. Shaw, A.M. Gibson, D.C. Mugford, An earthy off-flavour in wheat flour: geosmin produced by *Streptomyces griseus*, *Chem. Ind.*, p. 841, 1991.
139. Buttery, R.G., D.G. Guadagni, L.C. Ling, Geosmin, a musty off-flavor of dry beans, *J. Agric. Food Chem.*, 24(2), p. 419, 1976.
140. Hsieh, T.C.Y., U. Tanchotikul, J.E. Matiella, Identification of geosmin as the major muddy off-flavor of Louisiana brackish water clam (*Rangia cuneata*), *J. Food Sci.*, 53(4), p. 1228, 1988.
141. Marth, E.H., M. Constance, C.M. Capp, L. Hasenzahl, H.W. Jackson, H.V. Hussong, Degradation of potassium sorbate by Penicillium species, *J. Dairy Sci.*, 49, p. 1197, 1966.
142. Horwood, J.F., G.T. Lloyd, E.H. Ramshaw, W. Stark, An off-flavor associated with the use of sorbic acid during Feta cheese maturation, *Aust. J. Dairy Technol.*, 36, p. 38, 1981.
143. Daley, J.D., G.T. Lloyd, E.H. Ramshaw, W. Stark, Off-flavours related to the use of sobic acid as a food preperation, *CSIRO Food Res. Q.*, 46, p. 59, 1986.
144. Crowell, E.A., J.F. Guymon, Wine constituents arising from sorbic acid addiction, and identification of 2-ethoxyhexa-3,5-diene as source of geranium-like off-odour, *Amer. J. Viticult.*, 26, p. 97, 1975.
145. Saxby, M.J., A survey of chemicals causing taints and off-flavours in foods, in *Food Taints and Off-Flavors*, M.J. Saxby, Ed., Blackie Acad. Prof., London, p. 41, 1996.

146. Carey, R., Natural cork: the new closure for wine bottles, *Vineyard Winery Management*, 14, p. 5, 1988.
147. Whitfield, F.B., J.H. Last, C.R. Tindale, Skatole, indole and p-cresol: components in off-flavoured frozen french fries, *Chem. Ind.*, p. 662, 1982.
148. Miller, A., R.A. Scanlan, J.S. Lee, J.M. Libbey, M.E. Morgan, Volatile compounds produced in sterile fish muscle (*Sebastes melanops*) by Pseudomonas perolens, *J. Appl. Microbiol.*, 225, p. 257, 1973.
149. Strauss, C.R., T. Heresztyn, 2-Acetyltetrahydropyridines — a cause of the mousy taint in wine, *Chem. Ind.*, p. 109, 1984.
150. Mottram, D.S., R.L.S. Patterson, E. Warrilow, 2,6-Dimethyl-3-methoxypyrazine: a microbiologically-produced compound with an obnoxious musty odor, *Chem. Ind.* (London, U.K.), p. 448, 1984.
151. Buttery, R.G.H., W.F. Haddon, R.M. Seifert, J.G. Turnbaugh, Thaimin odor and bis-(2-methyl-3-furyl) disulfide, *J. Agric. Food Chem.*, 32, p. 674, 1984.
152. Forss, D.A. Review of the progress of dairy science: mechanisms of formation of aroma compounds in milk and milk products, *J. Dairy Res.*, 46, p. 691, 1979.
153. Fan, L.L., J.-Y. Tang, A. Wohlman, Investigation of 1-decyne formation in cottonseed oil fried foods, *J. Amer. Oil Chem. Soc.*, 60, p. 1115, 1983.
154. Borle, F., R. Sieber, J.O. Bosset, Photo-oxidation and photoprotection of foods, with particular reference to dairy products: an update of a review article (1993–2000), *Sciences des Aliments*, 21(6), p. 571, 2002.
155. Hsieh, O.A.L., A.S. Huang, S.S. Chang, Isolation and identification of objectionable volatile flavor compounds in defatted soybean flour, *J. Food Sci.*, 47(1), p. 16, 1981.
156. Al-Shabibi, M.A., E.H. Langner, J. Tobias, S.L. Tuckey, Effect of added fatty acids on the flavor of milk, *J. Dairy Sci.*, 47(3), p. 295, 1964.
157. Kinsella, J.E., What makes fat important in flavors, *Amer. Dairy Review*, 31(5), p. 36, 1969.

# Part II

## Flavor Technology

# 8 Flavoring Materials

## 8.1 INTRODUCTION

The flavor industry has a large number of flavoring materials available to it for use in the creation of a flavoring. These materials consist of various plant and animal materials (and derivatives thereof), products of fermentation and enzymology, as well as synthetic chemicals. There has been a gradual change in materials used over time due to the discovery of new materials (e.g., exotic plants), developments in processing technology (e.g., distillation, extraction, and adsorption), biotechnology (enzymology and fermentation), process chemistry (process flavors), and organic synthesis (e.g., chiral syntheses). This chapter will present an overview of these materials as well as their means of manufacture.

### 8.1.1 DEFINITIONS

#### 8.1.1.1 Flavoring

The Code of Practice of the International Organization of the Flavor Industry (IOFI), [1] of which the Flavor and Extract Manufacturers Association of the United States (FEMA) [2] is a member, defines flavorings as "Concentrated preparations, with or without flavor adjuncts [Food additives and food ingredients necessary for the production, storage, and application of flavorings as far as they are nonfunctional in the finished food] required in their manufacture, used to impart flavor, with the exception of salt, sweet, or acid tastes. They are not intended to be consumed as such." IOFI further defined the flavor ingredients that were permitted to be used in flavorings in 1990 as presented in Table 8.1 [3].

#### 8.1.1.2 Natural Flavoring

The definition of natural flavoring in the U.S., which can be found in the Code of Federal Regulations (CFR) [4], is: "The term natural flavor or natural flavoring means the essential oil, oleoresin, essence or extractive, protein hydrolysate, distillate, or any product of roasting, heating or enzymolysis, which contains the flavoring constituents derived from a spice, fruit or fruit juice, vegetable or vegetable juice, edible yeast, herb, bark, bud, root, leaf or similar plant material, meat, seafood, poultry, eggs, dairy products, or fermentation products thereof, whose significant function in food is flavoring rather than nutritional. Natural flavors include the natural essence or extractives obtained from plants listed in Secs. 182.10, 182.20, 182.40, and 182.50 and part 184 of this chapter, and the substances listed in Sec. 172.510 of this chapter." The sections referred to list Generally Recognized as Safe (GRAS) substances (Sections 182.xx, Generally Recognized as Safe) and "Natural flavoring

**TABLE 8.1**
**Definitions of Flavoring Substances Permitted in Flavorings**

*Flavor substance:* Defined chemical component with flavoring properties, not intended to be consumed as such.
*Natural flavoring substance:* Defined substance obtained by substance appropriate physical, microbiological, or enzymatic processes from a foodstuff or material of vegetable or animal origin as such or after processing by food preparation processes.
*Nature identical flavoring substance:* Flavoring substance obtained by synthesis or isolated through chemical processes from a natural aromatic raw material and chemically identical to a substance present in natural products intended for human consumption, either processed or not. (This category of flavoring substance does not exist in the U.S.)
*Artificial flavoring substance:* Flavoring substance, not yet substance identified in a natural product intended for human consumption, either processed or not.
*Flavoring preparation:* A preparation used for its flavoring properties that is obtained by appropriate physical, microbial, or enzymatic processes from a foodstuff or material of vegetable or animal origin, either as such or after processing by food preparation processes.
*Process flavor:* A product or mixture prepared for its flavoring properties that is produced from ingredients or mixtures of ingredients, which are themselves permitted for use in foodstuffs, or are present naturally in foodstuffs, or are permitted for use in process flavorings, by a process for the preparation of foods for human consumption. Flavor adjuncts may be added. This definition does not apply to flavoring extracts, processed natural food substances, or mixtures of flavoring substances.
*Smoke flavoring:* Concentrated preparation, not obtained from smoked materials, used for the purpose of imparting a smoke type flavor to foodstuffs. Flavor adjuncts may be added.

*Source:* From Ziegler, E., H. Ziegler, *Flavourings: Production, Composition, Applications and Regulations*, Wiley-VCH, New York, 1998. With permission.

substances and natural substances used in conjunction with flavors"(Section 172.510).

### 8.1.1.3 Artificial Flavoring

The definition of artificial flavoring in the U.S. can be found in the CFR [4] and is: "The term artificial flavor or artificial flavoring means any substance, the function of which is to impart flavor, which is not derived from a spice, fruit or fruit juice, vegetable or vegetable juice, edible yeast, herb, bark, bud, root, leaf or similar plant material, meat, fish, poultry, eggs, dairy products, or fermentation products thereof. Artificial flavor includes the substances listed in Sec. 172.515(b) and Sec. 182.60 of this chapter except where these are derived from natural sources." (Note: When the preceding quotation from the CFR refers to "this chapter," it means the chapter in the CFR, not in this book.) This definition basically states that if a flavoring substance is not natural, it is artificial.

## 8.2 NATURAL FLAVORING MATERIALS (PLANT SOURCES)

Plant materials, which are classified as foods themselves (e.g., raspberries, grapes, apples) may serve as flavorings substances; however, they tend to be quite weak in flavor strength and thus have limited application for food flavoring purposes unless they are a major component of the final food product (e.g., a fruit beverage, apple pie, etc.). Occasionally these "foods" are used to make an alcoholic infusion or extract of some type that provides a concentrated flavor form, but they are generally costly since the initial material is inherently low in flavor. Thus, these materials see limited application as food flavorings. The more common plant sources of flavoring materials include:

1. Herbs and spices, which are intrinsically highly aromatic and flavorful and may be used singly or as seasoning blends
2. Other aromatic plant materials used mainly as a source of essential oils (e.g., citrus peels)
3. Vanilla, which only develops its characteristic flavor profile after postharvest fermentation and curing
4. Coffee, tea, and cocoa, all of which require postharvest treatment to achieve an acceptable flavor profile and which are largely used to make beverages

These materials may be used as such (whole or ground/crushed) or processed to make a different flavor form such as an essential oil, oleoresin, or tincture.

## 8.2.1 Herbs and Spices

### 8.2.1.1 Definitions and Markets

Botanically, herbs are soft-stemmed plants the main stem of which dies down to ground level and either does not regrow (annuals), grows again in the following year only (biennials) or regrows each year (perennials). Culinary herbs are a restricted group of such plants together with some traditionally used nonherbs such as sage (sub-shrub) and bay laurel (the leaves of a large tree). All have been used for centuries in the seasoning of foods as well as many other domestic and commercial outlets.

Herbs may be used fresh or after dehydration. In the latter case it is usual to separate the leaves, floral parts, and seeds from the heavier, harder, and less aromatic stems by screening. Such herbs are called broken or rubbed, and it is in this form that they are generally sold for domestic and commercial use. Those of greatest commercial value are listed in Table 8.2. One can gain an appreciation for the volume of herbs used in the U.S. from Figure 8.1.

Spices are all other aromatic plant materials used in the flavoring or seasoning of foods, although this very broad definition has several exceptions (e.g., vanilla) which are not generally regarded as spices. Spices are usually only parts of plants. The use of spices in the U.S. has increased substantially in the past few years (Figure 8.2). This reflects the growth in sales of ethnic foods (e.g., Mexican, Asian, and Middle Eastern) in both the retail and food service areas. The major spices used in the food industry are listed in Table 8.3. Of these spices, pepper, salt, mustard, horseradish, and ginger are the most common. The top ten spices (dollar volume) are as follows: dehydrated onion and garlic, mustard seed, red peppers (except paprika), sesame seed, black pepper, paprika, cinnamon, cumin seed, white pepper, and oregano.

**TABLE 8.2**
**Principle Culinary Herbs**

| Herb | Botanical Source | Family | Part of Plant Used |
|---|---|---|---|
| Basil | *Ocimum basilicum* L. and other subspecies | Labiatae | Leaves and flowering tops |
| Bay laurel | *Laurus nobilis* L. | Lauraceae | Leaves |
| Lovage | *Levisticum officinale* Koch | Umbelliferae | Roots, leaves, and seeds |
| Marjoram | *Majorana hortensis* Moench. (= *Origanum majorana* L. ) | | Leaves and flowering tops |
| Mints | | | |
| Cornmint | *Mentha arvensis* L. var. *piperascens* Holmes and other varieties | Labiatae | Leaves distilled for essential oil |
| Peppermint | *Mentha piperita* L. | | Leaves distilled for essential oil |
| Spearmint | *Mentha spicata* (L) Huds var. *tenuis* (= M. *viridis* L.) and other varieties | | Leaves distilled for essential oil |
| Scotch mint | *Mentha cardiaca* Gerard & Baker | | |
| Oregano | | Labiatae | |
| Mexican sage | *Lippia* spp. usually *Lippia graveolens* HBK. | | Leaves |
| Greek | *Origanum* spp. mostly *Origanum vulgare* L. | | |
| Origanum (wild marjoram) | *Coridothymus capitatus* Richb (= *Thymus capitatus* Hoffm. & Link.) or *Origanum vulgare* L. or *Thymus masticina* L. and several other spp. dependent on source | | Leaves and floral parts |
| Rosemary | *Rosmarinus officinalis* L. | Labiatae | Leaves |
| Sage | | Labiatae | |
| Dalmatian | *Salvia officinalis* L. | | Leaves |
| Greek | *Salvia triloba* L. | | |
| Spanish | *Salvia lavandularfolia* (L) Vahl. | | |
| Savory | | Labiatae | |
| Summer | *Satureia hortensis* L. | | Leaves and flowering tops |
| Winter | *Satureia montana* L. | | |
| Tarragon (estragon) | *Artemisia dracunculus* L. | Compositae | Leaves |
| Thyme | *Thymus vulgaris* L; *T. zygis* L. and other spp. dependent on source | Labiatae | Leaves and floral parts |

*Source:* From Heath, H.B., G.A. Reineccius, *Flavor Chemistry and Technology*, Van Nostrand Reinhold, New York, 1986. With permission.

### 8.2.1.2 Historical Associations

The use of herbs and spices in the flavoring of foods is almost as old as man himself. Few other commodities have had such a profound effect upon world history, and their role in shaping the destiny of nations is a long and fascinating story. The spice

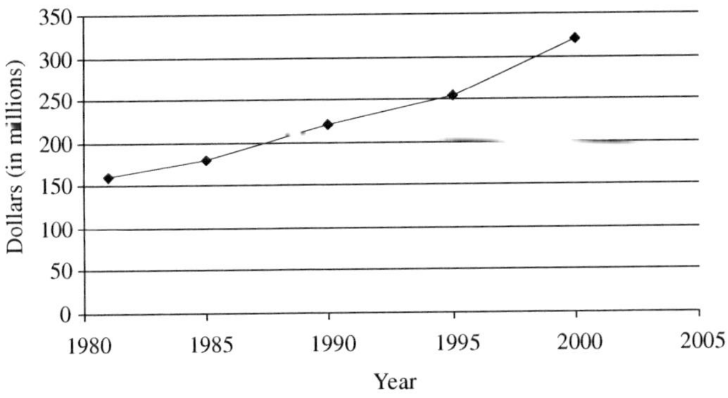


**FIGURE 8.1** Imports of major herbs into the U.S. (From Furth, P., *Summary of Market Trends and Herbs Consumption in The United States*, FFF Associates, 2001. With permission.)

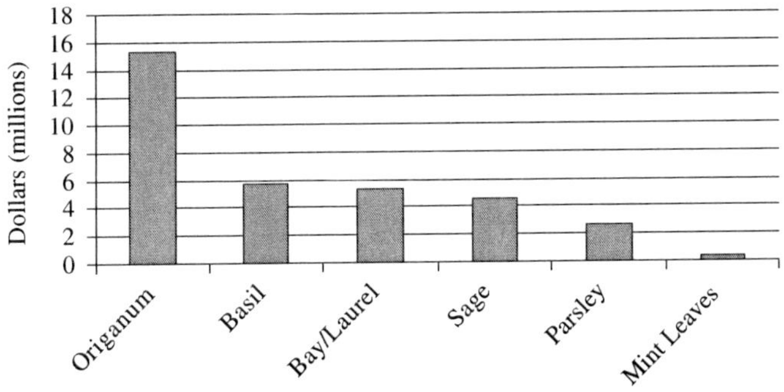


**FIGURE 8.2** Imports of major spices to the U.S. ($) from 1980 to 2000. (From Furth, P., *Summary of Market Trends and Herbs Consumption in The United States*, FFF Associates, 2001. With permission.)

trade is one of the oldest known to man and there is mention of spices such as cinnamon, cardamom, and ginger in the oldest manuscripts dating back to before 2000 B.C. The spice trade has always been extensive and profitable. The great wealth of King Solomon came from the traffic of the spice merchants, whose shrewd methods ensured the prosperity of the Middle East for centuries. By the tenth century the great cities of Alexandria and Venice were built and flourished. In the fifteenth century, great wealth was accrued from the commerce in spices, and bitter battles were fought to gain possession of valuable spice caravans and sea cargoes. The restless exploration of the Portuguese and the hazardous voyages of the great navigators such as Columbus and Magellan were all closely associated with the supply and commercial exploitation of these commodities. International rivalry reached unprecedented heights in the seeking out and annexing of spice growing areas and trading posts, for it was recognized that the country that controlled the spice trade was not only the richest but the most influential in the world.

The value of these natural flavoring materials, particularly black pepper, in converting otherwise monotonous diets into something appetizing was quickly

## TABLE 8.3
## Principle Spices Used in the Food Industry

| Spice | Botanical Source | Family | Part of Plant Used |
|---|---|---|---|
| Allspice (pimento Jamaican pepper) | *Pimenta dioica* (L) Merr. (= *Pimenta officinalis* Lindly) | Myrtacae | Fruit |
| Anise | *Pimpinella anisum* L. | Umbelliferae | Fruit |
| Anise; China star | *Illicium verum* Hook | Magnoliacae | Fruit |
| Capsicum | | | |
| Sweet peppers capsicum | *Capsicum annuum* L. | Solanaceae | Fruit |
| Cayenne (Tabasco chili) | *Capsicum minimum* Roxb. or C. | *frutescens* L. | |
| Caraway | *Carum carvi* L. | Umbelliferae | Fruit |
| Cardamom | *Elettaria cardamomum* (L) Maton var. *miniscula* Burkhill | Zingiberaceae | Seed |
| Cassia | *Cinnamomum cassia* (L) Blume | Lauraceae | Bark |
| Cinnamon | *Cinnamomum zeylanicum* Nees | Lauraceae | Bark |
| Saigon | C. *loureirii* Nees | | |
| Korintje | C. *burmanii* Blume | | |
| Clove | *Syzygium aromaticum* (L) Man. & Perryl. (= *Eugenia caryophylus* (Spreng) Bullock & Harrison; *E. caryophyllata Thunberg or E. Aromaticus* (L) Bailey) | Myrtaceae Flower bud | |
| Coriander | *Coriandrum sativum* L. Umbelliferae | Fruit | |
| Cumin | *Cuminum cyminum* L. | Umbelliferae | Fruit |
| Dill | *Anethum graveolens* L. | Umbelliferae | Fruit |
| Indian | *Anethum sowa* Roxb. | | |
| Fennel | *Foeniculum vulgare* Mill. | Umbelliferae | Fruit |
| Fenugreek | *Trigone/la foenum-graecum* L. | Leguminosae | Seed |
| Ginger | *Zingiber officinale* Roscoe | Zingiberaceae | Rhizome |
| Mace | *Myristica fragrans* Houttyn | Myristicaceae | Aril |
| Mustard | | Cruciferae | Seed |
| Black | *Brassica nigra* (L) Koch | | |
| White | *Sinapis alba* L. (= *Brassica alba* Boissier) | | |
| Indian | *Brassica juncea* (L) Cosson or Hook & Thorns. | | |
| Nutmeg | *Myristica fragrans* Houttyn | Myristicaceae | Seed |
| Paprika | *Capsicum annuum* L. | Solanaceae | Fruit |
| Parsley | *Petrosilinum sativum* Hoffm. or *P. crispum* (Mill) Nyman | Umbelliferae | Seed |
| Pepper black and white | *Piper nigrum* L. | Piperaceae | Fruit |
| Poppy | *Papaver somniferum* L. | Papaveraceae | Seed |
| Saffron | *Crucus sativus* L. | Iridaceae | Stigma |
| Mexican | *Carmanthus tinctorius* L. | Compositae | Florets |

**TABLE 8.3 (continued)**
**Principle Spices Used in the Food Industry**

| Spice | Botanical Source | Family | Part of Plant Used |
|---|---|---|---|
| Sesame | *Sesamum indicum* L. | Pedaliaceae | Seed |
| Turmeric (curmuma) | *Curcuma longa* L. | Zingiberaceae | Rhizome |

*Source:* From Heath, H.B., G.A. Reineccius, *Flavor Chemistry and Technology*, Van Nostrand Reinhold, New York, 1986. With permission.

appreciated and, in consequence, the European demand was almost insatiable; the prices were high and the profits enormous. In the days long before refrigeration and other modern methods for the preservation of fresh food commodities, what one could eat was dictated entirely by what was in season or could be preserved by the simple expedients of drying, cooking, or brining. The use of spices made such foods palatable, but there were then only a few who could afford the luxury.

Control of the spice trade passed through several hands over the ages. The initial supremacy of the Portuguese was replaced at the end of the sixteenth century by the Dutch who maintained their control of the spice trade until well into the eighteenth century. By the early nineteenth century Great Britain dominated this position, but today the U.S. is the major importer of these materials with New York as its trading center.

### 8.2.1.3 Classification of Herbs and Spices

It is possible to classify herbs and spices, and indeed other natural flavoring materials, in different ways depending upon the value of the classification to its application. Four methods used to classify herbs and spices are (a) botanical, (b) agronomic, (c) physiological, and (d) organoleptic. One could, of course, add alphabetical, which at least facilitates quick reference to available data. Of these classifications, classification by plant part or sensory properties are most useful to the flavorist. Classification by plant part (Table 8.4) has implications in manufacturing of essential oils or oleoresins, while a flavor character classification (perhaps by major constituent [Table 8.5] or sensory character [Table 8.6]) aids the product developer in selecting similar spices. The complexity of the sensory classification system depends on the specificity of the sensory attributes selected. For example, Arctander [7] proposed 88 distinguishable groups of sensorily related natural raw materials used in food flavorings and fragrances; of these, 22 groups involved herbs and spices. This is a very effective grouping for the trained flavorist and perfumer, but for most general purposes a much simpler grouping is quite adequate as an aid to the use of herbs and spices in blended seasonings. Such classifications are given in Tables 8.5 and Table 8.6. However, few are in complete agreement owing to the personal nature of sensory responses and the difficulties associated with the communication of findings. In spite of this, the system is of value and can readily be adapted to suit individual requirements.

**TABLE 8.4**
**Classification of Spices Based on Part of Plant Used**

| Part Used | Spice | Part Used | Spice |
|---|---|---|---|
| Leaf | Bay laurel | Stigma | Saffron |
| Fruit | Allspice (Pimento) Anise; Capsicum; Caraway; Celery; Coriander; Cumin; Dill; Fennel; Paprika; Pimento (Allspice) | Seed | Cardamom; Fenugreek; Mustard; Poppy; Sesame |
| Part fruit (aril) | Mace | Flower | Safflower |
| Bark | Cassia; Cinnamon | Flower bud | Clove |
| Root | Horseradish; Lovage | Rhizome | Ginger; Turmeric |

*Source:* From Heath, H.B., G.A. Reineccius, *Flavor Chemistry and Technology*, Van Nostrand Reinhold, New York, 1986. With permission.

**TABLE 8.5**
**Classification of Herbs by Main Flavor Component in Essential Oil**

| Essential Oil Contains | | | | | |
|---|---|---|---|---|---|
| Cineole | Thymol/Carvacrol | Alcohols/Esters | Thujone | Menthol | Carvone |
| Herb | | | | | |
| Bay laurel | Thyme | Oregano | Sweet basil | Peppermint | Spearmint |
| Rosemary | | Origanum | Marjoram | Cornmint | Dillweed |
| Spanish sage | | Savory | Tarragon | | |
| | | Wild Marjoram | | | |

*Source:* From Heath, H.B., G.A. Reineccius, *Flavor Chemistry and Technology*, Van Nostrand Reinhold, New York, 1986. With permission.

**TABLE 8.6**
**Classification of Spices by Sensory Associations**

| Pungent Spices | Aromatic Fruits or Seeds | Aromatic Barks | Phenolic Spices | Colored Spices |
|---|---|---|---|---|
| Capsicum, ginger | Nutmeg, mace (aril) | Cassia | Clove bud, allspice (pimento) | Paprika |
| Pepper (black and white) | Cardamom, fenugreek anise, fennel, caraway dill, Indian dill, celery root, lovage, parsley, cumin, coriander | Cinnamon | West Indian bay | Saffron |
| Mustard | | | Oils of: Clove stem and leaf; Cinnamon leaf | Safflower turmeric |
| Horseradish | | | | |

*Source:* From Heath, H.B., G.A. Reineccius, *Flavor Chemistry and Technology*, Van Nostrand Reinhold, New York, 1986. With permission.

Depending on their purpose, most books on herbs describe more than 100 different items that may be used in foods, for medicinal purposes, or in the preparation of toiletries and cosmetics either in their fresh form, as dried broken herbs, or as essential oils and/or extracts. The food industry uses only about 20 such plants and products made from them (Table 8.1).

#### 8.2.1.4 Flavor Character of Herbs

Herbs, for the most part, have a light and very distinctive aromatic character although they contain relatively low levels of essential oil. For culinary use, the whole herbaceous tops are harvested and may be used fresh or after drying. In the latter case it is usual to remove by mechanical screening the leaves and any floral parts or seeds as may be present from the heavier, harder, and less aromatic stems.

The flavor impact of the freshly cut herb is appreciably higher and of a different quality from that of its dried counterpart. This is due to loss or modification of the low boiling fraction of the essential oil made up mostly of terpene hydrocarbons. These do not directly contribute much to the aroma profile, but their presence gives the herb a clean freshness that is markedly absent from the dried product, although freeze-drying techniques have reduced this difference to a large extent. The characteristic top-notes of the profile of the freshly cut herb are not only lacking in the traditionally dried material but may be overlaid by a dull hay-like aroma, which is not removed by rehydration, but is generally dissipated during cooking.

Within the several groups of herbs given in Table 8.5 there is a close similarity of sensory effect and, when blended together, these tend to reinforce each other in their main flavoring attributes while supplementing each other in their minor aromatic nuances. Individual profiles are, of course, different and often dictate the use of one herb rather than another in a seasoning blend.

**Herbs containing cineole** — Herbs containing cineole have a very distinctive note that can readily be recognized as eucalyptus-like. Although each has its own diagnostic profile, these herbs may be described as "light" or "fresh" and their odor may be "penetrating," initially warming but usually developing a more cooling character. Such herbs are useful in the seasoning of white meats (e.g., veal, pork, chicken, fish) and other protein sources which have a low intrinsic flavor. They give an attractive lift to the overall profile of the cooked product, but when used in excess they can over-power the more delicate natural flavors present. All of them can be described as having green, herbaceous notes usually with a marked bitter after-note, which makes them of value in the seasoning of the sweet meats such as pork.

**Herbs containing thymol/carvacrol** — Herbs containing thymol and/or carvacrol are all richly sweet and spicy, aromatic, usually with a distinctive phenolic character. Their odor is penetratingly sharp to the point of being unpleasant, and this largely determines the level of their effective use in seasonings. The aromatic character is carried into the flavor profile which may be described as "sweetly phenolic" with a "warm, richly herbaceous" fullness. Seasonings containing herbs such as thyme may leave a distinctly unpleasant dry, woody after-note and a certain lingering bitterness.

**Herbs containing alcohols/esters** — The so-called "sweet" herbs are characterized by the presence of various alcohols and esters in their contained essential oils. These give an attractive fragrance and generally well-balanced roundness to their odor and flavor profiles. It is this property which makes them of value in blended seasonings. Each has a quite distinctive profile which is considerably more pungently fragrant when freshly cut than after drying. The initial odor impact is not high but is usually penetrating and may be slightly irritating. Many of them show an initial cooling effect backed by a "green, herbaceous" nuance with a hint of balsamic, floral notes. The flavor effect is smoother than the odor and can be described as warmly spicy, but in spite of this they are all pleasantly fresh and "herby" probably due to a camphoraceous note that has a delayed and somewhat persistent cooling effect. The aftertaste in most cases is bitter but not unpleasant.

**Herbs containing phenolics** — The varieties of *Salvia,* which are collectively known as *sage,* are all highly aromatic herbs having a penetrating odor that is initially cooling but then "warmly spicy." Their essential oils contain 40 to 60% of thujone modified by about 15% of cineole so that the profile is at first camphoraceous with a musty, dull, sickly sweet impact. These herbs are not to everyone's taste, and regional preferences for them are very marked. The flavor of the sages is overpoweringly herbaceous with a sweet, eucalyptus-like back-note and a lingering and not very pleasant bitter-sweet after effect. Sage is a powerful and effective seasoning and is widely used in stuffings together with onion for such meats as chicken, veal, and fish.

**Herbs: Mints** — The mints form another group of herbs distinguished by their wide variety. Several species are of commercial importance as flavoring materials either in the form of the fresh or dehydrated leaves or as essential oil. Their odor and flavor characteristics display wide variation depending on source, growing conditions, harvest and postharvest handling, and drying and/or distillation techniques employed. All the mints have a clean cooling odor that is strongly "mentholic" and usually irritating. Their attractive flavor is very cooling and refreshing with a marked mentholic character that is somewhat pungent but not irritating. Initially the flavor effect is warming but at higher concentrations rapidly becomes cooling and even numbing. Some varieties as well as older samples of essential oil leave the mouth with a heavy "oily-buttery" note, which may even be somewhat coconut-like, with a slightly "earthy," "unclean" back-note. Spearmint differs completely from peppermint in that the oil contains 50 to 60% of carvone, which gives it a characteristically "balsamic," "creamy" odor, and a warm, smooth pleasingly aromatic flavor that is popular in chewing gum.

### 8.2.1.5 Preparation of Herbs for Market

The main forms in which herbs are handled in commerce are as follows:

1. Whole-cleaned but not otherwise processed.
2. Chopped, cut, or sliced-leaves, very coarsely broken into pieces about 3–6 mm across. This is a convenient form for bulk transportation prior to further processing.

3. Broken or rubbed-leaves and other soft herbaceous parts are coarsely screened to remove all hard stems. This is the form in which most domestic herbs are marketed.
4. Ground-milled herbs powdered to a specified sieve size.
5. Blended seasonings-broken or ground herbs blended together ready for use in foods. Composition can be varied to suit particular dishes.

Domestically, fresh herbs are widely used in the form of *bouquet garni, fine herbs,* etc., as these have a better aroma and flavor than the dried forms. Herbs are, however, seasonal and the difficulties associated with handling fresh produce result in the dried form being extensively used almost exclusively on a commercial scale and very widely on a domestic scale.

The quality of herbs depends on several variables but particularly on the following: plant variety, environmental conditions during growth and harvesting, period of harvesting, drying and postharvest processing methods, storage and transportation, packaging and shelf life. Users of herbs are most concerned with cleanliness, particularly freedom from adhering soil, flavor, aroma, and color. Loss of quality is closely related to loss and/or degradation of volatile oil content, which is at most risk during the drying process as the loss of moisture is accompanied by loss of the lighter fractions of the volatile oil. As any residual dampness results in caking and mold infestation, the removal of water is very important. The fresh herb (60 to 80% moisture) is generally reduced to between 5 and 10% moisture to achieve a stable product. If this is carried out under optimum conditions, the color of the herb is retained and the odor and flavor losses are reduced to a minimum. Excessive heat results in an unacceptable degree of browning, the loss of freshness, and the imposition of an unacceptable off-odor that detracts from the overall profile. Most herbs are dried naturally under shade, but artificial drying at a maximum of 40°C and also freeze-drying are increasingly being used. These are more expensive processes but the resulting products have a much higher quality.

#### 8.2.1.6 Introduction to Spices

It is useful to classify spices based on sensory associations, which are of value in the formulation of well-balanced seasonings (Table 8.6). Spices differ from herbs in being only parts of plants that are very aromatic and, with few exceptions (e.g., capsicum, fenugreek), contain relatively high percentages of volatile oil as well as other powerful, nonvolatile, flavoring and/or coloring components. They are normally derived from plants growing wild in semitropical or tropical regions and, even when cultivated, are often grown, harvested and sun-dried under far from ideal conditions, all of which inevitably influences the quality of the marketed product. Of the various factors involved, temperature has been shown to have the greatest significance as a determinant of quality and suitability for use, but insect infestation during storage and mold infection in any damp material may cause serious deterioration in appearance and aromatic profiles to the extent of making the material quite unacceptable for use in food products.

#### *8.2.1.6.1 Milling of Spices*

Because of their physical characters, few spices can be incorporated directly into food products; most require further processing. Of the several processes used, particle size reduction or milling is most widely used. The degree of fineness of the processed spice is determined by its ultimate use (e.g., distillation, extraction, blended seasonings).

It is well recognized that the process of milling spices can directly affect the quality and keeping properties of the resulting material in the following ways:

1. By exposing the material to excessively high temperatures during the actual grinding process
2. By exposing the volatile oil in the ruptured cellular tissues to loss by evaporation and/or oxidative change
3. By altering the physical character of the product (e.g., creating a pasty mass instead of a free-running powder as occurs with nutmeg and mace), thereby influencing its subsequent shelf life and value as a flavoring material.

Milling or comminution of dried plant materials involves subjecting the whole or coarsely broken feedstock to considerable shear forces in specially designed machinery with fixed and high speed moving parts with the inevitable production of heat. Friction between the parts of the milling head and the plant material itself usually results in temperatures of between 40° and 95°C, depending on the fineness of the grinding and the nature of the starting material. Generally, the finer the grind, the higher the temperature created in the mill head (and greater volatile losses), but this is offset by the more ready availability of the flavor and increased ease of incorporating the milled spice into a food mix. Finely milled spices generally have a very limited shelf life, and in manufacturing practice they should not be retained in store for longer than about three months before use.

The critical deterioration temperature at which undesirable changes occur in the product (i.e., volatile losses, oxidation, discoloration, or physical transformation) is only about 36°C, and in some spices such as nutmeg and mace may be as low as 32°C. Various techniques have been developed to limit the formation of heat during grinding; two of particular interest in commercial use are (a) prechilling of the feed material, and (b) water-cooling or refrigeration of the milling chamber (i.e., cryogenic grinding).

Certain spices require special treatment, notably the following:

1. Those having high volatile and/or fixed oil content (e.g., caraway, cumin, nutmeg, mace)
2. Ground white pepper if prepared from whole black pepper by a process known as *decortication*
3. Cardamoms, the seeds of which must be first removed from the tough flavorless capsule in which they normally remain until required

4. The blending of curry powders that calls for a multistage milling to achieve the desired color and texture
5. Mustard seeds that must be partially deoiled and desolventized prior to grinding for use in prepared mustard powders

Loss of volatiles and autoxidation of fixed oils are not the only problems arising from milling. Older mills are fabricated from iron and only the fast moving parts are made from various grades of toughened steel. Studies have shown that many ground spices contain relatively high levels of particulate iron presumably originating from this source.

#### *8.2.1.6.2 Microbiology of Spices*

Spices may carry fairly high levels of microorganisms, e.g., black pepper commonly has standard plate counts of >$10^6$ CFU/g. The contaminating organisms include thermophilic aerobic bacteria and spores, yeasts, and molds, a few of which may be pathogenic or toxigenic. As spices are extensively used in processed foods and in meat products which are sold fresh (e.g., sausages), it is essential that microbial contamination be minimized; this is often accomplished through a sterilization process. The sterilization processes used for this purpose include: (a) exposure of the spices to sterilant gases (ethylene oxide or propylene oxide), (b) exposure of the spices to $\gamma$-irradiation, (c) sterilization by heat processing, or (d) distillation and/or extraction techniques to isolate the flavoring components of the spice [8].

**Gas sterilization of spices** — Gas sterilization involves the following steps: (a) loading the spices, in suitable bags, into a specially designed retort; (b) evacuation of the chamber; (c) raising the temperature of the material to about 35°C (maximum 60°C) and then add humidity; (d) introduction of a predetermined quantity of the sterilant gas; (e) maintenance of conditions to affect sterilization; and (f) removal of the gas and sparging of the retort with air several times to remove all residuals. Ethylene oxide is most commonly used and for practical and safety purposes, is normally diluted with carbon dioxide (80%) [9].

Ethylene oxide does not penetrate some spices very well particularly when it is diluted in carbon dioxide so the effectiveness of the treatment is variable. Often a spice must be treated more than once to achieve the sterilization desired. After treatment, the spices should be allowed to outgas for a week (or until no residual ethylene oxide can be detected) to reduce the residual ethylene oxide levels. The process (less the out gassing time) takes generally 12–18 hours [10].

U.S. regulations limit the residual ethylene oxide to 50 ppm. Ethylene oxide (ETO) is banned in many countries (e.g., Japan, and some European countries) due to its reaction with spice components to form ethylene chlorohydrin and ethylene bromohydrin. Ethylene chlorohydrin is a known carcinogen that persists in the spice for many months. In Canada, ETO cannot be used on vegetable seasonings or spice mixtures containing salt (residues of 1500 ppm are currently allowed) [10]. Ethylene oxide also has been labeled as a carcinogen by the W.H.O. This greatly complicates its use and handling in food operations. The complications of using ethylene oxide and its low efficiency make alternative treatment highly desirable.

**Gamma irradiation** — Ionizing irradiation is produced either by mechanical acceleration of electrons to very high velocities and energies, by x-ray machines, or by emission of γ-radiation from radioactive isotopes such as Cobalt 60 or Cesium 137. Of these, Cobalt 60 is the most favored source for commercial processing. Irradiation is approved in the U.S. for the treatment of spices, wheat, flour, potatoes, pork, fruits and vegetables, poultry and red meat [11]. Dosages applied to spices and herbs range from 5 to 15 kGy (generally use 10 kGy, or 1.0 to 1.5 rads). By U.S. and Canadian law, foods that have been irradiated must be labeled with a special international logo and the words "treated with radiation" or "treated by irradiation" must appear on the package appropriately (Minnesota Dept of Health, 2003). In the U.S., foods that use irradiated ingredients do not have to be labeled as such. However, in Canada, foods containing irradiated ingredients must be so labeled [9,11].

Irradiation is very effective in the sterilization of spices and herbs and results in few changes in quality or functionality of these materials. It is the preferred method of treatment. This process takes from 5 to 15 hours to complete [10].

**Steam treatment** — According to Leistritz [10], steam treatment is gaining in usage. Steam, wet or "dry," has the advantage of being effective and without residue (unlike ethylene oxide) and is well accepted by the end user (unlike irradiation). Wet steam has the disadvantage of adding moisture to the spice that then must be redried involving extra process time and potentially reducing the quality of the spice. The treatment may affect the volatile profile, color, functionality, and physical state (caking may occur).

Dry (superheated to 108–125°C) steam is beginning to be used for microbial reduction. Superheated steam adds little moisture to the product and is very rapid [10]. Thus, the quality of the spice suffers little from the process.

**Distillation or extraction** — The products resulting from distillation (i.e., essential oils) or extraction (i.e., oleoresins) are essentially free of bacterial contamination and form the basis of a whole range of products for direct incorporation into food mixes. The manufacture of these products will be discussed later in this chapter. The advantages and disadvantages of natural spices in various forms are given in Table 8.7.

## 8.2.2 Derivatives of Spices

### 8.2.2.1 Essential Oils (Distillation)

The bulk of all spices and herbs is cellulose, which contributes nothing to the aroma and flavor of the material. The aromatic profile is largely determined by the essential oil content, quantitatively as a measure of aromatic strength, and qualitatively as a determinant of aromatic character. In most instances, the volatile oil preexists in the plant and is usually contained in special secretory tissues the nature of which is often associated with particular plant families (e.g., Zingiberaceae-oil cells, Umbelliferae-vittae or oil canals, Labiatae-secretory hairs). For this reason one cannot over generalize on distillation techniques as optimum conditions will vary between different spices. Generally, younger plants produce more oil than older ones but the quality of such oils may lack the fuller aromatic attributes normally associated with

**TABLE 8.7**
**Advantages and Disadvantages of Traditional Ground Spices**

| | |
|---|---|
| **Advantages** | |
| Slow flavor release in high temperature processing | No labeling declaration problems |
| Easy to handle and weigh accurately | Presence of natural antioxidants in many herbs |
| **Disadvantages** | |
| Variable flavor strength and profile | Unhygienic — Often contaminated by filth (microorganisms) |
| Ready adulteration with less valuable materials | Presence of lipase enzymes |
| Flavor distribution poor | Dusty and unpleasant to handle in bulk |
| Discoloration due to tannins | Dried herbs usually have unacceptable hay-like aroma |
| Flavor loss and degradation on storage | Undesirable appearance characteristics in end products |

*Source:* From Heath, H.B., G.A. Reineccius, *Flavor Chemistry and Technology*, Van Nostrand Reinhold, New York, 1986. With permission.

oils from mature plants. In plants, essential oils are still a functional enigma; some may act as insect attractants as an aid to pollination, others may protect the plant tissues against parasites and animal depredation. Current opinion suggests that they are merely the waste products of plant metabolism, any functional value being coincidental.

Although several thousand plants are known to contain volatile constituents that could be recovered as essential oils, only about 200 are processed commercially and of these only about 60 are produced in significant quantities. The distillation of herbs and spices for the recovery of their contained essential oil is the subject of several standard texts [12–15]. Articles on essential oils are published regularly by Lawrence [e.g., 16–17] and compilations are offered in book form [18].

#### *8.2.2.1.1 Manufacture of Spice Essential Oils*

The recovery of aromatic constituents from plant materials in the form of essential oils dates back to the ninth century. Although the actual process has not changed much over the centuries, the manufacturing equipment used has been constantly refined and modified to give optimum yields under conditions that can be precisely controlled. Actual yields and quality depend on the nature of the feed material, the methods used in its handling, pretreatment, and speed of processing after comminution. Yield figures and aromatic quality are most likely to vary as a result of differences in distillation time, rate of distillation, efficiency of vapor condensation, and method of separation from the distillation water.

Modern distillation techniques fall into the following three categories:

1. Water distillation: the plant material is loaded into a still fitted with a slow-speed paddle stirrer, and covered with water that is brought to a boil by submerged steam coils. Distillation is continued until all the essential

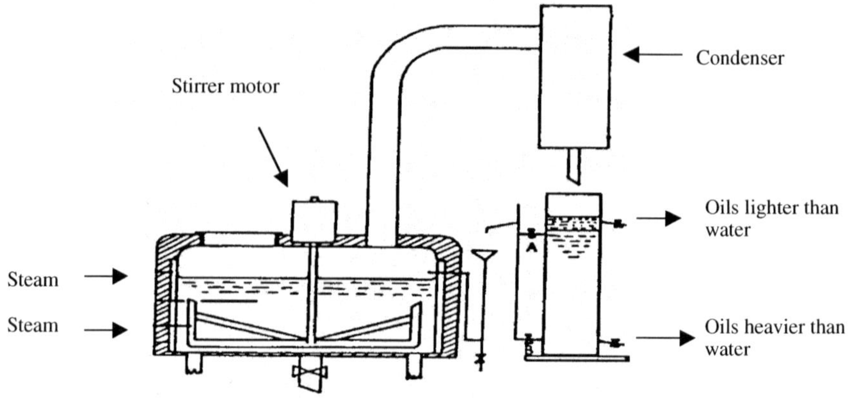


**FIGURE 8.3** Plant used for water distillation of spices. (From Heath, H.B., G.A. Reineccius, *Flavor Chemistry and Technology*, Van Nostrand Reinhold, New York, 1986. With permission.)

oil has been recovered, the condensed water being continuously returned to the still (a process called *cohobation)* (see Figure 8.3).

2. Water and steam distillation: The plant material is loaded onto a frame within the still body, fixed above a layer of water which can be brought to a boil by submerged steam coils. This technique is widely used for the distillation of green herbaceous materials (e.g., peppermint) (see Figure 8.4).
3. Steam distillation: The plant material is loaded into a suitable still through which steam may be injected from the base. A reboiler unit is often fitted to enable condensate to be recycled. The shape of the still body depends on the nature of the feed material. For example, the distillation of peppermint herb in the western states of the U.S. is carried out in stills formed from the back of container trucks that can be loaded directly in the fields; these trucks are driven to a center where a still head is bolted on and steam injected. Static steam stills tend to be tall and are often divided internally by grids to ensure uniform loading.

Most herbs and spices require some preparation prior to distillation. The unit operations involved include (a) cleaning; (b) comminution; (c) soaking, particularly of hard materials such as cinnamon bark; and (d) loading and charging, which requires close attention to avoid channeling or uneven exposure to the steam as this can have a very significant effect upon the sensory characters of the oil.

The method of distillation employed depends on the nature of the feed material. Over the years all the prime aromatic materials have been investigated to establish optimum conditions. Generally water distillation gives the finest quality oil as in this technique there is less chance of damage to sensitive components. Steam distillation may impart a burnt character, which can result from internal condensation and flowback of an aqueous extract onto the exposed steam injection coils.

The separation of the essential oil from the aqueous distillate can result in a reduction in quality depending on the following factors:

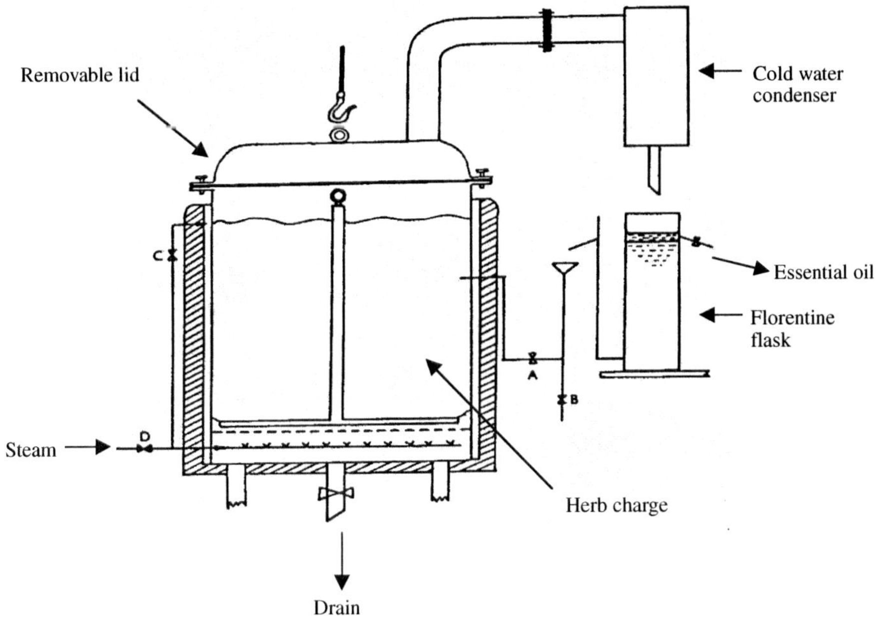


**FIGURE 8.4** Plant used for water and steam distillation of herbaceous materials. (From Heath, H.B., G.A. Reineccius, *Flavor Chemistry and Technology*, Van Nostrand Reinhold, New York, 1986. With permission.)

1. The chemical composition of the essential oil, particularly the ratio of low to high boiling constituents.
2. The temperature gradient of the condensate within the condenser, the outflow temperature of which determines the relative density of the oil and water phases. Optimally this should be between 45 and 50°C.
3. The flow rate that determines the time available to achieve effective phase separation.
4. The separator design, preferably allowing continuous cohobation of the separated water, which usually contains some residual unseparated oil globules.

Some essential oils are more effectively isolated by solvent extraction within the separator. For this purpose very pure grades of hexane and/or pentane are used. The essential oil is recovered by removing the solvent under high vacuum distillation [19].

The literature covering the distillation of essential oils is extensive if now somewhat out-of-date. However, techniques have not changed fundamentally. The texts quoted in the bibliography should be used as a source of further basic information on techniques and specific essential oils.

### *8.2.2.1.2 Use of Spice Essential Oils*

Essential oils offer a solution to a common criticism of ground spices, i.e., they have a uniform flavor strength and character across seasons. Most essential oils obtained

**TABLE 8.8**
**Advantages and Disadvantages of Spice Essential Oils**

| Advantages | Disadvantages |
|---|---|
| Hygienic, free from all microorganisms | Flavor good but often incomplete |
| Flavoring strength within acceptable limits | Flavor often unbalanced |
| Flavor quality consistent with source of raw material | Some readily oxidize |
| No color imparted to the end product | No natural antioxidants |
| Free from enzymes | Readily adulterated |
| Free from tannins | Very concentrated so difficult to handle and weigh accurately |
| Stable in storage under good conditions | Not readily dispersible, particularly in dry products |

*Source:* From Heath, H.B., G.A. Reineccius, *Flavor Chemistry and Technology*, Van Nostrand Reinhold, New York, 1986. With permission.

from materials from a designated source are reasonably constant in their flavoring effects and if used at a fixed rate, will give an end product with a consistent profile. However, in many instances, essential oils only represent a part of the total available flavor in the spice, and this can limit their usefulness. For example, in black pepper and ginger, the volatile oil only gives the characteristic odor of the spice; the pungent principles, being nonvolatile, are not present in the distilled oil. The advantages and disadvantages of using spice essential oils in food processing are listed in Table 8.8.

Essential oils are widely used in the making of plated spices most of which are formulated to replace the ground spice on a weight-for-weight basis. When making dispersions of essential oils onto an edible carrier it is necessary to reduce the dispersion rate to account for the proportion of oil, which is naturally retained within the unbroken cells of ground spices. The precise equivalence must be determined experimentally but may range within a 30 to 50% reduction. For example, if the yield of essential oil from 100 kg of allspice (pimento berries) is 4 kg, then a dispersion containing 2.5% oil of pimento, instead of 4%, will give an equivalent flavor when used weight for weight to replace the ground spice in a seasoning formulation.

The essential oil content of the most widely used culinary spices are listed in Table 8.9.

### 8.2.2.2 Oleoresins (Solvent Extraction)

Spice essential oils frequently lack the full, rounded profiles associated with the use of ground herbs and spices. In certain instances they are totally devoid of characteristic flavoring attributes contributed by nonvolatile components in the natural material. These can be recovered by extraction of selected, dried and ground herbs and spices. In practice, organic solvents are chosen that optimally dissolve both the essential

**TABLE 8.9**
**Volatile Oil Content of Spices**

| Spice | Volatile Oil Content (% v/v) | Spice | Volatile Oil Content (% v/v) |
|---|---|---|---|
| Allspice (pimento) | 3.0–4.5 | Dill | 2.0–4.0 |
| Anise | 1.5–3.5 | Fennel | 4.0–6.0 |
| Anise, China star | About 30 | Fenugreek | Trace |
| Capsicum | Nil | Ginger, Jamaican | 1.0–3.0 |
| Caraway | 2.7–7.5 | Ginger, African | 2.0–3.0 |
| Cardamom seed | 4.0–10.0 | Mace | 12–15 |
| Cassia (cinnamon) | 0.5–4.0 | Nutmeg | 6.5–15 |
| Celery | 2.5–3.0 | Paprika | Nil |
| Cinnamon | 0.5–1.0 | Pepper, black | 2.0–4.5 |
| Clove | 15–20 | Pepper, white | 1.5–2.5 |
| Coriander | 0.4–1.0 | Turmeric | 4.0–5.0 |
| Cumin | 2.5–4.5 | — | — |

*Source:* From Heath, H.B., G.A. Reineccius, *Flavor Chemistry and Technology*, Van Nostrand Reinhold, New York, 1986. With permission.

oil and the desirable nonvolatiles present. After extraction, the solvent is removed and the resulting concentrated extract is known as an oleoresin.

#### *8.2.2.2.1 Solvents*

Two main categories of solvents are widely used and these are listed in Table 8.10.

1. Polar solvents containing hydroxy or carboxy groups and are relatively reactive chemicals that are miscible with water. They are generally powerful solvents of a wide range of ingredients. Some pose fire hazards.
2. Nonpolar solvents are generally hydrocarbons or chlorinated hydrocarbons that are relatively unreactive and immiscible with water. They tend to have limited solvent properties, being better for oils and fats. The hydrocarbons may pose fire hazards while the chlorinated solvents are nonflammable but may be relatively toxic.

More recently, supercritical $CO_2$ has been used for this purpose. Extraction may be carried out at low temperature (0° to 10°C) and high pressure (8–80 atm). Under these conditions $CO_2$ liquefies to become a nonpolar selective solvent, which is odorless, tasteless, colorless, easily removed without residues and is nonflammable. Capital investment in the plant is high, but the extraction is claimed to be more effective with the added advantage that there are no solvent residues, no off-odors, low monoterpene hydrocarbon levels, more top-notes and more back-notes giving a fuller but *different* profile than with traditional extraction methods [20–25]. The higher cost and different flavor profile obtained with supercritical $CO_2$ has limited the use of this technique despite its inherent advantages.

**TABLE 8.10**
**Solvents Permitted for Use in Flavor Extraction**

| Solvent | BPt (°C)[a] 760 mm | Polarity[a] Log P | Permit[c] Residue (ppm) | Solvent | BPt (°C)[a] 760 mm | Polarity[c] Log P | Permit[a] Residue (ppm) |
|---|---|---|---|---|---|---|---|
| Hexane | 68.7 | 0.009 | 25 | Trichloroethane | 74.1 | | 30[b] |
| Methylene chloride | 40.0 | 0.309 | 30[b] | Ethanol | 78.4 | 0.654 | |
| Acetone | 56.5 | 0.355 | 30 | Supercritical $CO_2$ | –56.6 | | |
| Methanol | 64.7 | 0.762 | 50 | Isopropanol | 82.5 | 0.546 | 50 |
| Ethylene dichloride | 83.7 | | 30[b] | | | | |

[a] Source: From http://virtual.yosemite.cc.ca.us/smurov/orgsoltab.htm. With permission.
[b] The sum of all chlorinated solvents cannot exceed 30 ppm.
[c] (4)

### *8.2.2.2.2 Manufacture of Oleoresins*

The most widely used extraction process involves the following unit operations:

1. Preparation of the raw material. Dried herbs and spices must first be ground to ensure solvent penetration. The smaller the particle size, the greater the number of ruptured cells, the interfacial area, and the speed of achieving equilibrium in the extraction. Too fine grinding can lead to slow solvent penetration, slow percolation of the solvent through the mass, and difficulties in recovering solvent from the spent material.
2. Exposure of the material to solvent. This involves three phases: (i) addition of solvent to ensure even wetting of the dry mass, (ii) period of maceration to allow absorption of the solvent and equilibration between solutes and solvent, and (iii) continuous replacement of the miscella with fresh solvent. These conditions are influenced by the nature of the starting material, the design of the extractor, and whether the system is operated under gravity or forced-pump circulation. The system of recycling the solvent is economically important and is dictated by the solvent selected, the degree of heat treatment feasible without damage to labile constituents, the established flow rate and the optimum spice/solvent ratio. When the bulk quantity warrants it, a system of continuous countercurrent extraction offers many advantages.
3. Separation of miscella from extracted material and removal of solvent. The stripping of solvent from the desired extractives is critically important to the character and flavoring quality of the resulting oleoresin. Many spices contain highly volatile constituents so that the removal of solvent must be carried out under low temperature conditions so as to avoid loss or damage of these. The use of high vacuum is indicated but this can also

lead to volatile losses. For this reason, the lower the boiling point of the solvent the better, hence the attraction of supercritical $CO_2$ that does not have these problems.

Desolventization is a two stage process involving: (i) removal of about 98% of the solvent in a standard falling-film, rising-film, or calandria-type evaporator operated under carefully controlled vacuum conditions, and (ii) scrubbing, sparging with nitrogen, and high vacuum treatment of the concentrate to remove final traces of solvent down to the prescribed levels demanded by legislation (Table 8.10).

#### *8.2.2.2.3 Quality of Oleoresins*

The physical character of oleoresins ranges from viscous oils to thick, tacky pastes, making their direct incorporation into food mixes difficult. To assist in handling, many oleoresins may be admixed with a permitted diluent such as propylene glycol or a selected vegetable oil, such additions being clearly indicated in the product specification. It is the balance of volatile to nonvolatile resinous matter that is the best indication of quality and this should be closely similar to that in the original spice.

#### *8.2.2.2.4 Oleoresins and Related Products*

The advantages and disadvantages of using spice oleoresins in food processing are listed in Table 8.11. As oleoresins are not suitable for all food applications, various secondary products, suitable for direct incorporation into food products, have been commercially developed. These include the following:

- *Essences* are prepared by the maceration of ground spice materials with 70% ethanol. The resulting product may be fortified by the addition of top notes compounded from an essential oil and/or oleoresin. In the U.S. the term *extract* is often applied to a flavoring essence but has a defined meaning under FDA regulations.

**TABLE 8.11**
**Some Advantages and Disadvantages of Oleoresins**

| Advantages | Disadvantages |
|---|---|
| Hygienic, free from all microorganisms | Flavor quality good but as variable as the raw material |
| Can be standardized for flavoring strength | Flavor profile dependent on the solvent used |
| Many contain natural antioxidants | Very concentrated so difficult to handle and weigh accurately |
| Free from enzymes | Range from liquids to viscous fluids and pastes making difficult to incorporate into food mixes without "hot spots" |
| Long shelf life under good storage conditions | Tannins present unless specially treated |

*Source:* From Heath, H.B., G.A. Reineccius, *Flavor Chemistry and Technology*, Van Nostrand Reinhold, New York, 1986. With permission.

- *Emulsions* are liquid seasonings prepared by the emulsification of blended essential oils and/or oleoresins with gum Arabic (acacia) or other permitted emulsifying agents. A stabilizer may be added to prevent creaming.
- *Solubilized spices* are blended essential oils and/or oleoresins mixed with one of the polysorbate esters (e.g., Polysorbate 80) or other dispersing or solubilizing agents in such a concentration as to give a clear solution when mixed with water.
- *Dispersed, plated, or dry-soluble spices* are prepared by dispersing a standardized oleoresin or a blend of essential oil and oleoresin onto an edible carrier such as salt, dextrose, crackermeal, or rusk to give a product having a flavor strength, but not necessarily the same flavor character as that of a good quality freshly ground spice.
- *Encapsulated spices* are prepared by spray drying premade emulsions using gum Arabic (acacia) or one of the permitted modified starches. On drying, these become the encapsulant. Such products have a long shelf life and the flavor is only released when they are mixed with water.
- *Heat-resistant spices* are specialty encapsulated products in which the outer capsule is rendered water insoluble. Such products have a long shelf life and are designed to release their flavor only at high temperatures such as are achieved during baking.
- *Fat-based spices* are blends of essential oils and/or oleoresins in a liquid edible oil or hydrogenated fat. These latter products may be offered in bulk or as spray-cooled powders. They are designed for ready incorporation into high fat products such as mayonnaise or cream soups. Some of the advantages and disadvantages of using these secondary products in food processing are listed in Table 8.12.

### 8.2.2.3 Expressed "Essential" Oils (Citrus Oils)

Most essential oils are obtained by distillation using one of the processes already described in this chapter dealing with herbs and spices, but "essential" oils may also be obtained via other methods (albeit, essential oils to the purest are obtained *only* via distillation). The other methods include: (i) enfleurage-extraction of floral oils using fats, (ii) solvent extraction, and (iii) mechanical pressing, particularly applied to the recovery of citrus peel oils.

#### *8.2.2.3.1 Citrus Fruits*

The worldwide popularity of citrus fruits is based largely on their nutritional value, flavor, aroma, and other intrinsic attributes such as texture and color. Citrus fruits are an acknowledged source of vitamin C as well as of other dietary requirements. The characteristic color of the fruits is derived from carotenoids and flavonoids. Their characteristic odor is due to essential oil in the peel, and the flavor of the juice is determined by the ratio of sugars to organic acids, mainly citric acid, overlaid by the presence of low levels of aromatics. In this context we are only concerned with citrus fruits as a source of essential oils, which have a significant value as flavoring agents in several branches of food processing and soft drink manufacture.

**TABLE 8.12**
**Advantages and Disadvantages of Processed Spices**

| Advantages | Disadvantages |
|---|---|
| **Dispersed Spice Products** | |
| Standardized flavor profile by specifying source of raw material | Allowances necessary for the base used |
| Standardized flavoring strength | Volatiles lost on storage, particularly if ambient temperature is high |
| Hygienically acceptable | No resistance to high temperature processing under open conditions |
| Free from enzymes | |
| Readily handled and weighed with accuracy | |
| Readily dispersed in food mixes | |
| Usually free from dust during handling | |
| Many contain natural antioxidants | |
| Unwanted specks or color not contributed to the end product | |
| Water activity low | |
| **Encapsulated Spices** | |
| Aromatics fully protected from loss and degradation | Flavor profile may differ from that of equivalent natural spice |
| Where appropriate, natural color retained over long periods (e.g., paprika) | Concentration usually tenfold so that accurate weighing is difficult |
| Long shelf life under all conditions | Relatively expensive |
| Readily incorporated into food mixes | |
| Free from objectionable odors (applies particularly to garlic) | |
| Hygienically acceptable | |
| Free from enzymes | |
| Low water activity | |
| Do not contribute unwanted specks or color to the end product | |

*Source:* From Heath, H.B., G.A. Reineccius, *Flavor Chemistry and Technology*, Van Nostrand Reinhold, New York, 1986. With permission.

### *8.2.2.3.2 Citrus Essential Oils*

In all the citrus fruits, the essential oil is contained in numerous oval, balloon-shaped oil sacs or glands situated irregularly just below the surface of the colored portion of the peel (the flavedo). The white inner mesocarp (the albedo) does not contain any oil sacs but does carry the bitter glycosides such as hesperidin in lemon, orange, and tangerine, or naringin in grapefruit.

With the exception of lime oil, which is primarily recovered by distillation, the citrus peel oils are obtained by mechanical extraction that differs depending on the nature of the fruit to be handled. Other parts of the citrus plant also yield essential oils when distilled, and those from the flowers and leaves are of importance in fragrance compounding but of less value in flavorings. The oil from citrus leaves is called oil of petitgrain and that from citrus flowers, oil of neroli. The source is designated in the name, e.g., oil of neroli, Bigarade, is derived from the flowers of the bitter orange, C. *aurantium* L.

Commercially, the citrus oils are recovered directly from the peels of fresh fruit. The methods used include:

1. Sponge pressing. The separated half peels are hand pressed, the extruded oil being collected in sponges. This very laborious method is now hardly used but is still claimed to give the highest quality oil.
2. Machine pressing. Two main types of machines are used. (i) Sfumatrici machines do automatic pressing of the separated peels after removal of juice and pulp. Such machines can handle about 1 ton/h. (ii) Pellatrici machines are designed to handle whole fruit with the oil being released by rotary rasping of the fruit surface. Larger versions of these machines can handle 2 ton/h. The liberated oil is continuously removed by a jet of water and the oil separated from the coarse emulsion by standing. Additional oil is separated from the aqueous phase by centrifugation, and any intractable emulsion is distilled to recover an additional quantity of low grade oil.
3. Combined juice/oil presses. This machinery was developed in the U.S. and designed for the automatic extraction of juice and oil as a single operation without contact between the two products. Two principle systems are: (i) the fruit is halved, the juice reamed out and the cleaned peel is subjected to pressure between fluted rollers, and (ii) the surface of the whole fruit is first grated and the oil removed in a jet of water, while at the same time a hollow pipe is inserted into the base of the fruit and the juice extracted under pressure.

Lime oil is traditionally obtained as a by-product of lime juice manufacture and is recovered by distillation of the acid liquors obtained by crushing the whole fruit. Alternatively, a cold-pressed lime oil is also produced commercially; this has a much finer flavor quality than the more familiar distilled oil.

Citrus oils are characterized by having a high percentage of terpenoid hydrocarbons and a relatively low content of oxygenated terpenoid compounds that are mainly responsible for their aromatic profiles (Table 8.13).

The yield of oil obtained commercially depends on numerous factors other than the type of machinery used, but rarely does it equal that analytically determined in the fruit skin. The thickness of the fruit skin is a major determinant as thick-skinned fruits yield much less oil than thin-skinned varieties, owing to the absorption of oil by the spongy albedo. The delay between harvesting and processing is also important as flacid peel yields less oil. On average, yields vary as follows:

- Lemon 2–7 lb/t (ton)
- Orange 1-8 lb/t
- Grapefruit 1–2 lb/t

**TABLE 8.13**
**Composition of Citrus Oils**

| Oil | Terpenes (%) | Principle Oxygenated Compounds (%) | |
|---|---|---|---|
| Lemon oil | About 90 | Citral | 1.5–2.5 |
| Sweet orange oil | About 90 | n-Decanal | 0.9–3 |
| Bitter orange oil | 90–92 | n-Decanal | 0.8–1 |
| | | Linalyl acetate | 2–2.5 |
| Lime oil | 75–80 | n-Decanal | 1–5 |
| Tangerine oil | 95–96 | Methyl-N-methyl anthranilate | About 1 |
| Grapefruit oil | About 90 | n-Decanal | 1–2 |

*Source:* From Heath, H.B., G.A. Reineccius, *Flavor Chemistry and Technology*, Van Nostrand Reinhold, New York, 1986. With permission.

- Lime 0.1–0.3 lb/t
- Tangerine 1–2 lb/t

### *8.2.2.3.3 The Composition of Citrus Oils*

Citrus oils are characterized by the presence of large percentages of monoterpenes ($C_{10}/H_{16}$) and smaller amounts of sesquiterpenes ($C_{15}/H_{24}$). Both are the carriers of the oxygenated compounds comprising alcohols, aldehydes, ketones, acids, and esters, which are responsible for the characteristic odor and flavor profiles. The terpenoid composition of the various citrus oils is similar, the principle component being d-limonene. The terpenes possess little intrinsic odor or flavor value, but it would be incorrect to say that they have no flavoring effect. A citrus oil from which the terpenes have been removed is significantly flatter and lacks the characteristic freshness associated with a complete peel oil. The principle components of citrus peel oils are given in Table 8.14.

### *8.2.2.3.4 Processed Citrus Oils*

Terpene hydrocarbons are characterized by their relative insolubility in diluted alcohol and a ready tendency to oxidize and polymerize with an associated deterioration in the odor and flavor of the oil in which they are present. Cold-pressed citrus oils may be further processed to remove all or part of the monoterpenes and sesquiterpenes, the resulting products being known as *terpeneless* or *sesquiterpeneless* oils. As the content of the sesquiterpenes is relatively low, the term terpeneless has come to be used in a general sense to describe these concentrated products. Citrus oils from which terpenes have been removed are also called *folded oils* as the remaining flavorful oxygenated compounds are more concentrated. The degree of concentration is generally expressed as "fold" which reflects the volume reduction of the oil on concentration, i.e., a fivefold oil has one fifth the volume as the starting singlefold oil. While the oil becomes more potent with folding, potency is not exactly proportional to the fold. Also, the flavor character of the oil changes with folding beyond ca. fivefold (for orange oil).

**TABLE 8.14**
**Principle Components of Citrus Oils**

| | | |
|---|---|---|
| Monoterpenes | Oxygenated compounds | Aldehydes |
| α-Pinene | Acids | Acetaldehyde |
| β-Pinene | Acetic | Citral |
| γ-Terpinene | Capric | n-Decanal |
| α-Terpinene | Caprylic | Geranial |
| D-Limonene | Formic | Octanal |
| Myrcene | | |
| p-Cymene | | |
| Terpinolene | Alcohols | |
| Sabinene | Citronellol | |
| Camphene | Geraniol | |
| | Linalool | Esters |
| Sesquiterpenes | I-Nonanol | Geranyl acetate |
| Bisabolene | Octanol | Linalyl acetate |
| Cadinene | α-Terpineol | Methyl-N-methyl anthranilate |
| Caryophyllene | Terpinen-4-ol | Octyl acetate |

*Source:* From Heath, H.B., G.A. Reineccius, *Flavor Chemistry and Technology*, Van Nostrand Reinhold, New York, 1986. With permission.

In order to achieve maximum recovery of oil, the citrus processing industry applies several techniques that increase the yield. These include the following:

1. The distillation of the aqueous phase from which no more oil may be recovered by centrifugation
2. The steam distillation of the pressed residual peels
3. The superheated steam treatment of "citrus molasses" obtained from cured citrus wastes
4. The deoiling of canned single-strength citrus juices

All of these products are much inferior to cold-pressed citrus oils in both flavor and keeping properties.

### *8.2.2.2.5 Methods of Deterpenation*

Three methods widely used for the production of deterpenated oils are: fractional distillation under vacuum; selective extraction of the soluble oxygenated compounds in diluted alcohol, and chromatography using silica-gel.

1. Fractional distillation: The bulk of terpeneless oils are produced by fractional distillation under a high vacuum (1–2 mm Hg) so that the oil boils at a sufficiently low temperature to ensure the minimum damage to the heat sensitive constituents during the relatively long distillation period required to remove the large percentage of terpenes present. At these low pressures, the boiling points of the constituents are lowered making

efficient separation difficult, but as the terpenes are rarely individually separated, this is a problem of still design and reflux balance that can readily be solved should it be necessary. An alternative method to dry distillation is one in which steam injection is used. Unfortunately the distillation time is increased, leading to possible damage to the flavoring components. In addition, the presence of water may lead to hydrolysis of certain constituents with consequent damage to the aromatic profile of the end product.

2. Selective solvent extraction: Two methods used to produce terpeneless citrus oils having excellent solubility in diluted ethanol (30–35%) are the following:
   a. The whole oil is distilled with three to four times its volume of diluted alcohol in a still fitted with a separator and a return trap to permit recycling of the distillate. During the process, the terpenes distill over with the alcohol vapor and ultimately separate on the surface of the distillate; the required oxygenated fraction is progressively dissolved in the weak alcohol in the still. When no further separation of terpenes takes place, the recycling process is discontinued and the oxygenated constituents are recovered by fractional distillation.
   b. The whole oil is mixed with alcohol (95% ethanol by volume) and then with a calculated volume of water to reduce the alcohol strength to about 35%. The mixture is vigorously agitated and then allowed to stand to effect separation. The terpenes being insoluble in this strength of alcohol separate as an upper layer. The lower level may be drawn off and processed directly as a soluble essence for use in the soft drink industry, or it may be further diluted with water, saturated with sodium chloride to break any emulsion, and the terpeneless oil separated and dried. Such oils have a good bouquet but lack the freshness of the original oil.
3. Chromatographic separation: Braverman and Solomiansky [26] developed a chromatographic method of deterpenation for the production of high-quality terpeneless orange oil using silicic acid as an adsorbent for the oxygenated fraction. The technique involves passing the natural citrus oil through a column of silica gel followed by elution of the column with a nonpolar solvent (e.g., hexane). This removes any residual terpenes and allows the oxygenated fraction to be recovered from the silica gel by extraction with ethyl acetate or some other low boiling polar solvent that can then be removed by vacuum distillation. The process is exothermic so that the column must be cooled and precautions taken to ensure minimum heat damage to the flavoring constituents.

Terpeneless and concentrated citrus oils from which only a part of the terpenes have been removed are widely used as flavoring materials as they have improved stability and a longer shelf life, a lower usage rate, and improved solubility making them of particular value in the flavoring of soft drinks and liqueurs.

*8.2.2.3.6 Citrus Leaf and Flower Oils*

Although the essential oils obtained from citrus peels play a major role in food and beverage flavorings, oils obtained by steam distillation of citrus leaves, twigs, and flowers have an equal importance in fragrances. The oils obtained from the leaves and twigs are collectively known as oil of petitgrain; those obtained from the flowers are called oil of neroli. In each case the source is indicated in the precise designation.

The most widely used of these oils are derived from cultivated bitter orange trees *(C. aurantium* var. *amara)* and are designated "bigarade" from the French word for bitter orange tree "la bigaradier." The oils concerned are (i) oil of petitgrain bigarade — a pale yellow oil having a pleasantly sweet, fresh, floral odor with a woody herbaceous background character, and (ii) oil of neroli bigarade — a pale yellow oil, which darkens on keeping, has a very strong, fresh floral odor when newly distilled but loses its top-notes on storage. These oils have only limited value in flavorings and are used mainly to contribute interesting flavor nuances in imitation fruit flavorings.

### 8.2.2.4 Mint Oils

*8.2.2.4.1 Classification of Mint Oils*

The mints are a very complex and diverse group of perennial plants belonging to the family *Labiatae* which have hybridized to such an extent that their botanical parentage and classification is confusing, even to the experts. For practical purposes the following are recognized as of major commercial importance as flavoring materials either in the form of freshly cut leaves, dried and rubbed flakes, or as essential oils:

- Peppermint (*Mentha piperita* [L]), which is considered to be a natural hybrid of M. *aquatica,* L and M. *viridis* (L). It exists in two varieties: *vulgaris* (black or Mitcham mint) and *officinalis* (white mint).
- Cornmint *(Mentha arvensis,* L), characterized by having an essential oil containing a high percentage of menthol. Two varieties are of commercial importance: *piperascens,* Holmes (Japan and Brazil) and *glabrata,* Holmes (China).
- Spearmint *(Mentha spicata,* L or M. *cardiaca,* Ger.) — an essential oil rich in l-carvone.
- Garden or culinary mint, which represents a wide spectrum of species, varieties and cultivars including: M. *viridis* (garden mint), M. *rotundifolia* (applemint), M. *gentilis* (American applemint), and M. *spicata* var. *crispata* (curley mint).

*8.2.2.4.2 Peppermint*

Peppermint is cultivated almost exclusively for distillation of its essential oil. Although the so-called Mitcham mint is still regarded as having the finest flavor, this variety is now rarely grown in the U.K. where it originated. Plants of it were introduced into the U.S. in the early nineteenth century and prospered so well that this has become the world's biggest peppermint growing enterprise. Initially cultivation was carried out in the northwestern states but soon spread to the midwestern states of Michigan, Indiana, Ohio, and Wisconsin. Unfortunately the "white mint" variety developed various diseases, resulting in poor oil yields. This lead to the development of hardier "black mint" cultivars that were introduced into Oregon and Washington, which now form the prime center of peppermint oil production.

Significant peppermint is also cultivated in Argentina, Australia, Bulgaria, France, Germany, Hungary, India, Italy, Morocco, Netherlands, Poland, Spain, Russia, and Yugoslavia. Not all of these are of international importance. There are now so many varieties and strains that the commercially available oils display a wide spectrum of profiles and flavoring quality. Such wide differences are brought about by factors intrinsic to the plant and its environment and also to the methods of harvesting and distillation employed in the various growing areas.

Peppermint is probably one of the world's most popular flavors, being used in a wide range of sugar confectionery, chewing gum, chocolate centers, liqueurs, oral toiletries, and medicines. The fresh herb is popular in certain regions (notably western Germany and North Africa) for making a refreshing mint tea. For most purposes the essential oil is the preferred flavoring agent and is commercially available in the following forms:

1. *Natural:* The prime distillation of the leaves and flowering tops of the freshly harvested herb
2. *Rectified (or double distilled):* A redistillation of the prime oil in which the highest boiling fractions are rejected
3. *Double rectified (or triple distilled):* A rectification in which 5 to 10% of the lowest boiling fractions and 2 to 5% of the high boiling still residues are rejected

For most purposes the single rectified oil is used and is the one described in the Food Chemicals Codex and is official in many pharmacopoeias. In products where solubility is important, or where a clean peppermint flavor is essential (e.g., creme de menthe and chocolate fillings), the triple distilled oil is preferred.

**Cultivation and distillation** — Peppermint is propagated by root cuttings and, as the plants continue to thrive for about 5 years, it is necessary to select the cultivar very carefully. Under optimum soil and climatic conditions, the best yield and quality of oil is produced in the second and third years. The plants are normally replaced after the fifth harvest.

Harvesting takes place when the plants are in full flower, the precise time being critical to ensure maximum yield and a well-balanced oil. The yield decreases rapidly after full bloom due largely to leaf fall. If cutting is delayed, the oil loss can be significant but too early cutting results in a poor quality oil having an indifferent flavor due to an incorrect balance of menthol/menthone.

Distillation is carried out on the wilted or partially dried herbaceous tops. In most countries this is carried out in field pot stills, but in the main producing regions of the U.S., use is made of sophisticated mobile stills, the body of which is the truck used to carry the cut herb to the distillation center. In Oregon and Washington the entire peppermint operation is fully mechanized. The average yield of prime oil is upwards of 100 lb/acre. Percentage yields are rarely quoted as these depend on the state of dryness of the herb when distilled, hence it is more usual to quote yields as weight of oil per area of ground cut.

**Rectification** — Single distilled peppermint oils as received from the prime producers often have marked and objectionable sulfurous, "weedy" off-notes due to the presence of dimethylsulfide and other polysulfides. These undesirable taints, which detract from the clean, fresh, minty profile, are best removed by a process of rectification. Rectification of any essential oil is a process of redistillation in which high and/or low boiling fractions are separated to an extent necessary to provide an end product having desired physical and sensory properties. The process, although considerably improving the sensory attributes of an oil, cannot convert a poor quality oil into a good one. As redistillation involves some degree of concentration, the process may accentuate certain abnormalities or imbalance in the flavoring constituents rather than elimination of them. Rectification is carried out in one of two ways:

1. Dry distillation — The oil is redistilled under high vacuum (5–10 mm Hg) under which conditions the oil boils at a sufficiently low temperature to ensure minimum damage to any thermolabile constituents. The aromatic vapors rise through a suitable fractionating column which enables fractions of a defined boiling range to be separated and collected. Depending on the nature of the starting material and the desired character of the end product, the various fractions may be reblended as necessary and any unwanted fractions rejected.
2. Steam injection — This is the older method and is carried out in jacketed stills fitted with a suitable separator to permit either continuous recycling of the distillate where water is used in the still or its rejection where high-pressure live steam is injected. Contact of oils with boiling water and/or steam can lead to decomposition of certain constituents so that conditions must be very carefully controlled. Dry distillation *in vacuo* is now the more widely used technique.

As applied to peppermint oil, rectification results in two principle products:

1. *Oil of peppermint, single rectified* (also called *double distilled)* — A colorless oil prepared by rectification in which the last 2–5% of the oil remaining in the still is rejected. This eliminates the harsh, weedy resinous off-notes of the natural oil.
2. *Oil of peppermint, double-rectified* (also called *triple distilled)* — A water-white oil obtained by rectification in which the first fractions of distillate, up to whatever percentage is required, but usually within the range 5–10%, are rejected along with the last 2–5% of still residues. This process removes both the low-boiling terpenes and the high boiling odorous compounds. The resulting product has a much improved aromatic profile, better solubility, and increased shelf life.

### 8.2.2.4.3 Cornmint

Cornmint *(Mentha arvensis,* L) grows wild in China and in parts of Japan but is widely cultivated in both these countries as well as in Taiwan, Brazil, Argentina, India, and South Africa. The oil distilled from the harvested herb contains 60 to 80% of menthol and is usually solid at room temperature. This prime oil is not commercially available and is a rich source of natural menthol. The commercial oil is a dementholized product but still retains between 45 and 55% of total alcohols (as menthol). In the U.S. such oils are not recognized as peppermint oil and must be clearly labeled as mint oil, cornmint oil, or fieldmint oil. When used in any food product, such oils must be similarly declared. In international transactions, however, these oils are frequently called peppermint oil and usually are distinguished by their source (e.g., Brazilian peppermint oil, Chinese peppermint oil).

**Cultivation and Distillation** — Cultivation of these mints differs considerably with the growing regions, and one of the major texts on essential oils (reference 12) should be referred to for more specific information. The following notes merely indicate the differences:

*Japan.* Plants are raised from root cuttings and give their maximum yield of oil during the second and third years but must be renewed after the fourth year's cropping. In some areas, annual replanting is practiced. Harvesting, which varies between one crop per year in the north and three crops per year in the south, is carried out on bright, sunny days when the menthol content of the leaves is at its highest. The cut plant is bundled and hung up, under shade, to air dry for 10 to 30 days during which it loses about one third of its wet weight. The mint may be distilled locally or in larger centralized facilities. Oil yields vary considerably with location and season but average 50 lb/acre.

*China.* Mint is usually grown as an annual crop that is harvested twice per year, in June and September. Cutting takes place when the plants are about 3 ft high, the cut material being bundled and sun-cured prior to distillation. Field distillation is carried out in small but relatively modern stills, the product being sold to major central processors.

*Brazil.* Mint was introduced into the São Paulo region by Japanese settlers early in the 20th century. Propagation is by root cuttings planted in cleared ground in late August to early September. The plantations have to be renewed every 3 years, owing to limited soil fertility and the onset of rust diseases, which result in uneconomic yields. Two crops are harvested per year, with cutting taking place when the plants are in full bloom. The cut material is bundled and air dried prior to distillation. Prime oil produced in local field stills is transported to central processors for dementholization.

**Dementholization** — The oils derived from the distillation of the partially dried cut herb contain between 60 and 80% of total menthols. The dementholization process employed in the major processing centers consists of (a) drying the oil and filtering to remove particulate matter, (b) filling into metal containers, (c) immersing the containers into brine baths to reduce the temperature slowly to -5°C — this slow cooling encourages the growth of large menthol crystals, and (d) separating the dementholized oil by draining or centrifugation. The separated crystals are further purified by recrystallization.

#### 8.2.2.4.4 Spearmint

The designation spearmint is applied commercially to several species and varieties of *Mentha* possessing a characteristic odor profile that is quite different from either peppermint or cornmint, although the plant form, its cultivation, and harvesting are very similar. The main species involved are the following:

- *Mentha spicata* (L) Huds. var. *tenuis* Mich., which was formerly classified as *Mentha viridis* L. (U.S.)
- M. *spicata* var. *trichoura* Briq. (U.K.)
- M. *spicata* var. *crispata* Benth. (Germany and Japan).
- M. *verticillata* L. var. *strabola* Briq. (old U.S.S.R.)
- M. *cardiaca,* Gerard ex Baker, called Scotch mint (U.S.)

The cultivation of spearmint is centered in the western U.S. which produces ca. 80% of the total world production (2000 tons) [15]. The essential oil is extremely popular particularly in chewing gum, toothpaste, and some sugar confectionery. Most of the crop is distilled in the growing regions where yields of up to 40 lb/acre are obtained. There is a growing demand for the so-called Scotch spearmint, which is cultivated in the Midwestern states, as many consider this oil to have a sweeter and more rounded profile.

#### 8.2.2.4.5 Blended Peppermint Oils

The commercial designation of essential oils is not always an unequivocal indication of either source or quality. This is particularly the case with peppermint oils where there is not only a wide range of materials available but also a broad spectrum of application parameters. Peppermint oil may be single, double or triple distilled, and the quality and flavoring properties of the resulting products depend not only on the starting material but also on the processing techniques employed. Alternatives between *piperita* and *arvensis* oils are also available for use where price may be a significant factor. These oils may be offered in their genuine form or as blends formulated to satisfy particular manufacturing needs. In this specialized area of trading, it is not unusual to find essential oils which have been "sophisticated," even to the extent of the addition of synthetic versions of major components (adultered oils).

#### 8.2.2.4.6 Composition of Mint Oils

There are considerable differences in the composition of the various mint oils, determined by species and environmental conditions during growth, harvesting, and postharvest handling. It is convenient to consider these oils in two groups: (a) peppermint and (b) spearmint. Peppermint (M. *piperita)* and cornmint (M. *arvensis)* oils have much in common, the principle constituents of each being menthol, menthone, menthyl acetate, and other esters. It is the ratio of these components that determines their distinctive profiles.

**TABLE 8.15**
**Compositional Data for Mint Oils[a,b]**

| | *M. piperita* (%) | *M. arvensis* (%) |
|---|---|---|
| Terpenes | | |
| α-Pinene | 0.5–0.8 | 0.5–1.2 |
| Camphene | 0–0.2 | 0 |
| β-Pinene | 0.9–1.7 | 0.7–2.7 |
| Limonene | 3.2–4.3 | 4.9–11.2 |
| 1,8-Cineole and γ-terpinene | 6.4–13.5 | 0.2–0.1 |
| Oxygenated compounds | | |
| 3-Octanol and menthofuran | 3.2–9.4 | 0.3–3.6 |
| Linalool | 0–0.4 | 0.1–0.8 |
| Menthone | 8.9–23.8 | 21.5–29.7 |
| Isomenthone | 2.2–4.5 | 2.9–8.9 |
| Neomenthol | 2.8–4.2 | 2.8–5.0 |
| Menthol | 40.2–48.7 | 33.2–39.3 |
| Pulegone | 0.9–2.6 | 0.5–4.9 |
| Menthyl acetate | 4.4–11.6 | 0.9–6.2 |
| Piperitone | 0.7–3.1 | 1.8–11.6 |

[a] Minor components amount to about 1.5%.
[b] Precise composition is determined by variety and geographical source as well as by imposed environmental factors.

*Source:* From Smith, D.M., L. Levi, *J. Agric. Food Chem.*, 9, 1961. With permission.

Research into the biosynthesis of mint oil constituents has established that menthone is formed from piperitone as the plant matures and that this is further converted into menthol, the percentage of which continues to increase in the leaves until the plant is in full bloom. It is the final balance of menthol/menthone that is the major determinant of the flavoring quality of the distilled essential oil.

Spearmint (*M. spicata*) oil is characterized by having a high content of l-carvone, which is present at about 56%. Its sensory and physico-chemical characteristics are entirely different from those of peppermint. The compositional data of the major constituents of various mint oils is given in Table 8.15. An article by Lawrence provides a more detailed summary of mint oil composition [27].

#### *8.2.2.4.7 Other Commercially Important Sources*

Many other aromatic plant materials are collected or cultivated as a source of essential oils, many being of considerable commercial importance. Such essential oils are used as flavoring agents in a wide range of food, beverage, and confectionery products and also in numerous fragrance applications. These sources and oils have been fully described in several classical texts [7,12,13,29,30] and articles particularly those by Brian Lawrence (Lawrence has published three to four articles annually updating current findings on specific essential oils in *Perfumer and Flavorist*, e.g., [31–34]).

#### 8.2.2.5 Fruit, Fruit Juices, and Concentrates

Botanically, fruits are the matured ovary with, or without seeds, and sometimes with the remains of floral parts. The fruit wall, which may be either dry or fleshy is known as the pericarp and forms the edible part of most fruits. In the case of strawberries it is the swollen receptacle that forms the edible fruit. Most common fruits are used for dessert purposes, but what is not sold on the domestic market may be used as the starting material for several products of value in food processing. These include (a) fruit pieces, pulp, and puree; (b) dehydrated fruit; (c) fruit juices and concentrates; (d) fruit powders; (e) preserved fruits (canned or bottled); and (f) jams and jellies.

Nuts are also fruits but it is usual to treat these separately as their use and applications are very different.

Fruits are classified into seven categories, as follows [35]:

1. *Berries:* Fruits in which the edible part is pulpy and more or less juicy, containing numerous seeds: e.g., black currant, blackberry (bramble), blueberry, boysenberry, cranberry, gooseberry, huckleberry, kiwi fruit, loganberry, raspberry, red currant, and strawberry.
2. *Citrus:* Fleshy, edible portion is segregated into sections surrounded by a thick skin, which itself contains a strongly aromatic essential oil: e.g., bergamot, citron, grapefruit, kumquat, lemon, lime, tangerine (mandarin), orange, sweet, and orange, bitter.
3. *Stoned fruit:* Fruit contains a single seed (the stone or pit) surrounded by a fleshy pericarp, which is normally the edible portion: e.g., apricot, cherry, damson, peach, and plum.
4. *Grapes:* Small multiseed or seedless fruits that grow on vines in clusters.
5. *Melons:* Large fruit containing either a mass of seeds in the central cavity within a soft pulpy flesh or embedded throughout a watery cellular flesh: cantaloupe, honeydew, musk, and watermelon.
6. *Pomes:* Fruits in which the body is formed by an elongation and swelling of the receptacle surrounding the seed capsule: e.g., apple, pear, and quince.
7. *Tropical fruits:* Classified together for convenience but embracing many fruit forms: e.g., avocado, banana, date, fig, mango, pineapple, and pomegranate.

Most edible fruits consist of a skin or peel that encloses the soft pulpy tissues, and it is this fleshy pericarp, which is of most importance as a source of flavor. This tissue contains the juice comprising sugars (e.g., glucose, fructose, and sucrose), fruit acids (e.g., citric, tartaric, malic acids), coloring matter and, in some fruits, phenolic bodies, including tannins. There is a correlation between the sugar content, color, acidity, and flavor, and the balance changes as the fruit ripens. In general, the flavor is at a maximum when the sugar content is highest and the skin color at its brightest. The flavor changes and deteriorates significantly as the fruit becomes overripe.

#### *8.2.2.5.1 Fruit Juices*

A fruit juice is defined as the clear or uniformly cloudy, unfermented liquid recovered from sound fruits by pressing or other mechanical means. The precise technique employed in its manufacture depends on the nature of the fruit, and not all fruits are of equal value as a commercial source of juice [36,37]. The selection of fruit for juice processing calls for considerable expertise and an understanding of the balance between variety, ripeness, and physical quality of the fruit used and the character and stability of the separated juice. The expressed juice does not contain the flavor of the whole fruit as the characteristic volatile components (which are the source of the aromatic profile) are present in the freshly prepared juice in less than 0.07 part per thousand. The flavor normally associated with fresh fruit can only be achieved by blending juice with aromatics recovered from the rest of the fruit.

Fruit juices have been traditionally obtained by an extraction process that involved loading the soft fruit into a stainless steel vertical basket hydraulic press. As pressure was slowly applied, the juice was gradually expressed, passed through a coarse mesh screen and then collected. Such simple batch operations are necessarily slow, and continuous presses are increasingly employed. In these, the fruit is fed into a screw conveyor within a tapering perforated cage. The juice escapes as the pulp is increasingly compressed as it passes along the screw bed. After expression, the dry press-cake is usually discarded except in some instances when it is economically viable to remix the cake with water and repress, but such additional products are of very low quality and only fit for blending.

Fruits differ considerably in the ease with which they can be processed. The presence of pectins and proteinaceous matter may give rise to a viscous juice which is difficult to handle, slow to express, and poses problems when screening. Pectinase and other enzyme systems to break down the pectin result in better yields and a more fluid juice, but any such enzymes must be completely removed from the final product [38].

The processing of citrus fruit calls for special extraction equipment designed to minimize the content of peel oil in the resulting expressed juice. Two main types of extraction equipment are widely used throughout the industry:

1. Brown juice extractor. This simulates hand reaming of halved fruits. The fruit is cut in half by a sharp blade and each half is caught between the halves of a rubber cap and is then held against a rotating cone that removes the juice and a proportion of the cellular matter (made by Brown International Corporation).
2. FMC "in-line" extractor. The whole fruit is received in a row of serrated cups and squeezed by a similar cup that descends and meshes with the stationary one. The juice and cellular matter pass through a 1-inch hole cut in the bottom of the fruit by the sharpened end of a stainless steel tube. FMC (FMC Technologies, 2004) notes that 75% of the world's citrus juice is processed using their equipment.

In each case the initial pressings are passed through a fine screen to remove all pips and unwanted solids.

Limes, being very small fruits, are processed by crushing the whole fruit between rollers or in screw presses, the extracted juice is screened to remove solid matter and is then centrifuged to separate the peel oil. Freshly expressed citrus juices always contain extraneous matter comprising coarse cellular tissues and a fine suspension of pectins, gums, and proteinaceous substances, much of which may be colloidal in character. The nature of the finished juice depends on demand — orange and grapefruit juices desirably contain a uniform cloud of particulate solids; lemon and lime juices are usually presented as clear products, and the juice may have to be centrifuged to ensure this.

Fruit juices are very susceptible to deterioration and spoilage due to yeast growth hence products not intended for immediate use must be preserved to prevent fermentation. Throughout the processing and subsequent storage of fruit juices, rigid standards of cleanliness and hygiene are essential. The juice must be treated in a manner that will keep it from fermenting or undergoing any other form of deterioration. This is achieved in one of four ways:

1. Pasteurization. The juice is heated, either in batches or as a continuous process, at 60–88°C for periods of time ranging from 5 to 30 min. This inevitably leads to some changes in the sensory quality of the product, and the degree of heat treatment is limited to that necessary to destroy microorganisms in a relatively low pH medium and to inactivate any enzyme system that could lead to flavor deterioration and cloud instability during ultimate storage.
2. Freezing. The product is kept under refrigeration at below 4°C, under which conditions the flavor is retained almost indefinitely.
3. Chemical preservation. Where permitted by legislation (very limited), fruit juices may be preserved by the addition of sodium benzoate (0.1–0.3%) or sulfur dioxide (0.03–0.08%). In the latter case, there is a gradual loss of $SO_2$ during storage, and in bulk containers it is often necessary to add further quantities of sodium metabisulfite to maintain an adequate level of preservative. Ashurst [15] noted that sulfur dioxide finds its greatest use in the preservation of fruit juice concentrates (1500–2000 mg/kg).
4. Concentration. Juices that may be concentrated adequately (ca. 65° Brix) are stabilized by their low water activity. However, many juices undergo heat damage and/or become too viscous at this high solids level, and thus shelf-stable concentrates cannot be made.

#### *8.2.2.5.2 Fruit Juice Concentrates*

Most juices consist of about 85 to 90% water and have a flavor level that makes them acceptable to drink per se. However, as a source of flavor in beverages and other processed foods, most natural juices are much too weak and it is necessary to concentrate them, taking care that the delicate aromatic profiles and sugar/acid balance are not unduly changed.

Ideally, any concentration process should remove water without altering the characteristics of the natural juice, either by loss of essential volatiles or by the development of off-notes in either the aroma or flavor. Several methods have been developed to achieve this, including:

1. Vacuum distillation. This is carried out in heat-exchange units under carefully controlled conditions. This results in a concentration of soluble solids and extractive matter but with little aroma. The aromatic substances in the original juice are stripped out during the distillation, part being recovered by condensation and low-boiling fractions being lost. This technique leads to concentrates lacking much of the fresh character of the starting material. This can be partially replaced by recovery and incorporation of the volatiles from the aqueous distillate, but the end product will still be deficient in top-notes. Besides deterioration of aroma and flavor, the color of the juice may caramelize and darken with increasing concentration if the temperature is permitted to rise above 60°C.
2. Freeze concentration. When dilute juices are frozen, the ice crystals formed are pure water until the eutectic point is reached when the whole mass solidifies. As no heat is applied, using this technique results in a superior product having a well-balanced profile closely similar to that of the start material. The freezing temperatures employed range from –12° to –18°C. The concentrate is recovered either by allowing the slush-frozen mass to warm gently and running off the concentrated liquor as it thaws, or preferably, by centrifugation or vacuum filtration.
3. Continuous freeze concentration. This uses a highly efficient, low volume, scraped surface heat exchanger known as a votator and involves a continuous feed of prechilled natural juice, which is slush frozen and then centrifuged to separate the ice phase. The resulting concentrate is collected and may be recycled to effect further concentration until the processing cycle is steady. In general this system will handle juices containing 6 to 10% solids and will concentrate these to 40% solids in a single-stage operation.

The technique employed for the recovery of volatiles during concentration (distillation) was developed by the Eastern Regional Research Laboratory of the Department of Agriculture in 1944. The process has since been widely applied to improve the flavor of most fruit concentrates. The operation is carried out at atmospheric pressure, and heat damage is minimized by high velocity, turbulent flow in a preheater unit. Essence recovery is effected by condensing the fractionated vapors combined with scrubbing of the noncondensable gases before venting to atmosphere. It is usual to retain the recovered fruit essences separate from the main bulk of juice concentrate and to mix this in the correct proportion shortly before use.

Specifications for fruit juice products usually include a value for soluble solids expressed as *degrees Brix.* This figure is based on the relationship between the specific gravity of the product and the percentage of dissolved solids it contains. The Brix tables quoted in the literature are based on an accurate determination of these relationships using aqueous solutions of pure sucrose. It is assumed that all the dissolved solids in a juice will have a similar effect upon the specific gravity, but this is not strictly so, although it is sufficiently close for all practical purposes. In quality control, the specific gravity of the product is determined by using a hydrometer calibrated in degrees Brix or a refractometer, and the determined refractive index then being converted into degrees Brix from the appropriate tables. The two methods are not strictly comparable, but again, are sufficiently accurate for control purposes. The Brix value is really a measure of sucrose concentration, but since fruit juices contain more sugars than other soluble solids, the results act as a good ready guide.

#### *8.2.2.5.3 Blended Fruit Juice Products (WONF)*

Most natural and concentrated juices are either too low in available flavor or are too expensive to use in many food products. To cater to the needs of food processors and beverage manufacturers, the following two types of product have been developed and are widely used:

1. *WONF products* (with other natural flavors). These comprise blends of juices or juice concentrates so that at least 50% of the flavor is derived from the named fruit and not more than 50% from other natural flavors (to meet labeling requirements)
2. *Fortified concentrates.* These are products in which the flavor strength is fortified or the aromatic profile intensified by the addition of either other natural flavor constituents or of flavorings compounded from synthetics that are generally, but not necessarily, identical to those found in the natural fruit.

#### *8.2.2.5.4 Depectinized Juices*

All fruit juices contain pectic substances. These are hydrocolloids, which not only affect the viscosity of the juices but also stabilize the suspension of any insoluble solids or cloud. As many juices are required to be clear, these pectic substances must be removed during processing. This is achieved by enzymatic action of naturally occurring pectinases or by the addition of such enzymes (see reference 38). Depectinization has the added advantage that the juice or concentrate is miscible with diluted alcohol of the strength normally used in the manufacture of liqueurs whereas the natural juice would result in a heavy precipitate. In the case of orange and grapefruit juices, which are required to have a stable cloud, the pectinases must be inactivated otherwise the juice will separate into a sediment and a clear supernatant. The cloud of orange juice is generally stabilized by heating, the time and temperature being determined by the pH of the starting material (e.g., pH 3.8 — hold 2 min at 89°C or 15 sec at 94° C; pH 3.3–2 min at 85°C or 15 sec at 90°C). Depectinized juices are usually preserved by the addition of 0.05 to 0.06% benzoic acid.

#### *8.2.2.5.5 Dehydrated Fruit Juices*

There is an increasing demand for flavorings in powder form for use in a range of convenience foods and many attempts have been made to present natural fruit juices in this form, particularly for use in beverages and instant drinks. Dehydration as a unit operation is not difficult but the sugar/acids present in fruit juices tend to dehydrate as a plastic mass rather than as discrete granules, and the resulting products are invariably hygroscopic and must be packed in hermetic containers. It is usual to mix substances such as lactose with fruit juices in order to produce a free-flowing powder and to include a sachet of silica gel in the pack as a drying agent. The following processes have been developed with varying degrees of success:

1. Roller drying. This is the standard method of drying a film of the juice on heated rollers. It has a limited application as the resulting product develops an unacceptable cooked note and is abnormally dark in color.
2. Vacuum drying. This technique is similar to roller drying, but the rollers are operated under vacuum. This enables lower evaporation temperatures to be achieved and results in a product that is friable, easily reconstituted, and of good flavor quality.
3. Spray drying. This is carried out in a specialized facility and involves drying the juice, generally with the addition of a drying aid such as dextrose or maltodextrin, at a relatively high temperature for a short time. The stickiness of the juice solids under spray drier conditions makes them nearly impossible to produce without a drying aid as noted above.
4. Foam-mat drying. This involves the formation of a foam stabilized with proteinaceous matter or glyceryl monostearate. The foam is dried on a conveyor belt either in a stream of hot air or by blowing air through the foam. The product has a good fresh quality when first made but tends to oxidize on storage owing to its increased surface area.
5. Freeze drying. This operation involves the sublimation of ice directly from the frozen juice which is maintained at a very low temperature under high vacuum with just sufficient heat input to ensure the removal of the water. This results in products having an exceptional high flavor quality but are expensive as the process costs are relatively high.

Dried fruit juice products are difficult to handle owing to their hygroscopicity and the advice of the manufacturer should be sought to ensure optimum storage and usage conditions.

#### *8.2.2.5.6 Fruit Pastes and Comminutes*

For use in ice cream, frozen confectionery and certain types of beverages, citrus fruits may be passed through a colloid mill to form a homogenous paste. These finely ground products are made from the whole fruit and may be further blended with juice or natural peel oil emulsion in order to achieve a desired flavor profile and some degree of standardization. Such products are generally aseptically processed and packaged. The process inactivates any enzyme systems and results in a shelf-stable product.

### 8.2.2.6 Vanilla

Vanilla is an orchid that is indigenous to southeast Mexico and Central America where it grows wild in the moist forests. When the Spanish conquistadors under

Hernando Cortez invaded Mexico in 1520, they found the emperor Montezuma drinking *chocolatl,* a beverage based on ground cocoa beans and corn flavored with *tlilxochitl,* a blend of ground black vanilla pods and honey. Cortez was so impressed by its flavor that he sent the beans back to Spain. By the latter part of the sixteenth century, factories had been established in Spain for the manufacture of chocolate flavored with vanilla. The products soon gained wide popularity, and by the beginning of the seventeenth century vanilla was being enjoyed in England and many other European countries not only for its original use but also a flavoring agent in tobacco and medicines.

Attempts to grow the plant in the botanical gardens in Paris and Antwerp were successful, but the plants did not fruit owing to the absence of natural pollinators. It was not until 1841 that a practical method of artificial pollination, using a small bamboo stick to transfer the sticky pollen, was successfully developed, opening the way to commercial production in other suitable growing regions.

During the early nineteenth century vanilla plantings were established in Java, Reunion, Mauritius, the Malagasy Republic (Madagascar), the Seychelles, Tanzania (Zanzibar), Jamaica, and several other tropical Indian Ocean countries. Today, Madagascar and the islands of Comoro and Reunion grow about 90 percent of the world's crop of vanilla beans. The rest comes largely from Seychelles, Mexico, Central America, the West Indies, French Polynesia, Indonesia, and Uganda. The U.S. consumed 2.5 million pounds of vanilla beans (or vanilla made therefrom) in 2002, more than half of the world supply.

#### *8.2.2.6.1 The Vanilla Plant*

Vanilla beans are the fully grown fruits of the orchid, *Vanilla fragrans* (salisb.) Ames (= *V. planifolia,* Andrews), harvested before they are fully ripe, and then fermented and cured. Two other species are of some commercial value:

1. *Vanilla pompona* schiede is a wild vanilla found in southeast Mexico, Central America, the West Indies, and northern parts of South America. It is cultivated in Guadeloupe, Martinique, and Dominica, the products being sold under the name *vanillons.* These are thicker and more fleshy than genuine vanilla beans and are generally considered to be of poor quality for use in flavorings.
2. *Vanilla tahitensis* Moore is indigenous to Tahiti where it is cultivated. This species has a quite distinctive profile that differs from that of *V. fragrans.*

Vanilla is a fleshy, herbaceous, perennial vine that climbs to a height of 12 to 15 m. It thrives in a hot, humid climate up to about 2500 ft. The plant needs protection from wind. Under cultivation the supports are normally growing trees or shrubs planted at 8-ft intervals. The plant is propagated by cuttings that are planted close to the base of each support tree and regularly pruned to enable easy collection of the fruit. The plant flowers after 2 years, starts fruiting after 3 years and reaches a maximum yield in 10 to 12 years, after which time it is usually replaced by fresh stock. Vanilla flowers once a year over a period of about 2 months. The fully developed vine bears up to 1000 greenish yellow, waxy, fragrant flowers that bloom for 1 day only. The shape of the flower is tubular with an intricate structure so formed that self-pollination is impossible. Of these, only about 50 on each plant are selected for hand pollination, this being carried out using a needle or sharp pointed stick, early in the day on which the flower opens. The flowers occur in clusters and pollination is limited to about eight flowers in each to ensure adequate growth room as the pods develop. After successful fertilization, the pendulous, cylindrical pods take 6 to 9 months to mature and are harvested, while still unripe, as their color changes from green to yellow. At this stage, the beans are 12 to 24 cm long and about 2.5 cm in circumference and lack the characteristic odor and flavor that is only developed by postharvest processing. Yields are very variable and Purseglove et al. [39] reported yields of 450 to 710 lb per acre per year during a crop life of 7 years, but lower yields are not uncommon in some regions.

#### 8.2.2.6.2 The Curing Process

Although procedures differ between the various growing regions, the postharvest treatment of vanilla beans to produce a commercial product having an acceptable color and flavor is basically carried in four stages:

1. Killing or wilting, to stop further vegetative growth and to initiate enzymatic reactions
2. Sweating, to promote these enzymatic reactions, to stimulate the development of the dark brown color and to make the beans supple
3. Drying, to reduce the moisture content necessary for stable storage
4. Conditioning, to enable the flavor to develop fully

The postharvest fermentation and curing process is tedious and time consuming, and the methods used in the various producing regions significantly affect the quality and profiles of the resulting beans. The methods are described in detail by Muralidharana and Balagopol [40], Rossenbaum [41] and recently reviewed by Dignum et al. [42] and Havkin-Frenkel and Dorn, [43]. Briefly, the two main methods used are as follows:

1. Bourbon method. The sorted beans are first immersed in water at 65°C (150°F) for 2 to 3 min to rupture the inner cell walls and probably to stimulate the existing enzyme systems present. After draining, the wet beans are packed in cloth-lined boxes and allowed to sweat for 2 days. Thereafter, the beans are spread out on cloth-covered trays supported on a platform and sun-cured for about 6 h/day for about 6 to 8 days. The beans are covered with a blanket each night. This sun-curing process is uncertain and may take up to 3 months to complete, depending on prevailing weather conditions. Finally, the beans are slowly air dried in sheds over a period of 2 to 3 months or until they are judged to be in a correct condition for sorting and packing.

2. Mexican method. The harvested beans are first stored in sheds until they start to shrink, after which they are sorted and spread out on mats in the sun for about 5 h. As the beans begin to cool, they are rolled in a blanket and the rolls taken indoors and packed into air-tight boxes where they sweat. After 12 to 24 h the beans are removed and inspected and those which have changed color are removed for the next stage of processing. An alternative *oven-wilting* technique is also employed using a specially designed hot-room called a calorifico. This technique is gaining in favor as it gives greater control of temperature and humidity throughout the process. In both cases the next stage consists of alternate exposure to the sun and sweating, spread over a period of about 6 days. Some 20 to 30 days after the initial treatment, the beans will be sufficiently supple and have the characteristic odor and appearance. Very slow drying is then carried out indoors during which the beans are regularly inspected and sorted.

In Tahiti the method is quite different as the beans are collected when fully ripe since they do not split and dehisce like the Mexican species. Here the beans are stacked in piles until their color has changed to brown after which they are spread out on blankets resting on wooden platforms and exposed to the sun for 3 to 4 h each day. The beans are well wrapped in the blanket during the nights. This process is continued for about 20 days and is followed by slow air drying prior to sorting and packing.

When freshly gathered, vanilla beans are largely odorless and flavorless, having none of the characteristics that one associates with the commercial product. Vanilla beans of commerce are blackish brown in color and vary in length from 15 to 25 cm and from 6 to 12 mm in width. On storage, good quality vanilla beans may become frosted with a mass of glistening white, needle-like crystals of vanillin (the *givre* as it is sometimes called). This is not necessarily a criterion of quality as is so often assumed. The crystals do not appear in the producing areas where the climate is very hot and moist but develop only after shipment to cooler drier countries. The moisture content of most samples lies between 30 and 50% but batches dried to a much lower moisture content are available and are preferred for the production of vanilla extracts.

#### *8.2.2.6.3 Classification and Grading of Vanilla Beans*

Beans are generally graded into three categories: unsplit beans, split beans, and cuts (resulting from the chopping of beans prior to fermentation or the removal of damaged portions). Each producing region sets its own standards for grading, the details of which can be obtained from importers. Grades depend on bean length, overall appearance, particularly color, surface blemishes, moisture content, and aroma quality. They are usually tied into bundles, ranging from 70 to 130 beans depending on size, accordingly the weight of each bundle varies between 150 and 500 g. Bundles are packed into tin boxes lined with waxed paper, which usually contain 20 to 40 bundles in rows of ten. The tins are normally packed into wooden cases for shipment.

**TABLE 8.16**
**Comparative Profiles of Vanilla from Different Sources**

| Country of Origin | Sensory Profile |
|---|---|
| | **Bourbon Vanilla** |
| Malagasy Republic | Richly smooth, full bodied, sweet, spicy: less fine than Mexican vanilla |
| La Reunion | Closely similar to Madagascar beans |
| Comoro Islands | Smooth, balsamic with dry back-notes |
| Seychelles | Creamy, full bodied, slightly woody |
| Java | Deep, full bodied, very sweet (almost sickly) |
| | **Mexican Vanilla** |
| Mexico | Sharp, slightly pungent, sweetly spicy lacks body associated with Bourbon beans |
| | **Tahiti Vanilla** |
| Tahiti | Strongly perfumed, characteristically fragrant, heliotrope-like |
| | **Guadeloupe Vanilla** |
| Guadeloupe, Martinique, Dominica | Strongly perfumed floral, heliotrope-like, lacks body |

*Source:* From Heath, H.B., G.A. Reineccius, *Flavor Chemistry and Technology*, Van Nostrand Reinhold, New York, 1986. With permission.

### 8.2.2.6.4 *The Flavor of Vanilla*

Good quality vanilla beans have a quite distinctive aroma that is diagnostic for their country of origin. The differences in aromatic profiles are set out in Table 8.16. Common to all is an intrinsic profile that can be described as sharply acidic with a slightly bitter back-note and a pronounced pungency. This is not normally associated with vanilla preparations owing to the modifying effects of added sugar but it can be observed in vanilla oleoresins. The presence of nonvanillin volatile and nonvolatile flavor components is the reason for a marked preference for natural vanilla preparations over synthetic vanilla or compounded vanilla flavorings.

### 8.2.2.6.5 *The Chemistry of Vanilla Flavor*

Vanillin is the most abundant volatile aromatic constituent of cured vanilla beans. While the range in vanillin content of vanillas is often cited in the literature as 0.5 to 2.5%, Hartman (2003) found the level to be as low as 0.35% (a sample of Java vanilla). There is quite a range in vanillin contents of commercial vanillas depending upon source and manufacturing practice (Figure 8.5). While important, the quality of vanilla beans is not entirely defined by their vanillin content. Klines and Lamparsky [44] have identified 170 volatile components in Bourbon (Madagascar) vanilla and more recent work by Adedeji et al. [45] and Hartman et al. [46] reported 191 compounds, and Bala [47] ca. 250 in vanillas as a whole. Most of the other volatiles are minor in concentration relative to vanillin, but many have very low sensory thresholds which allows them to have an impact on sensory character. There is little

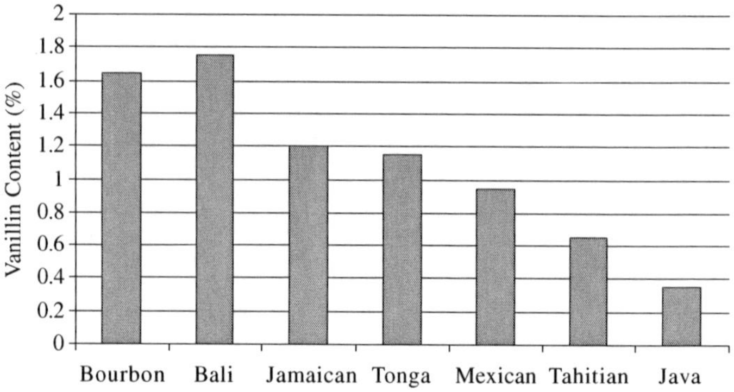


**FIGURE 8.5** Vanillin content of a vanillas from different origins. (From Adedeji, J., T.G. Hartman, C.-T. Ho, *Perfum. Flavorist*, 18(2), 1993. With permission. Also from Hartman, T.H., Composition of vanilla beans from different geographical regions, A presentation given at The First International Congress on the Future of the Vanilla Business, Princeton, NJ, Nov. 2003, http://www.aesop.rutgers.edu/~vanilla2003. With permission.)

question that the character of vanillas varies with origin and that this variation is due to the presence of unique components and amounts of the less abundant volatiles (Table 8.16). Quantitatively, the principle aromatic constituents of vanillas are as follows: vanillin (3-methoxy-4-hydroxy benzaldehyde), up to 2%; p-hydroxy benzaldehyde, about 0.2%; p-hydroxy benzyl methyl ether, about 0.02%; and acetic acid, about 0.02%. Other constituents include reducing sugars (about 10%), fats and waxes (about 11%), gums, and some resinous matter. Tahiti vanilla beans have a quite different aromatic composition and contain a higher percentage of anisyl alcohol, anisaldehyde, anisic acid, and piperonal (heliotropin).

When harvested, vanilla beans are totally lacking in what we recognize as vanilla flavor. The characteristic flavor only develops during postharvest fermentation and curing. The chemistry involved during this treatment has been known for many years. The presence of three glucosides, which yield vanillin and related phenols on hydrolytic cleavage, was confirmed in green vanilla beans as far back as 1947. Of these, the most abundant is glucovanillin with lesser quantities of glucovanillyl alcohol and the glucoside of protocatecuic acid (3,4-dihydroxy benzoic acid). The initial enzymatic reactions that take place during the curing process result in the formation of the respective phenol plus glucose (Figure 8.6). Further peroxidase and

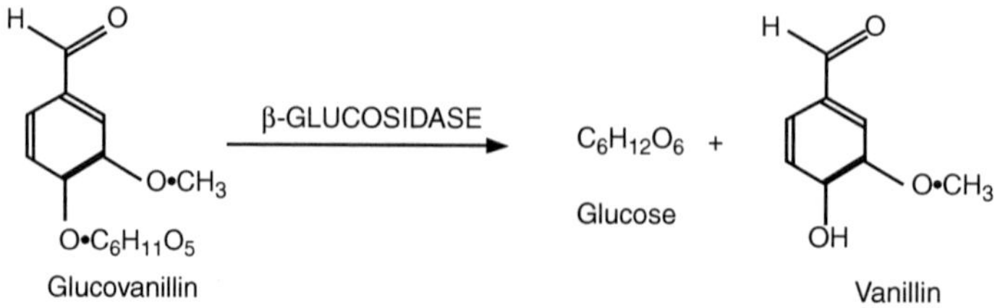


**FIGURE 8.6** The conversion of vanillin glucosides to free vanillin.

polyphenoloxidase reactions in the various substrates, as well as nonenzymatic transformations, result in a complex mixture of aromatic derivatives and the formation of brown pigments.

### *8.2.2.6.6 Vanilla Flavorings*

Before use, vanilla beans must be further processed as their form, color, texture, and fatty nature make direct incorporation into a food product of very limited applicability (often home use). A homogenized puree of vanilla beans has been used in the flavoring of ice cream, but the presence of unsaturated fatty matter poses problems of stability and shelf life. It is also very difficult to reduce the seeds so that they do not leave black specks. It is customary to use vanilla in the form of an alcoholic essence or extract and many such products, differing in flavor quality and strength, are available commercially.

**Vanilla extracts**. In the U.S., vanilla extract is defined in the legislation (CFR169.175) and is the only flavoring to have a standard of identity. U.S. law requires that the finished product shall be no less than 35% ethanol and contain the extractable matter from 283.85g of dry vanilla beans (either *V. fragrans (planifolia)* or *V. tahitensis*). The only other permitted ingredients are water and glycerin, propylene glycol, sugar, invert sugar, glucose, or corn syrup as appropriate to improve the sensory qualities and shelf life of the product. Such extracts are known as singlefold extracts. While U.S. law dictates the ingredients that may be used in making a vanilla, the processing equipment and protocols are left to the manufacturer, and thus there is considerable variation in vanillas available on the market. Extracts prepared from double the quantity of beans are known as double fold or twofold extracts. The maximum concentration that can be prepared directly from the beans is fourfold. Much stronger extracts can be made by dissolving a calculated weight of oleoresin vanilla in the appropriate strength alcohol or by solvent removal from an initial extract. Nonalcoholic extracts are also available; these are based on an acceptable solvent such as propylene glycol, according to the demands of the user industries.

The quality of vanilla extracts is determined by several factors, including the following: correct storage of the beans prior to extraction, blending of different quality beans as appropriate, degree of comminution of beans, method of extraction, correct storage of the extract, and aging for at least 90 days to ensure full flavor development.

Typically, vanilla beans are extracted as follows [49]: Vanilla beans are diced or minced and carefully blended with half their weight of sugar that helps with the extraction and ultimate maturation of the product. The mass is percolated in a suitable stainless steel extractor, using a mixture of ethanol and water of approximately 57% by volume, over a period of 3 to 4 weeks. The percolate is usually collected in fractions and later blended to produce the final product that is stored in stainless steel containers until sufficiently mature. After this period the extract is filtered or centrifuged and the alcohol content adjusted depending on analytical results.

**Oleoresin vanillin**. Oleoresin (or resinoid vanilla) is a dark brown, semifluid extract obtained by alcoholic or other solvent extraction of the finely chopped beans ("cuts" are often used). The process involves the complete removal of the solvent under vacuum. Yields depend on the strength of alcohol used, and Cowley [49] quotes a 40% yield with 50% ethanol, and only 30% with 65% ethanol. Purseglove et al. [39] has tabulated yields of oleoresin from beans of different origins. Inevitably the finer top-notes of the aroma are lost or modified by this heat treatment and diluted extracts made from oleoresin do not have the full flavor character usually associated with extracts made by direct percolation of the bean.

**Vanilla absolute**. In perfumery work it is desirable to have a vanilla extract that is completely soluble in ethanol and/or essential oils. Such products are made by selective solvent extraction and inevitably do not contain all the aromatic components present in the bean. The solvents used vary, but it is not uncommon to carry out a first extraction using a nonpolar solvent (e.g., hexane) followed by a polar solvent (e.g., ethanol). A high quality absolute is made using liquid $CO_2$ extraction.

**Vanilla sugar**. This food ingredient is prepared by mixing ground vanilla beans with sugar or by blending concentrated vanilla extracts with sugar. Imitation products are available; these comprise mixtures of vanillin and ethyl vanillin with dextrose and are often artificially colored. A minimum of 30% sugar is necessary to obtain a satisfactory product and up to 2% of a permitted anticaking agent may be incorporated.

#### *8.2.2.6.7 Adulteration of Vanilla*

Vanilla extracts are very expensive flavorants the composition of which is defined in the U.S. legislation. However, to satisfy a user demand, numerous products are offered that contain both natural extract and synthetic vanillin or other flavoring chemicals having a flavor resembling or compatible with that of vanilla. Such products are much cheaper than the authentic material and the problem of establishing the authenticity of vanilla extracts has been the subject of much research over the past 30+ years. Traditionally, tests concentrated on vanillin content, color, lead number and organic acid content, but these could quite easily be circumvented. This has lead to the development and use of very sophisticated methods of monitoring adulteration.

#### *8.2.2.6.8 Imitation Vanilla Flavorings*

Imitation vanilla flavorings that replace the natural extracts may be of excellent quality and provide a totally acceptable vanilla flavor profile in a wide range of food products. Their use can result in a considerable cost saving and an assurance of a stable of supply. It is usual for such imitation flavors to be offered in strength equivalent to a natural tenfold extract although other strengths can readily be obtained to suit manufacturers' needs. Often the choice between a natural vanilla and imitation vanilla is dictated by law (Standards of Identity) or a labeling preference (natural vs. artificial). These imitation vanillas are based on vanillin and ethyl vanillin with an assortment of other components to add sophistication.

Vanillin has been known since about 1816, and by 1858 the pure chemical had been obtained from alcoholic extracts of vanilla beans. It was not until 1872 that Carles established its correct formulation and in 1874, Tiemann and Haarmann reported it as 3-methoxy-4-hydroxy-benzaldehyde. Finally, Reimer synthesized vanillin from guaiacol and thus proved its chemical structure [5].

For very many years the most important source of synthetic vanillin was eugenol, present at about 80 to 90% in oil of clove, from which it was obtained by oxidation. Later processes involved first converting the eugenol to isoeugenol and this remained the standard method until the source material was changed to the sulfite waste liquors, some 2 million tons of which result from paper manufacture each year. Actual production of vanillin from this new source was first carried out in both the U.S. and in Canada in 1937 although the patented methods now used differ somewhat from those originally adopted. Today this is the prime source of vanillin.

Ethyl vanillin is a closely related compound, 3-ethoxy-4-hydroxy-benzaldehyde, which is not found in nature but is prepared synthetically from safrole. It has an intense vanilla-like odor and is about three to four times more powerful than vanillin as a flavoring agent. Like vanillin, it is widely used in the preparation of imitation vanilla flavorings but can give a somewhat harsh "chemical" character in higher dosage levels. In practice, a maximum of 10% of vanillin may be replaced by ethyl vanillin without this objectionable note being obvious.

### 8.2.2.7 Cocoa, Coffee, and Tea Flavorings

While each of these products are foods (beverages) in themselves, they also serve as a source of flavorings for other foods. This section will provide a brief description of these products as flavorings. (Note that cocoa/chocolate is also discussed in greater detail in Chapter 9.)

#### *8.2.2.7.1 Cocoa/Chocolate Plant and Processing*

The roasted seeds of the cacao tree have been used as the basis for a pleasant beverage since long before the Spanish conquest of Mexico in 1519. It was Cortez and the conquistadors who realized the potential of the drink called *chocolatl* and introduced the beans to Spain and ultimately Europe together with vanilla, which was used in the same preparation. In due course the source of cacao was studied and the tree classified as *Theobroma cacao* L.

To avoid confusion, it is necessary to note the following popular usage of well-known descriptions: (i) *Cacao* refers to the plant *Theobroma cacao* L., (ii) *Cocoa* refers to the beans and to the product made from them which is ready to mix into a beverage; and to its use as a powdered flavoring material or in the making of a chocolate beverage, and (iii) *Chocolate* refers to the solid product based on ground roasted cocoa beans.

The cacao tree grows in most countries within the tropical belt where temperatures range between 25 and 30°C associated with a high humidity (ca. 90%) and adequate ground water. It is a tall perennial evergreen bearing shiny leathery leaves and pod-like fruits on the main branches and along the trunk. Each fruit has a 2 to 4 in. diameter and is 7 to 12 in. long, with a dark green leathery skin that changes from yellowish orange to purple-red as the fruit ripens. Each contains up to 50 seeds embedded in a pinkish pulp. These seeds, after processing, become the cocoa beans of commerce. Cocoa beans are designated according to their country of origin and/or the port of export. Major sources are listed in Table 8.17.

**TABLE 8.17**
**Principle Sources and Types of Cacao Beans**

| Source | Commercial Designation |
|---|---|
| **Forestero-type beans** | |
| Brazil | Bahia |
| Cameroon | Douala |
| Congo | Vandeloo |
| Dominican Republic | Sanchez |
| Ecuador | Arriba, Guayaguil, Machala |
| Ghana | Accra |
| Nigeria | Lagos |
| Philippines | Maripipi |
| Trinidad | Port of Spain |
| **Crillo-type beans** | |
| Venezuela | Caracas, Maracaibo, Porto Cabello |

*Source:* Heath, H.B., G.A. Reineccius, *Flavor Chemistry and Technology*, Van Nostrand Reinhold, New York, 1986. With permission.

The ripe pods are harvested, split open, and the pulp and seeds removed. At this stage the seeds are very bitter due to the presence of tannins and are only made palatable by a process of fermentation, which not only removes the bitterness but also changes the color to a brownish red. Fermentation is carried out by covering the beans and allowing them to sweat for several days. This effectively kills the vegetative processes, softens the pulp and activates complex enzymatic reactions leading to the development of the characteristic flavor. The care with which this process is carried out has a significant effect upon the quality of the beans and their flavor when ultimately roasted prior to use. After fermentation, the beans still contain about 30% moisture and have to be dried to about 8% prior to storage and shipment. As marketed, cocoa beans comprise about 14% outer shell and 85% kernel, the so-called nib.

Raw cocoa beans do not have the distinctive odor and flavor of cocoa, this is only developed during a roasting process. There are two main types of cocoa beans, Crillo and Forastero, which produce beans having distinctly different flavor characteristics, although this is not now so pronounced as in the past owing to considerable cross-breeding and hybridization of plants. Commercially only two categories are recognized: *bulk or basic beans* and *flavor beans.* The former have a strong, harsh flavor character whereas the latter are smoother and considerably more pleasantly aromatic.

Chocolate is the product made by grinding freshly roasted and winnowed cocoa nibs. It contains 50 to 55% cocoa butter and when freshly made is liquid. This basic chocolate is usually cooled in molds to facilitate later handling. Commercial chocolate is prepared by blending the prime liquor with sugar, milk solids and flavorings, particularly vanilla.

#### 8.2.2.7.2 Cocoa/Chocolate Flavor

Considerable research has looked into the chemistry of cocoa flavor, i.e., the precursors responsible and the reactions involved at the various stages of production. Due to the wealth of information available on this product, the reader is encouraged to go to the following references for detail on cocoa processing [50–52] and flavor development [53–55]. The major reactions leading to flavor include:

1. *Fermentation* is when the flavor precursors are formed; the proteins present in the kernel are degraded and the level of free amino acids rises; sucrose is inverted to fructose and glucose, which in turn are oxidized to alcohol and various acids; some theobromine and most of the tannins are lost [56].
2. *Drying* reduces the moisture content to about 8% with loss of volatile acids and a rise in pH.
3. *Roasting* is when the precursors are converted into a complex mixture of some 800+ aromatic compounds, which qualitatively and quantitatively determine the profile of the roasted beans. The Maillard reaction plays a key role in flavor development.

Schnermann and Schieberle [54] reported 13 odorants with the highest flavor dilution factors in milk chocolate as: 3-methylbutanal (malty); 2-ethyl-3,5-dimethylpyrazine (potato chip-like); 2- and 3-methylbutanoic acid (sweaty); 5-methyl-(E)-2-hepten-4-one (hazelnut-like); 1-octen-3-one (mushroom-like); 2-ethyl-3,6-dimethylpyrazine (nutty, earthy); 2,3-diethyl-5-methylpyrazine (potato chip-like); (Z)-2-nonenal (green, tallowy); (E,E)-2,4-decadienal (fatty, waxy); (E,E)-2,4-nonadienal (fatty); δ-decalactone (sweet, peach-like); and 2-methyl-3-(methyldithio)furan.

#### 8.2.2.7.3 Coffee

Coffee has a very long history as a stimulating beverage. It is derived from an evergreen tree or shrub of which there are two main species, *Coffea arabica* L. and C. *robusta* L. A third species, C. *liberica* L., is of little importance. Coffee is native to Ethiopia from where it has spread first to India and then to Indonesia, Brazil, Colombia, El Salvador, and the Philippines. Presently the world crop (crop year 2002) is about 7.128 billion kg, the main commercial sources being Brazil (41%), Colombia (9.5%), Vietnam (8.4%), and Indonesia (4.8%) (International Coffee Organization [57]). These tropical plants grow between 2000 and 6000 ft above sea level and where the ground water is adequate. They bear clusters of small red, cherry-like fruits that contain two seeds. The fleshy berries are harvested almost continuously and, depending on the growing region, are either sun dried for 15 days prior to hulling or are hulled and then allowed to ferment prior to the separation of the greenish-yellow seeds. When dried these become the green coffee of commerce. As green coffee beans lack the characteristic odor and flavor normally associated with the beverage coffee, it is necessary to roast them. This induces a series of reactions resulting in significant changes to both the flavor and color of the beans.

#### 8.2.2.7.4 Coffee Flavor

The chemistry of coffee aroma and flavor has been extensively researched and a very comprehensive review published by Flament [58]. Flament [58] reviews the nonvolatile constituents of green coffee, including their structure and discusses their important contribution as flavor precursors during the roasting process. He also presents the chronological discovery of individual chemicals and critically examines the validity of their identification, highlighting the progress that has been made during the twentieth century. He notes that more than 300 volatiles have been identified in green unroasted coffee and in excess of 850 in roasted coffee.

In terms of key aroma components, Czerny et al. [59] conducted sensory studies using volatiles identified in previous studies as potential contributors to coffee aroma [60] to determine the odorants truly characterizing freshly brewed Colombian coffee. They found 2-furfurylthiol, 4-vinylguaiacol, several alkyl pyrazines, furanones, acetaldehyde, propanal, methylpropanal, and 2- and 3-methylbutanal had the greatest impact on the coffee flavor.

Coffee is a drink appreciated as much for its stimulating effect as for its attractive flavor. This is due to the presence of 1.2 to 1.9% caffeine that exists in the raw beans and survives any subsequent processing. Caffeine is an alkaloid, which is bitter in taste and is moderately soluble in hot water. To meet a demand for caffeine-free coffee, the alkaloid may be removed by extracting the dried beans using a nonpolar solvent (e.g., supercritical $CO_2$). This extraction process has a small effect on the flavor of the final beverage.

#### *8.2.2.7.5 Tea*

The beverage that we call tea is a hot water infusion of the dried broken leaves of *Thea sinensis* (L) sims. (= *Camillia sinensis*). This is an evergreen shrub, which grows throughout the tropics from sea level up to about 6000 ft, with the plants from higher altitudes producing what are considered to be the best quality leaves. In commerce, teas are designated by country of origin and often by growing region, but they are also classified by the method of postharvest handling and preparation for market. Three main types are as follows:

- Fermented teas (black tea) — India, Sri Lanka, Java, Sumatra, Pakistan, Japan, Taiwan, Malawi, and Kenya. These represent about 80% of the total tea consumption.
- Unfermented teas (green tea) — Japan, China, India, and Indonesia.
- Semifermented teas (Oolong or Pouchong tea) — Taiwan and Hong Kong.

**Black Tea.** The freshly harvested leaves are spread out on mats and allowed to dry slowly and wither, after which they are mechanically rolled and broken to liberate the oxidase enzymes present in the cell sap and ensure their close contact with precursors. The rolled leaves are then fermented for several hours, usually in baskets or spread out under damp cloths. During this stage the colorless tea tannins are partially oxidized, changing to a reddish brown color, and some essential oil is formed. This primary oxidation gives to the leaves a very agreeable odor and color and determines the flavor quality, strength, body, and color of tea made from them. The process is stopped by "firing" at about 85°C. This induces a secondary oxidation, a reduction in moisture content to about 5% and the full development of the characteristic tea odor and color. The whole process demands skill and experience to achieve a consistent quality.

**Green Tea**. Green teas are not fermented. After harvesting, the fresh leaves are steamed to prevent fermentation and associated color changes. This is followed by partial drying and rolling under pressure. A secondary drying then takes place, and the leaves are rolled and broken after which they are finally dried. The compounds present in the end product comprise tea tannins, largely in their original state. Very little essential oil is formed so that green teas lack the aromatic qualities of fermented teas.

#### *8.2.2.7.6 Tea Flavor*

The principle components of tea comprise (a) an essential oil (about 0.5%) formed during fermentation; (b) caffeine (1.8–5.0%); and (c) tannins (13–18%). The enzymatic formation of the black tea aroma has been investigated and is reviewed by Yamanishi [60,61] who established the involvement of the following biosynthetic pathways:

1. The main precursors are amino acids, carotenoids, including β-carotene, lutein, neoxanthin, and violaxanthin.
2. During fermentation a primary oxidation results in the significant reduction of carotenoids, particularly β-carotene resulting in the formation of ionone and terpenoid carbonyls.
3. A secondary epoxidation takes place during "firing" resulting in the formation of dihydroactinidiole, an epoxyionone, and two trimethyl substituted cyclohexanones.
4. By similar reactions, other carotenoids give rise to ionone, linalool and substituted hydroxy- and epoxy-ionones [63].

While there has been some research done on tea flavor, it is not nearly as comprehensive as that done on cocoa or coffee. Hara [64], Omori [65], and Kobayashi [66] may be consulted for more detail on the flavor of teas.

### 8.2.2.8 Aromatic Vegetables

Most vegetables are subtle in flavor and see little use in producing flavorings for the industry. However, their attractive flavors and bright colors make them valuable ingredients in cooking. The alliaceous vegetables (e.g., onion, garlic, leek, and chives) are popular because of their strong, aromatic character. They have been used domestically since the earliest times and, in spite of handling and processing problems, still form a major commodity in modern food processing. An increase in convenience food products has led to the extensive use of dehydrated forms (e.g., onion chips and powder) and of distilled essential oils. The development of very concentrated fluid extracts, which have the full aromatic profile associated with the freshly cut vegetable and are water miscible, has widened the field of application.

#### *8.2.2.8.1 Onion*

Onion, *Allium cepa* L. is one of the oldest known and most popular of the alliaceous vegetables. Considerable quantities are grown in the U.S. (California, Texas, and New York), Egypt, Japan, most European countries, Mexico, and Brazil. The plant is variably a biennial or perennial although normally grown as an annual either from seed or small bulbs called sets. It has four to six aromatic, cylindrical, hollow leaves. When flowers are present, they are greenish-white, small and in rounded umbels appearing in late summer. Egyptian onion (var. *vivaparum)* produces small bulbs or top-sets in the flower head. The leaves die down at the end of the growing season, leaving a hard-fleshed bulb having a thin outer scaly layer. The shape may be oblong, flat, globular, or oblate and the color brown, white, or red, depending on the variety and source.

Numerous cultivars exist and are classified according to their dry-solids content and pungency, most being raised to satisfy a particular market demand (i.e., for domestic use, pickling, canning, freezing, dehydration, etc.). Onions have almost no characteristic aroma until the tissues are cut or bruised, but once this occurs, flavor is produced very rapidly by the action of enzymes on odorless precursors that coexist in the cell sap. The resulting complex mixture of sulfides and other sulfur-containing compounds, can be recovered as so-called onion oil by distillation although the amount recovered is only in the order of 0.02 to 0.03% and frequently much less than this.

#### *8.2.2.8.2 Onion Flavor*

The biochemistry of onion flavor is now well known and has been extensively reviewed by Shankaranarayana et al. [67]. The chemistry involved has been covered in detail in an earlier chapter of this text. The distinctive flavor of onion and other alliaceous vegetables is due to S-alk(en)yl derivatives of L-cysteine sulfoxide, which are rapidly hydrolyzed by the enzyme alliinase to give an unstable sulfenic acid derivative together with pyruvic acid and ammonia. The sulfenic acid compound breaks down and rearranges to form the relatively more stable thiopropanal S-oxide, which has lachrymatory properties, or it reacts with other compounds to produce a complex mixture of di- or trisulfides which ultimately characterize the product.

In the case of onion, the principle substrate is *trans* (+)S-(l-propenyl)L- cysteine sulfoxide which, when acted on by alliinase, produces a distinctive odor profile, a marked pungency, a transient lachrymatory effect, a bitter aftertaste and a pink discoloration.

**Onion Oil**. The distillation of onion pulp that has been allowed to stand for 12 h results in a dark brown oil, the composition of which has been studied by Brodnitz and Pollack [68]. Later work by Galetto and Bednarczyk [69] established that three components (starred in following listing) are of significance in determining the sensory profile. The principle components were identified as: methyl-1-propyl disulfide*, di-1-propyl trisulfide, 3,4-dimethyl thiophene, cis-1-propyl-propenyl disulfide, cis-methyl-l-propenyl disulfide, methyl-l-propyl trisulfide*, trans-methyl-l-propenyl disulfide, and di-l-propyl trisulfide*. No allyl compounds were found to be present.

More complete listings have been published by Shankaranarayana et al. [67], Guntert and Losing [70], and Farkas et al. [71]. The flavor of onion oil is approximately 4000 times stronger than that of fresh onions and is obviously far too strong to be added directly to food products. Diluted versions, both liquid and dry, are available, and it is necessary to establish the relative flavoring power of these products from the supplier. It should be appreciated that, although such products provide an excellent source of onion flavor, they do not contribute anything of the textural quality provided by the raw vegetable. Furthermore, onion powder may significantly influence the water-holding capacity of some food mixes.

**Dehydrated Onion**. In many food and snack products it is necessary to use dehydrated ingredients and considerable tonnages of dehydrated onion and garlic are produced annually to satisfy this demand. Onions grown especially for dehydration are not generally the same as those used domestically. Optimally these should have a high dry-solids content and a good level of flavor and pungency. Those grown in California are special strains of white globe onions. These are grown intensively, are mechanically harvested, and are inspected in the fields prior to dispatch to the processing factory. Here they are flame peeled, washed to remove the burnt outer skin, and then mechanically sliced onto a perforated drying belt. Drying is carried out in a tunnel drier using hot air that circulates through the carrier belt. Initially the fresh onions have a moisture content of about 80%, and this is reduced to about 4% by the end of the drying cycle. An alternative fluid-bed technique is employed. The dehydrated product may be marketed as "kibbled" to various mesh sizes or "powdered." All of these products are prone to absorb atmospheric moisture and must be packed in well-closed containers and stored in an ambient temperature of above 15°C.

The above process results in some loss of the fresh odor and flavor and the relative flavor strength of fresh and dehydrated onion is difficult to determine owing to the differences in profiles. For general purposes it is reasonable to assume that dehydrated onion is about eight to ten times stronger than fresh onion. Missing top-notes may be added in the form of the fluid extract or onion oil.

#### *8.2.2.8.3 Garlic*

Like onion, garlic has a very long history and has been cultivated in Eastern countries for centuries. It is the compound bulb of *Allium sativum* L., which comprises 8 to 20 small bulblets, having a silky white to pink to mauve colored skin and generally known as cloves, enclosed within a white membranous outer casing. The plant is a hardy perennial with long flattened, pointed leaves arising from a crown. The unbranched stem is about 3 to 4 in (7.5–10 cm) long and carries an apical dense umbel of rose-white to greenish flowers often replaced by sterile pink colored bulbils. The U.S. is the biggest producer of garlic, but significant quantities are also available from Egypt, Bulgaria, Hungary, and Taiwan.

#### *8.2.2.8.4 Garlic Flavor*

Like onion, garlic is without odor until the tissues are cut or bruised, but once this occurs the odor is intensely strong and obnoxious (to some people). Garlic contains about 0.1 to 0.25% volatile constituents, formed enzymatically when garlic cloves are crushed in a manner similar to that of onion. This complex of aromatic sulfides can be recovered as so-called garlic oil by distillation.

The chemistry of garlic flavor parallels that of onion and Freeman and Whenham [72] established that any differences were due to qualitative and quantitative differences in the precursors present. The principle substrate is S-(2-propenyl)l-cysteine sulfoxide, which when acted upon by alliinase produces allyl thiosulfinate (allicin) which ultimately degrades into allyl disulfide, allyl thiosulfonate, and trace quantities of allyl trisulfide. The principle components of garlic oil are as follows [72]: allyl-(l-propyl) disulfide (2%), allyl-methyl disulfide (13.4%), and allyl-(2-propenyl) disulfide (84.6%).

Dehydrated garlic is a creamy-white powder prepared by the dehydration of selected garlic cloves. Like onion powder, it is hygroscopic and must be packed in well-closed containers. For general use a concentration of about 5:1 against fresh garlic is a good starting point though it should be recognized that the flavor profile lacks much of the fine quality associated with fresh garlic.

## REFERENCES


1. IOFI, International Organization of the Flavor Industry, *Code of Practice for the Flavor Industry*, Geneva, 1990.
2. FEMA, *Flavor Extract Manufacturers Assoc.,* Washington, D.C. http://femaflavor.org, accessed April 16, 2005.
3. Ziegler, E., H. Ziegler, *Flavourings: Production, Composition, Applications and Regulations*, Wiley-VCH, New York, 1998, p. 710.
4. CFR, *Code of Federal Regulations*, U.S. Government Printing Office, (http://www.access.gpo.gov), 2003.
5. Heath, H.B., G.A. Reineccius, *Flavor Chemistry and Technology*, Van Nostrand Reinhold, New York, 1986, p. 442.
6. Furth, P., *Summary of Market Trends and Herbs Consumption in The United States*, FFF Associates, Inc., 2001.
7. Arctander, S., *Perfume and Flavor Materials of Natural Origin*, Allured Pub. Corp., Carol Stream, 1960.
8. Tainter, D.R., A.T. Grenis, *Spices and Seasonings: A Food Technology Handbook*, 2nd ed., Wiley-VCH, New York, 2001, p. 249.
9. Jategaonkar, L., M. Marcotte, *Effects of Irradiation on Spices, Herbs and Seasonings — A Review of Selected Literature*, Nordion International, Kanata, Ont., Canada, 1993, p. 26.
10. Leistritz, W., Methods of bacterial reduction in spices, in *Spices: Flavor Chemistry and Antioxidant Properties*, S.J. Risch, C.-T. Ho, Eds., Amer. Chem. Soc., Washington, D.C., 1997, p. 7.
11. Marcotte, M. Irradiation as a disinfestation method — update on methyl bromide phase out, regulatory action and emerging opportunities, *Can. Radiat. Phys. and Chem.*, 52(1-6), p. 85, 1998.
12. Guenther, E., *The Essential Oils*, Vols. 1–6, Van Nostrand, New York, 1948–1952.
13. Gildermeister, E., F. Hoffman, *Die Aetherischen Oele*, 4th ed., Akademie-Verlag, Berlin, 1965-1967.
14. Dirschbaum, E., *Distillation and Rectification*, Chemical Pub. Co., Brooklyn, 1948, p. 426.
15. Ashurst, P.R., *Food Flavoring,* 2nd ed., Blackie Academ. Prof., New York, 1995, p. 332.
16. Lawrence, B.M., Progress in essential oils: Spanish marjoram oil, *Perfum. Flavorist*, 2004, 29(3), pp. 44–46, 48–54, 56–64, 66–67.
17. Lawrence, B.M., The essential oil of *Micromeria teneriffae* Benth, *J. Essent. Oil Res.*, 1(1), p. 43, 1989.

18. Lawrence, B.M., B.D. Mookherjee, B.J. Willis, Flavors and fragrances: a world perspective, *Dev. Food Sci.*, 18, p. 444, 1988.
19. Provatoroff, N., Some Details of the Distillation of Spice Oils, Proc. Tropical Prod. Inst. Conf., London, 1972.
20. del Valle, J.M., O. Rivera, O. Teuber, M.T. Palma, Supercritical $CO_2$ extraction of Chilean hop (*Humulus lupulus*) ecotypes, *J. Sci. Food Agric.*, 83(13), p. 1349, 2003.
21. Braga, M., P. Leal, J. Carvalho, M. Meireles, A. Angela, Comparison of yield, composition, and antioxidant activity of turmeric (*Curcuma longa* L.) extracts obtained using various techniques, *J. Agric. Food Chem.*, 51(22), p. 6604, 2003.
22. Povh, N.P., M.O.M. Marques, M.A.A. Meireles, Supercritical $CO_2$ extraction of essential oil and oleoresin from chamomile (*Chamomilla recutita* [L.] Rauschert), *J. Supercrit. Fluids*, 21(3), p. 245, 2001.
23. Calame, J.P., R. Steiner, $CO_2$ extraction in the flavour and perfumery industries, *Chem. Ind.*, (June 19), p. 399, 1982.
24. Brogle, H., Carbon dioxide as a solvent: its properties and applications, *Chem. Ind.*, (June 19), p. 385, 1982.
25. Moyler, D.A., Carbon dioxide extracted ingredients for fragrances, *Perfum. Flavorist*, 9(2), p. 109, 1984.
26. Braverman, J.B.S., L. Solomiansky, Separation of terpeneless essential oils by chromatographic methods, *Perf. Ess. Oil Rec.*, (48), p. 284, 1957.
27. Lawrence, B.M., Progress in essential oils, *Perfum. Flavorist*, 19(6), p. 57, 1994.
28. Smith, D.M., L. Levi, Treatment of compositional data for the characterizing of essential oils: determination of geographical origins of peppermint oils by gas chromatographic analysis, *J. Agric. Food Chem.*, 9, p. 230, 1961.
29. Formácek, V., K.H. Kubeczka, *Essential Oils Analysis by Capillary Gas Chromatography and Carbon-13 NMR Spectroscopy*, John Wiley, New York, 1982, p. 373.
30. Fenaroli, G., *Fenaroli's Handbook of Flavor Ingredients*, T.E. Furia, N. Bellanca, Eds., CRC Press, Boca Raton, FL, 1975.
31. Lawrence, B.M., Progress in essential oils, *Perfum. Flavorist*, 28(6), p. 56, 2003.
32. Lawrence, B.M., Progress in essential oils, *Perfum. Flavorist*, 28(4), p. 78, 2003.
33. Lawrence, B.M., Progress in essential oils, *Perfum. Flavorist*, 27(6), p. 46, 2002.
34. Lawrence, B.M., Progress in essential oils, *Perfum. Flavorist*, 27(1), p. 42, 2002.
35. Heath, H.B., *Flavor Technology: Profiles, Products, Applications*, Westport, AVI Pub. Co., 1987, p. 542.
36. Bates, R.P., J.R. Morris, P.G. Crandall, *Principles and practices of small- and medium-scale fruit juice processing*, Rome, Food Agriculture Organization United Nations, p. 226, 2001.
37. Naim, M., S. Wainish, U. Zehavi, H. Peleg, R.L. Rouseff, S. Nagy, Inhibition by thiol compounds of off-flavor formation in stored orange juice. I. Effect of L-cysteine and N-acetyl-L-cysteine on 2,5-dimethyl-4-hydroxy-3(2H)-furanone formation, *J. Agric. Food Chem.*, 41(9), p. 1355, 1993.
38. Novozyme, *Enzymes for juice extraction*, (http://www.novozymes.com), accessed 9/4/2004.
39. Purseglove, J.W., E.G. Brown, C.L. Green, S.R. Robbins, *Spices*, Vols. 1 and 2, Essex, U.K.; Burnt Mill, Harlow, 1981.
40. Muralidharana, A. and C. Balagopol, Studies in curing vanilla, *Indian Spices*, 10(3), p. 3, 1973.
41. Rossenbaum, E.W., Vanilla extracts, synthetic vanillin, in *Encyclopedia of Food Technology*, AVI Pub. Co., Westport, 1974.

42. Dignum, M.J.W., J. Kerler, R. Verpoorte, Vanilla production: technological, chemical, and biosynthetic aspects, *Food Rev. Int.*, 17(2), p. 199, 2001.
43. Havkin-Frenkel, D. and R. Dorn, Vanilla in *Spices: Flavor Chemistry and Antioxidant Properties,* S.J. Risch, C.-T. Ho, Eds., Amer. Chem. Soc., Washington, D.C., 1997, p. 29.
44. Klines, I. and D. Lamparsky, Vanilla volatiles — a comprehensive analysis, *Int. Flavours Food Addit.*, 7, p. 272, 1976.
45. Adedeji, J., T.G. Hartman, C.-T. Ho, Flavor characterization of different varieties of vanilla beans, *Perfum. Flavorist*, 18(2), p. 25, 1993.
46. Hartman, T. G., K. Karmas, J. Chen, A. Shevade, M. Deagro, H. Hwang, Determination of vanillin, other phenolic compounds, and flavors in Vanilla beans, in *Phenolic Compounds in Food and their Effects on Health,* C.-T. Ho, C. Lee, M. Huang, Eds., Amer. Chem. Soc., Washington, D.C., 1992, p. 61.
47. Bala, K., Natural Vanilla in America and Flavor Chemistry, A presentation given at The First International Congress on the Future of the Vanilla Business, Princeton, N.J., Nov. 2003, (http://www.aesop.rutgers.edu/~vanilla2003), accessed 9/4/04.
48. Hartman, T.H., Composition of Vanilla Beans from Different Geographical Regions, A presentation given at The First International Congress on the Future of the Vanilla Business, Princeton, N.J., Nov. 2003, (http://www.aesop.rutgers.edu/~vanilla2003), accessed 9/4/04.
49. Cowley, E., Vanilla and Its Uses, Proc. Tropical Prod. Inst. Conf., London, 1972.
50. Willson, K.C., *Coffee, Cocoa and Tea,* Oxford Publ., New York, 1999, p. 300.
51. Beckett, S.T., *Industrial Chocolate Manufacture and Use*, 3rd ed., Blackwell Science, Malden, 1999, p. 488.
52. Hilton, K., K. Lumpur, Proceedings: International Cocoa Conference, Challenges in the 90s, Kota Kinabalu (Sabah), Malaysian Cocoa Board, xxvii, pp. 611, 1994.
53. Maga, J., Cocoa flavor, *CRC Crit. Rev. Food Technol.*, 4, p. 39, 1973.
54. Schnermann, P., P. Schieberle, Evaluation of key odorants in milk chocolate and cocoa mass by aroma extract dilution analyses, *J. Agric. Food Chem.*, 45(3), p. 867, 1997.
55. Counet, C., D. Callemien, C. Ouwerx, S. Collin, Use of gas chromatography-olfactometry to identify key odorant compounds in dark chocolate: comparison of samples before and after conching, *J. Agric. Food Chem.*, 50(8), p. 2385, 2002.
56. Rohan, T.A., T. Stewart, The precursors of chocolate aroma, *J. Food Sci.*, 32, p. 395, 1967.
57. International Coffee Organization, London, U.K. (http://www.ICO.org, accessed 04/15/05).
58. Flament, I., *Coffee Flavor Chemistry,* New York, Wiley Publ., 2002, p. 424.
59. Czerny, M., F. Mayer, W. Grosch, Sensory Study on the character impact odorants of roasted Arabica coffee, *J. Agric. Food Chem.*, 47(2), p. 695, 1999.
60. Grosch, W., Flavor of coffee, *Nahrung*, 42(6), p. 344, 1998.
61. Yamanishi, T., The aroma of various teas, in *Flavor in Food and Beverages*, G. Charalambous, G. Inglett, Eds., Acad. Press, New York, 1978, p. 305.
62. Yamanishi, T., Tea, coffee, cocoa and other beverages, in *Flavor Research, Recent Advances*, R. Teranishi, R. Flath, H. Sugisawa, Eds., M. Dekker, New York, 1981, p. 231.
63. Sanderson, G.W., N.H. Graham, The formation of black tea aroma, *J. Agric. Food Chem.*, 21(4), p. 576, 1973.
64. Hara, Y., Flavor of tea, *Food Rev. Int.*, 11(3), p. 477, 1995.
65. Omori, M., Characterization of tea flavor, *Koryo*, 193, p. 59, 1997.

66. Kobayashi, A., Flavor precursors of green tea, oolong tea, and black tea, *Gendai Kagaku*, 297, p. 16, 1995.
67. Shankaranarayana, M.L., Volatile sulfur compounds in food flavors, *CRC Crit. Revs. Food Technol.*, 4, p. 395, 1974.
68. Brodnitz, M.H., C.L. Pollack, Gas chromatographic analysis of distilled onion oil, *Food Technol.*, 24(1), p. 78, 1970.
69. Galetto, W.G., A.A. Bednarczyk, Relative flavor contribution of individual volatile components of the oil of onion (Allium cepa), *J. Food Sci.*, 40(6), p. 1165, 1975.
70. Guntert, M., G. Losing, *Onion oil: Importance, Composition and Authenticity Tests,* Dragoco Report (Dragoco Co., Germany), 4, p. 165, 1999.
71. Farkas, P., P. Hradsky, Novel flavor components identified in the steam distillate of onion (Allium cepa), *Zeitschrift Lebensm.-Untersuch.–Forsch.*, 195(5), p. 459, 1992.
72. Freeman, C.G., R.J. Whenham, Nature and origin of volatile flavour components of onion and related species, *Int. Flavours Food Addit.*, 7, p. 222, 1976.

# 9 Flavoring Materials Made by Processing

## 9.1 INTRODUCTION

The term *processed flavor* is used in different contexts to describe: (a) products where the natural raw materials lack a characteristic flavor profile and the desired aromatic profile is achieved only by deliberate processing (e.g., coffee); (b) flavorings created as a result of Maillard and other related reactions between amino acids and sugars (e.g., meat-like flavors); (c) flavorings resulting from controlled enzymatic reactions (e.g., enzyme-modified dairy products); (d) products made by fermentation (e.g., wines, vinegar); and (e) products of thermal reactions of lipids (e.g. French fry flavor).

Some of these flavorings are produced as a result of a combination of these processing methods (e.g., coffee and cocoa, which involve fermentation and roasting; tea, which may or may not be fermented before roasting; vanilla, which must be fermented and cured; and thermally processed enzymatically modified dairy products). Some of these products (or reactions leading to products) have already been discussed in Chapter 5, but there will be a greater emphasis in this chapter in producing flavorings rather than just discussing the chemistry involved.

## 9.2 NATURAL PRODUCTS MADE BY ROASTING: COCOA/CHOCOLATE

The production of cocoa and chocolate from cocoa beans involves three prime processes: (a) fermentation, (b) drying, and (c) roasting, each of which has a significant effect upon the final flavor profile. The chemistry involved is complex [1–3] and the quantitative mix of aromatic constituents resulting from processing is determined by the precise conditions employed. Processing may be summarized as follows:

1. Fermentation facilitates separation of beans from the surrounding pulp and shell, inactivates the seeds, initiates the aromatic precursors and color development, and significantly reduces bitterness. During this stage the sucrose and proteinaceous constituents are partially hydrolyzed, any polyphenolic compounds are oxidized and glucose is converted into alcohol and then oxidized to acetic and lactic acids. Beans that have not been fermented do not develop a chocolate flavor when roasted [4].
2. Drying removes moisture to about 15%.
3. Roasting develops the desired flavor profile and completes the color changes, further reduces the moisture content to about 5%, conditions the

shell facilitating its removal, and makes the nib more friable. During roasting a portion of the initial fermentation products are converted by Maillard reactions into a complex mixture of volatile carbonyls, pyrazines, etc., which characterize the aroma and flavor of the final chocolate [5–7].

The temperature conditions encountered during the roasting process depend on the type of roaster employed, but they usually lie within the range 115 to 140°C. The precise conditions are still one of skill and judgment based on knowledge of the raw materials used and the ultimate product being made.

### 9.2.1 Production of Cocoa Powder

Having roasted the beans, it is necessary first to separate the shell from the kernel by passing it through rollers and then separating the broken shell and the cocoa nibs by winnowing, using graded screens and air elution. Final separation is carried out to ensure that the nibs are completely free from shell (shell is limited to 1.75% by law) prior to the next stage of processing.

Cocoa powder is manufactured by grinding the press cake after partial removal of cocoa butter. This product demands high roasted beans to give the end product a better flavor and darker color. Subjecting the separated nibs to a pressure of 7,000 lb/in$^2$ leaves a residual 20 to 28% cocoa butter in the press cake. On leaving the press, the cake is broken to form small pieces of about 1 inch size, which are then subjected to further milling under carefully regulated temperature conditions and cooled rapidly to 20 to 23°C immediately on discharge, to ensure a properly tempered cocoa butter content, the correct particle size, and an acceptable color.

Cocoa powder contains about 24% fat, 18% nitrogenous matter, 5.5% fiber, 10.5% starch, and about 4% moisture. When permitted, commercial products may contain additives such as sucrose esters, tripolyphosphate, disodium sulfosuccinate, or alginates in order to improve "wettability" as well as vanillin and other flavorings to give them a more rounded profile.

### 9.2.2 The Dutch Process

Dutching, or alkalizing, is carried out to modify the color and flavor of cocoa. Depending on the degree of alkalization, the color becomes darker and the flavor milder and much less harsh. This process involves the addition of alkali to the milled chocolate liquor accompanied by efficient mixing and the removal of water by heating.

### 9.2.3 Chocolate

Chocolate is a suspension of cocoa particles and sugar held in cocoa butter, the manufacture of which entails four unit operations:

1. Mixing is done with the correct proportions of sugar, cocoa butter, and cocoa liquor in a special plant known as a melangeur. This consists of two heavy rollers mounted over a heated rotating steel or granite bed, although other designs for continuous automatic operation are widely used

so that mixing and flavor development are optimized. The mixing operation is carried out at 50 to 60°C and usually takes 20 to 25 min, but in an automatic plant the dwell time may be only about 5 min. The aim of this stage is to reduce the particle size of the sugar and promote the loss of unwanted volatile acids. The product is a soft, plastic, pliable, non-flowing mass.
2. Refining is a size reduction process. This is carried out in a mill fitted with multiple, water cooled rollers [3–5]. Chocolate is usually refined to a particle size of 50–114 μm of which less than 10% are smaller than 5 μm and not more than 20% are larger than 22 μm. It is the distribution of particle sizes that determines the eating quality of the resulting chocolate [8].
3. Conching is necessary to ensure full flavor development by the elimination of the remaining volatile acids; it also removes moisture, smoothes angular sugar crystals, modifies the viscosity and alters the color of the product. The specially designed equipment produces efficient agitation and aeration to ensure good mixing and exposure of the product to air over a period of between 1 and 5 days at a controlled temperature below 100°C, depending on the type of chocolate being made. During conching, further oxidative changes occur together with some caramelization.
4. Tempering is necessary to induce crystallization of the cocoa butter in a stable form in the fluid chocolate mass. Cocoa butter can crystallize in six forms: I ($\gamma$, melts at 17°C and is very unstable), II ($\alpha$, melts at 21 to 24°C), III and IV ($\beta'$, melts at 25 to 29°C), V ($\beta$, melts at 34 to 35°C and is the preferred form), and VI (associated with chocolate bloom) [9,10].

The use of cocoa and chocolate products as flavoring materials is described by Minifie [9] and Beckett [10]. Key volatiles responsible for the characterizing aroma of chocolate have been reported by Schnermann and Schieberle [11] and Schieberle and Pfnuer [12] (briefly summarized in section 8.2.2.7.2 Cocoa/Chocolate Flavor).

## 9.3 PROCESS FLAVORS: MEAT-LIKE FLAVORS

The Maillard reaction can produce very many volatile compounds having strong and distinctive odors, depending on the reactants and the reaction conditions [13]. The classical Maillard reaction and the closely associated Strecker degradation have already been discussed in Chapter 5. Although these two reactions are only a part of the total chemical changes occurring during the heat treatment of foods, they generate more characteristic flavoring effects in processed foods than any other reaction.

The reproduction of meat-like flavors through process chemistry has been a primary target of the flavor industry for many years. Meats are expensive, and thus there is a strong financial incentive to develop substitute flavorings. The chemistry of raw and cooked meat flavor has been the subject of considerable research over the past 30 years, and this has provided the flavor industry with invaluable data for the recreation of process meat-like flavorings [14–18].

### 9.3.1 The Evolution of Process Meat-Like Flavorings

The development of meat-like flavorings has been an evolutionary process [19]. Early meat-like flavorings would probably more correctly be called meat extenders and were largely made from spice blends. The manufacturer would use spices normally associated with specific meats and sell them as meat enhancers or extenders. In the U.S. culture, one can envision sage being associated with pork, turmeric and celery with chicken, and onion and black pepper with beef. The use of these spice combinations did little to enhance the true meat flavor, but at least there was a flavor. Meat extracts have found some use in the industry and still do today. These byproducts of the meat processing industry (e.g., corned beef) provide little desirable flavor but contribute to a label statement of meat being present.

The beginning of thermal processes as a means to produce meat-like flavorings was the introduction of hydrolyzed vegetable proteins (HVPs). These products are thermally processed and bring both a characteristic flavor of their own (primarily from processing) and an abundance of flavor enhancers, particularly monosodium glutamate (MSG). Recent concern over the presence of monochloropropanol and dichloropropanol (suspected carcinogens) in HVPs has resulted in a new generation of HVPs produced through enzymatic action as opposed to heat and acid [20,21]. However, these new products do not offer the same flavor profile as the traditional products and thus are not as desirable.

The discovery and commercialization of savory flavor enhancers were a boon to these flavorings. The contribution of MSG and the 5′ nucleotides cannot be overestimated. While the consumer often states that these ingredients are not desired, they are intimately associated with the recognized meat-like flavorings and are indispensable. The final development in these flavorings has been the advent of topnoting, i.e., the addition of synthetic chemicals that add potency or specific sensory nuances to the base flavor. Aroma chemicals such as 2-methyl-3-furanthiol fortify the characteristic flavor (e.g., beef in this case) of the meat-like flavoring and thereby afford greater flavor potency, which is generally lacking in this flavoring group. Other chemicals may be added to provide a given sensory nuance such as a rare, bloody, or roasted note to the base flavor.

The advent of gas chromatography and interest by the academic community in flavor chemistry in the 1960s offered promise that the aroma of meat could be reproduced synthetically. It was proposed that if one could identify all of the aroma components of meat, the aroma of that meat could be reconstituted from pure chemicals to yield a quality, formulated meat flavoring. This promise has never been fully realized for several reasons. One was that meat aroma proved to be extremely complex, being made up of well over 800 aroma chemicals. A large number of these aroma compounds either are not commercially available or cannot be added to foods for legal reasons (i.e., a lack of toxicological data). Simply the challenge of reformulating 800+ aroma chemicals into a flavoring in a processing environment is daunting. These limitations resulted in efforts to determine which chemicals were truly *required* to provide a characteristic meat aroma. It has been proposed that 20 to 30 chemicals may be adequate to reproduce most food flavors so the question was which 20 to 30 chemicals? If these chemicals are available and permitted to be used in foods, then perhaps we can formulate synthetic meat-flavors. This line of research was pursued in the 1980s and '90s largely by Prof. Grosch and his research group in Garching. At this time, we have solid data on the chemicals needed to recreate meat aroma [22–32]. The primary hurdle today is that many of these chemicals are not available as natural compounds, and perhaps this blend of volatiles still lacks elements of carrying all of the characteristic meat flavor, i.e., aroma and taste.

The inability of the flavor industry to offer natural meat flavorings via chemical recombination spawned considerable research on reaction systems (meat flavor precursor systems). In the 1960s and '70s, a host of patents were issued outlining reaction formulations and processing conditions that created meat-like flavorings [19]. These flavorings did a reasonable job of providing both the taste and aroma of meats with the added benefit of carrying a natural label (provided the ingredients were natural and process conditions within approved guidelines). Much of this work was based on studies that pointed to the presence and consumption of certain flavor precursors in meats (primarily amino acids and reducing sugars) during cooking. It was assumed that a given flavor, e.g., roasted beef, could be formed in a reaction system from the same primary reactants and reaction conditions as exist in the product to be imitated, e.g., beef. This fundamental research, as well as a large helping of trial and error, has lead to an understanding of how to create meat-like flavorings through process chemistry. Processes for creating these flavorings will be described in the following sections. International Organization of the Flavor Industry (IOFI) guidelines for the creation of these flavorings are presented in Table 9.1.

## 9.3.2 The Creation of Process Flavorings

### 9.3.2.1 Reaction System Composition

The base ingredients used in process flavors generally include: (a) a protein nitrogen source, (b) a carbohydrate, (c) a fat or fatty acid, (d) water, (e) a pH regulator(s), and (f) various flavor enhancers. Depending upon the flavor and the creator, numerous other ingredients may be added to provide a given sensory note, e.g., smoke, or spice.

#### *9.2.2.1.1 Sulfur Sources*

If one is creating a meat-like flavor, a sulfur source is required in the reaction mixture. This sulfur source is most commonly cysteine and/or cystine (free acids or hydrochlorides). However, hydrogen sulfide has been used as a replacement (since the sulfur amino acid is typically considered a generator of hydrogen sulfide), but hydrogen sulfide is both corrosive and toxic in the workplace, and thus manufacturers prefer to work with the amino acids. Thiamin is an excellent source of hydrogen sulfide and will form various meat-like flavors by heating it alone (in water) at selected pHs, times, and temperatures. The aroma of heated thiamin initially is egg-like (hydrogen sulfide), but then it becomes more chicken-like and finally beef-like. While thiamin is an excellent precursor for the creation of meat-like flavorings, its cost, natural status, and vitamin activity (legal considerations) all are negative factors. Methionine is used to provide vegetable notes (potato due to its thermal degradation

## TABLE 9.1
## International Organization of the Flavor Industry (I.O.F.I.) Guidelines for the Production and Labeling of Process Flavorings

Process flavorings are produced by heating raw materials which are foodstuffs or constituents of foodstuffs in similarity with the cooking of food.

The most practicable way to characterize process flavorings is by their starting materials and processing conditions, since the resulting composition is extremely complex, being analogous to the composition or cooked foods. They are produced every day by the housewife in the kitchen, by the food industry during food processing and by the flavor industry.

The member associations of I.O.F.I. have adopted the following Guidelines in order to assure the food industry and the ultimate consumer of food of the quality, safety and compliance with legislation of process flavorings.

1. *Scope*
   1.1. These Guidelines deal with thermal process flavorings; they do not apply to foods, flavoring extracts, defined flavoring substances or mixtures of flavoring substances and flavor enhancers.
   1.2. These Guidelines define those raw materials and process conditions which are similar to the cooking of food and which give process flavorings that are admissible without further evaluation.

2. *Definition*
   A thermal process flavoring is a product prepared for its flavoring properties by heating food ingredients and/or ingredients which are permitted for use in foodstuffs or in process flavorings.
3. *Basic Standards of Good Manufacturing Practice*
   The chapter 3 of the Code of Practice for the Flavor Industry is also applicable to process flavorings.

4. *Production of Process Flavorings*
   Process flavorings shall comply with national legislation and shall also conform to the following:
   4.1. Raw materials for process flavorings. Raw materials for process flavorings shall consist of one or more of the following:
      4.1.1. A protein nitrogen source:
         • Protein nitrogen containing foods (meat, poultry, eggs, dairy products, fish, seafood, cereals, vegetable products, fruits, yeasts) and their extracts
         • Hydrolysis products of the above, autolysed yeasts, peptides, amino acids and/or their salts.
      4.1.2. A carbohydrate source:
         • Foods containing carbohydrates (cereals, vegetable products, and fruits) and their extracts
         • Mono-, di- and polysaccharides (sugars, dextrins, starches, and edible gums)
      4.1.3. A fat or fatly acid source: foods containing fats and oils
         • Edible fats and oils from animal, marine, or vegetable origin
         • Hydrogenated, transesterified, and/or fractionated fats and oils
         • Hydrolysis products of the above.
      4.1.4. Materials listed in Table 9.2.
   4.2. Ingredients of process flavorings
      4.2.1. Natural flavorings, natural and nature identical flavoring substances and flavor enhancers as defined in the I.O.F.I. Code of Practice for the flavor industry.
      4.2.2. Process flavor adjuncts. Suitable carriers, antioxidants, preserving agents, emulsifiers, stabilizers and anticaking agents listed in the lists of flavor adjuncts in Annex II of the I.O.F.I. Code of Practice for the flavor industry.

**TABLE 9.1 (continued)**
**International Organization of the Flavor Industry (I.O.F.I.) Guidelines for the Production and Labeling of Process Flavorings**

4.3. Preparation of process flavorings. Process flavorings are prepared by processing together raw materials listed under 4.1 and 4.1.2 with the possible addition of one or more or the materials listed under 4.1.3 and 4.1.4.

4.3.1. The product temperature during processing shall not exceed 180°C.

4.3.2. The processing time shall not exceed ¼ hour at 180°C with correspondingly longer times at lower temperatures.

4.3.3. The pH during processing shall not exceed 8.

4.3.4. Flavorings, flavoring substances and flavor enhancers (4.2.1) and process flavor adjuncts (4.2.2) shall only be added after processing is completed.

4.4. General requirements for process flavorings.

4.4.1. Process flavorings shall be prepared in accordance with the General Principles of Food Hygiene (CAC/Vol A-Ed. 2 (1985) recommended by the Codex Alimentarius Commission.

4.4.2. The restrictive list of natural and nature-identical flavoring substances of the I.O.F.I. Code of Practice for the Flavor Industry applies also to process flavorings.

5. *Labeling*

The labeling of process flavorings shall comply with national legislation.

5.1. Adequate information shall be provided to enable the food manufacturer to observe the legal requirements for his products.

5.2. The name and address of the manufacturer or the distributor of the process flavoring shall be shown on the label.

5.3. Process flavor adjuncts have to be declared only in case they have a technological function in the finished food.

to methional) to the flavoring. The patent literature abounds with suggestions for other sulfur sources, which include a host of plant extracts (e.g., extracts from the Allium family) as well as pure chemicals (e.g., methanethiol, ethanethiol, and dimethylsulfide). However, the amino acids mentioned earlier are the workhorses of this group of flavorings.

#### *9.3.2.1.2 Reducing Sugars*

The second critical component of meat-like flavorings is a reducing sugar. While amino acid choice strongly influences the sensory character of the final flavoring, carbohydrate choice has less influence on character. Reducing sugars cannot arbitrarily be interchanged in a reaction mixture, but this choice has a more subtle influence on flavor character.

Glucose is the most common sugar used for this purpose. Glucose is cheap and reasonably reactive. Arabinose, xylose, fructose, and lactose also find application in various meat-like flavorings. Of these latter reducing sugars, the use of xylose has increased in recent years. Xylose, being a pentose, is more reactive in the Maillard reaction and thus shorter process times are possible. The cost of xylose has until recently prohibited its use in process flavorings.

**TABLE 9.2**
**Materials Used in Procession**

| |
|---|
| Herbs, spices, and their extracts |
| Water |
| Thaimine and its hydrochloric acid salt |
| Ascorbic, citric, lactic, fumaric, succinic, and tartaric acids |
| The sodium, potassium, calcium, magnesium, and aluminum salts of the above acids |
| Guanylic acid and inosinic acid and its sodium, potassium and calcium salts |
| Inositol |
| Sodium, potassium, and ammonium sulfides and hydrosulfides, and polysulfides |
| Lecithin |
| Acids, bases and salts as pH regulators |
| Acetic, hydrochloric, phosphoric, and sulphuric acids |
| Sodium, potassium, calcium, and ammonium hydroxide |
| The salts of the above acids and bases |
| Polymethylsiloxane as an antifoaming agent (participating in the process) |

*Source:* From Manley, C.H., Flavor Interactions Workshop, IFT sponsored, Orlando, FL, 2003. With permission.

Ribose, ribose-5-phosphate, hydrolyzed polysaccharides, yeast, and yeast autolysates also appear in the patent literature as carbohydrate sources for these flavorings. There are no economical sources of ribose or ribose-5-phosphate. Yeast and yeast autolysates both will contribute some free carbohydrate.

Similar to the amino acids, the carbohydrate may be replaced by a pure chemical such as propanal or diacetyl. The Maillard reaction is broadly the reaction of an amine with a carbonyl, and thus the carbonyl does not have to be a reducing sugar.

#### *9.3.2.1.3 Acids*

Acids are typically used to adjust the pH of the reaction system. Meat-like flavorings are generally thermally processed at pHs around 5.2, similar to the pH of meat. Phosphoric acid is most commonly used since it is known to be a catalyst of the Maillard reaction. Other inorganic acids also may be used such as hydrochloric or sulfuric acids. Numerous organic acids appear in the patent literature. These acids typically contribute a flavor of their own and may be used to not only produce a given pH but also add a desired sensory note. Citric, lactic, acetic, propionic, malic, succinic, and tartaric acids appear in the patent literature.

#### *9.3.2.1.4 Lipids*

Some of the earlier research on meat flavors demonstrated that the common meaty characteristic of cooked meats can be reproduced by heating the water soluble components of the meats [33–35,16]. However, the differentiation amongst species was found when the lipid portion of the meat was heated. This has resulted in most meat-like flavorings using a similar base meat reaction system but then using added lipid to provide the unique species character of the meat, lamb or fish, for example.

The lipid is typically fat trimmings from the animal whose flavor is to be reproduced. While the industry is now able to add synthetic chemicals to create much of this species differentiation without using lipid in the reaction system, the chemicals are not natural, and the product must carry an artificial label.

The addition of lipids (as animal byproducts) to the reaction system is problematic. The issues resulting from this addition include:

1. Lipids are insoluble in the reaction system, so the flavoring either has a fat cap (not homogeneous) or has to include emulsifiers (label issue)
2. The lipids result in a cloudy product upon final use — clear stocks cannot be prepared
3. If a dry flavor form is desired, lipids make drying more difficult
4. The final product may have an oil slick (appearance)
5. Religious reasons (Kosher, Halal)
6. Fear of animal products (mad cow disease)
7. The lipids must be stabilized against oxidation (label issue)
8. The use of animal byproducts requires that the flavor company has a USDA compliant manufacturing facility including inspection

Probably the largest barrier to most companies using lipids (as animal byproducts) is this latter requirement. A flavor company uses very few materials that support microbial growth and, in fact, resembles a chemical company more than a food company. Thus, the need to operate the equivalent of a meat processing facility is outside their normal operating environment.

Some of the listed issues are irrelevant when nonanimal lipid sources are used. Thus one finds numerous patents proposing that plant oils or specific fatty acids may be used to create species character (the author is not aware of there being much success in this respect).

#### *9.3.2.1.5 Solvents*

The patent literature lists several solvents for use in the creation of meat-like flavorings. They include water, alcohol, propylene glycol, and glycerol. For meat-like flavorings, water is by far the most commonly used solvent. The only reason to consider a different solvent is to alter the water activity of the reaction system. For meat-like flavorings, this rationale is not compelling because meats have very high water activities, which must be duplicated in the reaction system. In the case of creating a chocolate or coffee process flavoring, nonaqueous solvents are commonly used since the cocoa and coffee beans are dry when roasted, and thus a low water activity must be obtained in the reaction system to duplicate this condition.

#### *9.3.2.1.6 Hydrolyzed Vegetable Proteins (HVP), Savory Enhancers (MSG and 5′-Nucleotides), and Autolyzed Yeast Extracts (AYE)*

These common components of meat-like flavorings are nearly universally used in the industry [19]. While there has been and continues to be a negative consumer attitude towards the use of MSG and HVP, their use continues since they are essential to the sensory character of meat-like flavorings. (MSG is associated with the Chinese

Restaurant Syndrome. HVPs are not desired because they may contain the earlier noted carcinogens [Section 9.3.1] and they are abundant sources of MSG). The historical, abundant use of the HVPs, MSG, and AYEs in these flavorings has resulted in their characteristic flavor being recognized as an integral part of the flavoring despite real meat having only subtle touches of these components. Some companies have chosen to offer meat-like flavorings without these components, but they have very little success in the marketplace. Since each of these ingredients will be discussed later in this chapter or in Chapter 11 (Flavor Potentiators), there will be no additional discussion here.

#### *9.3.2.1.7 Miscellaneous Components*

In addition to the components discussed above, numerous minor (in quantity) components may be added to the reaction system after thermal treatment. They include:

**Salt** — There may be more than enough salt in the flavoring already due to the use of HVPs or neutralization of the reaction mixture (the reaction mixture is normally neutralized after thermal treatment). However, in some applications, additional salt may be needed.

**Preservatives** — Preservatives (e.g., chelators or antioxidants) may be added after processing to inhibit the oxidation of lipids used in the reaction mixture.

**Thickeners** — Thickeners (e.g., gums or starches) may be added to the reacted mixture to stabilize the system against physical separation. The reaction mixture may contain various insoluble components (not only lipids) that will separate during cooling and storage. Increasing the viscosity will help provide a homogeneous product.

**Surfactants** — Surfactants (e.g., lecithins or monoglycerides) may be added after processing to emulsify the lipids into the reaction system. This again insures a homogeneous flavoring.

**Flavorants** — Numerous flavorants may be added after processing. They may be complete flavorings such as a smoke flavoring, various spice or herb essential oils, or oleoresins. The addition of top notes also may be put into this category. Top noting refers to the addition of individual pure chemicals that strengthen or modify the character of the process flavor. Top noting has become very common (as was mentioned earlier) and is a current area of significant research.

**Colorants** — There is little rationale to add colorants (e.g., caramel color) to a meat-like flavoring other than for the convenience of the final user. The colorant would be added to color the final product, thus the choice of the colorant and its usage level would be customer product dependent.

**Drying agents** — Drying agents (e.g., low DE maltodextrins) may be added to the processed reaction mixture to facilitate drying (provide a dry flavor form). Processed reaction mixtures are very hygroscopic and become sticky at elevated temperatures, and thus are difficult to dry without the addition of some drying aid. Low DE maltodextrins are commonly used for this purpose since they are cheap, bland, and dry well. Pan (or tray) drying may be accomplished with minimal addition of a drying aid. However, if one desires a spray dried product, it may be necessary to use nearly a 1:1 (solids basis) ratio of maltodextrin to reaction mixture solids. This results in a significant dilution of the flavoring. One of the common criticisms of this flavor category is that they lack flavor strength. Dilution of the flavoring to facilitate drying exacerbates this weakness.

**Free flowing agents** — As noted above, reaction products are very hygroscopic. This property means that they will very readily cake on storage unless some protection against moisture uptake is provided. Caking is minimized by adding drying aids to increase the glass transition temperature of the mixture, using a good vapor barrier packaging, and/or adding silicates. Adding drying aids dilute the product (as noted earlier), and thus this is not necessarily the desired choice. Applying packaging with high water vapor barrier properties adds cost and has minimal value once the package is opened. The use of free flowing agents is the common approach. Silicates will absorb considerable moisture thereby reducing the need for expensive packaging. It must be remembered, however, that silicates will not stop caking but only prolong the time to caking. The process flavor will cake at the same water activity with or without silicate, but the competitive absorption of water by the silicates simply prolongs the time until this water activity is reached.

#### 9.3.2.2 Reaction Conditions

Different blends of reactants may be heated under a wide range of conditions (including solids content and pH) to create a wide range of flavorings. The IOFI and Flavor Extract Manufacturers Association (FEMA) have established guidelines for the thermal processing of such products as not exceeding 15 min at 180°C or proportionately longer at lower temperatures. More recently, additional regulations have come into being that set minimum processing conditions. As a result of concern that some food processors may be adding water binding ingredients (e.g., proteins) to meat-products under the guise of natural flavorings, it is now required that these flavorings be processed minimally at 100°C for 15 min.

The time and temperature of heating strongly influence the flavor character of the process flavoring. It is generally acknowledged that heating temperatures of ca. 100°C yield cooked or stewed meat flavors and temperatures of ca. 120°C produce roasted meat flavors. Heating may be done in an open pan (which allows volatiles to escape) if the system is heated at temperatures equal to or below boiling, or in a closed vessel (maintains volatiles in the system) if desired, or higher temperatures are used. Capturing volatiles or allowing them to escape influences the flavor character. In some processes, the manufacturer may choose to heat a part of the formulation under one set of conditions and then add additional ingredients and further heat.

While the pH during heating is generally about 5.2, IOFI and FEMA guidelines state that pH should be between 3 and 8. After heating, the pH is generally adjusted to neutrality. Examples of meat-like flavoring formulations and processing were presented by Manley [36] and are summarized in Table 9.3.

**TABLE 9.3**
**Examples of Process Flavor Formulations and Processing**

| | |
|---|---|
| **Roast Beef Type Flavor** | |
| HVP | Sodium guanylate |
| Malic acid | L-methionine |
| Xylose | Water |
| Reflux at 100°C for 2.5 h | |
| **Chicken Type Flavor** | |
| L-cysteine HCl | Glycine HCl |
| Glucose | L-arabinose |
| Water | |
| Heated to 90–95°C for 2 h then adjust pH to 6.8 with NaOH | |
| **Pork Type Flavor** | |
| L-Cysteine HCl | Wheat gluten hydrolysate |
| D-xylose | |
| Heat at reflux temperatures for 60 min | |
| **Bacon Type Flavor** | |
| L-cysteine HCL | Thiamin HCL |
| Liquid soy hydrolysate | D-xylose |
| Smoke flavor | Bacon fat |
| Cook at reflux temperatures for 90 min | |

*Source:* From Manley, C.H., Flavor Interactions Workshop, IFT sponsored, Orlando, FL, 2003. With permission.

### 9.3.2.3 Final Flavor

The process flavor may be sold as a paste or as a powder. The paste often separates into a bottom layer of insoluble solids, a middle section of solution/suspension, and a top layer of solid fat. This striation makes it difficult to obtain a uniform sample/aliquot from the container. The preferred product form is a powder. Powders are easier to handle and store, have a longer shelf life, and are easier to apply in dry applications. As noted earlier, however, these materials are difficult to dry and prevent from caking during storage.

One can see that this overall process is very much an art and that the exact duplication of a competitive flavor is extremely difficult. It generally is difficult to determine initial reaction system composition due to the consumption of the reactants, plus there is little information that can be gathered about processing solids content, pH, times, or temperatures of heating, all critical factors in determining flavor, from examining the finished flavoring.

### 9.3.3 Hydrolyzed Vegetable Proteins (HVP)

The production of high quality hydrolyzed plant proteins and yeast autolysates has served as the basis of most commercial process meat-like flavorings for many years [37]. These products are offered as single components for direct use in food mixes or blended with flavor potentiators (e.g., MSG, 5′-nucleotides) and, often, spices. Currently, there is a wide range of such hydrolyzates produced in the market, each claiming specific color and flavoring characteristics, which in turn are dependent on the source protein (e.g., wheat gluten, maize gluten, soybean flour, etc.). The biggest proportion of HVP is based on soybean protein [38].

The three available industrial methods for the production of HVPs are (i) enzymatic: a slow process that may result in the formation of bitter peptides and typically lacks the desired aroma/taste profile [20,21,39]; (ii) alkaline hydrolysis: which typically results in unacceptable flavor profile and an unbalanced amino acid content, and (iii) acid hydrolysis: the most preferred method that is cost effective and yields a range of good flavors (see Chapter 11, section 11.4.1.2 Hydrolyzed Vegetable Proteins).

Complete hydrolysis of the protein is desirable using a high hydrochloric acid to protein ratio to ensure minimal amino acid degradation and a better flavor profile (as well as minimize any allergy issues). However, a large excess of acid has to be avoided as its neutralization would result in an abnormally high salt content in the finished product. Charcoal filtration of the neutralized aqueous solution results in the partial removal of certain amino acids as well as some unwanted impurities. The degree of filtration has to be carefully controlled as this determines the level of color in the end product.

If the starting material is high in glutamic acid, then the resulting HVP will have marked flavor potentiating properties. Some commercial HVPs are essentially free from reducing sugars so that further heating as part of a food system does not significantly enhance any Maillard reaction that may take place (other than contributing amino acids). Others are high in polysaccharides and give a decided roast meat or even burnt character on further heat processing. These differences underline the need to evaluate HVPs under actual processing conditions so as to determine their precise flavor contribution. The manufacturer's literature and specifications should be carefully studied before using these flavoring ingredients.

### 9.3.4 Autolyzed Yeast Extracts (AYE)

This is the product of the self-digestion of yeast cells, achieved by heating an aqueous slurry of active yeast at 45°C that effectively kills the cells without damaging any enzymes present. These enzymes are mostly proteases, which attack the cell components, partially solubilizing them. The resulting low molecular weight compounds include peptides, amino acids, and nucleotides. The degree of autolysis depends on the temperature/time conditions and the pH of the medium. The process is terminated by raising the temperature in order to inactivate the enzyme system. The degree of protein and carbohydrate breakdown can be varied to satisfy particular flavor and handling characteristics.

The flavor of the end product is determined, to a large extent, by the choice of yeast and the nature of the substrate. Aerobically grown *Saccharomyces cerevisiae* gives a much stronger, more beef-like flavor than anaerobically grown *Saccharomyces carlsbergensis*.

Hydrolysis with hydrochloric acid at 100°C is a more effective method of solubilizing yeast than autolysis. The conditions used are similar to those employed for the production of HVP (see Chapter 11, section 11.4.1.2 Hydrolyzed Vegetable Proteins and section 11.4.1.1 Yeast Extracts). When the hydrolysis has reached the desired level, the reaction liquor is neutralized to pH 5.5, clarified to remove all solid matter and partially concentrated. In some grades of yeast extract, the cell wall tissue is retained, the higher glycan content making these effective emulsifiers and thickeners. Further purification involves using either adsorption or ion-exchange techniques to remove undesirable components. The purified liquor is then finally concentrated and may be spray dried. Autolyzed yeast and hydrolyzed yeast extracts are excellent savory flavoring agents as well as contributors to the nutritive value of an end product. The industry has developed a line of such products that have been thermally processed to provide base meaty flavors.

## 9.4 ENZYMATICALLY DERIVED FLAVORINGS

### 9.4.1 INTRODUCTION

Enzymes find use in several areas of the flavor industry, including processing aids (e.g., free essential oils from plant cell structures, e.g., Sakho et al. [40]), for the liberation of flavor compounds from nonvolatile precursors (e.g., glycosides or cysteine conjugates, e.g., Winterhalter and Skouroumounis [41]), and the conversion of a precursor to a desired flavorant (biotransformations, Berger et al. [42]). In this chapter, enzymes will initially be discussed as they are used to enhance the potency of dairy flavorings. Later in this chapter (in the biotechnology section, 9.7) their use in producing pure aroma chemicals (catalysts for biotransformations) will be discussed.

Dairy products are typically low in flavor and are a high-cost ingredient. Thus, there is a strong incentive for the development of more potent dairy flavors to provide more cost-effective ingredients to the food industry. Increased flavor potency is generally achieved through one, or a combination of, the following processes:

1. Enzymatic modification (EM) of the base dairy product (controlled lipolytic and/or proteolytic enzymic action on fresh milks, creams, butters and butterfats, and cheeses)
2. An initial fermentation of the product (e.g., EM sour cream vs. EM fresh cream) followed by enzyme modification
3. The addition of other natural flavors (e.g., starter distillate, natural butyric acid, or methyl ketones)
4. Controlled high temperature processing (thermally processed to obtain browned flavor) [43]

Enzyme modified dairy products as a group are typically used to enhance or strengthen a dairy flavor as opposed to completely replacing the dairy product. They are approximately 15 to 30 times more intense than the corresponding non-EM product, and thus find usage at 0.1 to 2% in finished applications [44]. Unfortunately, EM products lack a complete flavor profile, and thus become harsh or unnatural when used at high levels. In EM processes, manufacturers primarily use lipolytic enzymes, which produce free fatty acids. These acids provide only a part of the characteristic flavor profile of a dairy product and thus become unnatural at high levels. While more complete flavor profiles may be obtained by using additional enzymes (e.g., proteases, peptidases, and esterases), the flavor profile is still lacking.

Some basic properties of enzymatic reactions will be discussed as will some of the existing EM dairy products and modifications thereof. Reviews by Kilcawley et al. [44], Freund [45,46], and Klein and Lortal [47] on EM dairy products are recommended for detail.

### 9.4.2 Properties of Enzyme Catalyzed Reactions

Enzymes are complex proteins that act as catalysts in various biochemical interactions. They generally function at ambient temperatures, are specific in action, have an optimum pH, and are effective at very low concentrations. A brief discussion of some of the general properties of enzyme catalyzed reactions follows below.

The optimum temperature for the activity of an enzyme as well as lower and upper temperature limits are unique to each enzyme. Some enzymes are extremely heat stable while others are easily denatured. The plot shown in Figure 9.1 illustrates a typical relationship between enzyme activity and reaction rate. Initially, the reaction rate increases with temperature, but at some temperature, the enzyme starts to denature, thereby losing activity. If one is using enzymes to generate a flavor, one chooses to use the optimum reaction temperature to minimize processing time.

Each enzyme also has a unique optimum pH for activity (see example in Figure 9.2). There is an optimum because enzyme structure is pH dependent and outside a

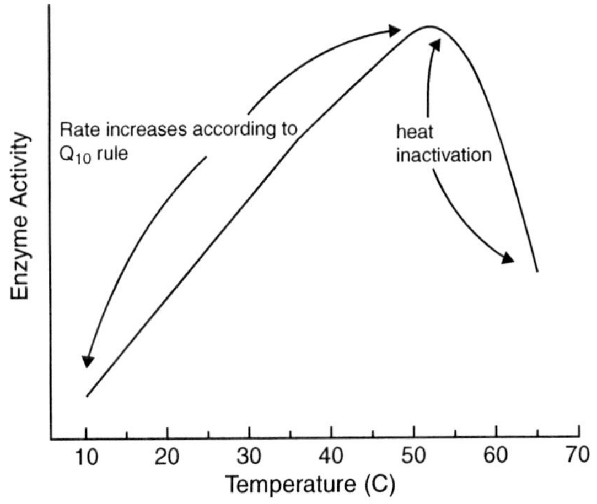


**FIGURE 9.1** The influence of temperature on reaction rate of a typical enzyme catalyzed reaction. (From Mathewson, P.R. *Enzymes*, Eagan Press, St. Paul, 2004. With permission.)

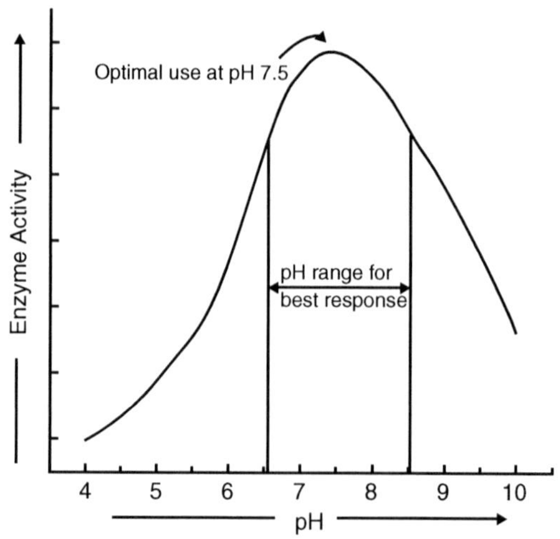


**FIGURE 9.2** The effect of pH on reaction rate of a typical enzyme catalyzed reaction. (From Mathewson, P.R. *Enzymes*, Eagan Press, St. Paul, 2004. With permission.)

given pH range, the enzyme begins to denature and thereby loses its activity. One often illustrates the activity of an enzyme in terms of a lock and key drawing. The substrate is the key that must fit (or orient) itself into the enzyme (the lock) in the proper manner or the enzyme cannot act upon it. Denaturation is viewed as a change in structure of the enzyme resulting in a loss in enzyme activity. The pH optimum as well as working pH range is again enzyme dependent. Enzymes derived from acid sources (e.g., bromelin, a protease from pineapple) will tend to have very low pH optima while those isolated from a neutral or alkaline environment (lower intestine, egg white, or fish muscle) would have optimal pHs in the alkaline region.

Many factors can denature, i.e., inactivate, enzymes. Shear can be very effective in denaturing an enzyme. The plot shown in Figure 9.3 illustrates how an enzyme may lose activity due to high shear mixing. Thus, if one is using an enzyme to produce an EM dairy product, one should apply high shear only before the enzyme is added. The purpose of the shear may be to dissolve or emulsify the base dairy product to increase the surface area of the substrate (e.g., butter fat). This would enhance the reaction rate and thereby reduce processing time.

The influence of water activity on enzyme activity is shown in Figure 9.4. The reaction rates of all enzymatic reactions slow as water activity becomes limiting. This may be due to reduced enzyme and substrate mobility at low water activities. If water is needed for the reaction, e.g., a hydrolysis reaction, limited water availability will further slow the reaction rate. One should note that in the example, water was added to all samples on day 48 (to increase the aw to 0.75). It is relevant that the enzyme activity of all the lower water activity samples increased back to their full activity. This has relevance to EM dairy products in that a dry EM product may appear to be stable, i.e., exhibit little change with time, even if it has active enzymes. However, if the active dry product is given the opportunity to gain moisture, e.g., added to a high moisture food, the enzymes may continue to attack the dairy product

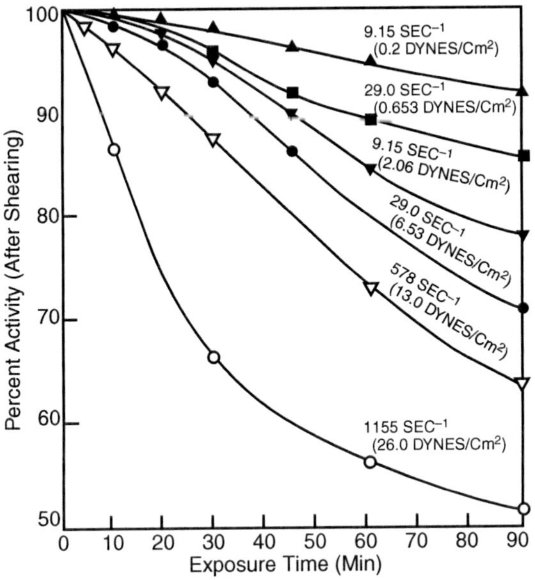


**FIGURE 9.3** The effect of shear on the inactivation of a typical enzyme. (From Richardson, T., D.B. Hyslop, *Food Chemistry,* 2nd ed., O.R. Fennema, Ed., Dekker, New York. With permission.)

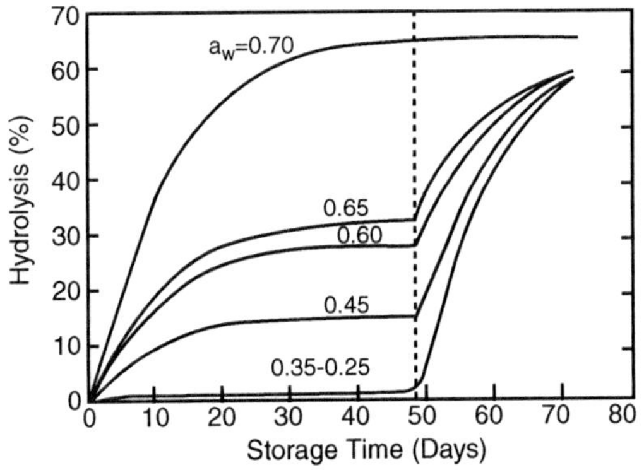


**FIGURE 9.4** The effect of water activity on reaction rate of a typical enzyme catalyzed reaction. (From Richardson, T., D.B. Hyslop, *Food Chemistry,* 2nd ed., O.R. Fennema, Ed., Dekker, New York. With permission.)

base or the foodstuff resulting in off-flavors. It is critical that all enzymes used in making an EM product be completely inactivated before it is used in a finished application.

Enzymes used in making EM products are most commonly denatured by heating. It is critical to know the heat tolerance of the enzymes used in a product so that an adequate heat treatment may be applied. The plots provided in Figure 9.5 show how

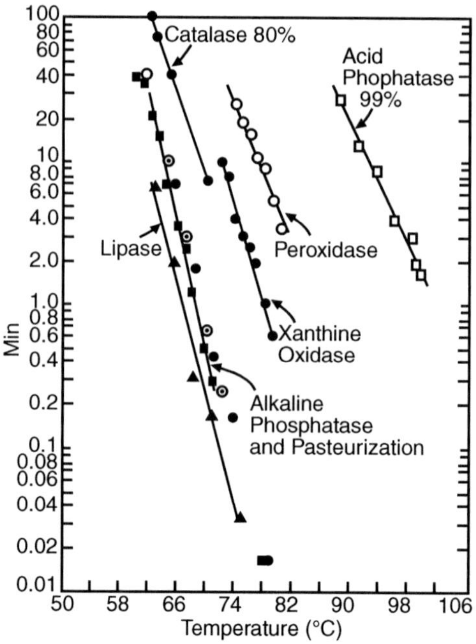


**FIGURE 9.5** The effect of temperature/time on the inactivation of a typical enzyme. (From Richardson, T., D.B. Hyslop, *Food Chemistry,* 2nd ed., O.R. Fennema, Ed., Dekker, New York. With permission.)

greatly enzyme activity varies with enzyme type. However, it must be further recognized that enzymes within a given class of enzymes will also vary greatly in heat stability. For example, a heat treatment that denatures one proteolytic enzyme, e.g., from a microbial source, may not have any effect on another proteolytic enzyme, e.g., one found in pineapple. One must check EM products for residual enzyme activity to insure there is no residual activity. This often done some period (e.g., 72 h) after manufacturing to allow any enzyme regeneration to occur [48].

### 9.4.3 Enzyme Modified Butter/Butter Oil

Butter oil is a commercial product made from melted butter by either centrifuging or directly making it from cream by patented processes by which the unwanted curd is removed by running the nearly churned cream into hot water. This causes the curd to precipitate, and the butter oil can then be recovered by centrifuging. Butter oil can be used directly in many food products or may be treated with selected enzymes (or organisms), which bring about lypolysis. The resulting very flavorful product is known as EM butter oil (or lipolyzed butter oil, LBO).

The commercial production of EM butter oil has been described by Andres [49] and the making of a dry-powdered version by low-temperature, spray-mat, drying is discussed by Farnham et al. [50]. These products offer richer, more intensified butter flavor profiles that have wide applications, particularly in baked goods and

Anhydrous milkfat or butter

↓

Melt anhydrous milkfat or butter, and add water and emulsifiers

↓

Add lipase enzyme and react for 24 h or more at 30°C or above and then heat inactivate added enzyme

↓

Further process to obtain desired consistency and/or composition

↓

Fill into containers and store refrigerated or frozen

**FIGURE 9.6** Flow chart for the formation of enzyme modified butter or butter oil. (From Freund, P., Flavor Workshop II: Reaction Flavors, Encapsulation and Analysis, University of Minnesota, St. Paul, September 2003. With permission.)

spreads. Very little information has been published on the precise organisms or enzymes employed, but the use of the microorganisms *Trametes sanguinea* has been patented [51]. The efficacy of *Streptococcus lactis* subsp. *diacetylactis* in the production of the butter flavor has been discussed by Piatkiewicz et al. [52], and Olson et al. [53] have described the microencapsulation of multienzyme systems. Aqueous cell free extracts of S. *lactis* subsp. *diacetylactis* are emulsified in milk fat and extruded into cold water to produce capsules. It is claimed that these products are effective in significantly increasing, up to eightfold, the production of diacetyl and acetoin in young cheese.

Commercially, butter oil is modified by controlled enzyme action that releases the volatile and flavorful short chain fatty acids, C4–C10 (e.g., butyric, caproic, caprylic, and capric), and also the nonvolatile long-chain fatty acids. The enzyme system is heat inactivated when the process has reached the desired end point. Such products contain about 48% fat (Figure 9.6).

It is usual for these products to be tempered at 20 to 25°C for 12 h prior to use to make them easier to incorporate into food mixes and to allow flavor equilibrium to be achieved. End products containing them should not be submitted to sensory assessment for a period of 24 to 48 h after manufacture. Manufacturers recommend that LBOs should be stored at 2 to 6°C and be used within 6 months from date of manufacture.

### 9.4.4 Enzyme-Modified Cheese (EMC)

Cheese flavor is very important in many savory food and snack products. In some instances an adequate flavor level can be achieved by the careful selection and use of natural cheeses, using the powder form if necessary, but in many instances the percentage needed to attain the correct flavor is not technologically acceptable and frequently is uneconomical. In such products, more intensely flavored cheese flavorings are required, and the industry has produced these by the application of enzyme technology.

EM Cheddar, Swiss, Romano, Mozzarella, Provolone, Feta, Parmesan, Blue, Gouda, Gruyere, Colby, and Brick cheeses are commercially available [44]. Kilcawley et al. [44] have provided a thorough review of this topic and it is highly recommended for further reading.

#### 9.4.4.1 Enzymes Used

The key to producing Enzyme-modified cheeses (EMCs) is the proper choice of enzymes and their application under optimal conditions. The majority of research in this area has been on the lipase enzymes. A wide variety of lipase enzymes are commercially available. These enzymes come primarily from animal or microbial sources. Lipase choice is critical because each lipase has a specificity for hydrolysis thereby determining the free fatty acid profile (a key determinant of flavor ) in the EMC [54].

Animal lipases are derived primarily from bovine or porcine pancreatic sources, and the pregastric tissues of kid goat, lamb, and calf [55]. The pregastric (i.e., oral) lipases (PGE) have a high specificity for the short chain fatty acids that characterize most of the natural cheeses and thus are preferred enzymes. Most PGEs have an optimum pH ranging from 4.8 to 5.5 and optimum temperature ranging from 32 to 40°C. Birschbach [56] has described the flavor character generated by different PGEs as follows:

Calf PGE – buttery and slightly peppery flavor
Kid PGE – sharp peppery often called piccante flavor, which quickly clears from the palate
Lamb PGE – peccorino flavor described as "dirty sock"

The use of any animal source enzyme brings concern for religious (Kosher or Halal) or dietary choices (e.g., vegetarian). These concerns as well as cost issues have promoted the development of microbial lipases and esterases for this purpose. If these enzymes are approved as GRAS (Generally Recognized as Safe), they can be used in this application.

As one would expect, microbial enzymes range in specificity. For example, lipases from *Candida cylindracea* and *Staph aureus* have little specificity while those from *Geotrichum candidum* are specific for long chain acids [57], which are of little value in this application. Fortunately, most microbial lipases show specificity for the one and three positions of the triglyceride where the short chain fatty acids are found. While substantial information exists in the literature characterizing lipase enzyme activity to help in enzyme choice [58], enzymes must be evaluated for function in a model system. This often is done in a curd slurry system, which is produced by blending two parts of fresh cheese curd plus one part sterile dilute salt solution (5.2%

NaCl), aspirating with nitrogen (5 min) and then inoculation with enzyme [59]. The system is then incubated at 30°C for 4 to 5 days, with the flavor periodically evaluated.

While lipases play a major role in determining the flavor of many cheeses, proteolytic activities are often considered more important to the flavor of some hard cheeses, e.g., Cheddar. The use of proteinases becomes problematic since these enzymes often produce bitter peptides. Thus, the manufacturer typically has to use a balance of proteinases and peptidases: the peptidases are necessary to break down the bitter peptides [59].

#### 9.4.4.2 General Processes Employed

EMCs are generally manufactured from cheese pastes that are made from the cheese of the same type. Additional components such as butterfat or cream may be added to add extra precursors when appropriate. Noncheese ingredients such as MSG, yeast extract, diacetyl, or other flavorants may also be added, but they may have to be declared on the label of the final product. Consistency in this base material is critical to the production of a standardized EMC product. Off-flavors may develop during incubation of the paste/enzyme slurry since the conditions are optimal for microbial growth. Equipment must be sterilized and precautions taken to prohibit microbial contamination. Bacterial inhibitors such as nitrates, sorbate, or nicin may be used. Free fatty acids generated by lipase enzymes afford some inhibition. Incubation time and temperature influence enzyme action and must be carefully controlled.

As noted earlier, the inactivation of residual enzymes to terminate the reaction is critical to flavor stability. The heat treatment must be adequate to inactivate the enzymes but not cause undesirable flavor changes. One should note that commercial enzyme preparations used in manufacturing may contain enzymes other than the primary enzyme(s): these are crude preparations as opposed to reagent grade enzyme preparations. For example, West [48] noted that pancreatic lipase preparations may contain amylases that can cause undesirable changes in flavor and texture in the finished product if not inactivated.

Kilcawley et al. [44] have outlined several industrial procedures for the production of EMCs. One example is presented in Figure 9.7.

It is claimed that the use of EMCs can result in considerable raw material cost savings since up to 50% of cheese solids in many product formulations can be replaced by a small quantity of EMC. Their use also enables food processors to adjust any variable flavor in natural cheese included in their product. They are of particular value in (a) processed cheese, (b) cheese powders, (c) cheese snacks, (d) pizza sauce, (e) cheese biscuits, (f) cheese spreads, (g) quiche, and (h) dips. The rate of addition is directly related to the profile of the natural cheese present and the strength required in the end product. On average, the recommended use rate is 1 to 3% of the total recipe weight. EMCs should be stored at 2 to 6°C and used within 6 months from date of manufacture.

Similar to the process flavors discussed earlier, EM dairy products are the result of the application of science plus substantial trial and error (i.e., art). Flavor character is dependent the starting materials, the enzymes added, the incubation conditions,

Fresh Cheese Curd

↓

Cut and grind the cheese, add water and emulsifiers to make a cheese slurry, and pasteurize (85°C for 30 min)

↓

Homogenize (7,000 psi)

↓

Add lipase and/or protease enzymes and/or starter cultures (incubate 37-40°C with mixing)

↓

Heat inactivate added enzymes/culture (85°C for 30 min)

↓

Paste — Further process to obtain desired consistency and/or composition

↓

Fill into containers and store under refrigeration or freeze

Powder —Mix with drying aid (e.g., maltodextrin and whey)

↓

Spray dry

↓

Package

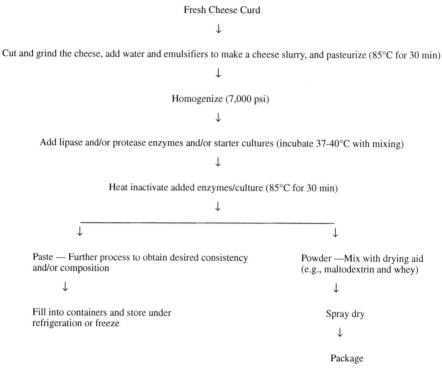

**FIGURE 9.7** Flow chart for the formation of enzyme modified cheese. (From Kilcawley, K.N., M.G. Wilkinson, P.F. Fox, *Int. Dairy J.*, 8, 1998. With permission.)

and virtually all steps of processing. This makes them extremely difficult for competitors to duplicate.

### 9.4.5 Further Processed EM Dairy Products

The flavor of an EM dairy product can be further manipulated to provide other sensory profiles. A simple way to do this is through an initial fermentation, i.e., prior to EM. A fermentation can strengthen the flavoring plus add other notes not added by the enzyme treatment. Thus a more complete flavor profile can be achieved. The cultures used in fermentation are typically the same as those used in a cultured dairy product.

Many of these products can be further enhanced by the use of top noting. This is similar to the process as it is applied to meat-like flavorings. Pure chemicals are added to the EM dairy product to provide notes not obtained in subsequent EM treatment. This process can greatly strengthen and round out an EM dairy flavoring. The availability of numerous natural dairy flavor chemicals often permits these top noted products to carry a natural label.

The final process applied to EM dairy products is thermal processing. Essentially, one applies a process similar to that of making a process flavoring (meat, chocolate, or coffee) to dairy products. The EM dairy product may have added ingredients prior to thermal processing to promote certain reactions, e.g., the Maillard reaction. Since dairy products are widely used as ingredients for the baking industry, it is logical that heated EM dairy products are very useful ingredients for the baking industry.

## 9.5 FLAVORS MADE BY FERMENTATION

The word *fermentation* (from the Latin *fermentare* meaning to cause to rise) was first applied to the making of wines and beer, processes whose origins are lost in history. It is now used to describe enzyme-catalyzed reactions in which compounds such as carbohydrates are broken down anaerobically by bacterial or yeast enzymes, leading to the formation of pyruvic acid (and other end products). In the case of yeast, this intermediate is split into carbon dioxide and acetaldehyde, which in turn is converted into ethanol. Industrially, fermentation is carried out using carefully selected microorganisms appropriate to the substrate, under precisely specified conditions. The products are wide ranging and include alcohol, glycerol, butyl alcohol, acetone, as well as lactic, acetic, citric, gluconic, and glutamic acids, many of which are important in food flavorings.

### 9.5.1 YEASTS

Yeast is a microscopic, unicellular organism of which there are several hundred species, only a very few of which are of commercial importance, namely, *Saccharomyces cerevisiae, Saccharomyces carlsbergensis, Candida utilis* (formerly *Torula utilis), Candida tropicalis,* and *Kluyveromyces fragilis.* Selected *Saccharomyces* spp. are used for the production of wines and other alcoholic beverages.

The choice of yeast depends on the nature of the substrate in which it is to be grown. S. *cerevisiae* and S. *carlsbergensis* utilize glucose, galactose, maltose, and sucrose; *Candida utilis* (the so-called torula yeast) prefers pentose sugars such as arabinose and xylose and is widely used in the fermentation of molasses and sulfite waste liquors; *Kluyveromyces fragilis* grows on lactose contained in dairy byproducts.

### 9.5.2 VINEGAR/ACETIC ACID

Vinegar has a long and useful history both as a condiment and as a food preservative. It is the product of the alcoholic and acetic fermentation of a sugar-containing solution without any intermediate distillation except in the production of spirit vinegar. Numerous vinegars are available for use in food processing. These include (i) wine vinegar, the alcoholic fermentation being from grapes; (ii) malt vinegar, made from malted barley; (iii) spirit vinegar, prepared by the acetic fermentation of distilled alcohol, itself made by fermentation; (iv) fruit wine vinegars, made from grape and nongrape fruit wines; (v) cider vinegar, made from fermented apple juice; and (vi) distilled vinegar, obtained by the distillation of malt vinegar. In recent years, the variety of vinegars has grown immensely [60]. The flavor characters of these vinegars vary considerably.

Two entirely different fermentations are involved in the manufacture of vinegar:

1. Fermentation of sugar by yeast:

$$\text{hexose} \rightarrow \text{ethanol} + 2\ CO_2$$

2. Conversion of alcohol to acetic acid by the action of acetobacter organisms:

$$\text{ethanol} + 2O_2 \rightarrow \text{acetic acid} + 2\ \text{water}$$

The process is usually carried out within the temperature range 24 to 27°C. The alcoholic fermentation is induced by a strain of *Saccharomyces cerevisiae* var. *ellipsoides,* which has been grown in the presence of a minimum level of sulfur dioxide to eliminate extraneous organisms.

Using a substrate of 8 to 20% fermentable sugars, the primary fermentation is carried out in open vats fitted with a cooling system to regulate the temperature. The added yeast, during its normal reproduction, converts the sugars to alcohol and carbon dioxide and, in the case of malt vinegar, similarly converts dextrins obtained from the malted barley. Fermentation takes about 7 days to completion at which stage the product contains 6 to 7% alcohol, which represents about 90% of theoretical conversion of the sugars. After sedimentation of the yeast cells, the supernatant alcoholic liquor is filtered and is ready for the second acetic oxidative fermentation.

This secondary fermentation requires considerable oxygen and is carried out in one of two ways:

1. Packed generator systems. The traditional acetifier comprises a tall vat packed with an inert material (usually birch twigs or wood shavings) which act as the carrier for the acetobacter culture. The alcoholic substrate is sprayed on the top of the column and allowed to flow down against a stream of air, introduced from the bottom of the vat, into a false bottom and is recirculated for several days until most of the alcohol has been oxidized to acetic acid.
2. Fring's acetator. This is a submerged culture fermentation, involving vigorous stirring of the substrate to induce adequate aeration. Several purpose-designed acetators are available, but in each the acetobacter organisms are introduced into the substrate and grow in a suspension of fine air bubbles produced by the considerable upwards turbulence. The process is continued until the desired level of acetic acid is achieved. This type of plant requires less space than the traditional vats and gives a more rapid conversion but is more energy consuming. The resulting vinegar is filtered and pumped into storage vats where it is allowed to mature for several months during which time it develops a delicate and characteristic bouquet.

Federal standards require vinegar to contain not less than 4 g of acetic acid per 100 ml.

Distilled vinegar is prepared by distilling the acetic acid and volatile constituents from a normally fermented vinegar. It is colorless and retains much of the characteristic aroma and flavor of the original vinegar. It is of particular use in products where color is an important attribute (e.g., pickles, mayonnaise, tomato ketchup).

Concentrated vinegar, containing up to about 40% acetic acid, may be prepared by freeze concentration using a scraped-surface Votator heat exchange unit operating at –8°C. The water is removed as ice and most of the aromatics remain in the concentrated product.

Some of the newer vinegars such as balsamic vinegar are made by combining processes. For example, balsamic vinegar is made by cooking down the must (juices) of unfermented grapes (red Lambrusco [used extensively on the Emilian plains], white Spergola and Occhio di Gatta, and/or Berzemino grapes), adding a culture to the reduced must (or back inoculating), allowing fermentation and then aging in selected wood barrels [61,62]. The characteristic flavor of this product is due to a combination of thermal processing (cooking of the must), fermentation, and wood aging — somewhat analogous to the heating of EM dairy products to create new flavorings.

### 9.5.3 Dried Inactive Yeast Powder

An inevitable and commercially important byproduct of fermentation processes is yeast. This can be recovered, dried, and inactivated to form a valuable food ingredient. Although the brewing industry is the major source, the demand is sufficient to warrant the growing of selected yeasts primarily for this purpose. Henry [63] has described the production of dried Torula yeast, resulting from the fermentation of the waste sulfite liquors from paper production.

The yeast slurry, containing about 1% solids, is continuously separated from the substrate by centrifuging, the collected yeast cells are washed to remove all adhering chemicals and then dried to a moisture content of <7% to give the commercial product. This product has a very high protein content (min. 45%) and substantial amounts of B vitamins. The USF XIII confines the source to either *S. cerevisiae* or *Candida utilis.*

Dried inactive yeasts are used not only as valuable nutrient and vitamin supplements but in the flavoring of dry savory seasonings, dip mixes, gravy and sauce mixes, snack products, etc., to which they give a rich, full-bodied flavor character. Because of the many variables that affect the flavor profile of inactivated yeast, the products of different manufacturers have quite distinctive flavoring properties. The profile may be characterized by such nuances as sweet, caramel-like, roasted, beefy, cheesey, or creamy, and this implies that such products must be individually assessed prior to use.

## 9.6 FLAVORS MADE BY PYROLYSIS: SMOKE FLAVORS

### 9.6.1 The Smoking of Foods

The process of curing and smoking food, which has evolved slowly over thousands of years, was probably discovered by accident when it was observed that meats and fish hung up to dry in a smoky atmosphere remained edible for longer periods and also acquired a pleasant taste. For many hundreds of years the process has been carried out as a means of preserving both meat and fish, although the emphasis has gradually changed from preservation to the imparting of a very pleasant flavor to the food product.

Initially, the smoking process was carried out in a kiln with little or no control over the smoking process. Today, smoke generators are designed to ensure the burning of hardwood sawdust under controlled conditions of forced air recycling and temperature. Even so, the slow burning of wood demands considerable skill and may give rise to products of variable flavor quality. The direct use of wood smoke is relatively inefficient, results in messy tarry deposits throughout the smoke generating plant, and often results in the deposition of unsightly particulate matter on the surface of the smoked product. In spite of these problems the smoking of meats and fish is still extensively used, but it is for these reasons that the use of natural liquid smoke flavors have gained wide acceptance.

### 9.6.2 Natural Liquid Smoke Flavorings

The Code of Practice issued by the IOFI [64] specifies smoke flavors as follows:

"Smoke flavors are concentrated preparations, not obtained from smoked food materials, used for the purpose of imparting smoke type flavor to foods. They are prepared according to one or more of the following methods:

1. By subjecting various untreated hardwoods to processes described as (a) controlled burning, or (b) dry distillation at appropriate temperatures, usually between 300 and 800°C, or (c) treatment with superheated steam at temperatures usually between 300 and 500°C, and condensation and capturing of those fractions which have the desired flavor potential
2. By applying further isolation techniques to the fractions obtained under 1. [item] 1 in order to retain only the flavorwise important fractions or components
3. By compounding chemically defined flavoring substances. Smoke flavors should contribute no more than 0.03 ppb 3,4-benzopyrene to the final food product."

### 9.6.3 Pyroligneous Acid

When wood is heated without an excess of air to temperatures in excess of 250°C, it decomposes into charcoal and a volatile fraction, which condenses on cooling. The aqueous phase of this condensate is known as pyroligneous acid. The chief component is water containing about 10% of acids, mainly formic, acetic and propionic.

### 9.6.4 Smoke Condensates

These are prepared by drawing the distillation and combustion products from damp hardwood sawdust into a condenser fitted with an electrostatic precipitator. This removes all particulate matter from the condensate, which is collected and settled to allow separation into aqueous and tarry phases. The clear aqueous phase is the desired product.

Modern generators produce smoke under carefully controlled conditions of combustion temperature and air flow. Sawdusts of hickory, maple, and oak woods are used as the source material. The smoke passes through a settling chamber to remove heavy tars and wood ash and is then directed upwards through a column of inert ceramic material countercurrent to a steady flow of water. This is recycled as necessary until the titratable acidity is 5 to 12%, as acetic acid. The product is then filtered and set aside to age, during which time it loses its unpleasantly sharp flavor character and additional tarry matter deposits.

More recent manufacturing processes consist of inclined rotary calciners. Sawdust is fed continuously into one end of the inclined tube and discharged at the other end as charcoal. The vapors are handled as already described. The rate of feed, inclination of the tube, and rate of air flow are optimized to give a product with the best smoke flavor and the lowest content of polycyclic aromatics.

## 9.7 BIOTECHNOLOGY TO PRODUCE FLAVORING MATERIALS

### 9.7.1 Introduction

The flavor market in the U.S. (and world) has been heavily weighted toward natural flavorings for the last 30+ years. Artificial flavorings find use today in lower cost food products or in situations where natural flavoring counterparts do not exist (e.g., top noting of meat-like flavorings). The market demand for natural flavorings has resulted in a great deal of research directed at developing processes for making natural flavoring chemicals. Gatfield [64] estimated that approximately 100 natural flavor components are being produced through biotechnology.

The classification of a flavoring (or flavor chemical) as being natural has been defined by the FDA and published in the CFR (presented earlier). It is adequate here to note that any product of fermentation or enzymatic activity starting with a natural precursor(s) yields a natural product according to this definition. Gatfield [65] commented that it would also seem reasonable that only those compounds found in nature could be considered natural for it may be possible to make compounds not found in nature through these approved techniques. In this chapter we will focus on examples of using enzymology and fermentation to yield natural flavoring chemicals or flavorings. Texts including those edited by Parliment and Croteau [66] and authored by Berger [67] as well as a textbook chapter by Berger et al. [42] are all highly recommended. Several comprehensive reviews are also available [68–70].

### 9.7.2 Production of Natural Flavoring Materials by Enzymatic Action

There is little question that enzymes exist in nature to accomplish nearly any chemical conversion that one would desire. Thus, theoretically, one could find an enzyme (or series of enzymes) to convert some precursor to any related natural flavor compound. However in practice, enzymes have found only a few commercial applications for this type of task (despite an abundance of publications in the literature). Enzyme catalyzed reactions suffer from several practical limitations, including:

1. Instability and often very short life times in application
2. Often require cofactors, e.g., NAD or ATP
3. Are slow and offer a lower yield than a chemical reaction
4. Are expensive, especially when batch processes are employed
5. Catalyze only a single step of a reaction

### 9.7.2.1 Ester Formation

The most widely used enzymatic bioconversion used in the flavor industry is the condensation of natural acids (generally acetic or other short chain acids) and alcohols (generally ethanol or fusel alcohols) to yield a wide range of esters. It is recognized that enzymes typically are reversible and thus, lipases (or esterases) normally associated with the *hydrolysis* of esters, can be used to *synthesize* esters under certain conditions.

Numerous researchers have published on the formation of esters using enzymes in nonpolar solvents (see Gatfield [71,72]). The diagram in Figure 9.8 illustrates how conducting enzymatic reactions in an organic solvent can reverse its action [73]. In this case, the enzyme is hydrated and thus presents an aqueous environment that accumulates high concentrations of free acids and alcohols due to their solubility in aqueous systems. The high concentration of the acids and alcohols favors their condensation into esters. The esters are less polar and thus go into the organic phase depleting the ester in the enzyme environment. This drives the equilibrium toward the further formation of esters. As one might anticipate based on this partitioning, the use of nonpolar solvents often will also increase reaction rate and yield.

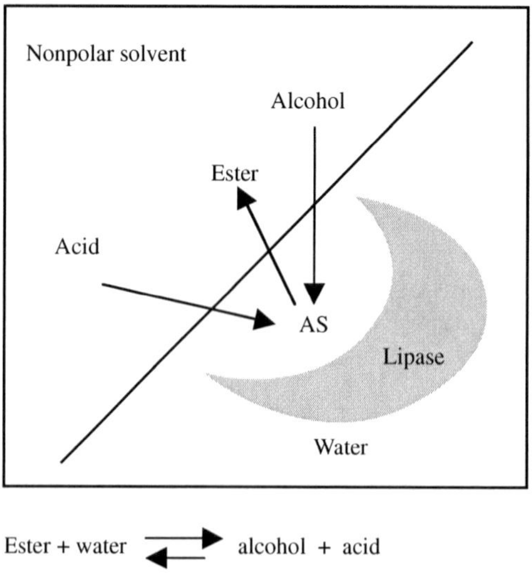


**FIGURE 9.8** Enzymatically catalyzed synthesis of esters. (From Armstrong, D.W., B. Gillies, H. Yamazaki, *Flavor Chemistry: Trends and Developments*, R. Teranishi, R.G. Buttery, F. Shahidi, Eds., Amer. Chem. Soc, Washington, D.C., 1989. With permission.)

A very similar process is used to catalyze the oxidation of aromatic and aliphatic alcohols to their corresponding aldehydes. An alcohol oxidase enzyme isolated from *Pichia pastoris* placed in a two-phase system will readily produce benzaldehyde from benzyl alcohol. While this oxidation is slow and inefficient in aqueous systems, oxidation in a two-phase system increased yield by a factor of nine and reaction time was shortened greatly [74].

It is interesting that industrial byproducts such as fusel oils (distillation byproducts of alcoholic beverages when making high proof products, Table 9.4) were once a disposal problem but now have become valuable starting components for enzymatic conversions. There are few good sources of branched chain alcohols other than the fusel alcohols. The bioconversions of selected amino acids will yield branched chain alcohols and acids, but this is a tedious process in itself. The bioconversion of a fusel alcohol to an acid, and ultimately its esterification to form cis-3-hexenyl-2-methyl butyrate, is shown in Figure 9.9. In this case, the oxidation of 2-methyl-1-butanol is likely catalyzed by immobilized cells of certain Gluconobacter strains as opposed to being enzyme catalyzed. Gatfield et al. [75] have shown that this organism is an effective oxidant of fusel alcohols. Gatfield [65] has also noted that some yeasts can also catalyze such oxidations [76].

Welsch et al. [69] have provided a very comprehensive review of the enzymatic synthesis of flavor molecules. They have provided extensive referencing as well as

**TABLE 9.4**
**Typical Fusel Oil Composition**

| Alcohol | % Composition |
|---|---|
| Ethyl alcohol | 5 |
| Propyl alcohol | 12 |
| Isobutyl alcohol | 15 |
| Amyl alcohol isomers | 63 |
| Residue | 5 |

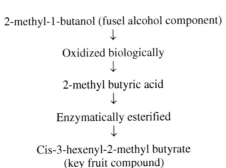

2-methyl-1-butanol (fusel alcohol component)
↓
Oxidized biologically
↓
2-methyl butyric acid
↓
Enzymatically esterified
↓
Cis-3-hexenyl-2-methyl butyrate
(key fruit compound)

**FIGURE 9.9** Enzymatic conversion of a fusel alcohol to a fruity ester. (From Candella, C., Personal communication, Robertet Flavors, S.A., Grasse, F.R., 1995. With permission.)

detailed discussions of some fundamental considerations in making these bioconversions. This reference is highly recommended.

#### 9.7.2.2 Resolution of Racemic Mixtures

Enzymatic conversions have also found application in the resolution of racemic mixtures. When an organic synthesis involves a chiral center, one typically obtains a racemic mixture of the end products. Very commonly, chiral isomers differ in sensory properties, and one desires one enantiomer as opposed to the racemic mixture, e.g., L-menthol is the desired form of menthol. It is very difficult to separate enantiomers by chemical means on a commercial basis (cyclodextrin-based columns have some utility in this application). However, a characteristic of enzymatic reactions is their stereospecificity, i.e., they will act on one enantiomer but not another. Thus, enzymatic processes may be incorporated into chemical synthesis to obtain a pure optical isomer. This may be done in either of two ways.

In one approach, an ester of one chiral isomer (e.g., L-menthol) may preferentially be formed via enzymatic action [78]. Then one only has to separate the initial alcohol form (e.g., D-menthol — not esterified) of the target compound from its ester form (L-menthol ester), which is reasonably easy. The alternative approach is to chemically form esters of both enantiomers and use an enzyme to cleave the ester from one chiral form. This again leaves a mixture that is an ester and an alcohol that are readily separated. Berger et al. [42] have illustrated this process for the separation of enantiomers of karahanaenol (Figure 9.10). Gatfield et al. [79] have a recent patent on this method for the isolation of pure L-menthol.

### 9.7.3 Production of Natural Flavoring Materials by Microbial Action

Unlike enzymatic reactions, microorganisms have the ability to perform multiple reactions, and they do not require cofacors for regeneration (albeit they require nutrients). They may be used to generate a flavor compound from a nonvolatile precursor (e.g., produce a lactone from castor oil), to effect the bioconversion of one volatile to another (e.g., valencene to nootkatone), or effect a chiral resolution (a racemic mixture of menthol). The primary limitation of using microoganisms for

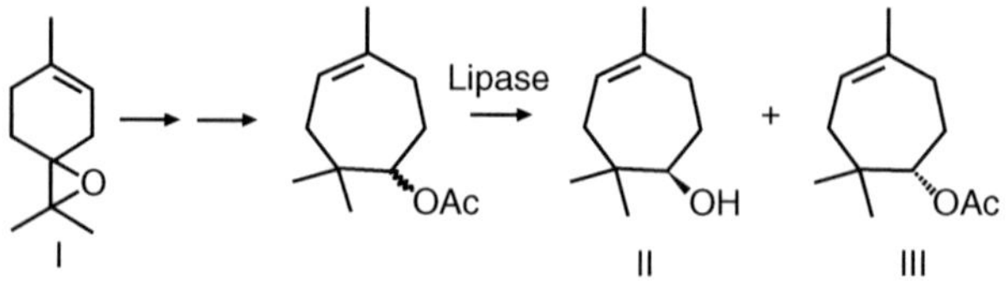


**FIGURE 9.10** Enantiospecific alcoholysis of karahanaenol acetate by *Pseudomonas cepacia* lipase. (I) terpinolene oxide; (II) (R)-karahanaenol; and (III) (S)-karahanaenol acetate. (From Roy, A. *J. Agric. Food Chem.*, 47, 12, 1999. With permission.)

these purposes is that either the precursor or end product may be toxic to the microorganism, and therefore the yields (g/L fermentation broth) may be low and cost of recovery may be high. Thus the use of fermentation to produce natural flavoring materials is largely an economic issue.

### 9.7.3.1 Fermentation to Produce Flavoring Materials

If one surveys the literature, it is clear that nearly any fruity type flavor could theoretically be produced by fermentation. However, as noted, few of these organisms produce aroma compounds at sufficient concentrations to be economically viable. Researchers try to overcome this problem through both genetic manipulation of the organism and media choice. Genetic manipulation is often done by classic methods such as UV irradiation or chemical mutation (n-nitrosoguanidine). By either method, random mutations occur and a selection is made of those mutants that produce higher levels of the desired end product. Because the mutations are random and hundreds of different mutants may be produced, one of the problems is the isolation and screening of the mutants for the increased production of the target molecule. To illustrate the problem and a possible solution, McIver [80] treated a strain of *Pseudomonas perolens* that produced low levels of 2-methoxy-3-alkyl pyrazines (bell pepper aroma) with n-nitrosoguanidine and selected high yielding mutants for further study. He first plated the organisms, selected pure colonies and grew them in nutrient broth. His screening process was simply smelling the nutrient broth. Those cultures that had an intense odor were analyzed by gas chromatography to determine the actual concentrations of the methoxy alkyl pyrazines. While smelling the tubes is not very scientific, it would have been impossible to analyze each of the cultures by any other method. The cultures possessing the most intense aromas were then grown on different media to enhance volatile production.

As can be seen in Table 9.5, nutrient source had a strong influence on the production of the methoxy pyrazines by a selected mutant of *Pseudomonas perolens*. The end result is that an organism that initially produced a few ppb of the methoxy alkyl pyrazines was converted into an organism that produced ppm levels of the target compounds. While one may not think that ppm levels of an aroma compound is significant, these pyrazines have sensory thresholds in the pptr levels.

Similar approaches (UV irradiation) have been used to enhance the yield of diacetyl in the production of starter distillates [82]. This method yielded cultures that produced a 4.5-fold increase in total acetoin + diacetyl production. Starter distillates are very broadly used in the reproduction of dairy flavors.

One of the earliest and most commercially successful examples of producing flavoring materials by fermentation is the production of 4-decalactone from castor oil (Figure 9.11, [83]). Castor oil is unique in that it is made up of nearly 80% ricinoleic acid (12-hydroxy-9-octadecenoic acid). *Yarrowia lipolytica* initially hydrolyses the ricinoleic acid from the triglyceride and then through β-oxidation, converts this acid to 4-hydroxydecanoic acid. This acid forms a lactone at low pHs to yield the γ-decalactone. The yield on this process is generally considered to be ca. 6 g/L which is very attractive.

**TABLE 9.5**
**The Production of Selected Methoxy Alkyl Pyrazines by Selected *P. perolens* Mutant on Different Growth Media**

| Carbon Source | Conc. (% w/v) | 2-Methoxy-3-isopropyl pyrazine | 2-Methoxy-3-secbutyl pyrazine |
|---|---|---|---|
| Glucose | 0.5 | 2,030[a] | 21[a] |
| | 1.0 | 960 | 11 |
| | 2.0 | 840 | 9 |
| Lactate | 0.5 | 6,280 | 68 |
| | 1.0 | 9,890 | 112 |
| | 2.0 | 4,730 | 48 |
| Pyruvate | 1.0 | 11,120 | 100 |
| Glycerol | No effect at | 0.5 or 1.0 | |
| Nutrient broth | No effect | | |

[a] ppb of compound in growth medium.

*Source:* From McIver, R.C., Ph.D. thesis, University of Minnesota, St. Paul, 1986. With permission.

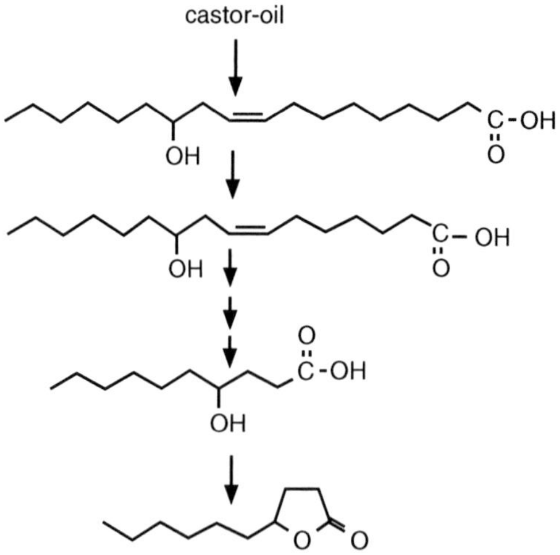


12-hydroxy-9-octadecenoic acid

10-hydroxy-7-hexadecenoic acid

4-hydroxydecanoic acid

4-decalactone

**FIGURE 9.11** The conversion of ricinoleic acid to γ-decalactone by *Yarrowia lipolytica*. (From Farbood, M., B.J. Willis, Production of γ-Decalactone, U.S. Patent 4560656, 1985. With permission.)

It is interesting that the δ-decalactone is formed by an entirely different biotransformation. This lactone is formed by the microbial reduction of an α,β-unsaturated lactone found in Massoia bark oil. Massoia bark oil contains about 80% 2-decen-5-olide, 7% 2-dodecen-5-olide and 6% benzyl benzoate. Several microorganisms

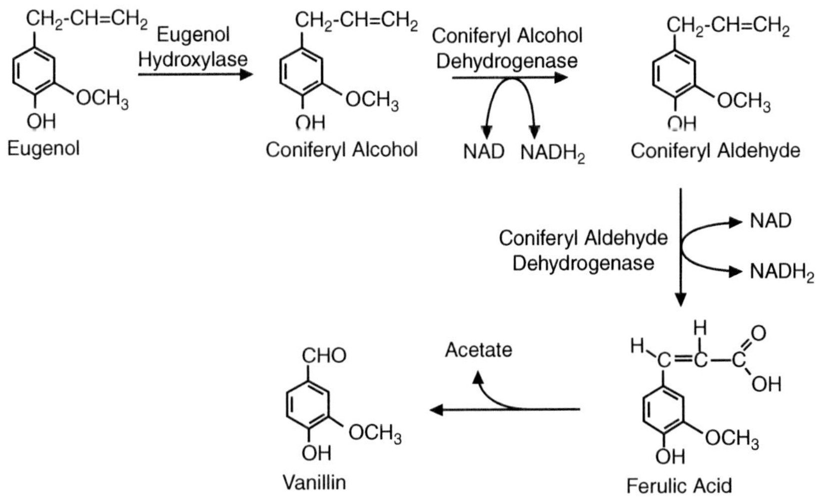


**FIGURE 9.12** The bioconversion of ferulic acid to vanillin. (From Gatfield, I.L., *Flavor Chemistry: Thirty Years of Progress*, R. Teranishi, E.L. Wick, I. Hornstein, Eds., Kluwer Acad. Publ., New York, 1999. With permission.)

have been found that will hydrogenate the ring double bond to form the corresponding saturated δ-lactones [84].

Natural butyric acid is produced by fermentation by growing *Clostridium butyricum* or *C. acetobutyricum* on glucose under anaerobic conditions [85,86]. Through the right choices in pH and growth media, concentrations may range from 1 to 2% in the fermentation broth.

### 9.7.3.2 Bioconversions Via Microorganisms

There are numerous references to the use of microorganisms for bioconversions [42,67,69]. While many such conversions produce too low yields to be of commercial viability, some have found commercial application. Bioconversions can be very efficient since the microbes generally only have to accomplish a limited number of steps in conversion as opposed to multiple steps in a *de novo* synthesis.

An example of this approach was outlined by Gatfield [65] on the conversion of ferulic acid to vanillin (Figure 9.12). Ferulic is an abundant, inexpensive, natural vanillin precursor recovered from waste streams or from the fermentation of eugenol [87]. While many microbial systems have been found that accomplish the desired bioconversion, Gatfield [65] has suggested that Amycolatopsis species is the most efficient providing vanillin concentrations up to 10g/L in a fermentation period of only 36 h [88,89]. This is an excellent yield and few side products were found.

### 9.7.3.3 Resolution of Racemic Mixtures

Just as enzymes find utility in the resolution of racemic mixtures, so do microorganisms. They both show very strong chiral preferences in their activities. The resolution of menthol racemates is an example of commercial application of this approach [68].

In this example, synthetic D,L-menthol racemate is initially chemically esterified with acetate, formate, propionate, myristate, benzoate, or succinate. The ester is then selectively hydrolyzed by steriospecific microbial enzymes to produce L-menthol that is readily isolated from the D-menthol ester. The low water solubility of the menthyl esters and L-menthol has made this process particularly suitable to non-aqueous systems. Omata et al. [90] immobilized cells of *Rhodotorula minuta* in photo-cross-linked or polyurethane resin gels and used a water saturated hexane eluant to obtain a product of 100% optical purity. The immobilized cells had an estimated half-life of 55 to 63 days in this solvent.

### 9.7.4 Economics of Biotechnology

It is generally impossible for biotechnologically produced aroma chemicals to compete economically with their chemically synthesized counterparts. Synthetic aroma chemicals often sell for $5 to 50 per kg while the equivalent chemicals produced via biotechnology generally cost 10 to 100 times more. The cost of fermentation, low yields and subsequent recovery costs (e.g., from a fermentation broth) drive these costs up greatly. The justification for paying this premium in cost is being able to use a natural label on the product. The industry appears willing to pay this premium in many cases and thus, biotechnology continues to be applied in the flavor industry.

## REFERENCES


1. Buyukpamukcu, E., Understanding cocoa flavor: where does it all come from?, *Chem. Rev.,* 11, 4, p. 8, 2002.
2. Chen, C.Q., E. Robbins, Chocolate flavor via the Maillard reaction, in *Caffeinated Beverages.* T.H. Parliment, C.-T. Ho, P. Schieberle, Eds., Amer. Chem. Soc., Washington, D.C., 2000, p. 286.
3. Martin, R.A., Jr., Chocolate. *Advan. Food Res.,* 1987, 31, p. 211.
4. Ikrawan, Y., S. Chaiseri, O. Vungdeethum, Effect of fermentation time on pyrazine concentration of Thai Forasero beans. *Kasetsart J. Natural Sci.,* 1997. 31, 4, p. 479.
5. Mermet, G., E. Cros, G. Georges, Preliminary study to optimize cocoa roasting parameters: aroma precursor consumption, pyrazine development, organoleptic quality. *Cafe Cacao,* 1992. 36, 4, p. 285.
6. Cros, E., Effect of roasting parameters on the development of cocoa aroma. *Ind. Aliment. Agric.,* 1993. 110, 7-8, p. 535.
7. Hashim, L., Flavor development of cocoa during roasting, in *Caffeinated Beverages.* T.H. Parliment, C.-T. Ho, P. Schieberle, Eds., Amer. Chem. Soc., Washington, D.C., 2000. p. 276.
8. Allerton, J., Chocolate and cocoa products, in *Encyclopedia of Food Technology,* A.H. Johnson, M.S. Peterson, Eds., AVI Pub. Co., Westport, 1974, p. 195.
9. Minifie, B.W., *Chocolate, Cocoa and Confectionery: Science and Technology,* 3rd ed. Van Nostrand Reinhold, New York, 1989, p. 904.
10. Beckett, S.T., *Industrial Chocolate Manufacture and Use,* 2nd ed., Blackie Academ. Prof., New York, 1994, p. 408.

11. Schnermann, P., P. Schieberle, Evaluation of key odorants in milk chocolate and cocoa mass by aroma extract dilution analyses, *J. Agric. Food Chem.*, 45, 3, pp. 867–872, 1997.
12. Schieberle, P., P. Pfnuer, Characterization of key odorants in chocolate, in *Flavor Chemistry: Thirty Years of Progress*, R. Teranishi, E.L. Wick, I. Hornstein, Eds., Kluwer Acad. Publ., New York, 1999, p. 147.
13. Hodge, J.E., F.D. Mills, B.E. Fisher, Compounds of browned flavour derived from sugars-amine reactions, *Cereal Sci. Today*, 17, p. 34, 1972.
14. Melton, S.L., Current status of meat flavor, in *Quality Attributes of Muscle Foods*, Y.L. Xiong, C-T Ho, F. Shahidi, Eds., Kluwer Academ. Publ., New York, 1999, p. 115.
15. Mottram, D.S., Flavor formation in meat and meat products: a review, *Food Chem.*, 62, 4, pp. 415–424, 1998.
16. Pearson, A.M., J.I. Gray, C.P. Brennand, Species-specific flavors and odors [of meat], in *Quality Attributes and Their Measurement in Meat, Poultry and Fish Products*, A.M. Pearson, T.R. Dutson, Eds., Blackie Academ., London, 1994, pp. 222–49.
17. Reineccius, G., Flavor and aroma chemistry [of meat], in *Quality Attributes and Their Measurement in Meat, Poultry and Fish Products*, A.M. Pearson, T.R. Dutson, Eds., Blackie Academ., London, 1994. p. 184–201.
18. Farmer, L.J., Meat flavor, in *Chem. Muscle-Based Foods* (Special Publication), Royal Soc. Chem., 106, pp. 169–82, 1992.
19. Nagodawithana, T., *Savory Flavors*, Esteekay Assoc., Milwaukee, 1995, p. 468.
20. Aaslyng, M.D., J.S. Elmore, D.S. Mottram, Comparison of the aroma characteristics of acid-hydrolyzed and enzyme-hydrolyzed vegetable proteins produced from soy, *J. Agric. Food Chem.*, 46, 12, pp. 5225–5231, 1998.
21. Wu, Y.-F.G., K.R. Cadwallader, Characterization of the aroma of a meat-like process flavoring from soybean-based enzyme-hydrolyzed vegetable protein, *J. Agric. Food Chem.*, 50, 10, p. 2900, 2002.
22. Gasser, U. and W. Grosch, Identification of volatile flavor compounds with high aroma values from cooked beef, *Zeitschrift Lebens. Untersuch. Forsch.*, 186, 6, p. 489, 1988.
23. Gasser, U., W. Grosch, Primary odorants of chicken broth: a comparative study with meat broths from cow and ox, *Zeitschrift Lebens. Untersuch. Forsch*, 190, 3, p. 3, 1990.
24. Gasser, U., W. Grosch, Aroma extract dilution analysis of commercial meat flavorings, *Zeitschrift Lebens. Untersuch. Forsch*, 190, 6, p. 511, 1990.
25. Grosch, W., G. Zeiler-Hilgart, Formation of meat-like flavor compounds, in *Flavor Precursors,* R. Teranishi, G.R. Takeoka, M. Güntert, Eds., Amer. Chem. Soc., Washington, D.C., 1992. p. 183.
26. Cerny, C., W. Grosch, Evaluation of potent odorants in roasted beef by aroma-extract dilution analysis, *Zeitschrift Lebens. Untersuch. Forsch*, 194, 4, p. 322, 1992.
27. Cerny, C., W. Grosch, Quantification of character-impact odor compounds of roasted beef, *Zeitschrift Lebens. Untersuch. Forsch*, 196, 5, p. 417, 1993.
28. Grosch, W., et al., Studies on the formation of odorants contributing to meat flavors, in *Progress Flavour Precursor Studies*, P. Schreier, P. Winterhalter, Eds., Allured Publ., Carol Stream, 1993, p. 329.
29. Cerny, C., W. Grosch, Precursors of ethyldimethylpyrazine isomers and 2,3-diethyl-5-methylpyrazine in roasted beef, *Zeitschrift Lebens. Untersuch. Forsch*, 198, 3, p. 210, 1994.
30. Kerler, J., W. Grosch, Character impact odorants of boiled chicken: changes during refrigerated storage and reheating, *Zeitschrift Lebens. Untersuch. Forsch.*, 205, 3, p. 232, 1997.

31. Kerscher, R., W. Grosch, Comparative evaluation of potent odorants of boiled beef by aroma extract dilution and concentration analysis, *Zeitschrift Lebens. Untersuch. Forsch.* 204, 1, p. 3, 1997.
32. Kerscher, R., W. Grosch, Comparison of the aromas of cooked beef, pork and chicken, in *Frontiers of Flavour Science*, P. Schieberle, K.H. Engel, Eds., Deutsche Forschung. Lebensmit., Garsching, 2000, p. 17.
33. Wasserman, A.E., N. Gray, Meat flavor. I. Fractionation of water-soluble flavor precursors of beef, *J. Food Sci.,* 1965. 30, 5, p. 801.
34. Zaika, L.L. Meat flavour: method for rapid preparation of the water-soluble low molecular weight fraction of meat tissue extracts, *J. Agric. Food Chem.*, 17, 4, 893, 1969.
35. Wasserman, A.E., A.M. Spinelli, Effect of some water-soluble components on aroma of heated adipose tissue, *J. Agric. Food Chem.*, 20, 2, p. 171, 1972.
36. Manley, C.H., Flavor Interactions Workshop, IFT sponsored, Orlando, FL, 2003.
37. Blake, A., The world of meat flavours, *Food Manuf.*, 57, 9, p. 65, 1982.
38. Prendergast, K., Protein hydrolysis — a review, *Food Trade Rev.*, 44, p. 14, 1974.
39. Roozen, J.P., The bitterness of protein hydrolysates, in *Progress in Flavour Research*, J. Adda, Ed., Applied Sci. Publ., London, 1979, p. 321.
40. Sakho, M., Enzymatic maceration: effects on volatile components of mango pulp, *J. Food Sci.*, 63, 6, p. 975, 1998.
41. Winterhalter P., G.K. Skouroumounis, Glycoconjugated aroma compounds: occurrence, role and biotechnological transformation, *Adv. Biochemical Engin./Biotechnol.*, 55, p. 73, 1997.
42. Berger, R., G., U. Krings, H. Zorn, Biotechnological flavour generation, in *Food Flavour Technology*, A.J. Taylor, Ed., Sheffield Academ. Press, Sheffield, U.K., 2002, pp. 60–104.
43. Freund, P., Flavor Workshop II: Reaction Flavors, Encapsulation and Analysis, University of Minnesota, St. Paul, September, 2003.
44. Kilcawley, K.N., M.G. Wilkinson, P.F. Fox, Enzyme-modified cheese, *Int. Dairy J.*, 8, p. 1, 1998.
45. Freund, P.R., Cheese flavours ripen into maturity, *Int. Food Ingred.*, 5, p. 35, 1995.
46. Freund, P.R., Speciality milks and milk derived flavors in confections, *Manuf. Confect.*, 75, 6, p. 80, 1995.
47. Klein, N., S. Lortal, Attenuated starters: an efficient means to influence cheese ripening — a review, *Int. Dairy J.*, 9, p. 751, 1999.
48. West, S., Flavour production with enzymes, in *Industrial Enzymology*, T. Godfrey, S. West, Eds., Macmillan Publ., London, 1996, p. 211.
49. Andres, C., Butter flavor variations meet specific usage demands, *Food Process,* 1981. 42, 5, p. 62.
50. Farnham, J.G., R. Jordan, H. Saal, Spray/mat effectively dries tough enzyme modified butter-oil, cheese, *Food Process*, 42, 3, p. 104, 1981.
51. Hori, S., H. Shimazono, Intensification and supplementation of the flavor of dairy products, GB Patent #1075149, 1967.
52. Piatkiewicz, A.E.A., Selection and mutation of *Streptococcus diacetylactis* for butter production, in *Advances in Biotechnology Vol. II. Fuels, Chemicals, Foods and Waste Treatment*, A. R. Liss, New York, 1981, p. 491.
53. Olson, N.E.A., Microencapsulation of multi-enzyme systems, in *Advances in Biotechnology: Vol. III. Fermentation Products*, A. R. Liss, New York, 1981, p. 361.
54. Kilara, A., Enzyme modified lipid food ingredients, *Process Biochem.*, 20, p. 35, 1985.
55. Birschbach, P., Origins of lipases and their action characteristics, in Bulletin 294, *Int. Dairy Fed.*, Brussels, p. 7, 1994.

56. Birschbach, P., Pregastric lipases, in Bulletin 269, *Int. Dairy Fed.*, Brussels, p. 36, 1992.
57. De Greyt, W., A. Huyghebaert, Lipase catalyzed modification of milk fat, *Lipid Technol.*, 7, p. 10, 1995.
58. Kwak, H.S., I.J. Joen, S.K. Perng, Statistical patterns of lipase activities on the release of short chain fatty acids in Cheddar cheese slurries, *J. Food Sci.*, 54, p. 1559, 1989.
59. Kristofferson, T., E.M., J.A.G. Mikolejeik, Cheddar cheese flavor: 4. Directed and accelerated ripening process, *J. Dairy Sci.*, 50, p. 292, 1967.
60. Harvest Fields, *Vinegar Types, Methods of Making Vinegar & Recipes,* The Harvest Group, http://www.harvestfields.netfirms.com, accessed 04/19/05.
61. Napoleon Company, *Balsamic vinegar*, http://www.napoleon-co.com, accessed 9/13/04.
62. Zingermans Co., *The Truth About Balsamic Vinegar*, http://www.zingermans.com/Category.pasp?Category=balsamic%5Fvinegar, accessed 04/19/05.
63. Henry, B.S., The Production, Properties and Applications of Inactive Yeast Powders and Extracts, Proc. Symp. British Soc. Flav., London, June, 1982.
64. I.O.F.I. (International Organization of the Flavor Industry) http://www.iofi.org, accessed 04/19/05.
65. Gatfield, I.L., Biotechnological production of natural flavor materials, in *Flavor Chemistry: Thirty Years of Progress*, R. Teranishi, E.L. Wick, I. Hornstein, Eds., Kluwer Acad. Publ., New York, 1999, p. 211.
66. Parliment, T.H., R. Croteau, Eds., *Biogeneration of Aromas*, Amer. Chem. Soc., Washington, D.C., 1986, p. 397.
67. Berger, R.G., *Aroma Biotechnology*, Springer Verlag, Berlin, 1995, p. 240.
68. Schreier, P., Enzymes and flavour biotechnology, in *Advances in Biochemical Engineering*, R.G. Berger, Ed., Springer Verlag, Berlin, 1997, p. 51.
69. Welsch, F.W., W.D. Murray, R.E. Williams, Microbiological and enzymatic production of flavor and fragrance chemicals, *Crit. Rev. Biotechnol.*, 9, 2, p. 105, 1989.
70. Janssens, L., H.L. De Pooter, N. Schamp, E.J. Vandamme, Production of flavours by microorganisms, *Proc. Biochem.*, 27, p. 195, 1992.
71. Gatfield, I.L., The enzymatic synthesis of esters in non-aqueous systems, *Ann. N.Y. Acad. Sci.*, 436, p. 569, 1984.
72. Gatfield, I.L., The enzymatic synthesis of esters in non-aqueous systems, *Lebensm. Wiss. Technol.*, 19, p. 87, 1986.
73. Armstrong, D.W., B. Gillies, H. Yamazaki, Natural flavors by biotechnological processing, in *Flavor Chemistry: Trends and Developments*, R. Teranishi, R.G. Buttery, F. Shahidi, Eds., Amer. Chem. Soc., Washington, D.C., 1989, p. 105.
74. Duff, S.J.B., W.D. Murray, Oxidation of benzyl alcohol by whole cells of *Pichia pastoris* and by alcohol oxidase in aqueous and nonaqueous reaction media, *Biotechnol. Bioeng.*, 34, p. 153, 1989.
75. Gatfield, I., T. Sand, Enzymic production of esters and lactones, U.S. Patent 4451565, 1982.
76. Whitehead, I.M., B.L. Muller, C. Dean, Industrial use of soybean lipoxygenase for the production of natural green note flavor compounds, in *Bioflavour 95*, P. Étiévant, P. Schreier, Eds., Institut. National de la Recherche Agronomique, Paris, 1995, p. 339.
77. Candella, C., Personal communication, Robertet Flavors, S.A., Grasse, France, 1995.
78. Shimada, Y., et al., Enzymatic synthesis of L-menthyl esters in organic solvent-free system, *J. Amer. Oil Chem. Soc.*, 76, 10, p. 1139, 1999.
79. Hilmer, J.M., U. Bornscheuer, R. Schmidt, S. Vorlova, I.L. Gatfield, Process for the preparation of L-menthol, U.S. Patent 6706500, 2004.
80. Roy, A. Chemoenzymatic synthesis of homochiral (R)- and (S)-karahanaenol from (R)-limonene, *J. Agric. Food Chem.*, 47, 12, p. 5209, 1999.

81. McIver, R.C., *Factors Influencing the Synthesis of Methoxy Alkyl Pyrazines by Pseudomonas Perolens,* Ph.D. thesis, University of Minnesota, St. Paul, p. 88, 1986.
82. Kuila, R.K., B. Ranganathan, Ultraviolet light-induced mutants of *Streptococcus lactis* subspecies diacetylactis with enhanced acid- or flavor-producing abilities, *J. Dairy Sci.*, 61, 4, p. 379, 1978.
83. Farbood, M., B.J. Willis, Production of γ-Decalactone, U.S. Patent 4560656, 1985.
84. Van der Schaft, P.H., N. Ter Burg, S. Van den Bosch, A.M. Cohen, Microbial production of natural δ-decalactone and δ-dodecalactone from the corresponding α, β-unsaturated lactones in massoi bark oil, *Appl. Microbiol. Biotechnol.*, 36, 6, p. 712, 1992.
85. Dziezak, J.D., Biotechnology and flavor development: an industrial research perspective, *Food Technol.*, 40, 4, p. 108, 1986.
86. Sharpell, F., C. Stegmann, Developments of fermentation media for the production of butyric acid, in *Advances in Biotechnology*, M. Moo-Young, Ed., Pergamon Press, Toronto, 1981, p. 71.
87. Hopp, R., P, J. Rabenhorst, Process for the preparation of vanillin and microorganisms suitable therefor. U.S. Patent 6133003, 2000.
88. Priefert, H., J. Rabenhorst, A. Steinbuchel, Biotechnological production of vanillin, *Appl. Microbiol. Biotechnol.*, 56, 3–4, p. 296, 2001.
89. Rabenhorst, J., R. Hopp, Process for the preparation of vanillin, U.S. Patent 5017388, 1991.
90. Omata, T., Stereoselective hydrolysis of D,L menthyl succinate by gel-entrapped *Rhodotorula minuta* var. texensis in organic solvent, *Appl. Microbiol. Biotechnol*, 11, p. 199, 1981.
91. Mathewson, P.R. *Enzymes*, Eagan Press, St. Paul, 2004, p. 109.
92. Richardson, T., D.B. Hyslop, Enzymes, in *Food Chemistry,* 2nd ed., O.R. Fennema, Ed., Dekker, New York, 1985, p. 371.

# 10 Artificial Flavoring Materials

## 10.1 ARTIFICIAL FLAVORINGS

The flavor industry initially put great effort into attempting to duplicate nature through the use of synthetic chemicals. Natural flavorings derived only from natural sources had severe limitations, particularly when the flavoring was obtained from fruits, due to low flavor levels, seasonal availability issues, and high material costs. Also, very often flavorings isolated from nature did not perform well due to undesirable components indigenous to the flavoring. The composition of artificial flavorings could be precisely controlled, their aromatic profiles and physical form designed to meet specified consumer requirements or manufacturing parameters, and their consistent supply assured at a reasonable cost. Thus, it appeared to the industry that artificial flavorings were the future of the industry. This prospect drove the industry to broadly develop synthetic sources of chemicals that were found to define most of today's foods or food ingredients.

However, despite the sound reasons for artificial flavorings dominating the market, natural flavorings comprise the major portion of the market, and artificial flavoring materials find limited use. Today, artificial flavorings tend to be used when there is no natural flavoring material counterpart (e.g., most meat top notes, some components of fruit flavors, ethyl vanillin, ethyl maltol, etc.) or when the consumer product is of low cost and will not carry the additional expense of a natural flavoring. Thus, the market for synthetic flavor chemicals has shrunk over time.

Artificial flavorings are mixtures of synthetic compounds recognized as safe for use in foods and approved solvents (e.g., water, ethanol, propylene glycol, triethyl citrate, benzyl alcohol, and triacetin) or carriers (e.g., salt, maltodextrin, dextrose) as required to produce the flavoring in the desired form and concentration. In the U.S. the list of synthetic chemicals permitted for use is regulated by law and is discussed in Chapter 16 as are the labeling requirements for any flavoring containing synthetic chemicals.

## 10.2 SYNTHETIC FLAVORING MATERIALS

### 10.2.1 Introduction

Researchers, using the analytical techniques available, are now able to separate and identify most of the compounds responsible for the aroma of foods [1]. From this vast list of aroma compounds (>7,000), a more limited subset of compounds, i.e., those that have significant flavoring properties, has been selected for manufacture by synthetic means. To be of any value to the flavorist or flavor manufacturer, a synthetic organic chemical must be readily synthesized from available starting materials in sufficient yield to make it economically viable. The product must be adequately pure and stable to achieve consistency of effect in the end product and have physical character, which enables its quality to be controlled within acceptable limits.

The purity of synthetic flavorants is a very important quality attribute. Chemicals are rarely so pure that they have simple odor and flavor profiles. Most contain sufficient trace impurities, stemming either from the starting materials or processing conditions used in their manufacture, to display a spectrum of perceptible character notes. Sometimes it is the presence of these impurities, which characterize the compound and make it far more acceptable as a flavorant than the pure material itself!

From a quality point of view, the purchasing of synthetic flavoring materials calls for precise analysis of their physical characters, particularly the refractive index and specific gravity, coupled with a critical appraisal of their odor and flavor under controlled conditions against an acceptable reference sample. This latter evaluation is of far greater value in determining suitability for use than insistence on rigid compliance with specified physical standards — the two methods of assessment should always be considered together and the organoleptic results given the greater emphasis in deciding acceptance or otherwise.

### 10.2.2 Consumer Attitudes Toward Synthetic Chemicals

Not surprisingly, consumers express considerable concern about the indiscriminate use of chemicals in foods although Stofberg and Stoffelsma [2] have shown that the amounts used and consumed as flavorings are infinitesimal compared with the input from natural sources the safety of which is not questioned. Nevertheless, in most countries, the laws governing safety of foodstuffs recognize this consumer attitude and thus laws are in place that reflect this consumer concern. The U.S. laws and labeling regulations will be discussed in Chapter 16 of this text.

## 10.3 CLASSIFICATION OF AROMA COMPOUNDS BY MOLECULAR STRUCTURE

Naturally occurring aroma and flavoring components embrace the whole spectrum of organic chemistry and the most important groups of compounds are listed in Table 10.1. Parliment [3] has pointed out the wide range of structures, thresholds, and characters of aroma chemicals present in any one plant source. Although the functional group of an organic compound has a quite significant affect on its aroma and hence flavor profile, there is not much correlation between chemical structure and organoleptic properties, and there are literally thousands of different odor qualities. Some compounds of similar structure do have broadly similar odors but there are very many exceptions and even where the odors are similar the intensities may vary considerably. Macleod [4], in reviewing the chemistry of odors, gave examples of the following categories of aromatic chemicals: (a) compounds of similar structure with similar odors (Figure 10.1); (b) compounds of different structure with similar

**TABLE 10.1**
**Chemical Nature of the Major Aromatic Compounds Used in Flavorings**

| | |
|---|---|
| **Hydrocarbons** | **Esters** |
| Aliphatic | Aliphatic |
| Aromatic | Aromatic |
| Cyclic terpenes (mono- and bicyclic terpenes) | Terpene esters ethers |
| Sesquiterpenes (mono-, bi-, and tricyclic sesquiterpenes) | Heterocyclic compounds ketals |
| Alcohols | Ketones |
| Aliphatic (saturated, unsaturated, and terpenic alcohols) | Aliphatic aromatic |
| **Aromatic** | **Cyclic terpene ketones** |
| Cyclic terpene alcohols | Ionones |
| Sesquiterpene alcohols | Irones |
| Aldehydes | Lactones |
| Aliphatic | Oxides |
| Aromatic | Phenols |
| Cyclic terpene aldehydes | Phenol ethers |
| Heterocyclic aldehydes | Compounds containing nitrogen and sulfur |
| Acetals | |
| **Acids** | **Amines** |
| Aliphatic | Amino compounds |
| Aromatic | Imino compounds |
| Anhydrides | Sulfides |
| | Pyrazines |

*Source:* From Parliment, T.H., *Chem. Tech.*, March, p. 284, 1980. With permission.

odors (Figure 10.2); (c) compounds of similar structure with different odors (Figure 10.3); and (d) stereoisomers with different odors (Figure 10.4).

These examples show that structural relationships are not sufficiently precise to allow predictability of odor or flavoring effects either qualitatively or quantitatively. Numerous attempts have been made to provide a theory that will explain structure-activity relationships but most of these apply only to a narrow range of examples. Macleod [4] concluded that the most important factors affecting this relationship would seem to be overall shape and size of the molecule, its detailed stereochemistry, and certain chemical properties (e.g., polarity, nature of functional groups) and physical properties (e.g., volatility and solubility). (Interestingly, recent advances in perception support his hypothesis — Chapter 1.) Such considerations are of little importance when a characterized aromatic component is being synthesized but are of real concern to the organic chemist aiming to create an organic compound having a defined odor and/or flavor character. At present, organic chemists are largely working on an empirical basis observing the effects of substitution, etc., by sensory analysis as the work progresses.

Certain relationships are worthy of comment. Within very broad limits, the flavoring strength of an organic compound, as opposed to its aromatic quality, is

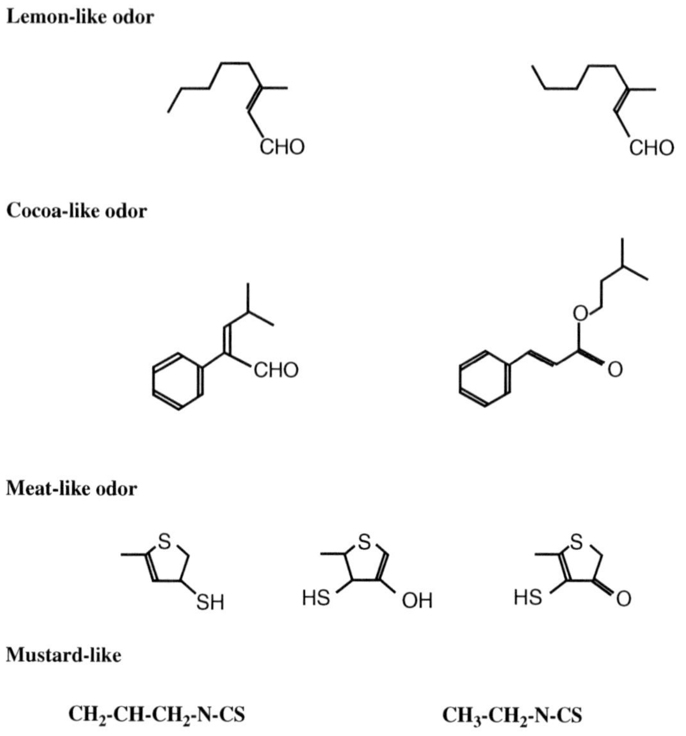


**FIGURE 10.1** Compounds having similar structures and similar odors. (From Macleod, A.J., *Proc. Symp. Zool. Soc.*, 45, p. 15, 1980. With permission.)

inversely related to its vapor pressure: the lower the vapor pressure, the greater the aromatic impact. (This seems counterintuitive but relates to mechanisms of perception.) For this reason, branched chain substitution is likely to result in compounds having a higher flavoring value than the equivalent straight chain compound (e.g., isoamyl butyrate has a considerably lower threshold and hence greater flavoring strength than n-amyl butyrate). Unfortunately the effect is not quantitatively predictable. The substitution of groupings within the molecule may also have a profound effect on its aromatic attributes. The degree of effect depends to some extent on the size of the molecule; in small molecules the substitution of one functional group for another may radically change its whole chemistry whereas in larger molecules, where functional groups form a much lesser part of the total structure, the effect is not so pronounced.

The modification of simple molecules such as $CH_3OH$, $CH_3SH$, and $CHNH_2$ demonstrates this effect whereas in the much larger vanillin molecule, the substitution of an ethoxy group for the normal methoxy group results in a compound having three times the flavoring power and only a subtle change in the aromatic profile. Structural relationships between synthetic organics used as flavorants are listed in Table 10.2.

**Musk-like**

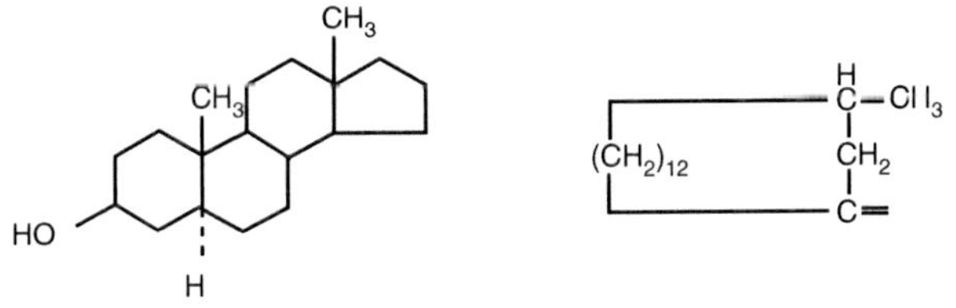


**Camphoraceous-like**

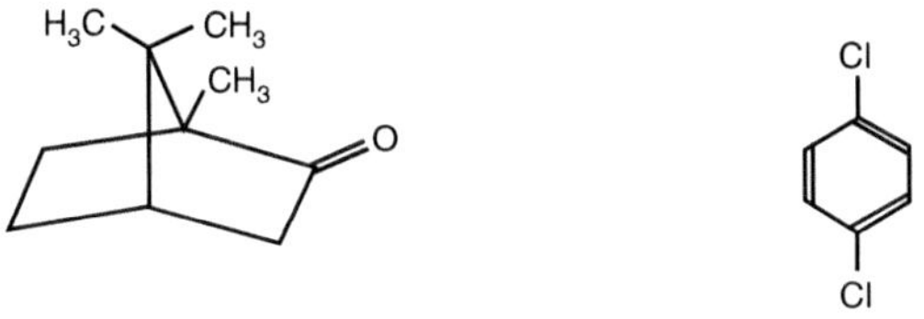


**FIGURE 10.2** Compounds having different structures and similar odors. (From Macleod, A.J., *Proc. Symp. Zool. Soc.*, 45, p. 15, 1980. With permission.)

**Grapefruit-like** **Woody odor**

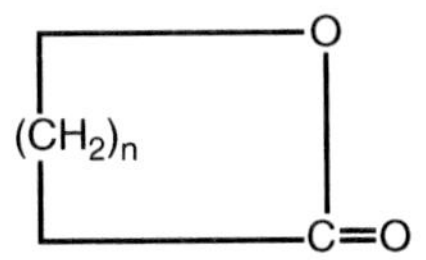

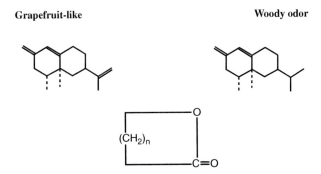


**FIGURE 10.3** Compounds having similar structures and different odors. (From Macleod, A.J., *Proc. Symp. Zool. Soc.*, 45, p. 15, 1980. With permission.)

## 10.4 SENSORY THRESHOLD VALUES

One of the principal properties of an organic compound which is to be used in flavorings is its sensory threshold value(s). The American Society for Testing and Materials (ASTM) has provided the following definitions:

1. Detection threshold: Minimum physical intensity detectable where the subject is not required to identify the stimulus
2. Difference threshold: Smallest change in concentration of a substance required to give a perceptible change

**Similar odors**

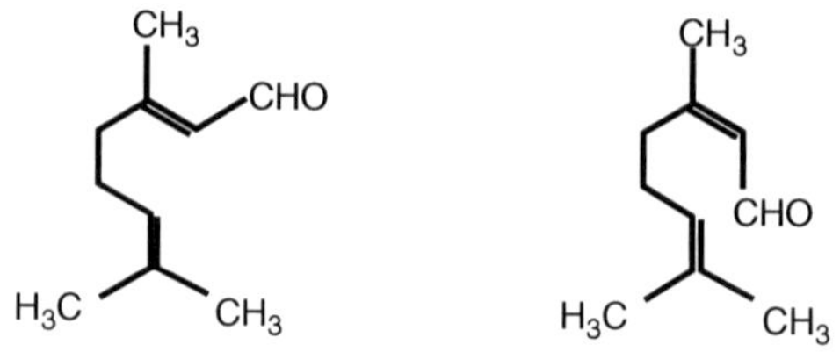


**Different odors**

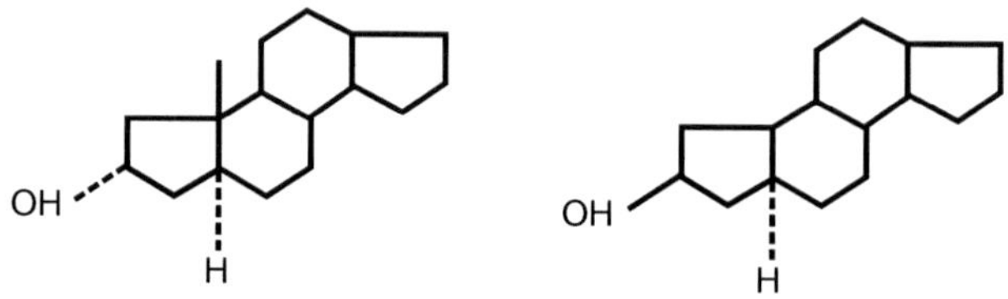


**FIGURE 10.4** Isomers and sterioisomers. (From Macleod, A.J., *Proc. Symp. Zool. Soc.*, 45, p. 15, 1980. With permission.)

## TABLE 10.2
## Structural Relationships of Flavoring Organics

| Class of Compound | Distinguishing Systematic Suffix or Prefix | Functional Group | Typical Structure | |
|---|---|---|---|---|
| | | | Aliphatic | Aromatic |
| Alcohols | -ol | –O–H | R–OH | $C_6H_5R$–OH |
| Phenols | -ol | –O–H | | $C_6H_5$–OH |
| Carboxylic acids | -ic acid or -oic acid | –COOH | R–COOH | $C_6H_5$–COOH |
| Esters | -ate | –COO– | R–COO–R′ | $C_6H_5$–COO–R |
| Aldehydes | -al | –CHO | R–CHO | $C_6H_5$–CHO |
| Ketones | -one | –CO– | R–CO–R′ | $C_6H_5$–CO–R |
| Ethers | -ether | –O– | R–O–R′ | $C_6H_5$–O–R |
| Lactones | -lactone or -ide | –COO– | | |
| Amines | -amine or amino | $–NH_2$ | $R–NH_2$ | $C_6H_5–NH_2$–R |
| Nitro-compounds | nitro- | $–NO_2$ | $R–NO_2$ | |
| Sulfides | -ide | –S– | R–S–R′ | |
| Disulfides | di- | –S–S– | R–S–S–R′ | |
| Thio-alcohols | -mercaptan | –SH | R–SH | |

*Source:* From Heath, H.B., G. Reineccius, *Flavor Chemistry and Technology*, Van Nostrand Reinhold, New York, 1986, p. 442.

3. Recognition threshold: Lowest concentration at which a substance is correctly identified.
4. Terminal threshold: The concentration of a substance above which changes in concentration are not perceptible.

The ASTM has provided standard protocols for the determination of these thresholds as well as a compilation of sensory thresholds for some compounds [6,7]. Such protocols are also available in several sensory texts/reference books [8–10].

Sensory threshold data (and descriptions) can be found in numerous places. For example, the Flavor-Base — 2004 [11,12] database provides a list of ranges in sensory thresholds of 762 GRAS and European Union flavor chemicals in several media (e.g., water, orange juice, beer, wine, oil, etc.). A series of articles has appeared over the years in *Perfumer and Flavorist* [13–16] that characterize the sensory properties and potential product applications of a large number of aroma chemicals used in the industry. Van Gemert [17] has provided a literature compilation covering the sensory thresholds of over 1,100 compounds in air and water. The World Wide Web (WWW) has provided a wealth of such information either in compilation form [14] or as general searches on compounds of interest. General searches will typically provide substantial information on individual aroma chemicals (e.g., supplier, sensory character and threshold, and chromatographic and physical properties).

There is little question that sensory thresholds are dependant upon the medium they are tested in, the testing protocol, and the ability of panelists. Thus, one always finds a range in threshold values occurring in any literature compilation. This range may be as much as a factor of $10^3$ for a given compound and thus such data must be used with care.

## 10.5 SENSORY CHARACTERS OF ODOR COMPOUNDS

As noted above, there is substantial information available describing the sensory properties of aroma compounds. The classic references in this respect were not mentioned above but are the works of Fenaroli [18] and Arctander [19]. In addition, most suppliers of aroma chemicals provide a characterization of their product. However, those directly concerned with creating flavoring compositions have to develop their own characterizations based on their personal evaluation of these chemicals. The descriptions provided in the literature are based on an individual's vocabulary, experience, and perception which vary as widely from individual to individual as do the sensory threshold data. Thus, those coming into this field must keep records of their own evaluations of odor and flavor characters of all aroma chemicals. This whole process is what differentiates one flavorist's creation from another.

With that said, the remainder of this chapter will present a discussion of the sensory characters of numerous aroma compounds. In some cases, the sensory descriptors are fairly constant across the population. This is generally the case for character impact compounds that singly define commonly known foods or food ingredients. For example, most people know eugenol as clove-like or vanillin as vanilla-like. The disagreement comes when the aroma chemical is not clearly defined as a common food or food ingredient. If a group of individuals has experienced a chemical from this group, it may have been in widely different foods or experiences and thus, their characterizations will differ greatly.

### 10.5.1 Hydrocarbons

The simplest of the organic chemical groupings is the hydrocarbons comprising only carbon and hydrogen. The saturated paraffin hydrocarbons, having the general formula $C_nH_{2n+2}$, are of little importance in a flavor context although some of the higher members of the series ($C_5$–$C_8$) are of some value as extraction solvents. Aromatic hydrocarbons, based on a single (benzene), double (naphthalene), or triple (anthracene) ring structure are more aromatic but again are of no direct use in flavorings.

The unsaturated hydrocarbons may be significant contributors to desirable aroma. For example, Schieberle and Steinhaus [20] have found (E)3,(Z)5-undecatriene and 1,(E)3, (Z)5,9-undecatetraene to be significant contributors to the aroma of hops essential oil.

In this group of volatile compounds, the terpenoid hydrocarbons are of greatest importance. These compounds, which are found in so many of the essential oils, form a very distinctive group of chemicals that have carbon skeletons comprising isoprene units joined together in a regular head-to-tail configuration in accord with what is called the *isoprene role*. They may be open or closed chain, or cyclic compounds of the general formula $C_{5n}H_{10n-4}$. Where n = 2 they are called terpenes; where n = 3, diterpenes; and where n = 4, sesquiterpenes. The compounds may be saturated or unsaturated. Typical examples of these compounds are given in Figure 10.5.

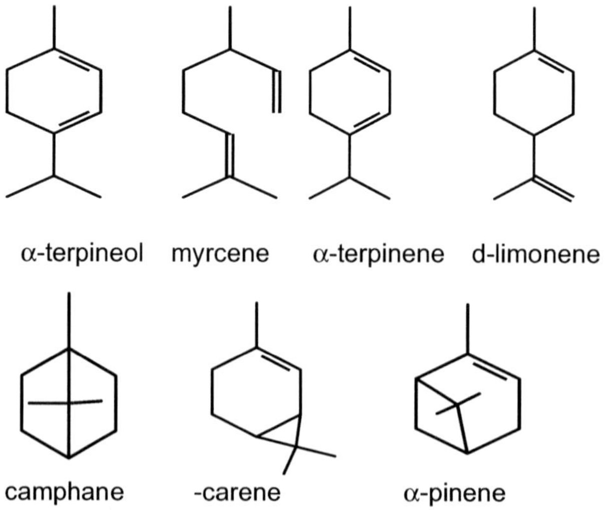


**FIGURE 10.5** Examples of Terpenoids. *Source:* From Heath, H.B., G. Reineccius, *Flavor Chemistry and Technology*, Van Nostrand Reinhold, New York, 1986, p. 442.

Terpene hydrocarbons act primarily as diluents and carriers for other, usually oxygenated, flavoring constituents although their presence adds a considerable degree of freshness to the odor and flavor profiles. Essential oils, which have been deterpenated, are flat and lack this fresh attribute [21].

### 10.5.2 Carboxylic Acids

The carboxylic acids are characterized by having a carboxyl group attached to either an alkyl group (R·COOH) or an aryl group (Ar·COOH). The presence of a hydroxyl group tends to suppress the odor (e.g., propionic acid $CH_3CH_2COOH$ has a pungent odor whereas lactic acid $CH_3CH(OH){\cdot}COOH$ is odorless). The di- and tri-carboxylic acids are also odorless.

In foods, acetic acid $CH_3COOH$ is by far the most important of the carboxylic acids and typifies the very short chain saturated fatty acids ($C_1$–$C_3$) which are generally unpleasant with strongly pungent, penetrating odors. In the range $C_4$–$C_8$, this character is progressively replaced by a rancid buttery/cheesy notes and above $C_9$ the odor becomes unpleasantly fatty/aldehydic. Acids above $C_{14}$ are waxy solids at room temperature and have little or no odor. However, the $C_{12}$ fatty acid is well known for its soapy taste. The lower members of the unsaturated oleic series are pungent and more acrid, their character changing, with increasing molecular weight, to a slightly spicy, aldehydic, and tallow-like quality. Those above $C_{11}$ are odorless.

The aromatic acids do not have a well-defined pattern of odors; all are faintly balsamic with various spicy, floral, or even fruity overtones. Benzoic and cinnamic acids, and the higher members of the series are odorless, although compounds based on a saturated cyclic structure (e.g., hexahydrobenzoic acid) have an unpleasant rancid character.

As a group, the organic acids impart a sharp, sour taste to a product. The di- and tri-carboxylic acids are of particular value in this respect as they occur naturally in many fruits. These compounds are odorless and are not used directly in flavoring formulations, but citric, malic, and tartaric acids are often used in end products to improve and even enhance added fruit flavorings which would otherwise be atypical of the natural fruit.

### 10.5.3 Acetals

Alcohols and glycols may react with the carbonyl group of aldehydes in the presence of an anhydrous acid to form mixed ethers called acetals. It is suggested that, in alcoholic solution, aldehydes really exist in equilibrium with compounds called hemiacetals but that these are too unstable to be isolated. In the presence of acids, the hemiacetals act as an alcohol and react to form the acetal. The reaction is reversible, and in the presence of water, acetals are readily hydrolyzed back to the parent aldehyde and alcohol.

The acetals are liquids having a wide range of odors but are not extensively used in flavorings. Two are of particular interest: (a) acetaldehyde diethyl acetal — pleasantly nutty and (b) cinnamaldehyde ethylene glycol — warm, sweetly spicy. The product of glycol with hydrotropic aldehyde (1-oxo-2-phenyl-2-methylethane) has a strong odor reminiscent of mushrooms. The formation of acetals in finished flavorings often adds unpleasant notes to the product. The alcohol may be the flavor diluent (ethanol or propylene glycol) and the aldehyde may be a part of the flavor. This reaction removes a desirable flavor component (the aldehydes) and forms acetals that have unpleasant notes resulting in an off-flavored product.

### 10.5.4 Alcohols

Alcohols are compounds having the general formula R-OH where R is any alkyl or substituted alkyl group. Alcohols may be grouped in several ways, including (a) primary (R–CHOH), secondary (R–R′–CHOH), tertiary (R–R′–R″–OH); (b) open chain or cyclic; (c) saturated or unsaturated; (d) containing a halogen atom; and (e) containing an aromatic ring.

In the aromatic series where the hydroxyl group (-OH) is substituted directly on the benzene nucleus, the compounds are known as phenols. These differ significantly from the alcohols and are considered separately. The alcohols as a group are important flavoring materials and are extensively found throughout nature. All contain a hydroxyl group, which largely determines their characteristics.

In the lower alcohols, where the hydroxyl group represents a large part of the molecule, they are polar solvents and readily miscible with water. The change in solubility with increasing carbon number is a gradual one and for normal primary alcohols those above $C_5$ are generally regarded as water insoluble. Almost all alcohols can be distilled, and this forms a convenient method for their preparation and purification. Alcohols of lower molecular weight have a sweet, spiritous odor that intensifies up the series to $C_{9,}$ which has an unpleasant oil-note. They are excellent and widely used solvents in the preparation of liquid flavorings. With increasing molecular weight the physical form changes and they become viscous fluids, the pleasant odor being progressively replaced by a fatty and irritating odor character. The higher alcohols are waxy solids having almost no odor.

Unsaturated short chain alcohols, in which the double bond is close to the hydroxyl group, have a harsh, penetrating and irritating odor. However, mid chain length, unsaturated alcohols such as cis-3-hexenol are very important contributors to the "green" note used in many foods. 1-Octene-3-ol is characterizing of mushrooms. The aromatic alcohols are mostly pleasant and have well-rounded odor profiles. The substitution of two or more hydroxyl groups removes the odorous character of the compound in both the aliphatic and aromatic series (e.g., glycerol).

The terpenoid alcohols represent a very distinct and important group of flavoring compounds found widely in essential oils. Two examples are of interest: (a) neraniol (trans-3,7-dimethyl-2,6-octadien-l-ol), which occurs in rose, lemongrass, and geranium oils and has a characteristic rose-like odor, and (b) linalool (3,7-dimethyl-l,6-octadien-3-ol), which occurs widely in either its optically active or inactive forms and has a typical floral odor. Aromatic alcohols in which the hydroxyl group is in the side chain have relatively weak but usually pleasant odor characters.

### 10.5.5 Carbonyls

These are compounds characterized by the presence of a carbonyl group (C=O) and may be either aldehydes, ketones, or diketones.

#### 10.5.5.1 Aldehydes

The aldehydes are intermediate between the alcohols and acids, being more reactive than the alcohols because of their double bond linkage with oxygen. Aldehydes are particularly susceptible to oxidation to acids. For example, it is common to find a white crystalline material in an old bottle of benzaldehyde. This crystalline material is benzoic acid, which if used in a product in place of benzaldehyde, will impart a bitter "bite" to the back of the throat.

Aldehydes of lower molecular weight are characterized by their unpleasant, sharply pungent, irritating odors. With increasing molecular weight, there is a progressive change in the profile to a pleasing fruity character though the penetrating nature of the odor remains apparent throughout the series, hence the true aromatic profile of aldehydes is better appreciated in dilution where this attribute is mostly eliminated. Aldehydes $C_8$–$C_{10}$ are markedly floral and most attractive on dilution. They are best used at low concentrations in compositions in which a floral nuance is required. As the molecular weight increases further, the compounds lose their odor and become somewhat waxy.

Unsaturated aldehydes have particularly irritating odors, and this is further increased by the presence of a triple bond. Like alcohols, some of the mid chain length unsaturated aldehydes such as trans-2-hexenal have very desirable green but more spicy odors. The di unsaturated aldehydes, e.g., 2,4-decadienal and 2,6-nonadienal, are important aroma contributors to meaty and fried aromas, and cucumber aroma, respectively. The aromatic aldehydes are powerful and display a wide spectrum of profiles, depending on their complexity.

By long tradition certain compounds are commonly known by names such as aldehyde $C_{16}$ or strawberry aldehyde, aldehyde $C_{14}$, or peach aldehyde. While the sensory description is accurate, the chemical designation is not. Aldehyde $C_{16}$ is in fact ethyl 2,3-epoxy-3-phenyl-butanoate whereas aldehyde $C_{14}$ is 4-hydroxy undecanoic acid, a $\gamma$-lactone.

#### 10.5.5.2 Ketones

A ketone molecule contains two substituent carbon radicals on the carbonyl group. These may be either similar or different. Those of low molecular weight are of little value as flavorants although acetone is used in the flavor industry as a valuable, if somewhat flammable, extraction solvent.

The lowest molecular weight ketone of real value as a flavoring material is methyl amyl ketone, which has a strong fruity character that persists through the next members of the series. With increasing molecular weight, the fruity character is replaced by floral notes, but the effects are not sequential and the ketones display a wide range of aromatic profiles and strengths. For example, the character impact compound in blue cheese is 2-heptanone and that of mushrooms is 1-octene-2-one.

### 10.5.6 Esters

Esters are formed by the reaction of alcohols or phenols with acids and their derivatives:

$$R \cdot COOH + R' \cdot OH \rightarrow R \cdot COO \cdot R' + H_2O$$

where R may be alkyl or aryl and where R′ is usually alkyl. To the flavorist, esters form the principal source of a very wide spectrum of odor and flavoring effects, and there are over 200 of these compounds permitted for use in foods. Esters are widely distributed in the essential oils and in some instances represent the major constituent (e.g., oil of wintergreen is about 90% methyl salicylate).

Although some general structural/sensory relationships can be found within this group, particularly their overall fruity character, these are by no means sufficiently well defined to form the basis of a worthwhile classification. From the broad similarities which exist between them, it is preferable to consider esters according to their acid constituent (e.g., acetates, butyrates, isobutyrates, benzoates, etc.) rather than on their alcohol or phenol radical (e.g., methyl, ethyl, benzyl, eugenyl, etc.), although this moiety does have an increasingly significant effect on the profile as it becomes larger and/or more unsaturated. The profiles of the constituent acids and alcohols have almost no sensory relationship with that of the resulting ester.

### 10.5.7 Ethers

Ethers are compounds of the general formula R-O-R; R-O-Ar or Ar-O-Ar and are named from the two groups that are attached to the oxygen atom followed by the name *ether* (e.g., diethyl ether). Some have common names which are more widely used (e.g., anisole, which is methyl phenyl ether). If one group has no simple name, the compound may be called an alkoxy derivative (e.g., 2-ethoxyethanol).

Ethers of lower molecular weight are extremely volatile and have a sweet, spiritous odor whereas the higher ethers have very interesting profiles which may be ethereal and fruity. The ether group does not impose much sensory impact, and most ethers have only light odors influenced more by the substituents in the molecule. There is little relationship within the group and individual characters must be experienced at first hand.

### 10.5.8 Heterocyclic Compounds

Heterocyclic compounds are those compounds that contain a ring made up of more than one kind of atom. The most common from a flavor chemistry point of view are:

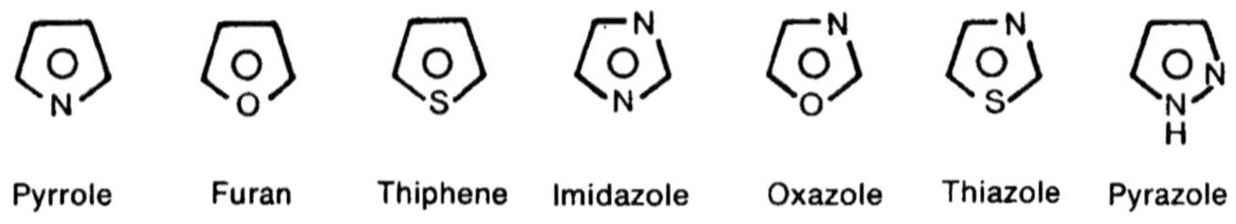


(note that variations in number of double bonds in these structures occur to give a broad range of compounds.)

The heterocyclic compounds containing O, N, or S, typically have very potent and often characteristic odors [22–24]. They are essential to the flavor of some plants and are key volatiles in defining the flavor of many thermally processed foods. Examples of members of this class of aroma compounds have been presented in Chapter 5.

### 10.5.9 Lactones

Internal esterification resulting in the loss of one molecule of water from the γ- and δ-hydroxyacids yields a class of cyclic compounds called lactones. The five-membered ring system is readily formed and the resulting γ-lactones are very stable.

As a group, the lactones are powerful and distinctive flavorants, which are responsible for many of the characteristic notes of fruits, e.g., peaches and apricots. A comparison of the odors of δ- and γ-lactones indicates that the increase in the size of the ring in the δ-compound results in an increased odor intensity but that the γ-lactones have in addition some "green" nuances not present in the δ-lactones. Many lactones have a strong musk-like odor that is very persistent. In the aromatic series, the best known lactone is coumarin, but this is no longer permitted for use in foods. The part played by lactones in foods has been reviewed by Maga [25] and Dufosse et al. [26].

### 10.5.10 Nitrogen-Containing Compounds

#### 10.5.10.1 Amines

Amines are substituted ammonia compounds having the general formula $RNH_2$ (primary); $R_2NH$ (secondary), and $R_3N$ (tertiary), where R is any alkyl or aryl group. Aliphatic amines are named according to the alkyl group(s) attached to the nitrogen following these with the word *amine.* Sometimes the prefix *amino* is added to that of the parent chain. Aromatic amines in which the nitrogen is attached directly to the aromatic ring are named as derivatives of aniline.

All amines of low molecular weight have strong and very distinctive profiles, which on dilution, have an ammoniacal character. The profile becomes increasingly "fishy" in the secondary and tertiary amines, although this objectionable odor is gradually lost with increasing molecular weight.

In the aromatic amines, an amino group substituted on the side chain results in a marked ammoniacal odor nuance whereas direct substitution of the amino group onto the nucleus results in a very weak and undefined odor character (e.g., anthranilic acid is odorless whereas methyl anthranilate has a strong orange-flower-like profile). The amines are of very limited use in flavorings as most of them have an objectionable odor but they do play an important part in some foods [27].

#### 10.5.10.2 Amides

The hydroxyl group of carboxylic acids may be replaced by the amino group to give compounds celled amines. These are chemically inert compounds, and as they are odorless, they are of no value as flavorants.

#### 10.5.10.3 Imines

In flavor chemistry one sometimes comes across compounds called imines. These are usually only intermediates in complex reactions and have the general formula RCH=NH or $R_2C$=NH.

#### 10.5.10.4 Amino Acids

Although not directly used as flavorants, the amino acids, as the prime constituents of proteins, not only play an essential role in body nutrition but also enter into the Maillard reaction when heated with reducing sugars. This initiates a series of interrelated reactions resulting in a complex mixture of highly aromatic compounds and a characteristic brown color. The chemistry of this reaction is dealt with in greater detail elsewhere (see Chapter 5).

#### 10.5.10.5 Isothiocyanates

The isothiocyanates contain the -N:C:S group and are found in several pungent vegetables, particularly, (a) allyl isothiocyanate, produced by the enzymatic action of myrosin on the glycoside sinigrin in black mustard seed; (b) benzyl isothiocyanate; found in garden cress; and (c) phenyl ethyl isothiocyanate, responsible for the sharp pungency of horseradish.

### 10.5.11 Phenols

Phenols are compounds of the general formula ArOH where Ar is phenyl, substituted phenyl or one of the other aryl groups. They are characterized by having the hydroxyl group attached directly to the aromatic ring. Phenol has a very distinctive odor long associated with disinfectants, but substitution of methyl groups into the benzene nucleus reduces the sharp, acidic character associated with phenol so that the cresols and xylenols have much weaker odors. Within the cresols, *meta* substitution relative to the hydroxyl group gives a lower odor impact than either the *ortho* or *para* positions. The introduction of two or more hydroxyl groups totally suppresses the odor character. Phenols are important in food flavors and have been reviewed by Maga [28].

### 10.5.12 Sulfur-Containing Compounds

Sulfur and oxygen, being chemically related, form similar compounds but have very different sensory properties. The -SH group is commonly known as mercapto-, thiol-, or sulfhydryl. Sulfur-containing compounds play an important role in natural flavor chemistry as they are not only responsible for the objectionable odors associated with rotting vegetable matter but also contribute, and often characterize, the desirable aroma of many plants and thermally processed foods [23,29,30]. The contribution of sulfur-containing volatiles to the aroma of foods is only recently being recognized due to their extremely low sensory thresholds [32]. Prior to the advent of sophisticated analytical instrumentation, these compounds went unrecognized and unidentified (see Chapter 5).

Sulfides, disulfides, and polysulfides (also called thioethers) are the characterizing components of the alliaceous vegetables (e.g., onion, garlic, leek), although the more subtle attributes of these profiles are contributed by substituted thio compounds.

The preceding sections on sensory characters of organics cover only the major organic groupings widely encountered in flavor work and is by no means exhaustive. Discussing each compound class in detail would require a book in itself. It is intended merely to draw attention to the enormous range of aromatic effects that can be derived from synthetic chemicals and in no way is a substitute for practical exposure to the odor and flavor characteristics of the chemicals permitted for use in foods.

## 10.6 NOMENCLATURE OF ORGANIC CHEMICALS

One cannot read any research paper involving flavor chemistry without encountering innumerable compounds having almost impossibly long chemical names. These are generally systematic and give a direct indication of the molecular structure of the compound concerned but unfortunately some confusion may arise from the use of different naming systems. Trivial or common names are most likely to be used in a commercial laboratory and in some older references, these designations are used almost exclusively. The International Union for Physics and Chemistry (IUPAC) system is by far the most preferred, and reference should be made to standard texts on organic chemistry to find the rules pertaining to the various types of organic compounds involved. In general, the system is based on the identification of the longest chain containing the substituent group and applying to it the name of the parent hydrocarbon or basic compound named on a strictly systematic basis. The position of the substituent group or groups is then indicated by sequential numbers using the lowest possible combination. It is a good practice to adopt the designation used in the official GRAS listings and to maintain an alphabetical listing of synonyms for ready reference [31].

## REFERENCES


1. TNO Nutrition and Food Research, Volatile compounds found in foods, http://www.voeding.tno.nl/vcf/VcfNavigate.cfm?CFID=1321050&CFTOKEN=33202682, accessed 04/20/05.
2. Stofberg, J. and J. Stoffelsma, Consumption of flavoring materials as food ingredients and food additives, *Perfum. Flavorist*, 5, p. 19, 1980.
3. Parliment, T.H., The chemistry of aroma, *Chem. Tech.*, March, p. 284, 1980.
4. Macleod, A.J., Chemistry of odors, *Proc. Symp. Zool. Soc.*, 45, p. 15, 1980.
5. Heath, H.B. and G. Reineccius, *Flavor Chemistry and Technology*, Van Nostrand Reinhold, New York, p. 442, 1986.
6. ASTM, Sensory testing protocols, http://www.astm.org, accessed 04/20/05.
7. ASTM, DS 48A, Compilation of Odor and Taste Threshold Values Data, West Conshohocken, p. 508, 1978.

8. Deibler, K.D. and J. Delwiche, Eds., *Handbook of Flavor Characterization: Sensory Analysis, Chemistry, and Physiology*, Marcel Dekker, New York, p. 493, 2004.
9. Lawless, H.T. and B.P. Klein, Eds., *Sensory Science Theory and Applications in Foods*, Marcel Dekker, New York, p. 441, 1991.
10. Carpenter, R.P., D.H. Lyon, T.A. Hasdell, Eds., Guidelines for sensory analysis, in *Food Product Development and Quality Control Guidelines*, Aspen Publishing, Gaithersburg, p. 210, 2000.
11. Leffingwell & Assoc., Flavor Base, http://www.leffingwell.com, accessed 04/20/05.
12. Leffingwell & Assoc., Odor Thresholds, http://www.leffingwell.com/odorthre.htm, accessed 04/20/05.
13. Mosciano, G., M. Fasano, J.C. Cassidy, K. Connelly, P. Mazeiko, A. Montenegro, J. Michalski, S, Sadural, Organoleptic characteristics of flavor materials, *Perfum. Flavorist*, 20, 6, p. 49, 1995.
14. Mosciano, G., W. Pohero, J., Michalski, C., Holmgren, D. Young, Organoleptic characteristics of flavor materials, *Perfum. Flavorist*, 25, 4, p. 71, 2000.
15. Mosciano, G., W. Pohero, J. Michalski, C. Holmgren, D. Young, Organoleptic characteristics of flavor materials, Consensus of Bedoukian GRAS 19, *Perfum. Flavorist*, 25, 5, p. 72, 2000.
16. Mosciano, G., Organoleptic characteristics of flavor materials, *Perfum. Flavorist*, 26, 6, p. 58, 2001.
17. van Gemert, L.J., *Compilations of Odour Threshold Values in Air and Water*, TNO Nutrition and Food Research Institute, Zeist, The Netherlands, 2003.
18. Fenaroli, G., *Fenaroli's Handbook of Flavor Ingredients*, 4th ed., G.A. Burdock, Ed., CRC Press, Boca Raton, 2001.
19. Arctander, S., *Perfume and Flavor Materials of Natural Origin*, Allured Pub., Carol Stream, 1960.
20. Schieberle, P. and M. Steinhaus, Characterization of the odor-active constituents in fresh and processed hops (variety Spalter Select), in *Gas Chromatography-Olfactometry: The State of the Art*, J.V. Leland, P. Schieberle, A. Buettner, T.E. Acree, Eds., Amer. Chem. Soc., Washington, D.C., p. 23, 2001.
21. Heath, H.B., *Flavor Technology: Profiles, Products, Applications*, AVI Pub. Co., Westport, CT, 1978.
22. Schieberle, P. and T.H. Hofmann, Flavor contribution and formation of heterocyclic oxygen-containing key aroma compounds in thermally processed foods, in *Heteroatomic Aroma Compounds*, G.A..Reineccius, T.A. Reineccius, Eds., Amer. Chem. Soc., Washington, D.C., 2002, p. 207.
23. Mottram, D.S. and H.R. Mottram, An overview of the contribution of sulfur-containing compounds to the aroma in heated foods, in *Heteroatomic Aroma Compounds*, G.A. Reineccius, T.A. Reineccius, Eds., Amer. Chem. Soc, Washington, D.C., 2002, p. 73.
24. Ho, C.T. and Q.Z. Jin, Aroma properties of some alkylthiazoles, *Perfum. Flavorist*, 9, 6, p. 15, 1985.
25. Maga, J.A., Lactones in foods, *CRC Critical Rev. Food Technol. Nutr.*, 8, p. 1, 1976.
26. Dufosse, L., A. Latrasse, H.E. Spinnler, Importance of lactones in food flavors: structure, distribution, sensory properties and biosynthesis, *Sciences des Alim.*, 14, 1, p. 17, 1994.
27. Maga, J.A., Amines in foods, CRC *Critical Rev. Food Technol Nutr.*, 10, p. 373, 1978b.
28. Maga, J.A., Simple phenols and phenolic compounds in food flavor, *CRC Critical Rev. Food Technol. Nutr.*, 10, p. 323, 1978a.

29. Shankaranarayana, M.L., B. Raghavan, K.O. Abraham, C.P. Natarajan, Sulfur compounds in flavors, in *Developments in Food Science, 3A Food Flavours, Part A*, I.D. Morton and A.J. McLeod, Eds., Elsevier Scientific Publ., New York, 1982, p. 169.
30. Sneeden, E.Y., H.H. Harris, I.J.P. Pickering, C. Roger, S. Johnson, X. Li, E. Block, G.N. George, The sulfur chemistry of shiitake mushroom, *J. Amer. Chem. Soc.*, 126, 2, p. 458, 2004.
31. Heath, H.B., *Source Book of Flavors*, AVI Pub. Co., Westport, CT, 1981.
32. Rowe, D., K. Auty, S. Jameson, P. Setchell, G. Carter-Jones. High impact aroma chemicals Part III, *Perfum. Flavorist*, 29, 7, 42, 2004.

# 11 Flavor Potentiators

## 11.1 INTRODUCTION

Flavor potentiators (or enhancers), by the strictest definition, are compounds that have no flavor of their own (at effect levels) but yet intensify or enhance the flavor of a food. However, the industry may use these terms rather loosely at times and include any compounds that make a flavor taste/smell better. For example, the addition of ethyl butyrate to methyl anthranilate enhances the grape character of the methyl anthranilate. This latter usage of the term is not appropriate in the context of this chapter.

Traditionally, the only compounds considered to be true flavor potentiators are table salt, monosodium glutamate (MSG), and certain nucleotides, e.g., adenosine-5′-monophosphate (ADP), which is not used commercially due to low potentiating effect, guanosine-5′-monophosphate (GMP), and inosine 5′-monophosphate (IMP). It is noteworthy that these compounds have very characteristic tastes of their own at typical industry usage levels (above sensory thresholds) contributing the umami taste (translated as the delicious) sensation. Monosodium glutamate and the 5′-nucleotides were originally identified by Japanese researchers as occurring naturally in food ingredients used to enhance the flavor of traditional recipes. Ikeda [1] isolated and identified MSG from sea tangle (*Laminiaria Japanica*) and Kodama [2] identified the histidine salt of 5′-inosinic acid in dried bonito (type of fish). In 1960, Kuninaka, identified guanosine monophosphate as the sensory enhancer in shiitake mushrooms (*Lentinus edodus*). MSG went into commercial production nearly immediately after its discovery. It was not economically feasible to produce the 5′-nucleotides at the time they were first identified, and commercial production was not begun until the early 1960s. These potentiators are now broadly used to enhance savory flavors.

Two other amino acids (tricholomic and ibotenic acids) have been identified as having flavor potentiator properties. Tricholomic and ibotenic acids were found originally in Japanese mushrooms (*Tricholoma muscarium* and *Aminita strobiliforms,* respectively) by Takemoto et al. [3–5] and Takemoto and Nakajima [6,7]. These amino acids were of particular interest since they were found to be effective in killing the common house fly. While both amino acids have lower sensory thresholds for potentiating effects than either MSG or the 5′-nucleotides, they have not found commercial application.

There has been substantial research interest in finding new flavor potentiators. This is partially driven by the negative image the public has of MSG, the key member of this group of compounds. Also, all of these traditional potentiators are useful only in savory products. It is desirable to find new potentiators that will also function in other food categories. For example, there is great interest in finding an enhancer of the sweet sensation. It is obvious that any compound that enhances sweetness without adding significant calories is desired. Thus, market needs have resulted in several new flavor potentiators being identified.

This chapter will present an overview of flavor potentiators. While focus will be on the traditional flavor potentiators since they have commercial application, recent additions to this category of flavor ingredients will be presented. Maga [8] has provided a very comprehensive but now somewhat dated review of this subject while Yamaguchi [9] has focused his review on the sensory aspects of MSG and the 5′nucleotides, and Ninomiya [10] has focused on the sensory and safety aspects of MSG only. Nagodawithana [11] has provided a less detailed review of the traditional flavor potentiators than Maga [8] but it is comparable in breadth.

## 11.2 CHEMICAL PROPERTIES OF L-AMINO ACIDS AND 5′-NUCLEOTIDES

### 11.2.1 Structures

The structures of the naturally occurring flavor potentiators are presented in Figure 11.1 (salt will be discussed separately). The commonly available naturally occurring flavor potentiators are either 5-carbon L-amino acids or purine ribonucleoside 5′-monophosphates having a 6-oxy group.

It is of interest that flavor potentiating effect is very dependent upon structure of the molecule. For MSG, it is only the L form of the amino acid that has potentiating properties. The D form has no activity. Also, the ionic form is very important. The monosodium form (rather than diacidic or dibasic salt) is active while the other ionic forms have little activity.

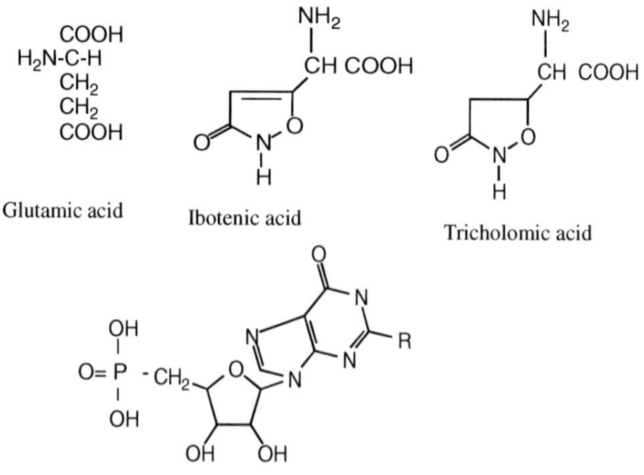


5'-Nucleotides: IMP R = H; GMP R = $NH_2$; XMP = OH

**FIGURE 11.1** Structures of traditional flavor potentiators. (From Nagodawithana, T.W., *Savory Flavors*, Esteekay Assoc., Milwaukee, 1995. With permission.)

In the case of the nucleotides, numerous isomers have been examined as possible flavor potentiators. Of the 2′, 3′, or 5′ isomers, only the 5′-nucleotides have been shown to have potentiating activity [12,13]. With the 5′ structure, a hydroxy group in the 6 position is required to generate flavor potentiation.

A large number of studies investigating related structures for potentiator activity have been reported [8]. Other than ibotenic and tricholomic acids, none of the amino acids or amino acid combinations was found to be as effective as MSG. Investigations of related nucleotides for potentiator activity by Imai [14] demonstrated that sulfur-substituted (2 position) nucleotides were more effective than the original nucleotides. For example, 2-furfurylthio IMP had a flavor potentiating activity approximately 17 times stronger than IMP. However, the cost of synthesis and necessity of obtaining food additive status has prohibited the sulfur-substituted nucleotides from becoming commercially available.

### 11.2.2 Stability

As was pointed out earlier, the monosodium salt of MSG is the only form of L-glutamic acid that has flavor potentiating properties. As expected then, pH has a pronounced effect on potentiating properties. Monosodium glutamate is most effective in the pH range of 5.5 to 8.0. The data of Fagerson [15] suggest that it is the R3 form (see Figure 11.2) of glutamic acid that is responsible for flavor potentiation.

MSG is unstable during thermal processing. Glutamic acid may lose a molecule of water to form its lactam thereby losing potentiating properties. Glutamic acid may also enter into the Maillard reaction. Free amino acids typically are substantially more reactive than protein bound amino acids. Participation in the Strecker degradation or any of the numerous browning pathways would result in loss in potentiating activity.

The 5′-nucleotides are most likely to be destroyed by hydrolysis of the ribose linkage to the purine base [16]. This linkage is weaker than the sugar phosphate bond. Destruction of the 5′-nucleotides during thermal processing shows a slight pH dependency. Product literature [17] has shown that 40 min of heating (115°C) in solution at pH 3 resulted in 29% loss of IMP and GMP while equivalent heating at pH 6.0 resulted in 23% loss. IMP and GMP are both quite stable when in the dry form and when exposed to light. Studies by the Ajinomoto Company [18] have shown that deep fat frying also produces minimal destruction of the 5′-nucleotides.

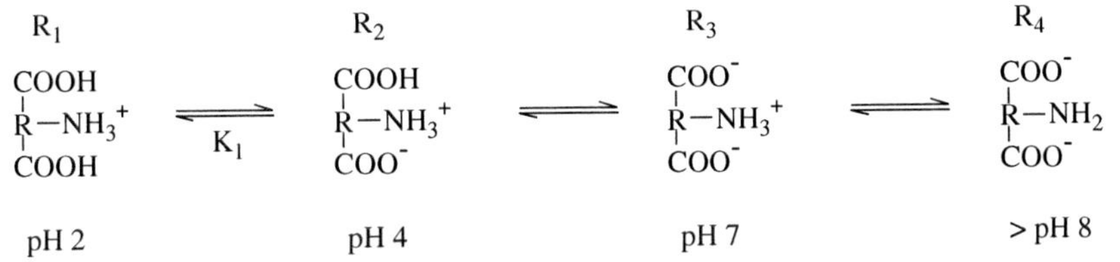


**FIGURE 11.2** Dissociation form of glutamic acid as a function of pH. (From Fagerson, I.S., *J. Agric. Food Chem.*, 2, 1954. With permission.)

It appears that canning is the major food process that results in significant losses in the 5′-nucleotides. Hashida et al. [19] have shown that canned short-necked clams lose from 39 to 48% of their 5′-nucleotides due to canning and 6 months storage at room temperature. Canned boiled lobster was found to lose 35 to 39% after 3 months of room temperature storage and 39 to 42% following 8 months storage. Corned beef lost 12% of the 5′-nucleotides when processed at 110°C for 120 min and 22% when processed at 120°C for 60 min. It appears that losses due to canning can be quite significant. Kinetic studies have shown IMP decomposition to be of first order [20]. Davidek et al. [21] have provided kinetic data over a temperature range of 60 to 98°C.

Other means of losing nucleotide activity are through water loss, enzyme activity, and irradiation. Since the 5′-nucleotides are water soluble, any loss of moisture from the product (e.g., canning or packaging in water, and drip losses following freezing) will result in reduced potentiator activity.

Enzymes can cause either an increase or a decrease in potentiator concentrations in a food. In meats, IMP results from the decomposition of AMP that came from ATP [22]. Therefore, IMP levels in the meat will typically increase shortly after death and then gradually decrease as the IMP is enzymatically degraded to inosine and then hypoxanthine. Enzymes may also be present in foods that hydrolyze the phosphate from the 5′-nucleotides [16]. This would also result in the loss of potentiator effects.

While irradiation has not become a common means of food preservation (albeit some ground beef is being treated to kill *E. Coli*), studies have shown that the 5′-nucleotides are unstable to irradiation. Terada et al. [23] found room temperature irradiation of 5′-IMP in distilled water resulted in a 40% loss at pH 7.0 and greater losses at acid pHs. The 5′-IMP is converted to hypoxanthine. The degree of loss is obviously dependent upon the dosage of radiation given the meat. Many of the early studies considered dosages required for sterilization which are very severe.

## 11.3 SENSORY PROPERTIES OF MSG AND 5′-NUCLEOTIDES

### 11.3.1 Influence of MSG and 5′-Nucleotides on Taste

The influence of flavor potentiators on the basic taste sensations is controversial. Early studies suggested that MSG intensified the sweetness or saltiness of a food when these tastes were near their optimum level while sourness and bitterness were found to be suppressed in some food systems [24]. However, Lockhart and Gainer [25] found no enhancement of either sweet or salt tastes by MSG. Mosel and Kantrowitz [26] also found no enhancement of either sweet or salt tastes but substantial reductions in the sensory thresholds of sour (2×) and bitter (30×). Van Cott et al. [27] found the opposite: that sweet and salt thresholds were lower, and sour and bitter showed no change. Yamaguchi and Kimizuka [28] found no significant influence of MSG on any of the basic tastes.

Yamaguchi [9] provided a summary of the literature on MSG and 5′-nucleotide interaction with the basic tastes both at the thresholds and at supra-thresholds. The effects of MSG on the basic tastes at threshold concentrations are shown in Table 11.1. It appears that MSG may:

**TABLE 11.1**
**Detection Thresholds of the Five Tastes (% (w/v)**

| Solvent | Sucrose | NaCl | Tartaric Acid | Quinine Sulfate | MSG |
|---|---|---|---|---|---|
| Water | 0.086 | 0.0037 | 0.00094 | 0.000049 | 0.012 |
| 0.094% (5 mM) MSG solution | 0.068 | 0.0037 | 0.0019 | 0.000049 | — |
| 0.25% (5 mM) IMP solution | 0.068 | 0.0037 | 0.03 | 0.0002 | 0.00019 |

*Source:* From Nagodawithana, T.W., *Savory Flavors*, Esteekay Assoc., Milwaukee, 1995. With permission.

1. Slightly reduce the sensory threshold of sweetness (sucrose)
2. Have no effect on salty tastes (NaCl)
3. Greatly increase the sensory thresholds of acids (tartaric acid)
4. Have no effect on bitterness (quinine sulfate)

IMP had a similar effect but increased the bitter threshold even more. The effect of MSG on the basic tastes at supra-threshold levels is similar except MSG was found to contribute to saltiness [9]. This may be partially due to the Na ion in the MSG.

The flavor potentiators themselves have a taste at high enough concentrations. Maga [8] noted that MSG has been reported to have a taste threshold ranging from 0.00067 to 0.067%. This range reflects the effects of panel number, age, sex, chemical purity, etc., on threshold determination. Yamaguchi and Kimizuka [28] reported an absolute threshold of 0.010% for MSG. The taste at recognition concentrations has been described as being salty-sweet. The sensory thresholds of IMP and GMP have been reported to range from 0.01 to 0.025% and from 0.0035 to 0.02%, respectively [16]. The sensory properties of IMP have been described as "beefy" while the GMP has been described as being "oaky-mushroom" [11].

### 11.3.2 Influence of MSG and 5′-Nucleotides on Aroma

The influence of flavor potentiators on aroma perception is also controversial. Wagner et al. [29] concluded that IMP could affect the aroma of some foods. They found that 0.01% IMP increased aroma strength of certain notes in beef noodle soup. Similar effects have also been reported by Caul and Raymond [30] and Kurtzman and Sjostrom [31]. However, Yamaguchi [32] reported that MSG has no effect on the aroma properties of 16 foods used in her study. This study is quite contrary to much of the other literature. While these studies involved sensory evaluation, Maga and Lorenz [33] reported on the effects of flavor potentiators on headspace volatiles of beef stock as measured by gas chromatography. They found similar trends for 0.05% MSG and 0.05% 5′-nucleotides with both increasing the volatiles in the headspace. However, the 5′-nucleotides increased overall peak areas by 2.3 times as compared to 1.7 times for the MSG.

The mechanism of enhancing aroma concentration in the headspace is unknown and such data need to be viewed with care. One can conceive how interactions

**TABLE 11.2**
**Taste Intensities of MSG-IMP and MSG-GMP Combinations**

| Ratio of MSG | Ratio of IMP | Relative Flavor Intensity | Ratio of MSG | Ratio of GMP | Relative Flavor Intensity |
|---|---|---|---|---|---|
| 1 | 0 | 1 | 1 | 0 | 1 |
| 1 | 1 | 7 | 1 | 1 | 30 |
| 10 | 1 | 5 | 10 | 1 | 18.8 |
| 20 | 1 | 3.5 | 20 | 1 | 12.5 |
| 50 | 1 | 2.5 | 50 | 1 | 6.4 |
| 100 | 1 | 2.0 | 100 | 1 | 5.4 |

*Source:* From Johnson, K.R., *Encyclopedia of Food Science*, M.S. Peterson, A.H. Johnson, Eds., AVI Pub. Co., Westport, CT, 1978. With permission.

between volatiles and nonvolatiles in a food may reduce the concentration of volatiles in the headspace, but it is difficult to conceive a mechanism for their increase. Most commonly researchers suggest a "salting out" effect, but the levels of potentiators used are well below the levels needed to produce salting out effects. The only other alternative is if an additive alters the polarity of a solution such that the air:water partition of an aroma compound changes significantly. This occurrence seems unlikely.

### 11.3.3 Synergism Between MSG and the 5′-Nucleotides

Synergism may be defined as the cooperative action of two components of a mixture whose total effect is greater than the sum of their individual effects [34]. MSG and the 5′-nucleotides show pronounced synergistic effects. Combinations of the 5′-nucleotides and MSG produce intensity effects as well as taste quality effects. Synergistic effects are illustrated in Table 11.2. A 1:1 ratio of MSG-IMP resulted in a sevenfold increase in flavor intensity (maximum effect).

It was of interest that a 1:1 mixture MSG-GMP produced a thirtyfold increase in flavor intensity. Therefore, it would appear that the most effective flavor potentiator system to use would be a 1:1 mix of MSG and GMP. However, due to the relatively higher cost of 5′-nucleotides, they are seldom used at a 1:1 ratio with MSG. Typically a 95:5 ratio of MSG-IMP/GMP (1:1 IMP:GMP) is used in the industry. This combination yields approximately a sixfold synergistic effect [8].

## 11.4 TRADITIONAL FLAVOR POTENTIATORS IN FOODS

### 11.4.1 MSG and 5′-Nucleotides in Foods

Both MSG and the 5′-nucleotides are found very commonly in foods. Maga [8] has provided a tabulation of the MSG and 5′-nucleotide content of numerous food products. There is no question that many foods contain sufficient levels of the flavor

**TABLE 11.3**
**MSG and 5′-Nucleotides Found in Foods**

| Meat and Poultry | (mg/100g) | Vegetables (con't) | (mg/100g) |
|---|---|---|---|
| Beef | 10 | Green asparagus | 106 |
| Pork | 9 | Corn | 106 |
| Cured ham | 337 | String bean | 39 |
| Chicken | 22 | Green pea | 106 |
| **Seafood** | | Onion | 51 |
| Scallop | 140 | Potato | 10 |
| Snow crab | 19 | Champignon | 42 |
| Blue crab | 43 | Shiitake mushroom | 71 |
| Alaska king crab | 72 | **Cheese** | |
| White shrimp | 20 | Emmental | 308 |
| **Vegetables** | | Parmegiano reggiano | 1680 |
| Cabbage | 50 | Camembert | 40 |
| Chinese cabbage | 94 | Cheddar | 182 |
| Broccoli | 30 | **Milk** | |
| Cauliflower | 46 | Cow | 1 |
| Spinach | 48 | Goat | 4 |
| Tomato | 246 | Human breast milk | 19 |

Source: From Maga, J.A., *CRC Crit. Rev. Food Sci. Nutr.*, 18, 3, 1983. With permission.

potentiators to make a significant contribution to flavor perception. A sampling of Maga's data is presented in Table 11.3. It becomes rather clear that some traditional food ingredients (e.g., tomato paste and Parmesan cheese) are likely used in cooking not only due to their inherent flavor but also due to the flavor potentiators they contribute. The high quantities of the flavor potentiators in certain food ingredients (i.e., hydrolyzed vegetable proteins and some yeast extracts) warrant a focused discussion of these ingredients.

#### 11.4.1.1 Yeast Extracts

Yeast extracts have high levels of RNA, the precursor of the 5′-nucleotides. Thus, they may serve as a source of these potentiators if the RNA is hydrolyzed. The yeast extracts also contribute a flavor of their own that is generally considered to be supportive of meat-like, cheese, and other savory flavorings.

Yeast extracts are made by a process that parallels that of making a Hydrolyzed Vegetable Protein (HVP) [11]. The initial step is selection of the starting material, a yeast byproduct, (e.g., brewers yeast) or a primary yeast (i.e., grown specifically for this purpose). Yeast selection is analogous to selection of the protein source in HVP manufacture and is equally important in determining the final flavor profile. Brewer's yeast generally brings a robust flavor that may be bitter and have a "beerish" character. The primary yeasts typically offer a blander flavor character. The yeast is hydrolyzed by autolysis, plasmolysis, or acid. If autolysis is used, the cell is killed without denaturing its proteolytic enzymes. Under the proper pH and temperature, the proteolytic enzymes begin to solubilize cell protein and nucleases act upon the RNA to provide the flavor enhancing aspects of the product. As the process progresses, cell constituents (e.g., carbohydrates) are also freed and may be acted upon by indigenous or added enzymes. Following autolysis, the cell walls are removed by centrifugation, the supernatant is filtered (polished), concentrated, and sold as such (a paste), or dried to yield a powder. The paste and dry forms differ in sensory properties likely due to the lower heat treatment of the dry form. Spray dried yeast extracts are notoriously susceptible to caking and require the use of sound packaging and care in shipping and storage.

If the yeast extract is produced by plasmolysis, cellular degradation is initiated by high salt levels. The salt dehydrates the cell, eventually resulting in death and liberation of cellular enzymes. While Nagodawithana [11] noted that there are several advantages to this approach, he stated that the product is typically higher in salt than desired. This limits its use in the food industry.

The last method of producing a yeast extract is via acid hydrolysis. This yields a product more similar to an HVP. Like HVPs, a yeast is initially slurried at 65 to 80% solids, concentrated HCl added, and the mixture is held at 100°C (or higher, under pressure) for 6 to 12 hrs. The mixture is cooled once the desired hydrolysis level has been achieved, is neutralized with NaOH to pH 5-7, filtered, concentrated to 35 to 40% solids and finally spray dried. These products have considerable MSG contents but are more expensive than those produced by autolysis due to high capital costs (acid and pressure). They also are high in salt. These two latter factors limit the market for acid hydrolyzed yeast extracts.

#### 11.4.1.2 Hydrolyzed Vegetable Proteins (HVP)

HVPs have been used commercially in the flavor industry since about the 1930s [11]. They are used in savory flavorings due to their intrinsic flavor (that developed on their manufacture or subsequent thermal processing to develop specific notes), their further reaction during processing, and/or their high level of MSG. They contain 9 to 12% MSG (naturally present in protein source) and ca. 45% salt (this level may be decreased based on acid strength and neutralization). The high levels of both salt and MSG make HVPs excellent sources of flavor potentiators. While this group of flavoring materials has been discussed earlier in Chapter 9 in section 9.3.3, their manufacture and sensory properties will be discussed further here.

The initial step in manufacture is to add the protein source (ca. 32% individual source or blend of defatted flours), water (23%), and 32% food grade HCl to a tank (Figure 11.3). The mixture is heated at various times and temperatures to yield the desired degree of hydrolysis and flavor character (perhaps ambient pressure [100°C] for 11 hrs or 121°C for 1.5 hrs) [11]. The mixture is cooled and brought to pH 5 to 6 with NaOH. The neutralization of the HCl with NaOH results in the formation of salt (and water). The reaction mixture is very dark in color and contains insoluble materials. It is then bleached and passed through a carbon filter to remove some of the color, insoluble solids and undesirable flavor notes. The product traditionally

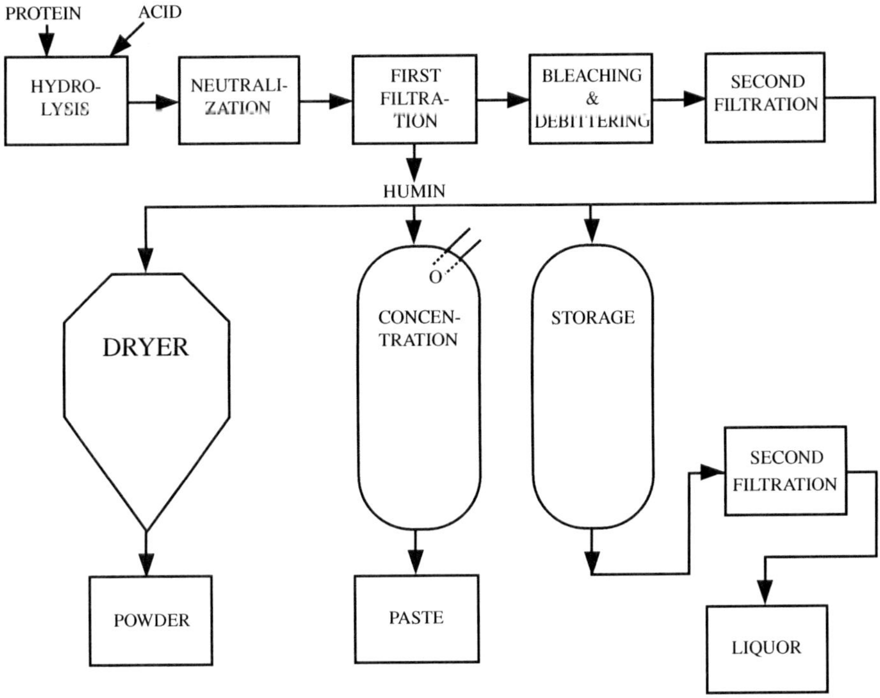


**FIGURE 11.3** Flow chart for the manufacture of HVPs. (From Nagodawithana, T.W., *Savory Flavors*, Esteekay Assoc., Milwaukee, 1995. With permission.)

was aged 6 to 9 months before further processing into powder form by oven or spray drying.

The sensory properties of a given HVP depend upon the starting protein source(s) as well as the processing conditions. As mentioned above, each HVP has a unique flavor character some portion of which arises due to the Maillard reaction during processing. The flours bring in carbohydrates (ultimately reducing sugars) and proteins (ultimately free amino acids) that may react to form flavor components. Since each protein source brings in unique sugars (or proportions of sugars) and amino acid profile, it is understandable that each HVP will have a unique flavor. HVP flavor may be enhanced by the addition of other reactants (additional amino acids or reducing sugars) and further thermal processing under specific conditions (not related to hydrolysis) to provide value added products.

### 11.4.2 MSG and 5′-Nucleotides Added to Foods

The application of flavor potentiators in the food industry is determined by the enhancing effects of the potentiator. While there is little agreement on what sensory properties are intensified or masked by flavor potentiators, there is no question that the flavor potentiators modify the flavor perception of some food systems. The influence of IMP on different flavor notes is presented in Table 11.4. These data again show that different flavor notes are affected to different degrees by IMP.

**TABLE 11.4**
**Influence of IMP on the Flavor, Aroma, and Texture Properties of Foods**

| Flavor/Testure Characteristic | Aroma | Flavor |
|---|---|---|
| Sweet | — | Increase |
| Salt | — | Decrease |
| Sour | No change | Decrease |
| Bitter | — | Decrease |
| Meaty | — | Increase |
| Brothy | No change | Increase |
| Hydrolyzed vegetable protein | Decrease | Decrease |
| Fatty | No change or decrease | No change or decrease |
| Buttery | No change or decrease | No change or decrease |
| Sulfury | Decrease | Decrease |
| Burnt | No change | No change or decrease |
| Starch | No change or decrease | No change or decrease |
| Herb/spice | No change | No change or decrease |
| MSG effect | — | Increase |
| Viscosity | — | Increase |
| Drying | — | Increase |
| Fullness | — | Increase |

*Source:* From Ajinomoto Company, Nucleotides. Part 3. Physical properties, *Ajinomoto Company Inc. Tech. Bull.*, with permission.

Flavor potentiators find wide usage in food products. Vegetables, sauces, soups, meats, and other savory foods constitute the major food applications. Typical food products and usage levels are presented in Table 11.5. The ability of the flavor potentiators to impart viscosity, drying, and fullness is a useful property in soups, gravies, sauces, and juices.

### 11.4.3 Sources of MSG and 5′-Nucleotides

Monosodium glutamate was originally produced as a byproduct of the sugar industry. However, today the majority of MSG is produced via fermentation [35]. Fermentation offers the advantage that only the L-isomer, which has the flavor potentiating effect, is produced. The organism used for MSG manufacture is a strain of *Corynebacterium glutamicum* (alternatively named *Micrococcus glutamicus*, [36]) that has an altered TCA cycle. This organism has reduced levels of α-ketoglutarate dehydrogenase that results in slow processing of glutamic acid to succinic acid. This results in the accumulation of glutamic acid (Figure 11.4). In order to allow the glutamic acid to leak into the fermentation medium, the biotin level in the growth medium must be kept low. Thus, the feed stock (cane molasses) is generally treated to reduce its biotin level. Under optimum growth conditions, this organism will produce 80 g of L-glutamic acid per liter of ferment.

**TABLE 11.5**
**Processed Foods' Flavor Potentiators and Usage Levels**

| | Usage Levels | |
|---|---|---|
| **Food** | **MSG (%)** | **5′-Nucleotides (50/50 IMP and GMP) (%)** |
| Dehydrated soups | 5–8 | 0.10–0.20 |
| Canned soups | 0.12–0.18 | 0.0022–0.0033 |
| Soup powder for instant noodles | 10–17 | 0.30–0.60 |
| Canned asparagus | 0.08–0.16 | 0.003–0.004 |
| Canned crab | 0.07–0.10 | 0.001–0.002 |
| Canned fish | 0.10–0.30 | 0.003–0.006 |
| Canned poultry, sausages, ham | 0.10–0.20 | 0.006–0.010 |
| Sauces | 1.0–1.2 | 0.010–0.030 |
| Dressings | 0.30–0.40 | 0.010–0.150 |
| Ketchup | 0.15–0.30 | 0.010–0.020 |
| Mayonnaise | 0.40–0.60 | 0.012–0.018 |
| Sausages | 0.30–0.50 | 0.002–0.014 |
| Snacks | 0.10–0.50 | 0.003–0.007 |
| Soy sauce | 0.30–0.60 | 0.030–0.050 |
| Vegetable juice | 0.10–0.15 | 0.005–0.010 |
| Processed cheese | 0.40–0.50 | 0.005–0.010 |

Source: From Maga, J.A., *CRC Crit. Rev. Food Sci. Nutr.*, 18, 3, 1983. With permission.

MSG is recovered from the ferment by first removing the cells followed by L-glutamate crystallization [11]. The crystals are washed to remove impurities, decolorized, filtered, and further purified to yield a very pure L-glutamate. The productivity of this organism, optical purity of the product, and ease of isolation from the medium all have made fermentation the production method of choice.

Maga [8] listed the following as commercial sources of the 5′-nucleotides: (1) fresh muscle of marine fish, (2) degradation of RNA by 5′-phospho-diesterase, (3) fermentation to produce nucleotides followed by chemical phosphorylation to produce the 5′-nucleotides, (4) direct fermentation to yield 5′-nucleotides, (5) chemical decomposition of RNA to produce nucleotides followed by phosphorylation to yield 5′-nucleotides, and (6) a combination of procedures.

Originally, fresh fish muscle served as a major source of the 5′-nucleotides. However problems in maintaining the fresh muscle (enzyme activity) and not being able to economically recover 5′-GMP from the fish muscle have made this process less important. Today the majority of 5′-nucleotides are obtained from either the enzymatic hydrolysis of RNA or fermentation. Fermentation is done by a mutant of *Brevibacterium ammoniagenes* that lacks feedback inhibition for the 5′-nucleotides. Magnesium is used to increase cell leakage of the nucleotides thereby limiting cell concentrations. Thus, one obtains high levels of the 5′-nucleotides in the fermentation broth.

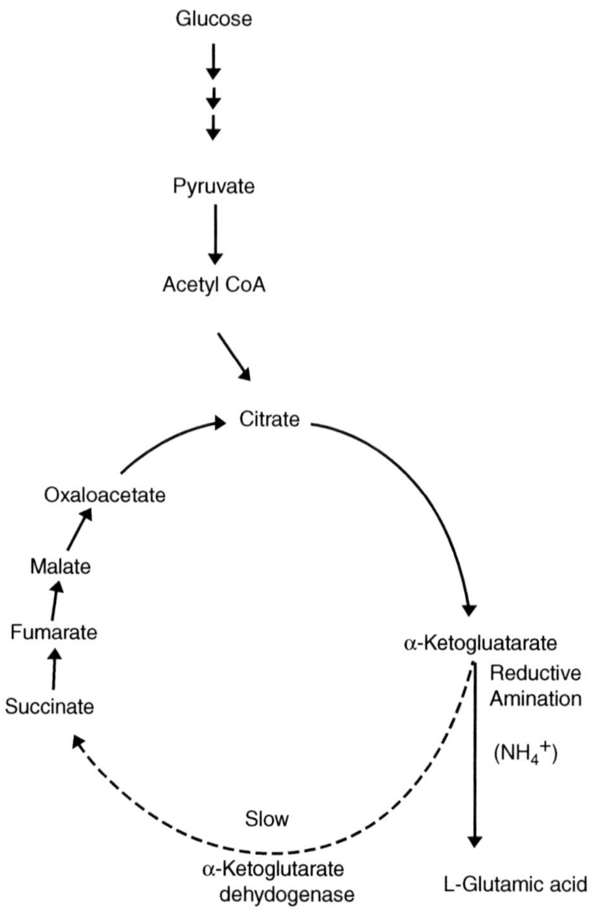


**FIGURE 11.4** Glutamic acid accumulation by a selected strain of *Corynebacterium glutamicum*. (From Nagodawithana, T.W., *Savory Flavors*, Esteekay Assoc., Milwaukee, 1995. With permission.)

Hydrolysis of yeast RNA via enzymatic processes is also a favorable process for 5′-nucleotides. The RNA content of yeast typically ranges from 2.5 to 15%, and thus yeast serves as a good source of the nucleotides [37]. The hydrolysis process proceeds in two steps, the first being a heat treatment to release the RNA from the interior of the cells. The second step is the enzymatic hydrolysis of RNA to 5′-nucleotides using a 5′-phosphodiesterase [38]. This enzyme is not indigenous to the yeast but is added due to its specificity to produce the 5′-nucleotides. It is derived from certain fungi, e.g., *Penicillium citrinum* or *Actinomycetes*, e.g., *Streptomycetes aureus*. The extract from this hydrolysis is isolated from the cell wall by centrifugation, filtered, concentrated, and dried to yield an extract high in the 5′-nucleotides.

### 11.4.4 Table Salt as a Flavor Potentiator

While table salt is generally classified as a flavoring agent (basic taste), it clearly functions as a flavor potentiator as well. There is little question that to most individuals, foods without salt are very bland and lack flavor. Salt contributes much more to sensory perception than simply adding a salty taste. Interestingly, this taste for salt is acquired.

Like MSG, salt has a generally unfavorable status with the consumer. Salt is associated with hypertension in some individuals, which contributes to stroke and some types of cardiovascular disease. There has been substantial research effort directed at finding salt substitutes, but other than $K^+$ chloride, little has been found that elicits a true salty taste. Unfortunately, $K^+$ is bitter and thus not well received. Some research has shown that the addition of MSG and/or 5′-nucleotides reduces the need for salt. There are other reports of L-lysine and L-arginine [39], L-ornithine-β-alanine [40,41], and trehalose [42] enhancing salt perception (thereby reducing usage level).

## 11.5 TOXICITY OF MSG AND 5′-NUCLEOTIDES

### 11.5.1 Monosodium Glutamate

MSG has low toxicity from an acute standpoint. An $LD_{50}$ of 19.9 g/kg body weight in mice has been reported by Ebert [43]. Long-term feeding (4% of diet) studies have failed to show any significant effects of MSG on several species of animals [44–46]. If any effects were seen, they were attributed to the large amount of sodium associated with MSG. Injection of MSG has been shown to induce rapid degeneration of neurons in the inner layers of the retina [47]. Degeneration of nerve cells in the brain due to subcutaneous injections of large amounts of MSG have also been observed in various species [48].

Monosodium glutamate has also been associated with the Chinese Restaurant Syndrome (CRS). Symptoms often associated with CRS are a burning sensation in the back of the neck, facial pressure, chest pain, and in some cases, headaches [49]. Additional symptoms provided by Ghadimi et al. [50] included sweating, nausea, weakness, and thirst. In fact there appears to be little agreement on the symptoms of CRS or even its existence in other than perhaps a very few, very sensitive individuals. Maga [8] provided a comprehensive review of this subject that concluded that there was no evidence for any negative impact on health or even response in the general population. More recent reviews of the safety of MSG have supported this conclusion [10].

Despite the overwhelming scientific evidence confirming the safety of MSG, there is a lingering public opinion that MSG may be responsible for some undesirable reactions upon ingestion. This public sentiment has resulted in the FDA requiring the labeling of MSG when added as such or when added as a part of a food ingredient that is a particularly rich in MSG (e.g., HVPs). In the past, it had been possible to add MSG to a food as a part of the natural flavoring declaration, i.e., not specifically list MSG on the label when it was added as a part of a flavoring. However, the Code of Federal Regulations, 21CFR101.22 Subpart B: Foods: Labeling of Spices, Flavorings, Colorings, and Chemical Preservatives, now states that the terms *flavors*, *natural flavors*, or *flavorings* may not include MSG, hydrolyzed proteins, and autolyzed yeast. Each of these must be declared on the label by its common or usual name rather than be covered under the blanket term. Thus, processed foods containing other ingredients with significant levels of free glutamate, such as hydrolyzed proteins, autolyzed yeast, and soy sauce, must declare these ingredients like any other ingredient on their labels. Interestingly, one does not have to acknowledge the presence of MSG when added as a food ingredient such as Parmesan cheese or tomato paste, also very rich sources of MSG.

### 11.5.2 5′-Nucleotides

The acute toxicities of the 5′-nucleotides have been found to range from 2.7 to 14.0 g/kg body weight, depending upon the species of animal and the means of administering the 5′-nucleotides [51]. While administration of near lethal doses of the 5′-nucleotides will cause temporary depression, clonic convulsion, and dyspnea, dietary intakes of 2% (of total diet) or less 5′-nucleotides have not been found to cause observable effects in any species of animals in long-term feeding studies [51,52]. Therefore, even though no human toxicity studies have been conducted on the 5′-nucleotides, the lack of observable effects in several species of animals at dietary levels approximately 1,000 times typical food applications strongly suggests that there are no significant toxicity concerns for 5′-nucleotides in the diet.

## 11.6 OTHER POTENTIATORS

As noted earlier in this chapter, there has been substantial research focused on finding new flavor potentiators to either replace MSG or provide more general flavor enhancement (not just savory foods). A few such compounds have been found and will be discussed below.

### 11.6.1 Beefy Meaty Peptide

The first new compound to be identified that appeared to offer flavor potentiation is the Beefy Meaty Peptide (BMP). This peptide was found in a papain digest of beef by Yamasaki and Mackawa [53]. The authors identified the peptide as an octapeptide (Lys-Gly-Asp-Glu-Glu-Ser-Leu-Ala). This peptide upon synthesis and sensory evaluation was reported to have beefy flavor properties [54]. Substantial additional research has been conducted on this peptide, and Spanier and Miller [55] have reviewed much of this work.

Confusion in the literature regarding the sensory character of the BMP and a new approach to peptide synthesis permitting larger scale synthesis, prompted van Wassenaar et al. [56] to do more a thorough sensory evaluation. They found that the BMP they synthesized and an equivalent commercial product lacked any taste, umami, or other. Additional work by Hau et al. [57] and Schlichtherle-Cerny [58] has supported the result that the BMP lacks any significant flavor character or enhancing ability. Thus, this compound does not appear to have any flavor potentiating properties.

### 11.6.2 Umami Tasting Glutamate Conjugates

Reaction products of glutamate have been reported to have an umami character. For example, a patent by Frerot and Escher [59] claims that the condensation product of glutamic acid and lactic acid (N-lactoyl-L-glutamate) provides a bouillon–like umami taste similar to MSG. This compound was found to stimulate the taste receptors responsible for umami perception. This finding prompted Beksan et al. [60] to study potential glutamate glycoconjugates (Maillard reaction products) for similar taste properties.

Beksan et al. [60] found that N-(glucos-1-yl)-L-glutamate and N-(1-deoxy-D-fructos-1-yl) )-L-glutamate both exhibited intense umami-like taste sensations at thresholds close to that of MSG. Applied in a savory application (bouillon base), they also gave a similar taste character as MSG. The N-(glucos-1-yl)-L-glutamate was unstable at neutral or low pH yielding glucose and glutamate; however, the N-(1-deoxy-D-fructos-1-yl) )-L-glutamate was stable under these conditions. While these compounds were found to elicit umami sensations, they were not investigated for potentiating effects. Thus, nothing can be said about these compounds in this respect.

### 11.6.3 Alapyridaine

N-(1-carboxyethyl)-6-hydroxymethyl-pyridinium-3-ol inner salt ([+]-[s] configuration) (Figure 11.5) was recently discovered as a taste enhancer [61]. Ottinger et al. [61] isolated this compound from heated glucose/alanine solutions and found that while it has no taste of its own, it enhances the sweetness of sugars, L-alanine and aspartame. Soldo et al. [62] have further investigated the taste enhancing properties of this compound and found that it has an unusually broad influence on tastes. It was found to enhance sweet (glucose and sucrose), umami (MSG and GMP), and salty tastes (NaCl) but had no effect on bitter (caffeine and L-phenylalanine) and sour tastes (citric acid). The sensory thresholds of these tastes were decreased by 2 to 32 fold in the presence of (+)-(s)-alapyridaine. It would appear that this compound could have substantial application in the food industry.

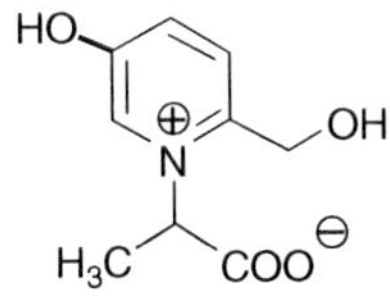


**FIGURE 11.5** Structure of alapyridaine. (From Soldo, T., I. Blank, T. Hofmann, *Chem. Senses*, 28, 2003. With permission.)

### 11.6.4 Sweetness Potentiators

#### 11.6.4.1 Maltol and Ethyl Maltol

Maltol and ethyl maltol are most commonly included in this category. Both of these compounds are used to enhance sweet products. The addition of 5 to 75 ppm maltol may permit a 15% reduction of sugar in some foods [63]. It is common to see maltol and/or ethyl maltol used in jams, jellies, fruit, or chocolate syrups, baked goods, sweet beverages, and other sweet foods. It is of interest that maltol has been found in nature but ethyl maltol has not.

#### 11.6.4.2 Cyclic Enolones

4-hydroxy-2-(or 5)ethyl-5-(or 2)-methyl-3(2H)furanone, 2-hydroxy-3-methyl-2-cyclopenten-1-one, and/or 3-hydroxy-4,5-dimethyl-2(5H)-furanone have been shown to enhance sweetness in foods. Namiki and Nakamura [64] noted that a 5% sucrose solution plus 15 parts per million 2-hydroxy-3-methyl-2-cyclopenten-1-one is as sweet as a 20% sucrose solution. The major problem with these compounds is that they all carry an aroma that is typically caramelic and seasoning-like and may not be desirable in the finished food product.

### 11.6.5 Other Potentiators

Other compounds included as flavor potentiators are dioctyl sodium sulfosuccinate (adds freshness), N, N′-di-O-tolylethylenediamine and cyclamic acid [8]. These compounds have found application in certain dairy products. It is questionable whether these compounds can be called *potentiators* since they do not potentiate a taste but add a sensory note.

## REFERENCES


1. Ikeda, K., On a new seasoning, 30, p. 820, 1909.
2. Kodama, S., On a procedure for separating inosinic acid, *J. Tokyo Chem. Soc.*, 1913. 34, p. 751, 1913.
3. Takemoto, T., T. Nakajima, R. Sukuma, Isolation of a flycidal constituent "ibotenic acid" from *Aminita muscaria* and *A. pantherina*, *J. Pharm. Soc. Japan* (Yakugaku Zasshi), 84, p. 1233, 1964a.
4. Takemoto, T., T. Nakajima, T. Yokobe, Structure of ibotenic acid, *J. Pharm. Soc. Japan* (Yakugaku Zasshi), 84, p. 1232, 1964b.
5. Takemoto, T., T. Yokobe, T. Nakajima, Studies on the constituents of indigenous fungi: 2. Isolation of the flycidal constituent from *Aminita strobiliformis*, *J. Pharm. Soc. Japan* (Yakugaku Zasshi), 84, p. 1186, 1964c.
6. Takemoto, T. and T. Nakajima, Studies on the constituents of indigenous fungi: 1. Isolation of the flycidal constituent from *Tricholoma muscarium*, *J. Pharm. Soc. Japan* (Yakugaku Zasshi), 84, p. 1183, 1964a.
7. Takemoto, T. and T. Nakajima, Structure of tricholomic acid, *J. Pharm. Soc. Japan* (Yakugaku Zasshi), 84, p. 1230, 1964b.
8. Maga, J.A., Flavor Potentiators, *CRC Crit. Rev. Food Sci. Nutr.*, 18, 3, p. 231, 1983.

9. Yamaguchi, S., Basic properties of umami and its effects on food flavour, *Food Rev. Int.*, 14, 2/3, p. 139, 1998.
10. Ninomiya, K., An overview of recent research on MSG: sensory applications and safety, *Food Aust.*, 53, 12, p. 546, 2001.
11. Nagodawithana, T.W., *Savory Flavors*, Esteekay Assoc., Milwaukee, 1995, p. 468.
12. Kuninaka, A., M. Kibl, K. Sakaguchl, History and development of flavor nucleotides, *Food Technol.*, 18, p. 287, 1964.
13. Kuninaka, A., *Symposium on Flavor Potentiation*, Arthur D. Little, Cambridge, 1964, p. 4.
14. Imai, K., Synthesis of compounds related to inosine 5′-phosphate and their flavor enhancing activity: IV. 2-substituted inosine 5′-phosphates, *Chem. Pharm. Bull.*, 19, p. 576, 1971.
15. Fagerson, I.S., Possible relationship between the ionic species of glutamate and flavor, *J. Agric. Food Chem.*, 2, p. 474, 1954.
16. Kuninaka, A., Flavor potentiators, in *The Chemistry and Physiology of Flavors*, E.A. Day, L.M. Libbey, Eds., AVI Pub. Co., Westport, CT, 1967, p. 515.
17. Takeda Chemical Industries, The synergistic effect between disodium 5′-inosinate, disodium 5′-guanylate and monosodium glutamate, *Product Bull.*, No.3,
18. Ajinomoto Company, Nucleotides. Part 3. Physical properties, *Ajinomoto Company Inc. Tech. Bull.*,
19. Hashida, W., T. Mouri, I. Shiga, Application of 5′-nucleotides to canned seafoods, *Food Technol.*, 22, p. 1436, 1968.
20. Lee, J.H., Studies on kinetics of inosine 5′-monophosphate instability, *J. Food Sci.*, 44, p. 946, 1979.
21. Davidek, J., J. Velisek, G. Janicek, Stability of inosinic acid, inosine and hypoxanthine in aqueous solutions, *J. Food Sci.*, 37, p. 789, 1972.
22. Terasaki, M., M., Kajikawa, E. Fujita, K. Ishii, Studies on the flavor of meats: I. Formation and degradation of inosinic acid in meats, *Agric. Biol. Chem.*, 29, p. 208, 1965.
23. Terada, J., K. Hata, T. Mouri, W. Hashida, I. Shiga, Effects of $\gamma$-irradiation on the stability of 5′-nucleotides in solution, *Shokuhin-Shosha*, 3, p. 160, 1968.
24. Cairncross, S.E., L. B. Sjostrom, What glutamate does in food, *Food Ind.*, 20, p. 982, 1948.
25. Lockhart, E.E., J.M. Gainer, Effect of monosodium glutamate on taste of pure sucrose and sodium chloride, *Food Res.*, 15, p. 459, 1950.
26. Mosel, J.N., G. Kantrowitz, The effect of monosodium glutamate on acuity to the primary tastes, *Amer. J. Psychol.*, 65, p. 573, 1952.
27. Van Cott, H., C.E. Hamilton, A. Littell, The Effects of Subthreshold Concentrations of Monosodium Glutamate on Absolute Thresholds, paper presented at 75th Annual Meeting Eastern Psychological Association, New York, April 10, 1954.
28. Yamaguchi, S. and A. Kimizuka, Psychometric studies on the taste of monosodium glutamate, in *Glutamic Acid: Advances in Biochemistry and Physiology*, L.J. Filer, Ed., Raven Press, New York, 1979, p. 35.
29. Wagner, J.R., D.S. Titus, J.E. Schade, New opportunities for flavor modification, *Food Technol.*, 1963, 17, p. 730.
30. Caul, J.F., S.A. Raymond, Home-use test by consumer of the flavor effects of disodium inosinate in dried soup, *Food Technol.*, 18, p. 353, 1964.
31. Kurtzman, C.H., L.B. Sjostrom, The flavor-modifying properties of disodium inosinate, *Food Technol.*, 18, p. 221, 1964.

32. Yamaguchi, S., The unami taste, in *Food Taste Chemistry*, J.C. Boudreau, Ed., Amer. Chem. Soc., Washington, D.C., 1979.
33. Maga, J.A. and K. Lorenz, The effect of flavor enhancers on direct headspace gas-liquid chromatography profiles of beef broth, *J. Food Sci.*, 37, p. 963, 1972.
34. Johnson, K.R., Synergism, in *Encyclopedia of Food Science*, M.S. Peterson, A.H. Johnson, Eds., AVI Pub. Co., Westport, CT, 1978, p. 722.
35. Ogata, K., S. Klnoshita, T. Tsunoda, K. Aida, Microbial production of nucleic acid-related substances, Kodansha Ltd., Tokyo, 1976.
36. Margalith, P.Z., Production of flavor and flavor enhancing compounds by microorganisms, in *Flavor Microbiology*, P.Z. Margalith, Ed., Thomas Publ., Springfield, 1981, p. 256.
37. Nakao, Y., Microbial production of neucleosides and nucleotides, in *Microbial Technology, Microbial Processes*, H.J. Peppler, D. Perlman, Eds., Academ. Press, New York, 1979.
38. Tanekawa, J., H. Takashima, Y. Hachiya, Production of yeast extract containing flavoring, U.S. Patent 4303680, 1981.
39. Guerrero, A., S.S.Y. Kwon, D.V. Vadehra, Compositions to enhance taste of salt used in reduced amounts, U.S. Patent 5711985, 1995.
40. Tamura, M., T. Nakatsuka, M.Tada, Y. Kawasaki, T. Kikuchi, H. Okai, The relationship between taste and primary structure of "Delicious Peptide" (Lys-Gly-Asp-Glu-Glu-Ser-Leu-Ala) from beef soup, *Agric. Biol. Chem.*, 53, 2, p. 319, 1989.
41. Seki, T., Y. Kawasaki, M. Tamura, M. Tada, H. Okai, Further study on the salty peptide ornithyl-b-alanine: some effects of pH and additive ions on the saltiness, *J. Agric. Food Chem.*, 38, p. 25, 1990.
42. Toshio, M., I. Satoshi, U. Yukio, Method for enhancing the salty-taste and/or delicious-taste of food products, Japanese Patent, EP 0813820, 1997.
43. Ebert, A.G., Dietary administration of monosodium glutamate or glutamic acid to C-57 black mice for two years, *Toxicol. Lett.*, 3, p. 65, 1979a.
44. Ebert, A.G., Dietary administration of L-monosodium glutamate and L-glutamic acid to rats, *Toxicol. Lett.*, 3, p. 71, 1979b.
45. Wen, C., K.C. Hayes, S.N. Gershoff, Effect of dietary supplementation of monosodium glutamate on infant monkeys, weanling rats, and suckling mice, *Amer. J. Clin. Nutr.*, 26, p. 803, 1973.
46. Huang, P., N. Lee, T. Wu, S. Yu, T. Tung, Effect of monosodium glutamate supplementation to low protein diets on rats, *Nutr. Reports Int.*, 13, p. 477, 1976.
47. Lucas, D.R. and J.P. Newhouse, The toxic effect of sodium L-g1utamate on the inner layers of the retina, *AMA Arch. Opthalmol.*, 58, p. 193, 1957.
48. Takasaki, Y., Studies in brain lesion by administration of monosodium-L-glutamate to mice: I. Brain lesions in infant mice caused by administration of monosodium-L-glutamate, *Toxicology*, 9, p. 293, 1978.
49. Schaumburg, H.H., R. Byck, R. Gerstl, J. H. Mashman, Monosodium-L-g1utamate: its pharmacology and role in the Chinese-restaurant syndrome, *Science*, 163, p. 826, 1969.
50. Ghadimi, H., S. Kumar, F. Abaci, Studies on monosodium glutamate ingestion: I. Biochemical explanation of Chinese-restaurant syndrome, *Biochem. Med.*, 5, p. 447, 1971.
51. Kojima, K., Safety evaluation of disodium 5′-inosinate, disodium 5′-guanlyate and disodium 5′-ribonucleotide, *Toxicology*, 2, p. 185, 1974.

52. Worden, A.N., K. F. Rivett, D. B. Edwards, A. E. Street, A. J. Newman, Long-term feeding study on disodium 5′-ribonucleotide on reproductive function over three generations in the rat, *Toxicology*, 3, p. 349, 1975.
53. Yamasaki, Y., K. Maekawa, A peptide with a delicious taste, *Agric. Biol. Chem.*, 42, p. 1761, 1978.
54. Yamasaki, Y., K. Maekawa, Synthesis of a peptide with a delicious taste, *Agric. Biol. Chem.*, p. 44, 1980.
55. Spanier, A.M., J.A. Miller, Roles of proteins and peptides in meat flavor, in *Food Flavor and Safety, Molecular Analysis and Design*, A.M. Spanier, H. Okai, M. Tamura, Eds., Amer. Chem. Soc., Washington, D.C., 1993.
56. van Wassenaar, P.D., A.H.A. van den Oord, W.M. M. Schaaper, Taste of "delicious" beefy meaty peptide, revised, *J. Agric. Food Chem.*, 43, p. 2828, 1995.
57. Hau, J., D. Cazes, L.B. Fay, Comprehensive study of the "beefy meaty peptide," *J. Agric. Food Chem.*, 45, p. 1351, 1997.
58. Schlichtherle-Cerny, H., Taste Compounds in Beef Juice (in German), Ph.D. thesis, Technical University of Munich,1996.
59. Frerot, E. and S.D. Escher, Flavoured products and method for preparing the same, Patent WO 97/04667, 1997.
60. Beksan, E., P.Schieberie, R. Fabien, I. Blank, L.B. Fay H. Schlichtherle-Cerny, T. Hofmann, Synthesis and sensory characterization of novel umami-tasting glutamate glycoconjugates, *J. Agric. Food Chem.*, 51, p. 5428, 2003.
61. Ottinger, H., A. Bareth, T. Hofmann, Discovery and structure determination of a novel Maillard-derived sweetness enhancer by application of the comparative taste dilution analysis (cTDA), *J. Agric. Food Chem.*, 51, p. 1035, 2003.
62. Soldo, T., I. Blank, T. Hofmann, (+)-(s)-Alapyridaine-A general taste enhancer?, *Chem. Senses*, 28, p. 371, 2003.
63. Sjostrom, L.B., Flavor potentiators, in *Handbook of Food Additives*, T.E. Funa, Ed., CRC Press, Boca Raton, FL, 1975, p. 513.
64. Namiki, T. and T. Nakamura, Enhancement of sugar sweetness by furanones and/or cyclotene, Japanese Patent, JP 04008264, 1992.

# 12 Flavorists and Flavor Creation

## 12.1 INTRODUCTION

One cannot consider the subject of flavor creation (imitation would perhaps be a better word as the term creation more correctly applies to abstract perfumes than to flavors) in the laboratory without first examining the type of persons who carry out this work (i.e., in the flavor industry: the flavorist; in food and related industries: the flavor technologist), the environment of a flavor laboratory, and the functional interfaces which influence the development of a successful flavoring product. This discussion will be centered on the flavor industry but much can be applied to technologists involved with the flavor of consumer products.

## 12.2 THE FLAVORIST

It is recognized that flavorists have to be multitalented and have a wide range of interests and expertise, not only within the flavor industry, but in the technological industries that it serves and the complementary areas of organic chemistry and process engineering that are fundamental to these industries. Not only must flavorists be scientists, they should have an artistic talent and a creative flair; they should not be bench bound but have a direct involvement in the commercial marketplace; they must be equally at home in the midst of research chemists and sophisticated instruments as with process technologists and the production plant as with potential customers; they must be informed about biochemistry, enzyme systems, sensory assessment, and a host of other related topics; and they should not be parochial but have a worldwide overview, particularly with regard to legislation, social responsibilities to consumers, and safety of their products in foods.

The core activities of a flavorist vary with company but within smaller companies, these activities fall within three broad areas: (a) the creation of flavorings that meet the requirements of present or potential end users of such products; (b) the participation in the marketing process for such flavorings, including technical presentation and collaborative work on specific applications; and (c) the advising on the legislative aspects of flavor usage worldwide. In larger companies, specialists handle most marketing and legislative responsibilities, and thus the flavorist is permitted to focus on creative endeavors.

### 12.2.1 Selection of Individuals to Become Flavorists

Any individual who has a normal sense of taste and smell and has interest in the work can be a flavorist. Most commonly flavorist trainees in the U.S. have come from within the company, having worked in another part of the company and found the opportunity to move into the creation laboratory. Today it is common that these individuals will have a four-year college degree in one of the basic sciences. However, few universities offer courses focused in this very specialized field, and none offer a complete program leading to a degree in flavor. Thus, most who choose to become flavorists are largely trained within the flavor industry on an apprenticeship basis. The process of becoming a flavorist involves a lengthy apprenticeship, and those who lack the needed skills are generally culled from the position over time. If they persevere and have the needed skills, they will ultimately take an examination offered by the Society of Flavor Chemists and if successful, they will be awarded the title of *flavorist*.

### 12.2.2 Training of Flavorists

While there are no schools in the U.S. specializing in training individuals for the flavor industry, there is a school in France devoted to the training of students for either perfumery or flavor professions (ISIPCA, International Institute of Perfumery, Cosmetics, and Food Flavourings, 36 rue de Parc Clagny, 78000 Versailles, France). This school offers two-year programs of study in perfumery, cosmetics, and food flavoring, including apprenticeships providing work experience. However, as noted above, most professional training in the U.S. is done in companies. The trainee generally will work under a senior flavorist, making trial formulations devised by the senior flavorist. This provides an opportunity for the trainee to learn materials and to see how they are used by the senior flavorist (master) in the art of flavor creation. The master may also suggest (or require) additional learning activities for the trainees. This may be reading, working in various positions within the company for variable periods (e.g., formulation, legal, quality assurance, research, etc.), and/or following prescribed exercises to learn how to create flavors. This entire process is very individual and company specific. Heath [1] and Jellinek [2] have suggested reading materials for this purpose. The diversity of approaches for training is evident in reading works by Fischetti [3,5], Di Genova [6], Perry [7], and Shore [4]. In all cases, a very significant amount of effort is devoted to learning materials. The flavor creation laboratory generally has several hundred different materials available to it, and the flavorist must be familiar with all materials used in creation of his or her specialty.

## 12.3 WORKING ENVIRONMENT

Within the flavor industry, laboratory facilities range from excellent to barely adequate, from purpose-designed and superbly equipped laboratory complexes (see Figure 12.1), to small all-purpose units providing only the minimum of necessities.

A laboratory suitable for flavor creation should include the following: (a) a clear run of work bench with an impervious surface for easy cleaning, (b) ambient and refrigerated storage to hold aroma chemicals typically used in flavor creation for that laboratory (a given laboratory is generally devoted to a specific flavor type, e.g., dairy, sweet, savory, seasonings, etc.), laid out systematically and within easy reach of the flavorist, (c) adequate cupboard storage space, (d) natural or indirect overhead

**FIGURE 12.1** A state of the art flavor creation laboratory (Photo courtesy of Robertet Flavors, Inc.).

lighting, (e) a sink fitted with hot and cold water, (f) air conditioning, (g) adequate venting or fume hood to handle extremely potent flavor compounds, and (h) a separate area for use as an office.

Due to the use of very potent aroma compounds, it is becoming more common to find one or two laboratories that are specifically designed to handle potent aroma chemicals without the contamination of a larger work area. This laboratory operates under negative pressure and separate venting to minimize the contamination of other creation laboratories.

The aim throughout should be to create a working environment for the flavorist that provides adequate room for several flavor projects to proceed concurrently without danger of cross-contamination, a work area free from distractions, and suitable facilities for the simple evaluation of flavorings that are being formulated.

The industry trend of forming close working relationships with given food companies (select or preferred suppliers) has resulted in some flavor companies having creation/evaluation laboratories isolated from the daily activities of the company as a whole, where the food company may send product development staff to work with the creative flavorists/application staff on flavor development. This laboratory must assure confidentiality both to the visiting company and other clients. The laboratory typically offers creation/evaluation facilities, office space, plus access to the outside world via phone and the Internet.

## 12.4 FLAVOR CREATION

### 12.4.1 IMITATION FLAVORINGS

The process begins with the customer. The customer must communicate information about the sensory character desired in the product. This is one of the most difficult tasks and often results in failure of the process. While a detailed description of the desired flavor may be given, differences in vocabulary between the customer and the flavorist make this approach less than ideal. It is most desirable to have a target sample. This may be a competing product or a natural product as a target for imitation. It is also critical that the flavorist be provided with substantial information regarding the customer's product to be flavored. Flavors must be designed for a given product: flavors are not created for all food products but a specific formulation of food ingredients (food product) that is processed in a specific manner, ultimately packaged, and stored under unique conditions until consumed. Ideally, the customer should give the flavorist a sample of the unflavored food product (or formulation) as well as processing conditions. Unfortunately, food companies are often reluctant to do so, and as a result, the ability to create the desired flavor profile in a customer's product is lessened. The role of flavor interactions in influencing flavor has been discussed earlier in Chapter 6, and this discussion should drive home the importance of supplying a base food or providing some other means of providing a realistic base food for flavor creation.

Furthermore, it is important that any flavoring constraints be clarified up front. The following questions should be addressed:

1. What label statement does the company wish to use on the product? Kosher? Halal? All natural? Natural/artificial? No added MSG? GM Free?
2. What are the cost constraints? Natural flavorings are more expensive than natural/artificial or artificial flavorings.
3. What is the degree of sophistication needed? An inexpensive flavor cannot contain expensive ingredients.
4. What market is being targeted? U.S.? European? Japan? The country of sale will determine the aroma chemicals that can be used in creation (legal constraints).
5. What flavor form is needed? That should be obvious if the company supplies the flavorist with an unflavored base, but unfortunately a company may use a liquid flavor in development and then demand a dry version of the product for manufacture (bad idea!).
6. If this is a matching project, how good does the match need to be? One will never make a perfect match of a flavoring or food — how close does the match to a target have to be?

Once these questions are dealt with, the flavorist can turn his/her attention to the task of creating the desired flavor. The flavorist cannot and should not endeavor to work in a vacuum as it is all too easy to become too close to the problem, to be oversensitive to constructive criticism/input and to have a closed mind to the suggestions of others. Success depends on the quality of interdisciplinary communications and collaboration with colleagues within the company. These functional interfaces are particularly necessary with (a) the research department of the company; (b) other flavorists within the company, in order to reduce bias and a too narrow view of the problem; (c) application technologists, in order determine the best form for the projected flavoring; (d) the purchasing department, to ensure commercial availability of proposed ingredients; (e) the quality control department, to agree on standards and methods of quality assurance as they apply to the new product; and (f) the applications laboratory for they will determine if a flavoring is acceptably close to the desired target. These interfaces have been critically examined by Heath [1] and Bonica [9] while behavioral and commercial aspects of flavors have been reviewed by Apt [8].

The two basic methods used in the creation or reconstruction of flavors are:

1. Traditional — based on the building up of the formulation by the use of known major constituents of the target flavor and the introduction of aromatic nuances by the use of chemicals having profiles reminiscent of the various attributes or notes detected in the target profile. The selection of aromatics is largely dictated by the experience of the flavorist. One might think that this method is hit and miss, but in reality, it is the expertise of the flavorist in problem solving, involving a flexible approach and an ability to apply different but associated sensory relationships that has made this method effective over many years.
2. Analytical — relying on the analysis of the target flavor by instrumental techniques, mainly GC/MS, and the use of the results to determine the composition of the imitation flavor. The quantitatively significant components of a flavor can often readily be identified, but experience has shown that frequently this knowledge is insufficient to compound a satisfactory flavoring that imitates the original. The flavor attribute that typifies a particular material is often due to compounds which are present in trace amounts. The threshold values of these compounds are usually very low so that their effective contribution to the odor and flavor profiles may be disproportionately higher than their quantitative presence would suggest.

In consequence, neither method is fully satisfactory — the traditional approach is too dependent on the subjective experience of the flavorist and is time consuming, while the purely analytical approach results in an incomplete or unbalanced profile due generally to the limitations of the analytical method. The most satisfactory approach is a combination of both approaches using all available analytical data and technological know how combined with an artistic interpretation of the flavor profile. This calls for the subjective judgment of the flavorist based on knowledge, practical experience of flavorants, and a creative flair.

The creation of a flavoring consists of several stages, which must be carried out systematically and each completed to the satisfaction of the flavorist before proceeding to the next stage. The steps vary with the individual, but Heath [10] has suggested the following:

1. Establishment of the target flavor. The target material may be a natural fruit, vegetable, or other flavorful product, an existing flavoring, or a competitive food product. Whatever its source, it must be critically examined in order to define its odor and flavor profiles, and the relative impact of the flavor attributes accessed. This initial sensory assessment may be carried out by the flavorist or by a team of trained assessors that should include the flavorist. Using the target profile as a guide, the flavorist critically reexamines the material, confirms each designated attribute, and associates each with known constituents or the known profiles of other flavorants not necessarily present in the target material. This reappraisal is very subjective to the flavorist who generally has his/her own system of equating odor and flavor categories with specific chemical entities. The result of this examination is a profile restated as key notes and nuances in terms of the chemical(s) directly associated with each [11].
2. The assembling of data on the target material and likely chemical components, essential oils, etc. that the flavorist may use in creation of the flavoring. This information will come from many sources including:
   - Existing well-proven formulations on file within the flavor company. These, generally, represent many years of experience, and although once totally acceptable, they may now leave something to be desired in terms of a match for the target profile but form an excellent starting point.
   - Research reports on the chemistry of flavoring components in the target material. The past four decades have seen an enormous increase in the identification of aromatic components found in foods. Today, most foods have been characterized in terms of the volatile components present and many have been studied adequately to know the key, or characterizing components, of the food [12].
   - Technical literature, particularly the published proceedings of symposia covering appropriate flavoring topics.
   - Personal records and practical experience of the odor and flavor profiles of permitted flavorants that are necessary to enable reported data to be interpreted in terms of practical aromatic effects.
   - Published GRAS (Generally Recognized as Safe) lists — basically all aroma compounds used in foods must appear on the GRAS list (albeit, sometimes compounds are used before they appear in the *published* list). A review of compounds on this list may make one aware of new compounds not previously known.
   - Services such as the Chemical Sources Association (CSA), flavor compound manufacturers, or in-house synthesis efforts, external organizations/companies, or internal synthesis programs may make new chemicals available for use. While those provided via in-house programs may not be on the GRAS list, compounds may be identified that are extremely valuable and thus an effort made to put them on the GRAS list.

3. Preparation of a trial blend. The flavorist prepares a first blend of ingredients. Fischetti [3,5] does this by breaking the flavor down into parts: The first being the character impact portion, the second being contributory compounds and the final being differential components. These parts may be further described as:
   - Character impact — These compounds convey the named flavor (see Table 12.1). They, by themselves, will create the named flavor, albeit, a cheap version of the desired flavor. For example, one may use methyl anthranilate to form the base of a Concord grape flavor, or citral to convey the central lemon character. The flavor will be recognized as such, but it will be of poor quality, lacking complexity.
   - Contributory — This group of chemicals reinforces or balances, adds a natural note through complexity, to the named flavor. An example would be the addition of ethyl butyrate to grape flavor. Ethyl butyrate is not grape, but it brings out the grape character.
   - Differential — This group of volatiles adds unique sensory notes. One may add garlic oil to a caramel flavoring to add warmth, or cis-3-hexenol to a fruit flavor to add a green or unripe fruity note, or compounds such as furaneol (or damascenone) to provide cooked or jammy notes. None of these compounds will be confused with the named flavor, but they add uniqueness to the flavoring.

   The first effort is unlikely to be sufficiently near the target profile and may indeed be something of a caricature as the main aromatic features are likely to be exaggerated and the minor nuances swamped or not even present. This first rough model must then be continuously modified by trial and error until a reasonably close match is achieved. Before the evaluation of any trial mixing, it is often desirable to allow the mixture to stand for a period in order to let the components blend together. If the assessment is carried out too soon, wrong judgments may result.
4. Subjection of the created flavor to appraisal by other flavorists against the target material in a suitable base material.
5. The preceding stages are repeated as often as is necessary to achieve an acceptable composition.
6. Establishment of application data. After an adequate period to permit the blend to mature, it is submitted for evaluation in either a nominated product or range of end products. This stage will prove the efficacy of the flavoring in various media and draw attention to any changes that result from processing. This work is typically done in an Applications Laboratory.
7. Commercial exploitation of the product. On the completion of the creative process, the product has to be offered for sale either to a specific customer or to food manufacturers generally. The flavorist may be called upon to make a presentation of his/her new product and to demonstrate its value in appropriate end products. The flavorist who created the new flavoring is also most likely to be involved in establishing quality norms and to be assisting in establishing its production [1].

**TABLE 12.1**
**Character Impact Compounds for Some Foods [14]**

| | | | |
|---|---|---|---|
| Almond | 5-Methylthiophen-2-carboxaldehyde | Anise | Anethole, Methyl chavicol |
| Apple | Ethyl-2-methylbutyrate Isoamylacetate | Banana | Isoamylacetate |
| Bergamot | Linalyl acetate | Blueberry | Isobutyl-2-buteneoate |
| Blue cheese | 2-Heptanone | Butter | Diacetyl |
| Caramel | 25-Dimethyl-4-hydroxy-3(2H)-furanone | Caraway | d-Carvone |
| Cassie | P-Mentha-8-thiol-3-one | Celery | 3-Propylidene phthalide |
| Cherry | Benzaldehyde, Tolyl aldehyde (para) | Chocolate | 5-Methyl-2-phenyl-2-hexenal, Isoamyl butyrate, vanillin, ethyl vanillin, methyl methoxypyrazine |
| Cinnamon | Cinnamic aldehyde | Coconut | γ-nonalactone, δ-octalactone |
| Coffee | Furfurylmercaptan Furfurylthiopropionate | Cognac | Ethyloenanthate |
| Clove | Eugenol | Coriander | Linalool |
| Cucumber | Nona trans-2-cis-6-dienal | Garlic | Diallyl sulfide |
| Grape (concord) | Methyl anthranilate | Grapefruit | Nootkatone |
| Green bell pepper | 2-Methoxy-3-isobutylpyrazine | Green leafy | Cis-3-hexenol |
| Hazelnut | Methylthiomethylpyrazine | Horseradish | Pentene-3-one |
| Jasmine | Benzylactetate | Lemon | Citral |
| Maple | Methylcyclopentenolone | Meat | Methyl-5-(hydroxyethyl) thiazole, 2-methyl-3-furanthiol |
| Melon | Methyl-3-tolylpropionaldehyde | 6 Dimethyl-5-heptenal, 2-Phenylpropionaldehyde | 2-methyl-3-(4-isopropylphenyl) propionaldehyde, Hydroxycitronellal dimethylacetal |
| Mushroom | 1-Octen-3-ol | Mustard | Allyl isothiocyanate |
| Peach | γ-Undecalactone, 6-Amyl-α-pyrone | Peanut | 2,5-Dimethylpyrazine |
| Pear | Ethyldecadienoate (cis-4 trans-2) | Peppermint | Menthol |
| Pineapple | Allyl caproate, Methyl-β-methylthiopropionate | Popcorn | 2-Acetyl-1-pyrroline methyl-2-pyridylketone |
| Potato | Methional,2,3-Dimethylpyrazine | Prune | Benzyl-4-heptanone, Dimethylbenzylcarbinylisobutyrate |
| Raspberry | α-Ionone, α-Irone, ρ-Hydroxyphenyl-2-butanone | Red currant | trans-2-Hexenol |

**TABLE 12.1 (continued)**
**Character Impact Compounds for Some Foods [14]**

| | | | |
|---|---|---|---|
| Seafood | Pyridine, piperidine, trimethylamine | Smoke | Guaiacol, 2,6-Dimethoxyphenol |
| Spearmint | L-carvone | Strawberry | Ethylmethylphenylglycidate, Ethyl maltol |
| Winter-green | Methylsalicylate | Vanilla | Vanillin |
| Vinegar | Acetic acid | | |

*Source:* From Fischetti, F., presentation at Flavor Workshop I: Flavor Creation, University of Minnesota, Dept. Food Science and Nutrition, St. Paul, MN, May 12, 2002. With permission.

### 12.4.2 Blending of Seasonings (Culinary Products)

In recent years the flavor industry has expanded to service the culinary trade. Traditionally, these needs were met by smaller, specialized companies that only serviced this market segment. However, the rapidly growing market for prepared meals (particularly frozen, take-out, or take-away), restaurant meals, and institutional feeding has made this market attractive to the larger flavor companies. This market opportunity has resulted in flavor companies hiring chefs who know the trade and can work with traditional ingredients. A combination of the chef's skills plus those of a flavorist to produce traditional culinary products (seasonings) has become an important business for many flavor companies. The blending of herbs, spices, and other food flavorings (e.g., salt, MSG, essential oils, oleoresins, process flavors, or chemical top notes) falls within the flavorists repertoire, but the chef brings the knowledge of traditional ingredients and products as well as needs of the industry.

The stages of creating a seasoning depend upon whether one is attempting to match an existing seasoning or formulating one for a designated end use. In the latter case, the process comprises the following:

1. Establishing the target profile. This implies a knowledge of the composition and processing parameters of the product in which the seasoning will be used. The flavor profile of the basic unseasoned product after cooking significantly determines the type of seasoning required. In general, the weaker the flavor of the basic product, the lower the level of added seasoning required (e.g., a fish product requires only a lightly balanced herbal seasoning; a product based on beef requires a heavy, spicy seasoning with a distinct pungency).
2. Deciding the balance required between the flavoring attributes of the seasoning ingredients. This subject has already been discussed, but ingredients should be assembled in the following flavor groups: (i) light, sweetly herbaceous, (ii) medium aromatic, (iii) heavy, full-bodied, spicy, and (iv) pungent (Table 12.2). These should be supplemented, as desirable, with nonspice materials such as MSG, hydrolyzed vegetable protein, yeast extract, salt, and dextrose.

**TABLE 12.2**
**Ingredients Used in Seasonings**

| | |
|---|---|
| Ground herbs and spices | Herb and spice essential oils (usually plated) |
| Encapsulated herb and spice extractives | Flavor enhancers |
| Preservatives (as permitted) | Food colors (as permitted) |
| Food acids (e.g., citric acid) | Imitation flavorings (e.g., smoke) (as permitted) |
| Salt, dextrose, sucrose, etc. | Artificial sweeteners (as permitted) |
| Milk whey | Sodium caseinate |
| Hydrolized vegetable protein | Yeast extract, etc. |

*Source:* From Heath, H.B. and G.A. Reineccius, *Flavor Chemistry and Technology*, Van Nostrand Reinhold, New York, 1986. With permission.

3. Preparation of the preliminary seasoning mix. This involves the creative abilities of the chef/flavorist team who decides on the most acceptable balance between the various flavoring categories.
4. Evaluation of the mix in a hot white sauce or neutral soup base [13] for its odor and flavor profiles.
5. Repetition of these stages until a well-balanced seasoning is achieved.
6. Evaluation of the seasoning in an end product processed under normal conditions. Seasoning effects are significantly altered by cooking and the presence of other materials, particularly meat. If the end product is to be deep frozen, then this test must include a period of deep freeze of at least 24 h.

The matching of an existing seasoning involves first an agreement on the target sample and its profile. This might sound obvious, but whereas liquid flavorings are very stable and show little change over long periods, blended seasonings have only a limited shelf life and lose their flavoring power. If the sample is old or has been incorrectly stored, the development of a matching product will be a waste of time. Once a target flavor profile is agreed upon, it is necessary to carry out some preliminary analysis to determine salt content, coloring matter, pungency level, MSG, etc. Since one can readily obtain this information via rather simple analytical testing, this is done to facilitate the matching process. Thereafter, the creation of the matching seasoning follows the usual lines of blending appropriate ingredients and repeated testing until a suitable, matching product is achieved. The nearness of trials to the target material is assessed in a white sauce or neutral soup base, adjustments being made until an acceptable product is obtained.

## 12.5 SENSORY ASSESSMENT

In any flavor development project it is necessary to carry out a continuous sensory assessment of trial samples in order to establish the nature and extent of differences between the trial and the target flavor profile. While the final evaluation of a flavoring is done in the Flavor Applications laboratory, there is a need to have a simpler, faster means of evaluating efforts during the creation process. This is typically done in the creation laboratory by the creative flavorist or chef, depending upon the product.

### 12.5.1 Sample Evaluation

The purpose of this section is to draw attention to the preparation of the material for examination, to the techniques of odor and flavor assessment, and to certain precautions that must be taken in the testing of highly flavorful materials.

In the case of concentrated preparations, it is necessary first to dilute these to an acceptable level to permit direct use. This is achieved in one of the following three ways: (a) admixture with a simple food ingredient (e.g., dextrose, sucrose, lactose, or salt) as appropriate; (b) dilution with an acceptable solvent (e.g., ethanol); or (c) incorporation into a neutral food product or carrier at the correct dosage rate. The neutral food products (or carriers) recommended for this purpose include the following:

a. Neutral soup comprising 60% corn starch, 22% caster sugar and 18% salt, used at 4.5% with boiling water and thickened by cooking for 1 min. The sample to be evaluated is added to the prepared soup base. This medium is particularly suitable for spices, culinary herbs, and blended seasonings.
b. White sauce, made with either butter or margarine, plain flour, and milk.
c. Sugar syrup — 10% sucrose in potable water. This medium is the standard one used for the evaluation of flavorings, essential oils, and sweet spices (e.g., ginger, cinnamon, cassia).
d. Reconstituted mashed potato, made in accordance with the manufacturer's instructions but omitting any butter. This is used in the evaluation of onion, garlic, paprika, and blended seasonings.
e. Fondant, using an instant fondant base reconstituted as required. The flavored base is formed in cornstarch molds.
f. Sugar boilings, using the following base: sucrose 120 g, water 40 ml, glucose syrup 40 g. The glucose and water are boiled and mixed until dissolved, the sugar is added and boiled to 152°C (300°F), the product then being removed from the heat and any flavor, color or citric acid stirred in. The mix is then poured onto a greased slab, allowed to cool and passed through drop rollers to form the product.
g. Pectin jellies, using the following base: sucrose 100g; water 112.5 ml; glucose syrup (43 DE) 100 g; citrus pectin (slow set) 6.25 g; citric acid solution 2 ml of a 50% solution. The pectin and sugar are mixed together and slowly added to the water at 77°C (170°F) with continuous stirring. The temperature is raised to 106°C (220°F) and the product removed from the heat; the flavoring, any color and the acid solution are quickly stirred in and the mass poured into cornstarch molds. The poured product is allowed to stand for 12 h before removing from the molds.
h. Carbonated soft drink base, using the following general formulation that may be modified to suit particular products: sodium benzoate solution, 0.25 ml of a 10% solution; citric acid solution, 1 ml of a 50% solution;

8-lb sugar syrup to make 160 ml. The flavor is stirred into this base at the appropriate dose level and the concentrate diluted 1 plus 5 with potable water prior to bottling and carbonation.

i. Milk, a useful carrier for the preliminary evaluation of flavorings for ice cream and other dairy products. The milk may be sweetened with 8% sucrose if desired. The aim is to present the assessor with the sample in the best form and under the best conditions to enable a reliable judgment to be made in a reasonable amount of time.

## 12.6 CONCLUSIONS

The creation of flavoring materials by flavorists and chefs involves a combination of science and art — neither science nor art alone will create a good flavoring material. The flavorist today typically has a college degree but learns his/her specific job skills on the job as an apprentice. The chef typically follows a course of academic training plus an apprenticeship although the academic training does not occur within a university but at a culinary school. The creative effort is a combination of trial and error with constant evaluation and refinement. Once a suitable flavoring is created, it goes to the Applications Laboratory for a formal evaluation, and if successful, it is sent to the customer. Flavor creation ultimately involves all aspects of the company from the instrumental laboratory to marketing. It is a fascinating task that offers the reward of seeing one's creation in the marketplace.

## REFERENCES


1. Heath, H.B., *Source Book of Flavors*, AVI Pub. Co., Westport, CT, 1981.
2. Jellinek, G., *Sensory Evaluation of Food: Theory and Practice*, Horwood, Chichester, 1985.
3. Fischetti, F.J., The training of a flavour chemist — an organised programme, *Flavour Ind.*, 5, 7/8, p. 166, 1974.
4. Shore, H., The training of a flavourist — one on one, *Flavour Ind.*, 5, 7/8, p. 165, 1974.
5. Fischetti, F., Flavors and spices, in *Kirk Othmer Encyclopedia of Chemical Technology*, John Wiley Sons, New York, 1994, p. 16.
6. Di Genova, J., The flavourist as an artist, *Flavour Ind.*, 5, 7/8, p. 174, 1974.
7. Perry, P., The flavourist as a technical man, *Flavour Ind.*, 5, 7/8, p. 171, 1974.
8. Apt, C.M. Flavor: Its chemical, behavioral and commercial aspects, *A.D. Little Symp.*, Westview Press, Boulder, 1977.
9. Bonica, T.J., The flavourist as a processor, *Flavour Ind.*, 5, 7/8, p. 172, 1974.
10. Heath, H.B. and G.A. Reineccius, *Flavor Chemistry and Technology,* Van Nostrand Reinhold, New York, 1986, p. 442.
11. Seidman, M., Sensory methods in the work of the flavour chemist, in *Progress in Flavour Research*, D.G. Land, H.E. Nursten, Eds., Appl. Sci. Publ., London, 1979, p. 15.
12. Peppard, T.L., How chemical analysis supports flavor creation, *Food Technol.*, 53, 3, p. 46, 1999.

13. Heath, H.B., *Flavor Technology: Profiles, Products, Applications*, AVI Pub. Co., Westport, CT, 1978.
14. Fischetti, F. Character Impact Compounds Found in Foods, presented at Flavor Workshop I: Flavor Creation, University of Minnesota, Dept. Food Science and Nutrition, St. Paul, MN, May 12, 2002.

# 13 Flavor Production

## 13.1 INTRODUCTION

Flavorings are commonly divided into classes based on the physical states, i.e., liquids, emulsions, pastes, or solids. Liquid- and emulsion-type flavors are used in liquid food products. These food products may be aqueous or fat-based with the corresponding flavor being either in an aqueous- (alcohol or propylene glycol) or fat-soluble (benzyl alcohol, triacetin, triethyl citrate, or vegetable oil) solvent. Paste flavorings are a mixture of components that typically are either very high in natural flavorings (e.g., botanical extracts) or components that differ in solubility. An example of a paste flavoring might be a Cheddar cheese flavor. The top-notes of Cheddar are composed of rather light carbonyls and fatty acids while the base-notes are made up of various salts and amino acids. Since these components are mutually insoluble, it is virtually impossible to make a complete Cheddar flavor without it being a paste.

Solid flavorings may be products very high in natural flavorings or flavor chemicals dispersed in a solid fat, dry gum, or starch-based matrix. This chapter will discuss the methods used to produce liquid, emulsion, and dry flavoring materials.

## 13.2 LIQUID FLAVORINGS

Liquid flavor production is basically a chemical blending operation. The production area consists of a store of raw materials and weighing, mixing, and packaging equipment. While some of the less stable chemicals are held under refrigeration (or freezing) and the very potent flavor chemicals (e.g., mustard oil) are held in well-ventilated rooms, the vast majority of chemicals are held at room temperature. The flavor production (compounding) area also contains a variety of mixing tanks being stirred with air-driven motors to minimize the danger of fire or explosion. These tanks will range in size from only 5 gallons up to several hundred gallons. The flavor compounder is responsible for this operation.

Recent innovations in electronic weighing and computerized controls have made this process much more foolproof (Figure 13.1 and Figure 13.2). This is very important since there are substantial costs associated with errors in flavor formulation. These costs include the loss of raw materials, disposal of errant mixtures, lost work time, and potentially a dissatisfied customer, or a consumer product recall if the product makes it to the final customer. Newer systems may be completely automated or require scanning a bar code of each ingredient used, weighing the ingredient in and computer checking of weighing against the formulation. The order of mixing is often important since solubility and reactivity may depend on the particular chemicals present and their concentration. Therefore, the solvent(s) are typically weighed in first so that these problems are minimized. The computer system

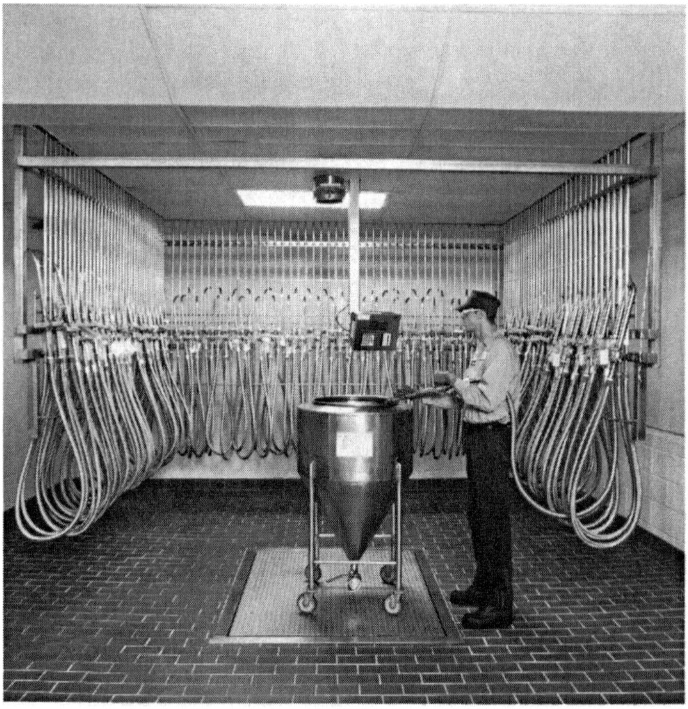

**FIGURE 13.1** Computerized system for the formulation of small batches of flavoring. (Courtesy of Robertet Flavors, Inc. With permission)

checks not only materials and their amounts but also order of addition. Once the formulation has been completed, a sample is taken for quality control. When quality control has given approval of the lot, it is packaged for transport to the customer.

It is of interest that sanitation is of little concern in this area. The products being handled will not support microbial growth and, more than likely, are toxic to microorganisms. The primary reason for cleanliness in a liquid formulation area is to avoid contamination of one flavor with another and to minimize worker exposure to organic vapors. Due to the very low flavor thresholds of some flavor constituents, cleanliness is essential.

## 13.3 EMULSIONS

Emulsions are typically used in the flavor industry to carry product flavor (e.g., beverage or baker's emulsions) or impart turbidity (cloud emulsion) to a product (Table 13.1). In some beverage applications, an emulsion may be used to impart both. They also are a part of the manufacturing process used in the production of dry flavorings (assuming the flavoring is oil soluble). Emulsions used in the flavor industry are generally oil (or flavor)(particulate phase) in water (continuous phase) emulsions. They offer a simple means of incorporating an insoluble flavor into an aqueous food system or carrier matrix for encapsulation.

**FIGURE 13.2** Computer controlled large scale weighing and mixing tanks for liquid flavorings. (Courtesy of Robertet Flavors, Inc. With permission.)

Emulsions offer cost advantages in the delivery of water-soluble (or dispersible) flavorings since water is a very inexpensive solvent. The primary disadvantages of emulsions are that they are physically unstable and require preservation from microbial spoilage. Physical stability is not an issue with baker's emulsions since they are stabilized by the system viscosity.

Emulsions are typically classified on the basis of particle (i.e., droplet) size. A microemulsion is one that has an average particle size $<0.1$ μm. This emulsion ranges from slightly hazy to clear in appearance in application depending upon particle size. This very small particle size is typically achieved only with the use of very effective emulsifiers (e.g., Tween 80) and the use of alcohol as a system solvent component (e.g., mouthwash applications). Microemulsions are used in rather highly flavored products such as toothpastes or mouthwashes since the high level of emulsifier often imparts an off-flavor to less flavorful products. Typical flavor and cloud emulsions have particle sizes ideally ranging from 0.5 to 2 μm. Smaller particle sizes result in a blue-hue to the emulsion, and larger particle sizes lack turbidity and stability.

### 13.3.1 Beverage Emulsions

Beverage flavor and cloud emulsions are composed of at least an oil phase (an oil soluble flavoring for a beverage flavoring and neutral terpenes or vegetable oil for a cloud), water, and emulsifier such as gum Arabic [2] or a chemically modified

**TABLE 13.1**
**Typical Formulations for Emulsions Manufactured by the Flavor Industry**

| | Emulsion Type (%) | | |
|---|---|---|---|
| **Ingredients** | **Beverage** | **Cloud** | **Baker's** |
| Xanthan | — | — | 0.70 |
| Gum arabic | 9.45 | 25 | 2 |
| Citrus oil | 4 | 10[a] | 10.8 |
| Sodium benzoate | — | [b] | 0.2 |
| Citric acid | — | [b] | 0.39 |
| Brominated vegetable oil | 0.5 | 0.45 | — |
| Glyceryl abietate[c] | 3.3 | 5 | — |
| Propylene glycol | 9.45 | — | — |
| Water | 73.3 | 59.55 | 85.91 |

[a] Vegetable oil or orange terpenes would be used.
[b] Optional ingredient.
[c] Also called ester gum.

*Source:* From Fischetti, F. Workshop in Food Flavors: Creation and Manufacturing, offered at the Dept. of Food Science and Nutrition, University of Minnesota, St. Paul, May 2002. With permission.

starch [3,4] (Table 13.1). Typically, a weighing agent and a preservative (e.g., sodium benzoate plus citric acid or propylene glycol) will be included in this formulation. A beverage emulsion is used to flavor a beverage (and add some turbidity) while a cloud emulsion is used specifically to impart turbidity to the beverage with minimal flavor contribution. A manufacturer will use a cloud either because the beverage contains too little real juice to impart the turbidity associated with the natural counterpart (e.g., lemonade with fresh lemon juice vs. a lemon oil product) or the manufacturer may choose to avoid the problems of natural clouds (i.e., use clear juices) and use a synthetic cloud to impart the expected turbidity. In some instances a beverage manufacturer will use a beverage emulsion to impart flavor and some turbidity and additional cloud emulsion to get the desired turbidity.

Creaming, coalescence, flocculation, and Ostwald ripening may all occur to varying extents in beverage and cloud emulsions [5,6]. In the industry, the common term for creaming in bottled soft drinks is *ringing* because the flavor emulsion separates from the soda, floats to the top, and shows a white creamy or oily ring in the neck of the bottle. Coalescence implies localized disruption of neighboring droplets in aggregates so that oil droplets merge together to form large ones. This obviously leads to a decreased number of droplets, enhances creaming, and eventually causes emulsion breakdown. Flocculation occurs when oil droplets of the dispersed phase form aggregates or clusters without coalescence. These aggregates behave as single large drops, and the rate of creaming is accelerated. In emulsion concentrates, a perceptible increase in viscosity can be observed when flocculation occurs. In a soft drink system, droplet concentration is so low that flocculation is often reversible. The aggregates can be readily redispersed by agitation when the interactions between droplets are weak [7]. Ostwald ripening is defined as the growth of larger droplets at the expense of smaller ones due to mass transport of soluble dispersed phase through the dispersing medium. Ostwald ripening is negligible unless the dispersed phase is at least sparingly soluble in the continuous phase. Since essential oils are somewhat soluble in water, beverage emulsions are prone to Ostwald ripening [8].

When we discuss approaches to stabilize emulsions, we first have to consider the problem of creaming. Factors determining creaming are defined by Stokes' law as follows:

$$V = \frac{2gr^2(\rho_2 - \rho_1)}{9\eta}$$

where $V$ is velocity of separation; $g$, acceleration of gravity; $r$, radius of droplet; $\rho_2$, density of continuous phase; $\rho_1$, density of dispersed phase and $\eta$, viscosity of continuous phase.

Stokes' law shows that viscosity is an important parameter influencing emulsion stability. Viscosity is of significance to the concentrated cloud or beverage emulsions, as well as to baker's emulsions since these products are quite viscous. However, in a beverage application, the finished beverage cannot be viscous, and thus the only parameters which can be used to maximize stability in beverage applications are the radius of the droplets and the density differences. The radius of the dispersed droplets should be as close to 1 μm as possible. This particle size yields a bright white and yet reasonably stable (if formulated and processed properly) emulsion. A small, uniform particle size is not easily produced with only gum Arabic or a modified starch as the emulsifier [9]. One typically reduces the particle size by passage through some form of homogenizer. This not only reduces the particle size of the dispersed phase, but also increases its surface area and uniformity (limits Ostwald ripening). While good homogenization contributes significantly to the stabilization of the resultant emulsion (note the $r^2$ term in Stokes Law), repeated homogenization (often tempting) may result in destabilization.

Homogenizers are of four main types:

*Valve homogenizers:* The crude emulsion is forced under considerable pressure through a small spring-loaded valve or annulus so the emerging high-pressure jet impinges directly onto the face of a ring set at right angle to the flow. The impact creates excellent shearing action which breaks oil droplets into those of smaller size [10] (see Figure 13.3).

*Rotary colloid mills:* The crude emulsion is fed into the center of a high-speed rotor and is thrown against the wall of a stator, having to pass through a gap of about 50 μm. The finer the clearance of this gap, the smaller the globule size, the smaller the volume throughput of the machine and the

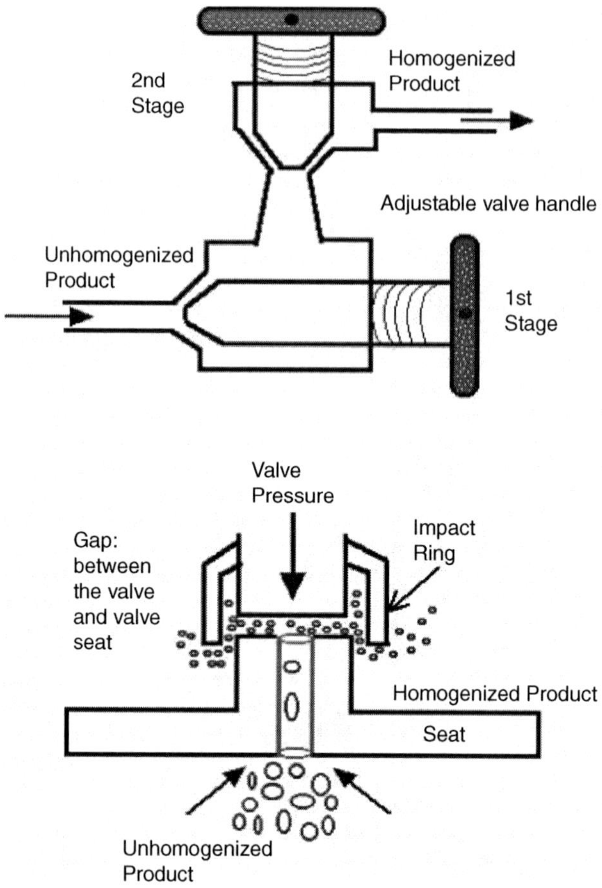


**FIGURE 13.3** Schematic of a valve homogenizer. (From Goff, D. Dept. of Dairy Science and Technology, University of Guelph, Guelph, ON, Canada, http://www.foodsci.uoguelph.ca/dairyedu/ homogenization.html. With permission.)

greater the heat generated — an important factor in flavoring emulsion production.

*Ultrasonic vibrators:* The prime emulsion is fed, usually by pump, into a tube containing a transverse blade vibrating with a frequency of about 30 kHz. This induces cavitation in the product and reduces the particle size of the dispersed phase very efficiently. These machines may be used for continuous in-line emulsification.

*Microfluidizer:* The Microfluidizer™ was originally designed by the Arthur D. Little Co., but was purchased by the Microfluidics Corp. (Newton, MA). This equipment has been traditionally used in the pharmaceutical industry to make emulsions and only recently has been used to produce flavor and cloud emulsions [11]. This equipment works on the principle of dividing a high pressure stream (in excess of 25,000 psi) into two parts, passing

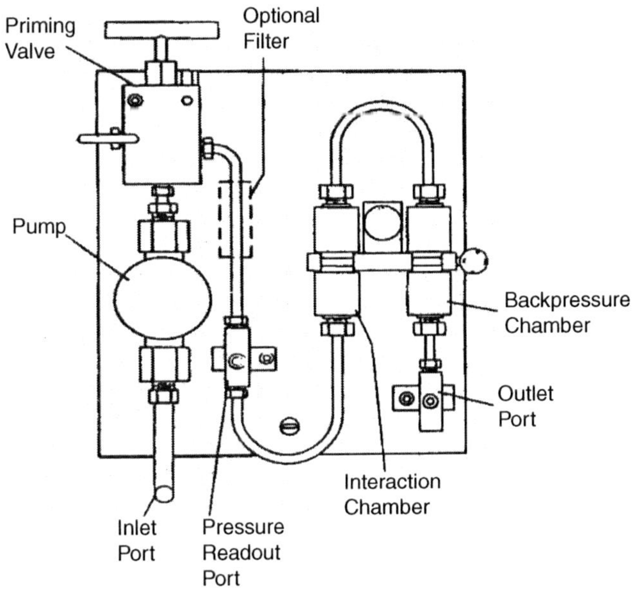


**FIGURE 13.4** Schematic of a Microfluidizer. (Courtesy of Microfluidics Inc. With permission.)

> each part through a fine orifice (in the μm range) and directing the flows at each other [12,13] (see Figure 13.4). This creates a tremendous shearing action, which provides an exceptionally fine emulsion. Studies by Tse [14] have shown that this type of homogenization is superior to the other types commonly used in the industry.

The data presented in Figure 13.5 illustrate the effect of homogenization (in effect, particle size) on emulsion stability. In this figure, the samples were centrifuged to speed their breakage, and emulsion quality was measured simply by turbidity. Two points are evident from the figure: 1) effectiveness of homogenization (particle size) influences emulsion stability, and 2) even a very fine emulsion will break if weighing agents are not used.

The finest emulsions are prepared by the gradual addition of the dispersed phase to the continuous phase with constant and vigorous agitation. Any other ingredients are usually dissolved in the appropriate phase before emulsification. Additions, particularly to the dispersed phase, are extremely difficult once an emulsion has formed. Success depends on the gradual addition of the dispersed phase in small successive quantities so that the continuous phase is always in large excess.

As noted earlier, particle size reduction alone does not produce stable emulsions; the density difference (particulate vs. continuous phases) must also be reduced. This is done by increasing the density of the particulate phase via the use of weighing agents. The commonly used weighing agents and legal limits are presented in Table 13.2. A flavoring such as orange oil or orange terpenes has a specific gravity of 0.84

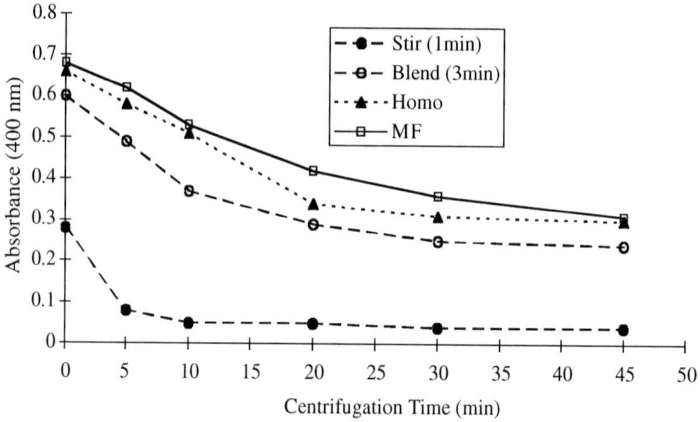


**FIGURE 13.5** Effect of type of homogenization on emulsion stability during centrifugation. (Stir-emulsion made using a wire whisk, Blend-emulsion homogenized with a rotor/stator high shear mixer, Homo-Manton Gaulin valve-type homogenizer used on emulsion, MF-emulsion was Microfluidized) (From Tse, A. and G.A. Reineccius, *Flavor Technology: Physical Chemistry, Modification and Process*, C.T. Ho, C.T. Tan, C.H. Hsiang, Eds., Amer. Chem. Soc., Washington, D.C., 1995. With permission.)

**TABLE 13.2**
**Weighing Agents for Flavor Oils Used in Emulsions**

| Ingredient | Specific Gravity | Legal Status[a] |
|---|---|---|
| Brominated vegetable Oil (BVO) | 1.33 | 15 ppm |
| Glyceryl abietate (ester gum) | 1.10 | 100 ppm |
| Sucrose acetate isobutyrate (SAIB) | 1.15 | 300 ppm |
| Glycerol tribenzoate | 1.14–1.2 | GMP[b] |
| Damar gum | 1.05 | [c] |
| Wood rosin | 1.06 | [c] |

[a] ppm in finished product.
[b] Good Manufacturing Practice: total amount of this and other weighing agents may not exceed 0.5%.
[c] Not permitted in the United States but permitted in some other countries.

*Source:* From Heath, H., G.A. Reineccius, *Flavor Chemistry and Technology*, AVI Pub. Co., Westport, CT, 1986. With permission.

while a vegetable oil will have a specific gravity of about 0.94. Since a carbonated beverage will have from 0 to 13% sugar, depending on the type of beverage and whether it is artificially sweetened, the specific gravity of the finished beverage will range from about 1.00 to 1.06. Traditionally, brominated vegetable oil (BVO) was used to increase the density of the oil phase, but in 1970, the use of this material became prohibited at levels exceeding 15 ppm in the finished product. BVO was extremely effective for this purpose due to its high specific gravity and solubility in an organic phase, and thus its loss (large reduction in permitted levels) required that manufacturers had to search for suitable substitutes.

Glyceryl abietate (ester gum), sucrose acetate isobutyrate (SAIB), and benzyl esters were made available for this purpose. As can be seen in Table 13.2, none of these substitutes has as high a specific gravity as BVO. Therefore, larger proportions of these weighing agents had to be used. In fact, not enough of these weighing agents can be used due to solubility and legal limits to bring the density up to the density of the typical sugar sweetened beverage. The manufacturer can generally add only enough weighing agents to bring the density of the oil phase up to about 1.02. While this is adequate for stability in artificially sweetened beverages, it is too low to obtain good stability in naturally sweetened products, which explains why emulsions continue to be problematic in the industry.

An example of how effective weighing agents are in stabilizing emulsions is shown in Figure 13.6. In this figure, emulsions with and without a weighing agent (ester gum) were prepared by both blending and Microfluidization. The emulsions were diluted to beverage strength and then centrifuged, monitoring turbidity during this time. It is clear that the use of ester gum both increased turbidity initially (partially due to increased amount of oil phase) and made the emulsion much more stable during centrifugation.

The propensity of an emulsion to undergo flocculation and coalescence are largely determined by the emulsifier used. Emulsifers that offer an electronic charge tend to minimize these phenomena since particles will be repelled from each other by the charge thereby limiting the opportunity for either coalescence or flocculation. The charge carried by the protein fraction of gum Arabic is partially responsible for its effectiveness as an emulsifier. Gum Arabic has another stabilizing property in that it readily accumulates at the oil/water interface. In effect, the gum forms a film around the oil, making either coalescence or flocculation less likely [7,16].

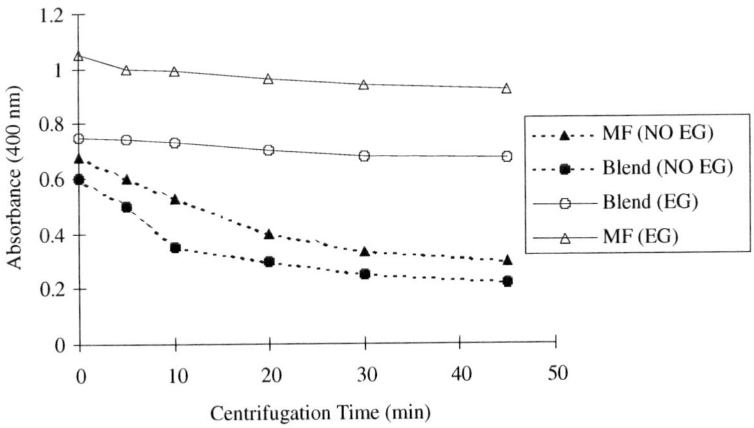


**FIGURE 13.6** The effect of using ester gum (EG) to stabilize a beverage emulsion. (MF sample was Microfluidized, Blend was homogenized with a high shear mixer.) (From Tse, A. and G.A. Reineccius, *Flavor Technology: Physical Chemistry, Modification and Process*, C.T. Ho, C.T. Tan, C.H. Hsiang, Eds., Amer. Chem. Soc., Washington, D.C., 1995. With permission.)

### 13.3.2 Baker's Emulsions

Baker's emulsions are very viscous macroemulsions. They are typically composed of the flavoring, gum Arabic, gum tragacanth (or more likely xanthan due to cost), propylene glycol, and water. The amount and proportion of gums are chosen to provide an emulsion, which is stable primarily due to its high viscosity. The viscosity facilitates incorporation of the flavor into a viscous filling or dough. Propylene glycol is added as a microbial preservative. A typical formula for a baker's emulsion is presented in Table 13.1. There are no significant problems with stability of these emulsions because of their final applications.

## 13.4 DRY FLAVORINGS

The food industry has need for flavorings in a dry form. These flavorings may be produced by plating or some form of encapsulation. Plating involves spreading a flavoring substance(s) *on* an edible food ingredient such as sugar, salt, or whey, while encapsulation involves incorporating flavor ingredients *into* a solid matrix of some food material. These materials may be starches or their derivatives, gums, proteins, lipids, cyclodextrins, or some combination thereof (e.g., Shahidi and Han [17], Risch and Reineccius [18], Gibbs et al. [19]).

Any of these processes may be used simply to convert a liquid flavoring into a more conveniently handled solid form. Alternatively, and especially in cases where the flavoring material will be stored for a substantial period of time before use, it can be to provide substantial protection for the material to be encapsulated. Such protection could be against loss by evaporation, the harmful effects of oxidation, exposure to light, moisture or acid pH, and/or reaction with other components. Finally, these processes may be employed in order to achieve controlled release in a food product. This is done primarily to prevent release of the component(s) of interest until an appropriate stage of processing (e.g., thermal release during the baking of cookies) or until the food is actually in the process of being consumed (e.g., release by physical fracture or wetting in the mouth).

Encapsulation is typically carried out in commercial practice using one of a number of processes, including: spray-drying, spray-cooling/chilling, freeze-drying, fluidized-bed coating, extrusion, coacervation, co-crystallization, and molecular inclusion [18–22]. Of these processes, all but molecular inclusion are macro processes; processes that typically result in particles having diameters in the range 3 to 800 μm. In some cases the particles comprise droplets of core material dispersed in a continuous matrix of carrier material, whereas in other cases, the core is continuous and surrounded by a shell of carrier. In contrast, the process of molecular inclusion occurs at the molecular level, whereby individual molecules of a flavoring are trapped or included within cavities present in individual molecules of carrier (most commonly a cyclodextrin).

### 13.4.1 Extended or Plated Flavors

The simplest process of producing a dry flavoring is to extend (or plate) the liquid flavor on an edible base. This base may be salt, sugar, silicates, dry whey, porous starch, etc., depending upon final application. For example, it is not uncommon for a sausage manufacturer to disperse essential oils and oleoresins on salt. The salt:flavoring blend is mixed into a sausage at the time of manufacture. A baker may choose to plate vanillin on sugar.

This method of producing a dry flavoring has a major advantage of being low in cost. Primary purposes are to insure uniform distribution of the relatively potent insoluble flavoring throughout the food material and facilitate weighing since the flavor oil is diluted by the carrier. A significant disadvantage is that the flavor is afforded little or no protection by the carrier. The flavor may evaporate from the product/package or become oxidized due to air contact if stored any significant time. Also, flavor load is limited to 2 to 7% (w/w) for at higher loadings it may become sticky and not be free flowing.

The exception to the above limitations is the use of silicates for this application. Silicates can absorb approximately an equal weight of flavoring and remain a free flowing powder. Surprisingly, the flavoring has considerable stability both to evaporative losses and oxidation once absorbed by the silicate. Bolton has written a M.S. thesis on this topic [23] and published some of her findings in Bolton and Reineccius [24].

Bolton [23] found that silicas differed greatly in their ability to protect flavoring materials both from evaporation and oxidation. Good flavor retention and flavor stability (to oxidation) were achieved with Syloid 74, Syloid 244, and (to a slightly lesser extent), Sylox 15 (W.R. Grace products). Poorer results were obtained by using Sylox 2 and Syloid 63 as flavor carriers: poor flavor retention was obtained using Sylox 2, and rapid oxidation of limonene occurred using Syloid 63 (see Table 13.3 for the properties of these silicas).

It was unexpected that a silica would protect a flavor compound (e.g., limonene) from oxidation. The ability of the silica to inhibit limonene oxidation was found to

**TABLE 13.3**
**Physical and Chemical Properties of the Selected Commercial Silicas**

| Product designation<br>Analysis[a] Physical Properties | SX 2 | SX 15 | SD 244 | SD 74 | SD 63 |
|---|---|---|---|---|---|
| Average bulk density (g/cc) | 0.110 | 0.128 | 0.120 | 0.152 | 0.300 |
| Average particle size (μm) Coulter 70 μm aper. | 3.53 | 4.05 | 3.87 | 6.65 | 8.54 |
| Average particle size (μm) Malvern | 7.80 | 8.40 | 6.60 | 7.70 | 10.20 |
| Surface area ($m^2/g$) Hg | 119 | 209 | 434 | 302 | 677 |
| Pore volume (cc/g) $N_2$ adsorption | 2.0–2.5 | 2.0–2.5 | 1.18 | 1.13 | 0.39 |
| Oil adsorption (g/100 g) | 203 | 219 | n.a.[b] | n.a. | n.a. |

[a] Data courtesy of W.R. Grace & Co., Technical Center, Baltimore, MD, 1990.

[b] n.a. = not analyzed.

*Source:* From Bolton, T.A., The Retention and Stability of Volatile Flavor Compounds Absorbed in Amorphous Silicon Dioxide, M.S. thesis, Department of Food Science and Nutrition, University of Minnesota, St. Paul, 1993. With permission.

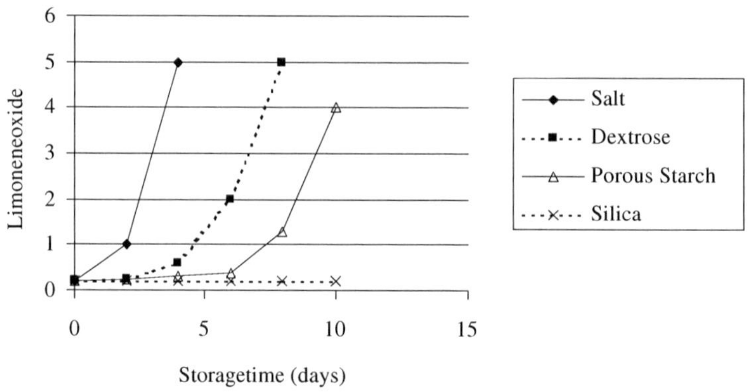


**FIGURE 13.7** Oxidation of orange oil when plated on various base materials (Limonene oxide is an oxidation product of limonene, values mg/g limonene). (From Bolton, T.A., G.A. Reineccius, The oxidative stability and retention of a limonene-based model flavor plated on amorphous silica and other selected carriers, *Perfum. Flavorist*, 17, 2, 1992. With permission.)

be dependent on the type of silica. This implies that differences in physical and chemical properties of the silicas play an important role in the mechanism(s) responsible for protecting limonene against oxidation. The amorphous silicas (Syloid 74, Syloid 244, and Sylox 15) were more effective flavor carriers in the plating process compared to a selection of traditional flavor carriers (Figure 13.7).

Flavor loss from the traditional carriers (salt, sugar, and carbohydrates) occurred by simple evaporation, whereas the adsorption and retention of flavor compounds in amorphous silica were found to be more complex. In general, lower molecular weight flavor compounds were lost to a greater extent than higher molecular weight flavor compounds. This is expected since these low molecular weight compounds are more volatile. Polar flavor compounds or compounds with unpaired electrons strongly interacted with the silica surface and were retained better than nonpolar flavor compounds. Flavor retention in the silicas was dependent on flavor load, and this suggests that the silicas had an optimum loading capacity with respect to flavor retention.

More recently Veith [25] has completed a thesis on silicas as potential flavor delivery materials. She chose to prepare emulsions of model aroma compounds in water and then allow silica to form a matrix around the flavoring droplets. The ability to control silica structure allowed her to study the relationship between silica structure and performance. The concept of forming silica particles around flavorings is novel. She is in the process of publishing her results in the *Journal of Agricultural and Food Chemistry* and the *Journal of Controlled Release*.

### 13.4.2 Inclusion Complexes (Cyclodextrins and Starches)

#### 13.4.2.1 Cyclodextrins

Cyclodextrins (CyDs) are a series of cyclic oligosaccharides that are enzymatically derived from starch employing CyD transglycosylase. α-, β-, and γ-CyDs comprise 6, 7, and 8 D-glucose units, respectively, connected through α-1,4 linkages. Each

CyD is more or less toroidal in shape, i.e., shaped like a (thick-walled) bucket, with a hydrophobic cavity and hydrophilic exterior. This unique structure enables CyDs to form an inclusion complex, entrapping the whole, or part, of a guest molecule inside its cavity, principally by means of weak forces such as van der Waals forces, dipole-dipole interactions, and hydrogen bonding. The dimensions of the inner cavity of α-, β-, and γ-CyDs are 5.7, 7.8, and 9.5Å, respectively, thus potentially accommodating a range of sizes of guest molecule [26].

For a substance included within a CyD, an equilibrium exists between free and complexed guest molecules; the equilibrium constant depends on the nature of the CyD and guest molecule, as well as on factors such as temperature, moisture level, and other aspects of food composition. Generally speaking, however, the presence of water or high temperature is required to liberate guest molecules once complexed. Accordingly there are published accounts of dry CyD inclusion complexes being stable for periods of up to 10 years at room temperature (e.g., [27,28]). In contrast, once dissolved in water, a portion of the complex will dissociate, with the degree of dissociation depending on the nature of the guest. The properties and performance characteristics of CyDs have been summarized by Pagington [29], Qi and Hedges [28], Hedges and McBride [30] and Del Valle [31]. The principal beneficial effects that can be achieved by complexation with CyDs include flavor modification (e.g., masking off-notes) as well as flavor stabilization and solubilization.

Pagington [29] listed several methods of preparing flavor/CyD inclusion complexes including: stirring or shaking a solution of CyD with the guest and filtering off the precipitated complex, blending solid CyD with the guest in a mixer and drying, and passing the vapor of a guest flavor through a CyD solution. Qi and Hedges [28] provide experimental details of a coprecipitation method deemed most suitable for laboratory evaluation. Large-scale production is generally accomplished by the paste method (may be called a slurry method) since less water must subsequently be removed during drying.

While CyDs have been known for more than a century and their ability to form inclusion complexes recognized for at least 40 years, their utilization in food applications did not commence until the 1970s, and then primarily in Japan and Hungary [17]. An important factor delaying the utilization of CyDs has been their regulatory status in the various markets; α- and β-CyDs have both received self-affirmed GRAS status (generally recognized as safe) in the U.S. while GRAS review is still underway in the case of γ-CyD [32]. β-CyD is generally approved for use in Europe (as food additive E459) whereas this is not the case for either α- or γ-CyD. CyD is permitted for use in Japan, though the regulations do not specify which members of this group.

A major barrier to the use of CyDs by the industry is their cost. CyDs are expensive relative to competing encapsulation technologies. Cost considerations will likely dictate that CyDs find niche uses, for example, as a means of protecting the most vulnerable component(s) of a flavor (one may put the unstable component or components of a flavoring into a CyD and then dry blend this into the remaining encapsulated flavor). CyD inclusion is the only process that results in the true isolation of one compound from all others. Thus, this process can confer stability upon a flavoring that no other process can and thereby offers unique functionality but at a cost.

There are numerous publications on the use of CyDs for the encapsulation of flavoring materials. A sampling of this literature follows. Reineccius and Risch [33] evaluated the use of β-CyD with artificial flavors; they found relatively poor retention for small molecules and warned of the danger of unbalanced flavors resulting from a lack of the fresh, light notes normally provided by low molecular weight volatiles. Kollengode and Hanna [34] were able to achieve enhanced flavor retention during operation of a twin screw extruder by adding flavors as CyD complexes. Labows et al. [35], studying flavor release in toothpaste, demonstrated that the slow release of citral from a CyD inclusion complex on water addition could be accelerated by employing a surfactant such as sodium lauryl sulfate. Molecular encapsulation in α-, β-, and γ-CyD was been used to prevent the loss of *l*-menthol from mixtures of maltodextrin and CyD during the drying of individual droplets; β-CyD appeared to perform best according to Liu et al. [36]. The decomposition of allyl isothiocyanate in aqueous solution was depressed in the presence of CyDs, with α-CyD being more effective than β-CyD [37]. Szente and Szejtli [38] reported the novel stabilization of a precursor of acetaldehyde, namely its diethyl acetal, by inclusion in 2-hydroxypropyl-β-CyD; free acetaldehyde was liberated upon hydrolysis at acidic pH.

We [39–43] have published five papers reporting on the encapsulation, release, and potential application of CyDs as flavor delivery systems. These papers consider the selectivity of CyD in including as well as releasing of flavor molecules. Their potential uses as flavor delivery systems in thermally processed foods and fat replaces are also reported.

#### 13.4.2.2 Starch

There is also some literature available reporting on the use of starch (e.g., potato starch) for the inclusion of flavoring molecules [44]. Starches have been shown to form helical structures that can bind flavor compounds. Examples include finding that decanal will form an inclusion complex with gelatinized potato starch in less than a minute while compounds such as limonene and menthone will require several days to accomplish any appreciable complexation at room temperature [45]. Encapsulation of acetaldehyde on starch hydrolysate has been shown to be possible by inclusion or adsorption [46]. This process involves spraying an aqueous solution of acetaldehyde solution on a starch hydrolysate (DE 10–13) at 5°C under constant agitation. A contact time of 15 min results in a product having 4.1% acetaldehyde. This product does not lose acetaldehyde when stored at 40°C for one week. The primary weakness of using starch for this purpose is that very low flavor loads are accomplished. Therefore, the economics of the process are unfavorable.

### 13.4.3 Phase Separation/Coacervation Processes

This technique employs a conventional three-phase system: the manufacturing vehicle (solvent), the flavor carriers (wall materials), and the flavor (core material). While there are several types of coacervation, complex coacervation is most commonly used in the food/flavor industry.

The basic principle involved in this method is to form an emulsion and then precipitate components of the continuous phase around the droplets of the discontinuous phase to form a wall (capsule). The mechanism of capsule formation by this process has been reviewed in the literature (e.g., [47]) and numerous patents exist further describing the manufacturing process in detail (e.g., [48,49]).

The details and importance of each of the stages of complex coacervation follow. Initially, two oppositely charged hydrocolloides (e.g., gelatin and gum Arabic) are dissolved separately in warm water (ca. 40°C). Flavoring is then added to one hydrocolloid dispersion with stirring to form an emulsion. The intensity of stirring determines the size of the oil droplets and thus, the resultant capsule size since the capsule wall forms around the emulsion droplet (large droplets yield large capsules while small droplets form small capsules). With time, the hydrocolloid accumulates at the oil:water interface, providing a weak wall structure around the oil droplets. At this point, the second hydrocolloid is added to the system and the pH is adjusted to ca. pH 4. This results in the hydrocolloides being oppositely charged (e.g., the gelatin has a net positive charge while gum Arabic has a net negative charge) and thus being attracted to each other at the oil droplet surface. This further thickens the particle shell and provides structure. The solution is diluted with water (1:2 or 3) and then allowed to cool thereby solidifying the gelatin around the oil droplet (if the solution were not diluted, the gelatin would form a solid gel in the mixing vessel). The pH is increased and then glutaraldehyde is added to cross-link (harden) the gelatin. The wall must be hardened to permit drying for producing a free flowing powder. There are patents on using other cross-linking agents (e.g., enzymes [50], but glutaraldehyde is the most effective material available. After adequate time, the capsules are recovered by centrifugation (or sieving) and dried. Drying is particularly problematic since the particles will generally be sticky and form a mat at normal drying temperatures. A schematic of this overall process is presented in Figure 13.8.

There is little information in the public domain on this process other than patent literature and a few overviews of the process. Many of the patents address the use of specific wall materials. For example, researchers at NIZO [52] have patented the use of whey proteins for this process while Soper [53] has patented the use of fish protein (kosher considerations with gelatin). Subramaniam and Reilly [49] address the use of a Type B gelatin to permit Halal certification. Bakker et al. [54] have worked with other hydrocolloides (e.g., carboxymethylcellulose and gelatin) in making coacervates. The primary requirement is that the hydrocolloid has an electric charge, and there are numerous hydrocolloides that fill this need (although the positively charged species is normally a protein).

A rather creative approach was taken by Soper et al. [55] in developing a system of forming neutral capsules (core is a neutral lipid) and then loading them with flavoring by increasing the moisture content of the wall thereby allowing the movement of flavoring into the neutral core material. This permits making a large quantity of unflavored capsules and then adding the flavoring desired.

While this process has been in existence since the early 1950s, it has found little application in the flavor industry. The reasons include material cost, time (overnight process), low yield (typically dilute solutions of hydrocolloides must be used), and

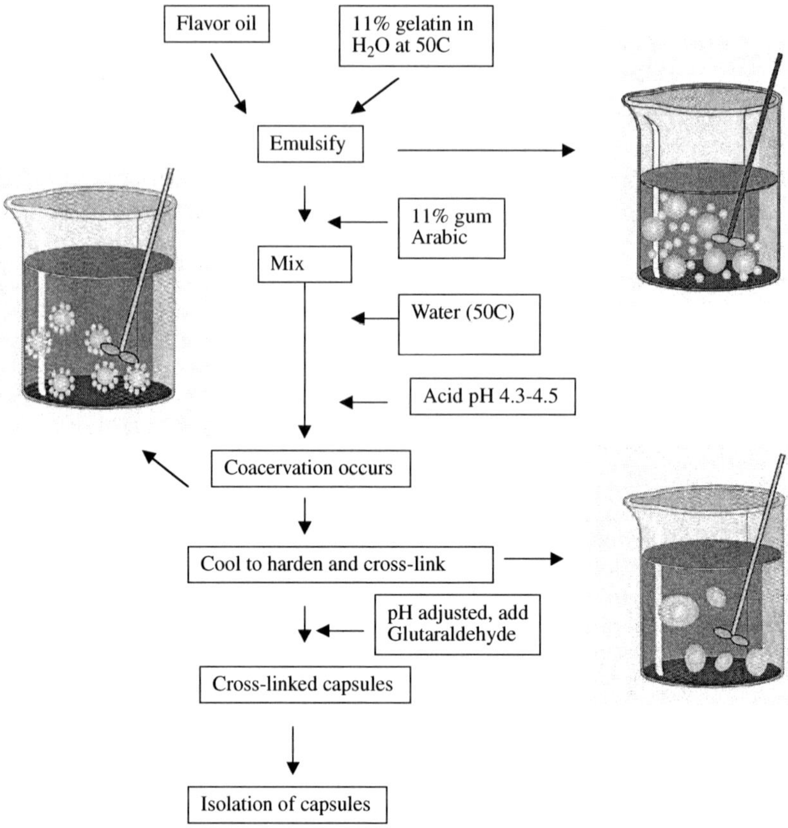


**FIGURE 13.8** Steps involved in the complex coacervation process. (From Vasishtha, N. 2003. Southwest Research Institute, San Antonio, TX. With permission.)

the need for glutaraldehyde (a toxic substance) in cross-linking. Yet the process produces capsules potentially of nearly any desired size (nm to mm), offers high loads (up to 80%), and uses a unique release mechanism (diffusion at low water activities or rupture when subjected to adequate shear stress). These unique properties are resulting in coacervated flavorings starting to be used in the market. The cost of materials and processing are likely to limit the potential for this process but it is finding commercial application in the food/flavor industry at this time.

### 13.4.4 Dehydration Processes

Dehydration processes offer an economical, simple, and flexible means of producing encapsulated flavorings. Dehydration methods yield a particulate powder that contains active flavor ingredients uniformly distributed in the carrier matrix (matrix encapsulation). While any method of dehydration (tray, vacuum tray, freeze, or drum) can and is used for some applications, spray drying is by far the major process used for flavor encapsulation.

#### 13.4.4.1 Spray Drying

Spray drying is the oldest commercial technique for producing encapsulated flavoring in volume. The accidental discovery that acetone, added to tomato puree to help maintain color and flavor of tomato powder, was not lost in the spray drying process started the development of encapsulated flavorings. Since A. Boake Roberts and Co. (N. Revie) made this discovery in 1937, spray drying has become the most important means of producing dry flavorings.

The spray drying equipment used for the production of dry flavorings is essentially the same as is used for the production of dry milk. We, therefore, find substantial equipment available for the manufacture of spray dry flavorings. Equipment availability has contributed toward making this process the dominant method for the production of dry flavorings.

##### *13.4.4.1.1 Encapsulation Matrices*

The initial step in spray drying a flavor is the selection of a suitable carrier (or encapsulating agent). The ideal carrier should have good emulsifying properties, be a good film former, have low viscosity at high solids levels (<300 cps at >35% solids levels), exhibit low hygroscopicity, release the flavor when reconstituted in a finished food product, be low in cost, bland in taste, stable in supply, and afford good protection to the encapsulated flavor [56]. Hydrolyzed starches, modified starches and gum Arabic make up the three classes of carriers in wide use today [57].

Gum Arabic (acacia) is the traditional carrier used in spray drying. It is a good, natural emulsifier and rates well on the other criteria used in evaluating a flavor carrier. Since beverage applications account for a large proportion of dry flavorings used, emulsion stability in the finished product is one of the most important criteria in carrier selection.

Gum Arabic is a natural exudate from the trunk and branches of leguminous plants of the family Acacia [58]. There are several hundred species of Acacia, however, only a few are gum producers, and these are located in the sub-desert region of Arabic. The main gum-producing countries are Sudan, Senegal, Mali, and Nigeria. Gum Arabic is collected during the dry season, which runs from November to May. The trees are tapped (injured) by villagers, exudation occurs, and then the exudate is hand collected. It is of interest that each tree will produce only about 300 g of gum/year.

As one might expect based on the growing region and somewhat primitive means of collecting gum Arabic, supply and quality are variable. The gum is often contaminated with foreign matter and may have high bacterial loads. Thus, manufactures must rehydrate the gum, filter it to remove foreign matter, and then pasteurize it to lower microbial counts. The gum is variable in functionality since it is collected in the wild from many different trees as well as species. Both droughts and political unrest in the growing regions occasionally result in short supplies of gum Arabic and increased prices.

Modified starches — Cost and limited supply of gum Arabic have led to the development of alternative carriers for spray drying. The chemically modified starches most closely reproduce the functional properties of gum Arabic. Native starches impart virtually no emulsification properties to an emulsion. This lack of emulsification creates two significant problems. The first is that poor flavor retention results. As discussed later in this chapter, fineness of the infeed emulsion has a strong influence on determining flavor retention during drying. The second problem relates to the stability of the flavor emulsion once reconstituted in the final product (a beverage, for example). If the carrier provides no emulsification to the flavor, the flavor rapidly separates from the product and forms a ring at the top of the container.

The most effective way to impart emulsifying properties to a starch that is FDA approved is to esterify the partially hydrolyzed starch with substituted cyclic dicarboxylic acid anhydrides. The FDA has approved the treatment of starch with a maximum of 3% octenylsuccinic acid anhydride. This corresponds to a degree of substitution of 0.02. The modified starch obtained from this treatment has been reported to be superior to gum Arabic in emulsification properties and in the retention of volatile flavors during spray drying [9,59]. However, the modified starches have some disadvantages: They are not considered natural for labeling purposes, and they may not offer adequate protection to oxidizable flavorings (albeit new generation products have been vastly improved in this respect) [48].

Hydrolyzed starches — The final group of carriers used in the flavor industry is the hydrolyzed starches [60]. These products range in dextrose equivalent from about 2 to 36.5. Hydrolyzed starches offer the advantages of being relatively inexpensive (approximately one third that of the modified starches), bland in flavor, low in viscosity at high solids, and they may afford good protection against oxidation (depending on dextrose equivalent). The major problem with these products is the lack of emulsification properties. It is, therefore, not uncommon to use blends of gum Arabic/hydrolyzed starches or modified starches/hydrolyzed starches.

#### *13.4.4.1.2 Flavor Retention During Drying*

When one considers the encapsulation of a flavoring by spray drying, one has to look very broadly at the product performance requirements. One has to consider the particle size and shape, absolute and bulk densities, flowability, dispersibility, moisture content, appearance, flavor load, shelf life, stability to caking, structural strength, release properties, and initial emulsion stability and stability once reconstituted. Neglecting any of these performance criteria will result in a flavoring that will not perform properly in a final application. In this chapter, spray drying will be considered fairly narrowly, i.e., only from a flavor retention and stability standpoint. The reader is encouraged to use other references, e.g., [61] for greater detail.

The primary considerations in optimizing the process of spray drying for flavorings are flavor retention during drying and in the final powder during storage, emulsion stability both during manufacturing and in a finished application, and protection afforded against deterioration (e.g., oxidation). The relative importance of each of these criteria depends upon the flavor being dried as well as the finished application. For example, one would not worry about long-term emulsion stability in a dry flavoring being produced for a Ready-to-Spread frosting, but would for a dry beverage flavor. However, one is concerned about emulsion stability *during* the drying process irrespective of the application. If the emulsion breaks during manufacture (i.e., drying), there is a potential for an explosion in the dryer later in the process when the oil phase of the broken emulsion is being dried (the dryer infeed is 100% organic components — the aqueous part of the infeed has been processed). Protection against oxidation would not be important in a flavor that contained no oxidizable constituents, but would for a flavoring containing citrus oils as another example. Thus, the discussion that follows on optimization is limited in perspective and must be considered in terms of the product and its application.

Flavor carrier choice has been shown to influence volatile retention during spray drying by numerous authors [19,62–66]. This influence can be indirect in the sense that some carrier materials become very viscous at relatively low solids contents. For example, Dronen [67] has shown that a soy protein concentrate infeed is limited to ca. 15% infeed solids or its viscosity prohibits effective atomization. Low solids means poor flavor retention. Dronen [67] found an average retention of a model volatile mixture to be only 13.1% when spray dried in a soy concentrate but 24.7% when dried in a whey protein isolate carrier (30% infeed solids). We assume that the two proteins would act reasonably similarly in terms of retaining volatiles and the difference in retention is primarily due to the low infeed solids of the soy protein infeed.

**Infeeds Solids Concentrations:** The effect of type of carrier on flavor retention can also be direct. Carriers, which are good emulsifiers and/or good film formers typically, yield better flavor retention than do carriers that lack these properties. As will be discussed later, emulsification ability is important for the retention of lipophilic volatiles but less so for hydrophilic volatiles.

Once a carrier (or blend of carriers) has been selected, it (or they) must be rehydrated in water. It is desirable to use a particular infeed solids level that is optimum for each carrier or combination of carriers. Research has shown that infeed solids level is the most important determinant of flavor retention during spray drying (Figure 13.9). While most research has suggested that one should use as high an infeed solids level as possible, other work has shown that there is an optimum infeed solids level for each carrier system [68]. The existence of an optimum in solids content may be due to either the possibility that adding solids beyond their solubility

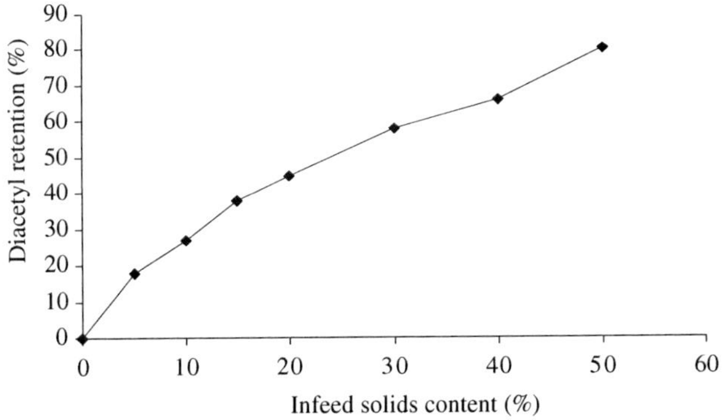


**FIGURE 13.9** Influence of infeed solids on the retention of diacetyl during spray drying. (From Reineccius, G.A. and S.T. Coulter, *J. Dairy Sci.*, 52, 8, 1969. With permission.)

level no longer benefits flavor retention, or perhaps the infeed viscosity becomes a problem resulting in poorer atomization and larger flavor losses.

It is postulated that increased infeed solids levels improve flavor retention by decreasing the drying time necessary to form a high solids surface film around the drying droplets. Once the droplet surface reaches about 10% moisture, flavor molecules cannot diffuse through this surface film, while the relatively smaller water molecules continue to diffuse through this surface film and are lost to the drying air (see [56] for a theoretical discussion of flavor retention and the supporting references).

**Flavor Concentration:** Once the carrier has been solubilized (with or without heating — heating is recommended for some carrier materials such as Capsul®, a chemically modified starch), the flavor system is added. A 20% flavor load based on carrier solids is traditional in spray drying. Higher flavor loads typically result in unacceptably high losses of flavors in the dryer. Emberger [70] has shown that about one third to one half as much flavor is retained during drying when a 25% flavor load is used compared to a 10% loading. Brenner et al. [71] obtained a patent for a process for the production of high-load spray dried flavorings. They claimed that high surface oils and poor flavor retention during drying are largely the result of particle shrinkage and cracking during drying. A cracked particle surface results in substantial flavor losses during drying and high levels of surface (or extractable) oil. Brenner et al. [71] used a combination of polysaccharides (e.g., gum Arabic, starch derivatives, and dextrinized and hydrolyzed starches) and polyhydroxy compounds (e.g., sugar alcohols, lactones, monoethers, and acetals) to form a carrier material that remained plastic during spray drying. Using this plastic carrier, they claimed to be able to spray dry infeed materials with a flavor load of up to 75% (based on dry solids). They reported oil recoveries (amount in/amount recovered in the dry powder) of 80% at this high loading. However, this author knows of no commercial products on the market using this patent.

**Infeed Emulsion Quality:** Emulsion size has a pronounced effect on volatile retention during spray drying [72–75] (Figure 13.10). In each of the cited studies, the retention of the model volatile (generally a lipophilic component) increased with decreased emulsion particle size. Soottitantawat et al. [75] noted that emulsion size had a profound effect on the retention of limonene (very lipophilic) in the finer emulsions (e.g., 0.5–2 μm) and became less dependent upon emulsion size at larger average particle sizes. They provided data also showing that the atomization resulted in a decrease in particle size of the coarser emulsions but had no effect on the finer emulsions. They hypothesized that the shearing effect during atomization broke the larger emulsion particles allowing them to evaporate during drying [75].

Soottitantawat et al. [75] also presented data on the effect of emulsion size on the retention of more water-soluble volatiles (ethyl butyrate and ethyl propionate), and they observed an effect different from that for limonene. There appears to be an optimum emulsion size for the retention of the more water-soluble volatiles, e.g., esters. The increased loss of the esters at small particle sizes was hypothesized to be due to the larger surface areas of the fine emulsions. The increased surface area would present a greater opportunity for diffusion into the matrix and loss from its surface during drying. Again at larger emulsion sizes, the droplets would be subject to shear losses. Thus, it appears that losses of flavorings that are predominantly

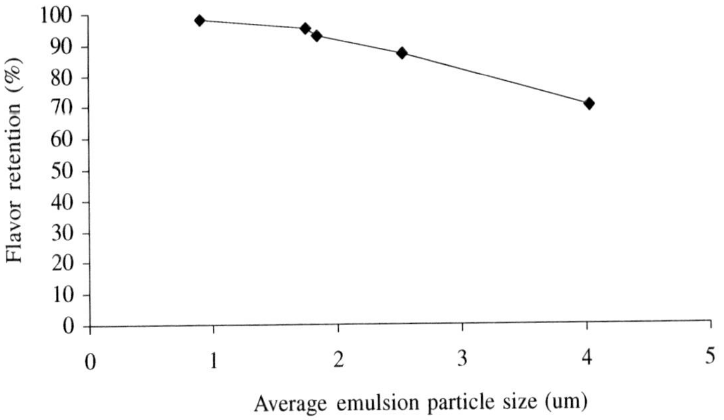


**FIGURE 13.10** Effect of emulsion size on flavor retention during drying. (From Risch, S.J., G.A. Reineccius, Eds., *Encapsulation and Controlled Release of Food Ingredients*, Amer. Chem. Soc., Washington, D.C., 1995, p. 214. With permission.)

composed of lipophilic components can be minimized by the use of carrier systems that promote the formation of fine, stable emulsions and good homogenization thereof. Those flavorings that are primarily composed of more hydrophilic components (perhaps fruity flavors) would not be as dependent upon the use of an emulsifying carrier or efficient homogenization.

**Infeed Temperature:** The dryer infeed may be kept at ambient temperature, or held hot (>155°F) to minimize the opportunity for microbial growth. The choice is company and product dependent. The influence of infeed temperature on flavor retention during drying can be considered in two different ways. Early work by Sivetz and Foote [76] indicated that chilling the dryer infeed (30% coffee solids extract) resulted in improved flavor retention. It was postulated that chilling the infeed increased viscosity, which in turn reduced the circulation currents within the drying droplets. Reduced circulation currents have been shown to improve flavor retention. Later work by Kieckbusch and King [77] has also shown increasing retention of volatiles as the infeed temperature was lowered. This effect was, however, attributed to the effect of viscosity on atomization properties. While this earlier work suggested that one should consider chilling the dryer infeed, Thijssen and coworkers [64,78–80] advocated the opposite: warm the dryer infeed. They pointed out that higher dryer infeed temperatures permit increased infeed solids, which thereby improves flavor retention and has the added advantage of increasing drying throughput.

**Dryer Operating Parameters:** Having considered the formulation and preparation of the material to be spray dried, one must now consider the effect of spray drier operating variables on flavor retention. One needs to choose operating temperatures, air flow rates, inlet air dehumidification etc.

Dehumidification of the dryer inlet air would be expected to have a beneficial effect on the retention of volatiles during spray drying. Lower air humidity would result in more rapid drying of the atomized infeed material, thereby shortening the drying time and the period available for the loss of flavors. However, the cost of dehumidifying large volumes of dryer air is prohibitive. Therefore, one seldom sees air dehumidification used in the flavor industry.

The influence of dryer inlet and exit air temperatures has received considerable attention [65,66,60,71,81–85]. It is desirable that a high inlet air temperature be used to allow rapid formation of a semipermeable membrane around the drying droplet but yet not so high as to cause heat damage to the dry product. Inlet air temperatures of 160 to 210°C have been reported as giving optimum flavor retention during drying [69,80–82]. Inlet air temperatures above 210°C have been found to decrease flavor retention for some types of carriers. However, spray dried flavorings have been successfully produced using inlet air temperatures from 280 to 350°C [56,86]. It is interesting that Reineccius and Coulter [69] did not find high inlet air temperatures to be beneficial to the retention of diacetyl. They suggested that dryer air humidity as a result of inlet/exit air temperature combinations is a governing factor as opposed to simply temperatures.

The influence of dryer exit air temperature on flavor retention is not as well documented. Reineccius and Coulter [69] have shown that the retention of small soluble flavorants such as diacetyl increases with increasing exit air temperatures. This is presumably due to the higher exit air temperatures (at a fixed inlet air temperature), giving the dryer air a lower relative humidity. Low humidity results in more rapid drying and, therefore, better flavor retention. There are other concerns for using high exit air temperatures, however. High exit air temperatures may result in heat damage to some flavoring materials (e.g., cheese and tomato). High exit air temperatures also decrease dryer throughput and reduce powder moisture content, which may be overriding considerations. The maximum retention of diacetyl occurs under operating conditions that result in very low manufacturing rates (low $\Delta T$ — inlet air minus exit air temperature) and therefore, one may choose to operate the dryer under less than maximum retention conditions for production or product reasons.

The influence of dryer air temperatures appears less significant when one is drying less volatile flavorants (e.g., orange oil) at higher concentrations (ca. 20% load) than diacetyl at low concentrations (ppm). It appears that increasing either inlet or exit air has a slight detrimental effect on volatile retention.

**Atomization:** During atomization, an infeed material is sprayed into very turbulent air (facilitates mass transfer from the infeed), is in a thin film (high surface area), and substantial infeed mixing occurs (lowers feed/air interface concentrations of volatiles) all enhancing volatile losses. Thus, the atomization process must be optimized for volatile retention.

In the case of pressure spray systems, King [87] has noted that high pressure operation enhances volatile retention (e.g., atomization at 1.83, 3.55, and 7.00 MPa resulted in 31, 45, and 53% retention of propyl acetate during drying) [88]. King [87] explained the improved volatile retention with high pressure atomization as being partially due to a higher pressure reducing the length of the sheet of infeed material emitted from the spray nozzle before breakup into spherical droplets, thereby reducing the length of time that the liquid is in a sheet (high loss rate period). High spray pressures also provide a greater momentum to the atomized material thereby drawing more hot air into the spray stream [89,90]. The increase in hot air incorporation results in more rapid drying and reduces time to the formation of a selective film around the drying droplet. King [87] commented that an analogous situation exists for centrifugal atomization — higher wheel speed would enhance volatile retention for similar reasons.

One would also expect the spray angle of the nozzle to impact volatile retention during the atomization and early drying period. King [87] suggests that as wide a spray pattern as possible (without wetting the dryer walls) should be used since this will increase the atomized particle/drying air contact thereby increasing drying rate. A dryer design that maximizes dryer air turbulence and infeed contact is also desirable.

**Air Flow Rate:** One would expect the air flow rate through the dryer to influence flavor retention. High air flow rates should improve retention since they would shorten drying time. However, Kieckbusch and King [91] found no influence of air flow rate on flavor retention. They suggested that the large amount of air entrapment by the expanding spray during atomization at high air flow rates offsets the benefits of decreased drying time.

#### *13.4.4.1.3 Flavor Stability*

The discussion to this point has focused on the influence of various infeed matrix and dryer operating parameters on the retention of volatile flavors during drying. While retention of the flavor during drying is the first concern, once dried, the flavor must exhibit good shelf life. When one discusses shelf life of an encapsulated flavoring, one is typically concerned about oxidative stability. Many flavors (especially citrus oils) are very prone to oxidation and readily develop off-flavors during storage [92]. The major determinant of oxidative stability is carrier choice [93].

**Carrier Effects:** As was noted earlier, gum Arabic has been the traditional carrier used for the spray drying of flavors. Gum Arabic has been shown to provide a variable amount of protection to encapsulated citrus oils depending upon the Arabic [94]. As can be seen in Figure 13.11, a particular gum Arabic provided about 40 days shelf life (generally consider shelf life to end at about 2 mg limonene oxide per g limonene although this value is not firm) to an encapsulated single-fold orange peel oil when held at 37°C (no antioxidant added). A modified starch product (Amiogum®, American Maize Products, Hammond, IN) tested under the same conditions provided a shelf life of only 20 days. The poor shelf life obtained by the product encapsulated using the modified starch has been observed using another commercial product (Capsul®, National Starch) both in unpublished studies in our laboratory and in work published by Baisier and Reineccius [95]. These results would appear to contradict the work of Trubiano and Lacourse [59], although too few data were presented in this latter publication to make a comparison of results.

As can also be seen in Figure 13.11, quite superior shelf life was observed for the orange oil encapsulated in 25 DE corn syrup solids (M-250, Grain Processing Corp., Muscatine, IA). The shelf life afforded by the corn syrup solids matrix was about 160 days. This is superior to either gum Arabic or an emulsifying starch.

Subramaniam [93] and Anandaraman and Reineccius [96] have shown that simple hydrolyzed starches can give excellent protection to spray-dried orange oils

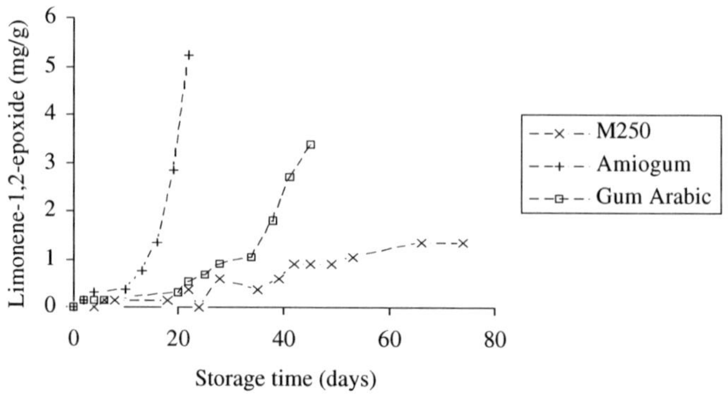


**FIGURE 13.11** The influence of carrier type on the shelf life of a single fold orange peel oil (storage temperature 37°C, M250–25 DE corn syrup solids, Amiogum®-chemically modified starch). (From Westing, L.L., G.A. Reineccius, F. Caporaso, *Flavor Encapsulation*, S.J. Risch, G.A. Reineccius, Eds., Amer. Chem. Soc., Washington, D.C., 1988. With permission.)

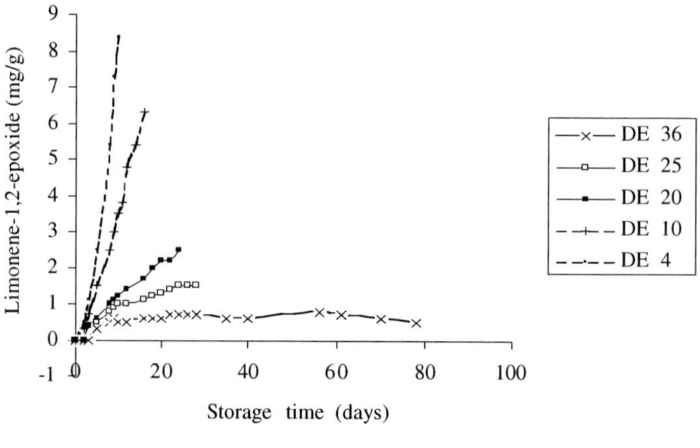


**FIGURE 13.12** The influence of starch dextrose equivalent (DE) on the shelf life of spray dried single fold orange peel oil. (From Anandaraman, S. and G.A. Reineccius, *Food Technol.*, 40, 11, 1986. With permission.)

and that the degree of protection is directly related to the dextrose equivalent of the hydrolyzed starch (Figure 13.12). There are, however, problems in using the simple hydrolyzed starches as flavor carriers. As was noted earlier, these products lack any emulsifying properties, typically result in poor retention of volatile flavors, dry poorly, and may result in caking during storage [94].

**Water Activity:** The role of water activity in determining shelf life of the spray-dried powders is of interest. It is well documented in the literature that lipid oxidation is slowest at the monolayer (water activity ca. 0.2) and will increase on either side of the monolayer. Anker and Reineccius [86] found that shelf life of orange peel oil encapsulated in gum Arabic increased with water activity within the range of water activity studied (0.0 to 0.536). Later work has shown that the type of encapsulation matrix determines the influence of water activity on shelf life [88]. Ma [97] confirmed the work of Anker [86] on gum Arabic and found that extending the water activity range to 0.75 resulted in decreased shelf life. Thus, it appears that there is an optimum water activity for the stability of materials encapsulated in gum Arabic (ca. 0.55). Ma [97] also looked at the role of water activity in influencing the shelf life of orange oil spray dried in Capsul and maltodextrin (M-100). She found that shelf life increased with water activity to 0.75 (upper limit of study) for the maltodextrin, and there was an inverse relationship between the two for Capsul. Thus, one will have to consider the matrix in determining the water activity (moisture content) that will give the optimum shelf life. However, it is clear that there is no shelf life rationale for producing powders in the traditional 2 to 5% moisture range. It is worthwhile to note that the proper choice of encapsulation matrix will yield shelf life in excess of one year, using typical spray drying processes without any antioxidant.

Today the industry largely uses blends of an emulsifying carrier (gum Arabic or modified starch) and hydrolyzed starches (10–25DE). Blends are used to reduce costs, and to impart both emulsifying properties and extend shelf life.

#### 13.4.4.2 Freeze, Drum, and Tray Drying

Alternative dehydration processes for the encapsulation of flavors include drum (contact) drying, tray drying, and freeze drying [98]. While these processes do not find major use in the industry, they are used in certain applications. These processes and their potential use for flavor encapsulation will be briefly discussed.

##### *13.4.4.2.1 Dryer Descriptions*

The basic configuration of a tray dryer is a chamber where the food is placed and a blower and ducts to provide hot air circulation around and across the food. Water is removed from the product surface and carried out of the dryer in a single operation. The air is heated while entering the dryer by means of heat exchangers or direct mixture with combustion exhaust gases. Typically, a tray drier operates in the batch mode. Its continuous counterpart is the tunnel dryer. The food stuff (flavor emulsion) is spread in thin layers, usually at 4.75 to 7.5 kg/m$^2$, on trays supported by angle slides mounted on the side of the cabinet. In larger operations, the trays are supported on trolleys, which can be pushed into and out of the cabinet and also turned around so that the air is not always leading on the same edge [99].

Contact dryers are dryers in which the wet material is dried by direct contact with a heated surface. Heat transfer to the wet material is mainly by conduction from this surface through the bed or layer of wet solids. This is in direct contrast to convection tray dryers: since no hot gas is required as a source of heat in contact dryers, gas flow through the system can be low and ultimately limited to the vapor evaporated from the wet material [100].

Specifically in drum drying, material is dried on the surface of an internally heated revolving drum. The material, in fluid, slurry, or paste-like form, is spread in a thin layer on the surface of the drum so that at no time is the drying rate governed by the diffusion of the vapor through the product layer. There are four variables involved in the operation of such a dryer:

1. Steam pressure or heating medium temperature, which governs the temperature of the drum surface
2. Speed of rotation, which determines the time of contact between the film and the preheated surface
3. Thickness of the film, which may be governed by the distance between the drums (double-drum) or a spreader (single-drum)
4. Condition of the feed material, that is, the concentration, physical characteristics, and temperature at which the solution/emulsion to be dried reaches the drum surface

In a double-drum type of operation, the feed level between the drums also determines the final concentration of the feed at the precise moment of contact with the hot drums [101].

Freeze drying is a multiple operation in which the material to be dehydrated is frozen by low-temperature cooling and then dried by direct sublimation of the water [102]. Freezing temperature and time are a function of the solutes in solution. Because the solutes become more concentrated in the unfrozen portion of the mix, the freezing temperature continually decreases until all the solution is frozen. At the end of the freezing process, the entire mass becomes rigid, forming an eutectic consisting of ice crystals and food components [103].

During the first drying step and as a result of the low pressure applied, the water vapor generated at the sublimation interface is removed through the outer porous layers of the product. The driving force for the sublimation is essentially the difference in pressure between the water vapor pressure at the ice interface and the partial pressure of water vapor in the drying chamber. The second drying step begins where no more ice (from unbound water) is in the product and the moisture comes from partially bound water in the material. Typically, this step takes up to a third of the total drying cycle to desorb the moisture from the internal surface within the dried product [104]. The structural rigidity afforded by the frozen material at the surface (where sublimation occurs) prevents collapse of the solid matrix remaining after drying. The result is a porous, nonshrunken structure that facilitates rapid and almost complete rehydration when water is added to the material at a later time. Major disadvantages of freeze drying are the high energy cost and the long drying time [105].

#### *13.4.4.2.2 Flavor Retention During Drying*

One might expect poor volatile retention during the tray or drum drying processes since there is no rapid (or early) formation of a selective membrane during the drying process. As described earlier, this phenomenon occurs in spray drying and essentially facilitates the retention of volatile organic compounds. However, as one can see in Table 13.4, orange oil was retained very well when dried in a tray dryer and poorly by both freeze and drum drying. If one considers the tray drying process, a film of dry solids appears on the drying surface quite early in the drying process. Apparently this dry film functions to slow the diffusion of organic volatiles from the drying surface thereby trapping the volatiles in the dry matrix. Drum drying involves

**TABLE 13.4**
**The Retention of Orange Oil during the Drying of a 30% Solution of Gum Acacia (20% Flavor Load) by Different Dehydration Processes**

| Drying Process | Flavor Retention (%) | Surface Oil (mg/100 g powder) |
|---|---|---|
| Spray dried | 89 | 3.4 |
| Tray dried | 95 | 75.5 |
| Drum dried | 71 | 8.8 |
| Freeze dried | 75 | 10.8 |

*Source:* Adapted from Buffo, R.A. and G.A. Reineccius, *Perfum. Flavorist*, 26, 2001. With permission.

vigorous boiling that would hinder the formation of a dry surface film and would not favor volatile retention (Table 13.4). Freeze drying fared little better than drum drying. While there is a drying interface that might retard volatile loss, apparently it forms too slowly to be effective in retaining volatiles.

In the industry, a large portion of the process flavorings (meat-like flavors) are dried by tray drying. This process involves a substantial amount of heat that continues the reaction process and also retains a high amount of aroma components. This drying process is well suited to thermoplastic/hygroscopic materials, which includes the majority of meat-like flavorings.

Table 13.4 also contains information on surface oil. Surface oil is generally considered detrimental to shelf life since it is readily available for oxidation (although there is often no relationship). The very high level of surface oil in the tray dried product is due to the need for grinding the dry cake. Grinding exposes a large amount of oil to extraction/oxidation. Spray drying results in the least amount of surface oil while freeze and drum drying have similar surface oil levels.

### 13.4.5 Extrusion

When one hears the word *extrusion*, one generally thinks of a high-temperature, high-pressure twin screw extruder that is used in the cooking and texturizing of cereal-based products. Extrusion as applied to flavor encapsulation may be used in a broader sense in that a molten flavor emulsion is forced through a die. However, unlike the high temperatures and pressures used in cereal processing, traditionally pressures were typically less than 100 psi and temperatures seldom exceeded 120°C.

#### 13.4.5.1 Traditional Processes and Formulations

Flavor extrusion originated from the basic ideas of Schultz et al. [106] who simply added citrus oils to a molten carbohydrate mass (like hard candy), agitated it to form a crude emulsion, and then allowed it to cool and solidify as a mass. The mass was ground to the desired particle size and sold as an encapsulated flavoring material.

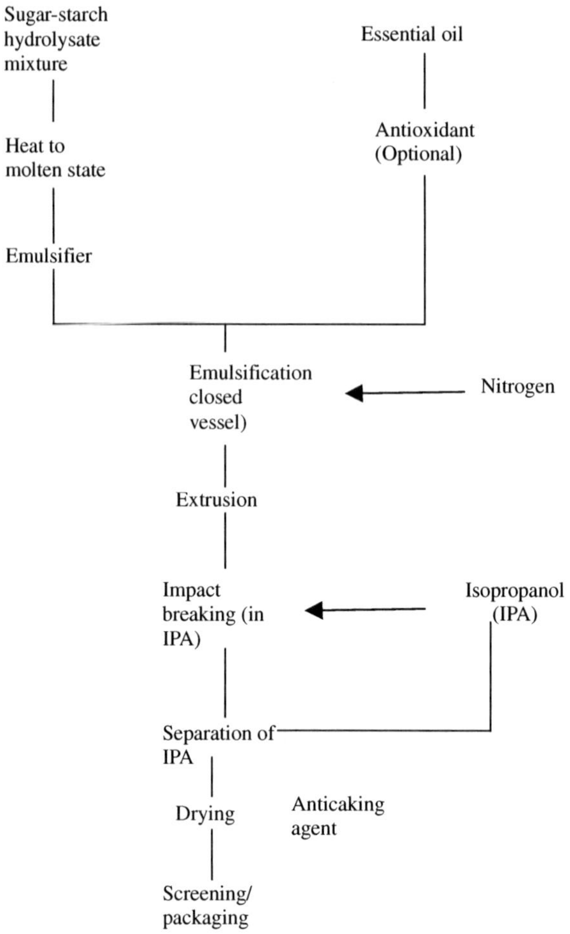


**FIGURE 13.13** Early batch extrusion processing of flavorings. (From Beck, E.E., Essential oil composition, U.S. Patent 3704137, 1972. With permission.)

Swisher [107] combined the basic formulation of Schultz et al. [106] with extrusion to produce a process similar to that used until only recently (Figure 13.13). Swisher [107] used a 42 DE corn syrup as the encapsulation matrix. Antioxidant was added to provide stability to the citrus oils during the high-temperature processing. Butylated hydroxyanisole and 4-methyl-2,6-ditertiarybutylphenol were mentioned by name in Swisher's patent and use level was suggested to be 0.05 percent of the flavor oil. Emulsifiers were also added in order to facilitate emulsion formation and promote stability. While numerous synthetic and natural emulsifiers were suggested in his patent, patent examples used a blend of monoglycerides and monoglyceride sodium sulfoacetate at 1% level (based on total emulsion weight). This corn syrup/antioxidant/emulsifier mixture contained from 3 to 8.5% moisture, and temperatures of about 120°C were necessary to keep the solution sufficiently low in viscosity to permit the incorporation of flavoring oils. Flavoring oils were incorporated at about a 10% level, and the mixture was agitated violently under nitrogen to form an oxygen-free emulsion. This emulsion was forced through a die into an immiscible liquid (e.g., mineral or vegetable oils), which was then rapidly cooled or simply extruded into pellets, which were allowed to solidify and then ground to the desired particle size. The ground material was washed with solvent (e.g., isopropanol) to remove surface oil and then dried under vacuum. The end product of this process was a free-flowing granular material containing 8 to 10% flavoring.

Swisher [109] was awarded a second patent on this process, which incorporated several improvements. He suggested incorporating 4 to 9% glycerol into the initial corn syrup solids. The glycerol functioned as a heat exchange medium, permitted solubilization of the corn syrup solids at a lower moisture content, and acted as a plasticizing agent in the finished product. The plasticizing effect minimized cracking of the product, which would permit oxygen to contact the flavoring material and thereby limit shelf life.

A second major innovation in this later patent was the extrusion of the hot flavor/carbohydrate melt through a die (approximately 1/64-in holes) into a cold solvent bath. While several solvents (e.g., kerosene, petroleum ether, methanol, acetone, methylethyl ketone, limonene, benzene, and toluene) were suggested in his patent, iso-propanol was used as the solvent of choice in patent examples and is used today in the industry. The extrusion into a cold solvent bath was a key innovation in that the solvent served to rapidly solidify the carbohydrate matrix into an impermeable amorphous structure, washed residual surface oil from the product that otherwise would have oxidized causing off-flavor problems and also served to extract moisture from the finished product. Swisher [109] found that soaking times ranging from 36 to 144 h reduced moisture content to ½ to 2%. When sufficient agitation was used in the bath, this step also served to reduce particle size (practical limit is when piece length equals diameter). Therefore, it was no longer necessary to grind the solidified flavoring and wash it since these two operations were now combined into the solidification step of the process.

The residual solvent was removed from the product through air drying. At this time, additional moisture could also be removed (if one chose to reduce soaking time) and various anticaking agents could be added. The combination of high dextrose equivalent corn syrup solids and glycol was quite hygroscopic, so anticaking agents such as tricalcium phosphate were recommended to maintain the product as a free-flowing particulate.

The process of encapsulation via extrusion remained largely unchanged for several years except for some changes in formulation. Beck [108] replaced the high DE corn syrup solids with a combination of sucrose and maltodextrin. He used a carbohydrate melt consisting of about 55% sucrose and 41% maltodextrin (10–13 DE). The remaining ingredients were moisture and additives. This low-DE maltodextrin/sucrose matrix was considerably less hygroscopic than that used by Swisher [107,109]. Beck [108], however, continued the use of anticaking agent and recommended pyrogenic silica rather than tricalcium phosphate. The flavor load obtained by Beck [108] generally ranged from 8 to 10%, with 12% considered as a practical maximum. Barnes and Steinke [110] were awarded a patent in which they used a modified food starch in place of the sucrose. The modified food starch selected was either Capsul (National Starch) or Amiogum® (American Maize). Both of these starches have been chemically modified to produce a product with emulsification properties. Barnes and Steinke [110] stated that the emulsifying starch, with its lipophilic characteristics, would absorb the flavor oils into the matrix. The maltodextrin was, therefore, used primarily to provide bulk and some viscosity control. Barnes and Steinke [110] claimed that the use of emulsifying starches in the encapsulation matrix would permit increasing the loading capacity to 40% flavoring (although these high loading have not been observed in commercial practice).

### 13.4.5.2 Modern Extrusion Processes

The fact that batch processes are typically more expensive and offer limited process control compared to continuous processes has motivated the flavor industry to develop methods of producing flavor capsules with continuous extrusion systems (e.g., see [111–114]). In addition to improved process control and reduced cost, the use of continuous extruders permits working with much lower moisture systems (higher viscosities). Modern twin screw extruders can work with high viscosity systems and still provide considerable mixing (emulsifying) compared to the traditional extrusion systems used in making this type of dry flavoring. The relevance of this ability is obvious in that the industry needs to make a *dry* product, so the less moisture in the formulation, the less drying is required in secondary steps.

A typical extrusion system used in making flavor capsules is presented in Figure 13.14. While cooking is often an important function of a cereal extrusion process, the primary functions of the extruder in flavor encapsulation is to melt, emulsify, and form. In the figure presented, the encapsulation matrix is dry blended in the feeder (10), and then water and blended matrix are each metered into the extruder (01 and 02). The extruder barrel is heated to melt the matrix and flavor is then injected into the barrel as late as possible (05). The desire is to minimize heat exposure of the flavoring, but there must be adequate time for emulsification. Some of the processes include a barrel venting step. The purpose of the venting step is to remove some of the water (plasticisor) added to the formulation thereby requiring little or no final drying. A patent example is presented below [112].

> A blend of "6% Amerfond (Domino Sugar, 95% Sucrose, 5% Invert sugar), 42% Lodex-10 Maltodextrin (American Maize, 10 DE), 2% Distilled monoglyceride (Kodak, Myverol 18-07) was fed at a rate of approximately 114 g/min into the continuous processor (Figure 13.14) with water at 2 g/min. The mixture was melted in the processor. The processor was maintained at 121°C. The processor screws were operating at 120 RPM. The molten mixture was discharged directly to the melt pump. Acetaldehyde was injected into the molten matrix on the discharge side of the melt pump using a piston metering pump. A static mixer was used to blend the matrix and flavor together. Immediately prior to flavor injection the temperature of the molten matrix was approximately 138°C. The matrix and acetaldehyde mixture was then delivered under pressure to one of the nozzle discharges for forming and subsequent collection. The flow system was arranged so that forming and solidification could take place under either atmospheric or pressurized conditions."

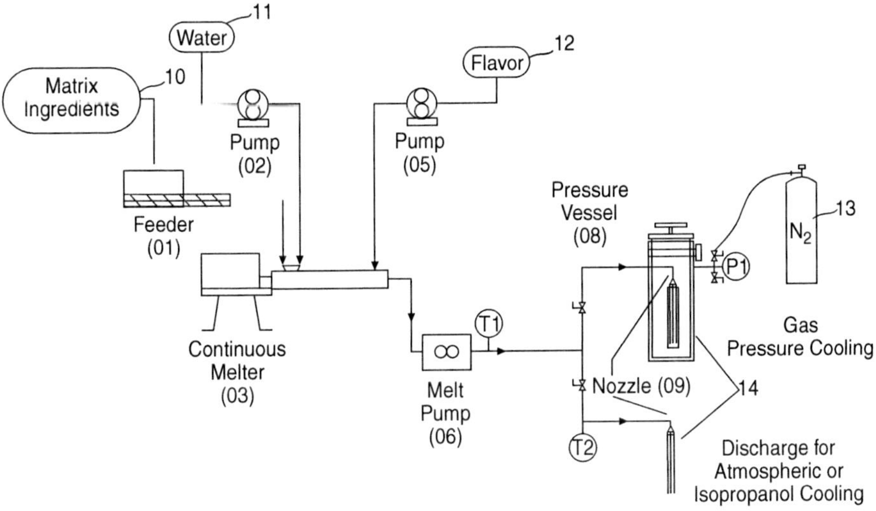


**FIGURE 13.14** Extruder system designed to produce flavorings. (From Fulger, C., L.M. Popplewell, Flavor encapsulation, U.S. Patent 5601865, Assigned to McCormick & Company, Inc., (Sparks, MD) 1997. With permission.)

In the example above, the authors set the system up to extrude under pressure (08 — potentially cold temperatures) so that highly volatile substances (such as acetaldehyde) would not volatilize and be lost thereby producing a porous capsule. Alternatively, the product discharge could be air cooled or injected into a cold isopropanol bath (T2). This process required particle size reduction as a final step. Size reduction by grinding potentially could damage the particle structure, reducing volatile retention and shelf life.

Benczedi and Bouquerand [111] approached the problem of particle formation by working both with formulation and processing conditions. By using the formulation and process conditions noted in the patent example below, they were able to form very uniform particles directly from the extruder.

Patent example:

**Formulation**

| **Ingredients** | **Parts by Weight** |
|---|---|
| Orange flavor[a] | 10 |
| Lecithin | 1 |
| Water | 7 |
| Maltodextrin 19 DE | 82 |
| Total | 100 |

[a] Origin: Firmenich SA, Geneva, Switzerland.

"The powder was extruded at a throughput of 50 kg/h through 1 mm die holes using a Clextral BC 45 twin-screw extruder equipped with a cutter knife allowing to chop

> the melt at the die exit while it is still plastic. At the low water content needed to guarantee a glass transition above 40°C at constant sample composition, the temperature of the melt in the front plate was of 105°C and the static pressure in the extruder was preferably kept below 1-20 × $10^5$ Pa." [111]

It appears that process development has progressed to the point where continuous extrusion processes can be used to produce encapsulated flavorings without the need for cooling, drying, or size reduction steps. These innovations all have contributed to reduced manufacturing costs for this product.

Historically, extrusion encapsulation has offered the advantages of yielding large particle sizes (0.25 mm to 2 mm), and extended shelf life (oxidation). Large particles are visible to the consumer (marketing?) and also do not fall through tea bags (as spray dried particles do). They may or may not classify during shipping, depending upon the product. The most touted strength is that the process offers exceptional shelf life to flavorings by providing an excellent oxygen barrier. The companies producing these products associate the exceptional shelf life with large particles (diffusion distances), high densities (lower diffusion rates), and the matrix being in the glassy state. However, it is this author's opinion that the shelf life is largely due to the matrix composition and that exceptional shelf lives can be obtained by other encapsulation processes if similar matrices are used.

The advent of continuous extrusion has also provided the opportunity to use a wider variety of matrix materials. In the past, viscosity has been a limiting factor, but this is no longer as critical when twin screw extruders are used. The wider range of matrix materials has provided the opportunity to use materials that can offer some controlled release properties, e.g., proteins or less soluble carbohydrates. This offers further advantage to this encapsulation process.

The primary weakness that still plagues this product is poor emulsion stability in final applications. The materials used in encapsulation generally have no inherent emulsification properties (exceptions would be when modified starches or gum Arabic is used as a wall component) and the emulsion must be formed in an exceptionally viscous matrix. Both factors work against forming a small enough emulsion size to yield a stable emulsion in a finished product, i.e., beverage. Nevertheless, major improvements have been made in this process that lower costs and improve performance. Thus, flavorings produced in this manner will likely have increased market share in the future.

#### 13.4.5.3 Shear Form Process

*Shear form extrusion* was a term coined by Fuisz Technologies (Chantilly, VA) to describe their process of encapsulation. Briefly, this process is similar to that of making cotton candy. The desired matrix and actives are dry blended together, and then the blend is passed through a heated spinner head. The matrix melts in the heated spinner head and is thrown out of the head through openings of the desired size and shape. The molten matrix/flavor mass congeals as it flies through the air. The cooled solid matrix is collected as the finished product. Fuisz Technologies owns 114 patents to date on this process as it is applied to a broad range of products (mostly food or pharmaceuticals). Patents 5427811 [115] and 6174514 [116] illustrate the process and its application. While there was substantial commercial interest in this process initially, it appears that it has limited utility for the encapsulation of volatile substances.

## 13.5 CONTROLLED RELEASE

Little will be said about controlled release flavorings in this chapter simply because there is limited space. However, there is substantial interest in imparting controlled release properties to encapsulated flavorings. Ideally, the flavoring would be protected from the environment until the final product is ready for consumption. Unfortunately, there are limited opportunities for imparting controlled release to flavorings due to both materials limitations (FDA approved) and cost considerations — cost of both materials and processing.

Encapsulated flavorings currently made generally use water as a manufacturing vehicle, and thus water (hydration) is the release mechanism. However, coacervation (when cross-linked) produces an insoluble wall, and thus release is via diffusion as opposed to dissolution. In high moisture systems, this slows release but does not stop it. The extrusion process also may provide some controlled release properties in that one may use less soluble matrices thereby reducing release rates.

Alternatively, one can apply secondary coatings to encapsulated flavorings to impart different release properties to flavorings, e.g., thermal release. This is most often done by fluidized bed coating or centrifugal suspension coating (both may use lipids). The primary issues with applying secondary coatings are: (1) the dilution of the flavorings (large amounts of coatings are required to coat small particles), and (2) the added cost. Often it is more cost effective to use more flavor in a product than pay for the additional processing costs of adding controlled release. The reader is encouraged to go to Risch and Reineccius [18], Reineccius [117], Li and Reineccius [118], and Heiderich and Reineccius [119] for detail on these processes and their application to food flavorings.

## 13.6 SUMMARY

The manufacture of liquid (including emulsions), paste, and dry forms of flavorings has been presented. Liquid flavorings are the simplest to assemble but are still problematic in terms of proper formulation and ultimately stability (physical and chemical). Often components will separate into distinct phases that complicates usage (pastes). During storage, chemical reactions may occur between flavor components as well as between flavor components and the system solvent or oxygen (air). These flavorings are not stable indefinitely and have a limited shelf life.

Delivering flavorings as emulsions is extremely problematic. Emulsions are inherently unstable and will readily undergo phase separation despite best efforts to add the optimal amounts of weighing agents and efficiently homogenize. Stability is very dependent upon the flavoring being used as well.

A wide array of manufacturing processes is available for the production of dry flavorings. Each of these processes results in a product that has unique strengths and weaknesses. The ultimate choice of process depends upon cost and performance needs.

## REFERENCES


1. Fischetti, F. Workshop in Food Flavors: Creation and Manufacturing, offered at the Dept. of Food Science and Nutrition, University of Minnesota, St. Paul, May 2002.
2. Buffo, R., G.A. Reineccius, Beverage emulsions and the utilization of gum acacia as emulsifier/stabilizer, *Perfum. Flavorist*, 25, 4, p. 24, 2000.
3. King, W., P. Trubiano, P. Perry, Modified starch encapsulating agents offer superior emulsification, film forming, and low surface oil, *Food Prod. Dev.*, 10, 10, p. 56, 1976.
4. Rutenberg, M.W., Starch and its modification, in *Handbook of Water Soluble Gums and Resins,* R.L. Davidson, Ed., McGraw Hill, New York, 1980.
5. Buffo, R., Optimization of the Emulsifying and Encapsulating Properties of Gum Acacia, Ph.D. thesis, University of Minnesota, St. Paul, p. 336, 1999.
6. Buffo, R., G.A. Reineccius, Shelf life and mechanisms of destabilization in dilute beverage emulsions, *J. Flavour Frag.*, 16, p. 7, 2001.
7. Tan, C.T., J. Wu Holmes, Stability of beverage flavor emulsions, *Perfum. Flavorist*, 13, p. 23, 1988.
8. Dickinson, E., *An Introduction to Food Colloids*, Oxford University Press, Oxford, U.K., 1992, p. 79.
9. King, W., P. Trubiano, P. Perry, Modified starch encapsulating agents offer superior emulsification, *Food Prod. Dev.*, 10, 10, p. 54, 1976.
10. Goff, D. Dept. of Dairy Science and Technology, University of Guelph, Guelph, http://www.foodsci.uoguelph.ca/dairyedu/homogenization.html, accessed 10/6/04.
11. Tse, K.Y., *Physical Stability of Flavor Emulsions,* M.S. thesis, Department of Food Science and Nutrition, University of Minnesota, St. Paul, p. 84, 1990.
12. Chadonnet, S., H. Korstredt, A. Siciliano, Preparation of microemulsions by microfluidization, *Soap/Cosmet./Chem. Specialties*, 61, 2, p. 37, 1985.
13. Washington, C., Dispersing problems of emulsion production, *Lab. Equip. Dig.*, Dec. 1987.
14. Tse, A. and G.A. Reineccius, Methods to predict the stability of flavor/cloud emulsions, in *Flavor Technology: Physical Chemistry, Modification and Process*, C.T. Ho, C.T. Tan, C.H. Hsiang, Eds., Amer. Chem. Soc., Washington, D.C., 1995, p. 172.
15. Heath, H., G.A. Reineccius, *Flavor Chemistry and Technology*, AVI Pub. Co., Westport, CT, 1986, p. 442.
16. Tan, C.-T., Beverage emulsions, in *Food Emulsions*, K. Larsson, E. Friberg, Eds., Marcel Dekker, New York, 1997, p. 491.
17. Shahidi, F., X.-Q. Han, Encapsulation of food ingredients, *Crit. Rev. Food Sci. Nutr.*, 33, 6, p. 501, 1993.
18. Risch, S.J., G.A. Reineccius, Eds., *Encapsulation and Controlled Release of Food Ingredients*, Amer. Chem. Soc., Washington, D.C., 1995, p. 214.
19. Gibbs, B.F., S. Kermasha, I. Alli, C.N. Mulligan, Encapsulation in the food industry: a review, *Int. J. Food Sci. Nutr.*, 50, 3, p. 213, 1999.
20. Sébastien, G., Microencapsulation industrial appraisal of existing technologies and trends, *Trends Food Sci. Technol.*, 15, 7–8, p. 330, 2004.
21. Arshady, R., B. Boh, Microspheres,microcapsule patents and products: the art and science of microcapsules, patents and patent databases, *Microcapsules Liposomes*, 6, p. 1, 2003.
22. Uhlemann, J., B. Schleifenbaum, H.J. Bertram, Flavor encapsulation technologies: an overview including recent developments, *Perfum. Flavorist*, 27, 5, p. 52, 2002.

23. Bolton, T.A., *The Retention and Stability of Volatile Flavor Compounds Absorbed in Amorphous Silicon Dioxide,* M.S. thesis, Department of Food Science and Nutrition, University of Minnesota, St. Paul, p. 182, 1993.
24. Bolton, T.A., G.A. Reineccius, The oxidative stability and retention of a limonene-based model flavor plated on amorphous silica and other selected carriers, *Perfum. Flavorist*, 17, 2, p. 1, 1992.
25. Veith, S.R., *Retention, Diffusion and Release of Flavor Molecules from Porous Silica Sol-Gel-Made Particles,* Ph.D. thesis, Swiss Federal Institute of Technology, Zurich, 2004.
26. Hedges, A.R., W.J. Shieh, C.T. Sikorski, Use of cyclodextrins for encapsulation in the use and treatment of food products, in *Encapsulation and Controlled Release of Food Ingredients*, G.A. Reineccius, S.J. Risch, Eds., Amer. Chem. Soc., Washington, D.C., 1995. p. 60.
27. Szente, L., J. Harangi, J. Szejtli, Long-term Storage Stability Studies on Flavor-β-Cyclodextrin Complexes, Proc. 4th Int. Symposium on Cyclodextrins, Kluwer, Dordrecht, p. 545, 1988.
28. Qi, Z., A. Hedges, Use of cyclodextrins for flavors, in *Flavor Technology: Physical Chemistry, Modification and Process*, C.-T. Ho, C.-T. Tan, C.-H. Tong, Eds., Amer. Chem. Soc., Washington, D.C., 1995, p. 231.
29. Pagington, J., Beta-cyclodextrin, *Perfum. Flavorist*, 11, 1, p. 49, 1986.
30. Hedges, A., C. McBride, Utilization of β-cyclodextrin in food, *Cereal Foods World*, 44, 10, p. 700, 1999.
31. Del Valle, E.M., Cyclodextrins and their uses: a review, *Process Biochem.*, 39, p. 1033, 2004.
32. Harrison, M., Wacker Biochem Corp., 3301 Sutton Rd., Adrian, MI, 2001.
33. Reineccius, G.A., S.J. Risch, Encapsulation of artificial flavors by β-cyclodextrin, *Perfum. Flavorist*, 11, 4, p. 1, 1986.
34. Kollengode, A., M. Hanna, Cyclodextrin complexed flavors retention in extruded starches, *J. Food Sci.*, 62, 5, p. 1057, 1997.
35. Labows J.N., J.C. Brahms, Cagan, R.H., Solubilization of flavors, in *The Contribution of Low- and Non-Volatile Materials to the Flavor of Foods*, W. Pickenhagen, C.-T. Ho, A. Spanier, Eds., Allured, Carol Stream, IL, 1996, p. 125.
36. Liu, X.D., T. Furuta, J. Yoshii, P. Linko, W.J. Coumans, Cyclodextrin encapsulation to prevent the loss of l-menthol and its retention during drying, *Biosci. Biotechnol. Biochem.*, 64, 8, p. 1608, 2000.
37. Ohta, Y., K. Takatani, S. Kawakishi, Kinetic and thermodynamic analyses of the cyclodextrin-allyl isothiocyanate inclusion complex in an aqueous solution, *Biosci. Biotechnol. Biochem.*, 64, 1, p. 190, 2000.
38. Szente, L., J. Szejtli, Cyclodextrin-complexed acetal as an acetaldehyde generator, *Perfum. Flavorist*, 20, 2, p. 11, 1995.
39. Reineccius, T.A., G.A. Reineccius, T.L. Peppard, Encapsulation of flavors using cyclodextrins: comparison of flavor retention in α, β, and γ types, *J. Food Sci.*, 67, 9, p. 3271, 2002.
40. Reineccius, T.A., G.A. Reineccius, T.L. Peppard, Comparison of flavor release from α, β, and γ -cyclodextrins, *J. Food Sci.*, 68, 4, p. 1234, 2003.
41. Reineccius, T.A., G.A. Reineccius, T.L. Peppard, Utilization of beta-cyclodextrin for improved flavor retention in thermally processed food applications, *J. Food Sci.*, 69, 1, p. 58, 2004.

42. Reineccius, T.A., G.A. Reineccius, T.L. Peppard, Beta-cyclodextrin as a partial fat replacer, *J. Food Sci.*, 69, 4, p. 334, 2004.
43. Reineccius, T.A., G.A. Reineccius, T.L. Peppard, The effect of solvent interactions on $\alpha$, $\beta$, and $\gamma$ -cyclodextrin/flavor molecular inclusion complexes, *J. Agric. Food Chem.*, 53, 2, pp. 388–392, 2005.
44. Korus, J., Microencapsulation of flavors in starch matrix by coacervation method, *Polish J. Food Nutr. Sci.*, 10, 1, p. 17, 2001.
45. Kitson, J.A., H.S. Sugisawa, Sugar captured flavor granules, *Can. Inst. Food Sci. Technol. J.*, 7, 1, A15, 1974.
46. Pitchon, E., M. Schulman, S.B. Chall, Solid Aroma Composition, in Ger. Offen. 2,840,511, *Chem. Abstr.*, 92, 196654h. 1980.
47. Thies, C., Microencapsulation, *Kirk-Othmer Encyclopedia of Chemical Technology*, John Wiley & Sons, New York, 2000, http://www3.interscience.wiley.com/cgi-bin/mrwhome/104554789/HOME, accessed 04/22/05.
48. Green, B.K., Oil containing microscopic capsules and method of making them, U.S. Patent Application 2800458, 1957.
49. Subramaniam, A. and A. Reilly, Flavor microcapsules obtained by coacervation with gelatin, Firmenich Company, Princeton, NJ, U.S. Patent Application 2004041306, 2004.
50. Soper, J.C., M.T. Thomas, Preparation of protein-encapsulated oil particles using enzyme-catalyzed crosslinking, Patent CODEN: EPXXDW EP 856355 A2, 1998.
51. Vasishtha, N. 2003. Southwest Research Institute, San Antonio, TX.
52. NIZO Food Research, Complex coacervates containing whey proteins, Patent CODEN: EPXXDW EP 1371410 A1, 2003.
53. Soper, J.C., Method of encapsulating food or flavor particles using warm water fish gelatin, and capsules produced from them, Patent CODEN: PIXXD2 WO 9620612 A1, 1996.
54. Bakker, M.A.E., S.A. Galema, A. Visser, Microcapsules of gelatin and carboxymethylcellulose, Patent CODEN: EPXXDW EP 937496 A2 19990825, 1999.
55. Soper, J.C., X. Yang, D.B. Josephson, Encapsulation of flavors and fragrances by aqueous diffusion into microcapsules, Patent CODEN: USXXAM US 6106875 A 20000822, 2000.
56. Bangs, W.E., Development and Characterization of Wall Materials for Spray-Dried Flavorings Production, Ph.D. thesis, Department of Food Science and Nutrition, University of Minnesota, St. Paul, 1985.
57. Reineccius, G.A., Carbohydrates for flavor encapsulation, *Food Technol.*, 45, 3, p. 144, 1991.
58. Thevenet, F., Acacia gums: Natural encapsulation agent for food ingredients, in *Flavor Encapsulation*, Risch, S.J., G.A. Reineccius, Eds., Amer. Chem. Soc., Washington, D.C., 1988, p. 37.
59. Trubiano, P., N.L. Lacourse, Emulsion-stabilizing starches: use in flavor encapsulation, in *Flavor Encapsulation*, Risch, S.J., G.A. Reineccius, Eds., Amer. Chem. Soc., Washington, D.C., 1988, p. 45.
60. Kenyon, M., R.J. Anderson, Maltodextrins and low-dextrose equivalent-equivalence corn syrups: production and technology for the flavor industry, in *Flavor Encapsulation*, Risch, S.J., G.A. Reineccius, Eds., Amer. Chem. Soc., Washington, D.C., 1988, p. 7.
61. Reineccius, G.A., The spray drying of food flavors, *Drying Technology,* 22, 6, pp. 1289–1324, 2004.

62. Leahy, M.M., S. Anandaraman, W.E. Bangs, G. Reineccius, Spray drying of food flavors. II. A comparison of encapsulating agents for the drying of artificial flavors, *Perfum. Flavorist*, 8, 5, p. 49, 1983.
63. Kerkhof, P.J.A.M. and H.A.C. Thijssen, Quantitative study of process variables on aroma retention during drying of liquid foods, *Amer. Inst. Chem. Eng. Symp. Ser.*, 73, 163, p. 33, 1977.
64. Thijssen, H.A.C. and W.F. Rulkens, Retention of aromas in drying food liquids, *De Ingenieur/JRG '80/NE*, 47, p. 45, 1968.
65. Buffo, R., *Optimization of the Emulsifying and Encapsulation Properties of Gum Acacia,* Ph.D. thesis, Department of Food Science and Nutrition, University of Minnesota, St. Paul, 1999.
66. Rosenberg, M., I.J. Kopelman, Y. Talmon, Factors affecting retention in spray-drying microencapsulation of volatile materials, *J. Agric. Food Chem.*, 38, p. 1288, 1990.
67. Dronen, D.M., *Characterization of Volatile Loss from Dry Food Polymers,* Ph.D. thesis, Department of Food Science and Nutrition, University of Minnesota, St. Paul, p. 637, 2004.
68. Reineccius, G.A., W.E. Bangs, Spray drying of food flavours. III. Optimum infeed concentrations for the retention of artificial flavours, *Perfum. Flavorist*, 10, 1, p. 27, 1985.
69. Reineccius, G.A. and S.T. Coulter, Flavor retention during drying, *J. Dairy Sci.*, 52, 8, p. 1219, 1969.
70. Emberger, R., Aspects of the development of industrial flavor materials, in *Flavour '81*, P. Schreier, Ed., de Gruyter, New York, 1981, p. 619.
71. Brenner, J., G.H. Henderson, R.W. Bergentsen, Process of encapsulating an oil and product produced thereby, U.S. Patent 3971852, 1976.
72. Risch, S.J., G.A. Reineccius, Effect of emulsion size on flavor retention and shelf-stability of spray dried orange oil, in *Flavor Encapsulation*, S.J. Risch, G.A. Reineccius, Eds., Amer. Chem. Soc., Washington, D.C., 1988, p. 67.
73. Sheu, T.Y., M. Rosenberg, Microencapsulation by spray drying ethyl caprylate in whey protein and carbohydrate wall systems, *J. Food Sci.*, 60, 1, p. 98, 1995.
74. Re, M.I., Y.J. Lui, Microencapsulation by spray drying: influence of wall systems on the retention of the volatile compounds, in *Drying-96-Proceed. of the 10th Intern. Drying Symp.*, Krakow, Poland, 1996, p. 541.
75. Soottitantawat, A., H.T. Yoshii, M. Furuta, P. Linko Ohkawara, Microencapsulation by spray drying: influence of emulsion size on the retention of volatile compounds, *J. Food Sci.*, 68, 7, p. 2256, 2003.
76. Sivetz, M., H.E. Foote, *Coffee Processing Technology*, AVI Pub. Co., Westport, CT, 1963.
77. Kieckbusch, T.G. and C.J. King, Losses in the nozzle zone during spray drying, *Pacific Chem. Engin. Confer. Proc.*, 2, 1, p. 216, 1977.
78. Thijssen, H.A.C., Effect of process conditions on in drying liquid foods on its aroma retention, *Proc. Nordic Aroma Symp.*, 3rd ed., 1972.
79. Thijssen, H.A.C., Effect of process conditions in drying coffee extract and other liquid foods on aroma retention, *Colloquium Int. Chimistrie Cafes*, 8, p. 222, 1973.
80. Thijssen, H.A.C., Optimization of process conditions during drying with regard to quality factors [Skim-milk], *Lebensmit. Wissen. Technol.*, 12 , 6, p. 308, 1979.
81. Blakebrough, N. and P.A.L. Morgan, Flavour loss in the spray drying of emulsions, *Birmingham Univ. of Chem. Eng.*, 24, 3, p. 57, 1973.

82. Bomben, J.L., S. Bruin, H.A.C. Thijssen, R.L. Merson, Aroma recovery and retention in concentration and drying of foods, *Adv. Food Res.*, 20, p. 1, 1973.
83. Brooks, R., Spray drying of flavoring materials, *Birmingham Univ. of Chem. Eng.*, 16, 1, p. 11, 1965.
84. Rulkens, W.H. and H.A.C. Thijssen, The retention of organic volatiles in spray-drying aqueous carbohydrate solutions, *J. Food Technol.*, 7, 1, p. 95, 1972.
85. Bradley, R.L. and C.M. Stine, Spray drying of natural cheese, *Manufact. Milk Prod. J.*, 54, p. 8, 1964.
86. Anker, M., G.A. Reineccius, Influence of spray dryer air temperature on the retention and shelf-life of encapsulated orange oil, in *Flavor Encapsulation*, Risch, S.J., G.A. Reineccius, Eds., Amer. Chem. Soc., Washington, D.C., 1988, p. 78.
87. King, C.J., Spray drying: retention of volatile compounds revisited, *Drying Technol.*, 13, 5–7, p. 1221, 1995.
88. Etzel, M.R. and C.J. King, Loss of volatile trace organics during spray drying, *Ind. Eng. Chem. Pro. Des. Devel.*, 23, 4, p. 705, 1984.
89. Papadakis, S.E. and C.J. King, Air temperature and humidity profiles in spray drying. 1. Features predicted by the particle source in the cell, *Ind. Eng. Chem. Res.*, 27, p. 2111, 1988.
90. Moor, S.S., C.J. King, Visualization of spray dynamics in a pilot plant dryer by laser-initiated fluorescence, *Ind. Eng. Chem. Res.*, 37, p. 561, 1998.
91. Kieckbusch, T.G. and C.J. King, Volatiles loss during atomization in spray drying, *Amer. Inst. Chem. Eng. J.*, 26, 5, p. 718, 1980.
92. Brenner, J., The essence of spray-dried flavours: the state of the art, *Perfum. Flavorist*, 8, 2, p. 40, 1983.
93. Subramaniam, A., *Encapsulation, Analysis and Stability of Orange Peel Oil,* Ph.D. thesis, Department of Food Science and Nutrition, University of Minnesota, St. Paul, p. 278, 1984.
94. Westing, L.L., G.A. Reineccius, F. Caporaso, Shelf life of orange oil: effects of encapsulation by spray-drying, extrusion, and molecular inclusion, in *Flavor Encapsulation*, S.J. Risch, G.A. Reineccius, Eds., Amer. Chem. Soc., Washington, D.C., p. 110, 1988.
95. Baisier, W., G. Reineccius, Spray drying of food flavors. V. Factors influencing shelf-life of encapsulated orange peel oil, *Perfum. Flavorist*, 14, 3, p. 48, 1989.
96. Anandaraman, S. and G.A. Reineccius, Stability of encapsulated orange peel oil, *Food Technol.*, 40, 11, p. 88, 1986.
97. Ma, Y.M., *Investigation of the Correlation between Glass Transition and Stability of Flavor Microcapsules,* M.S. thesis, Department of Food Science and Nutrition, University of Minnesota, St. Paul, 1991.
98. Buffo, R.A. and G.A. Reineccius, Comparison among assorted drying processes for the encapsulation of flavors, *Perfum. Flavorist*, 26, p. 58, 2001.
99. Brennan, J.G., *Dictionary of Food Dehydration*, Butterworth-Heinemann Ltd., Oxford, 1994.
100. Chirife, J., M. Karel, Volatile retention during freeze drying of aqueous suspensions of cellulose and starch, *J. Agri. Food Chem.*, 21, 6, p. 936, 1973.
101. Desobry-Banon, S.A., F.M. Netto, T.P. Labuza, Comparison of spray-drying, drum-drying and freeze-drying for beta-carotene encapsulation and preservation, *J. Food Sci.*, 62, 6, p. 1158, 1997.
102. Flink, J., The retention of volatile components during freeze drying: a structurally based mechanism, in *Freeze Drying and Advanced Food Technology*, Goldblith, S.A., L. Rey, W.W. Rothmayr, Eds., Academic Press, New York, 1975, p. 351.

103. Flink, J. and M. Karel, Effects of process variables on retention of volatiles in freeze drying, *J. Food Sci.*, 35, p. 444, 1970.
104. Bradley, R.L., Moisture and total solids analysis, in *Introduction to the Chemical Analysis of Foods*, S.S. Nielsen, Ed., Jones Barlett Publishers, Boston, 1994, p. 93.
105. Karel, M., Stability of low and intermediate moisture foods, in *Freeze Drying and Advanced Food Technology*, Goldblith, S.A., L. Rey, W.W. Rothmayr, Eds., Academic Press, New York, p. 351, 1975.
106. Schultz, I.H., R.P. Dimick, B. Mackower, Incorporation of natural fruit flavors into fruit juice powders, *Food Technol.*, 10, 10, p. 57, 1956.
107. Swisher, H.E., Solid essential oil containing compositions, U.S. Patent 2809895, 1957.
108. Beck, E.E., Essential oil composition, U.S. Patent 3704137, 1972.
109. Swisher, H.E., Solid essential oil flavoring composition, U.S. Patent 3041180, 1962.
110. Barnes, J.M., J.A. Steinke, Encapsulation matrix composition and encapsulate containing same, U.S. Patent 4689235, 1987.
111. Benczedi, D., P.E. Bouquerand, Process for the preparation of granules for the controlled release of volatile compounds, U.S. Patent 6607771, 2003.
112. Fulger, C., L.M. Popplewell, Flavor encapsulation, U.S. Patent 5601865, 1997.
113. Porzio, M.A. and L.M. Popplewell, Encapsulation compositions, U.S. Patent 6652895, 2003.
114. Levine, H., L. Slide, B. Van Lengerich, J.G. Pickup, Glassy matrices containing volatile and/or labile components, and processes for preparation and use thereof, U.S. Patent 5009900, 1991.
115. Fuisz, R.C., B.A. Bogue, Method and apparatus for spinning thermo-flow materials, U.S. Patent 5427811, Awarded to Fuisz Technologies Ltd., Chantilly, VA, 1995.
116. Cherukuri, S.R., A.L. Khurana, M.K. Jr. Schaller, T.L. Chau, M.J. Strait, Breath freshening chewing gum with encapsulations, U.S. Patent 6174514, 2001.
117. Reineccius, G.A., Controlled release techniques in the food industry, in *Encapsulation and Controlled Release of Food Ingredients*, S.J. Risch, G.A. Reineccius, Eds., Amer. Chem. Soc., Washington, D.C., 1995, p. 8.
118. Li, H.C., G.A. Reineccius, Protection of artificial blueberry flavor in microwave frozen pancakes by spray drying and secondary fat coating processes, in *Encapsulation and Controlled Release of Food Ingredients*, S.J. Risch, G.A. Reineccius, Eds., Amer. Chem. Soc., Washington, D.C., 1995, p. 180.
119. Heiderich, S. and G.A. Reineccius, The influence of fat content, baking method, and flavor form on the loss of volatile esters from cookies, *Perfum. Flavorist*, 16, 6, p. 14, 2001.

# 14 Flavor Applications

## 14.1 INTRODUCTION

Interestingly, there is little detailed information in print on flavor applications. Heath [1] offered the first book in this area, and it still serves as an excellent reference today. Ashurst [2] and Ziegler and Ziegler [3] devote significant space to this topic in their books, but it is not the focus of either book. Certainly, specialized books on a given food (e.g., baked goods authored by Matz [4]), include chapters on the flavoring of their product, but this information as a whole is then scattered among many sources instead of one. This topic is sufficiently important that it should be covered in a book on its own.

Flavor applications is a critical function both within a flavor company and the food industry. The Flavor Applications group brings expertise in how foods are formulated, processed, stored, and ultimately prepared by the consumer. A flavor must be designed for a given application, and it is this group that brings this expertise to the company. Furthermore, this group is responsible for the evaluation of flavorings for performance in the desired product. In the last 10+ years, flavor applications laboratories have been the fastest growing sections of most flavor companies. At one time, this group did not exist, and individual flavorists created a flavor, put it in a suitable base and they (perhaps with the salesman) decided it was good enough to submit to the customer or not. Things have changed greatly in recent years.

The trends in the food industry have been to pass much of the product development work back to suppliers in order to save money. Food companies often expect a flavor company not only to determine how to flavor their product, but to do additional product (not just flavor) formulation work, as well as market studies. It is interesting that one flavor company is planning on hiring marketing staff, and packaging materials and design expertise to their staff to service their customers. The widespread formation of strategic relationships (preferred suppliers, etc.) has coupled flavor companies and their customers much closer together than ever. Thus, flavor companies must have both the people and food processing equipment to make the range of products they wish to supply flavors to. In effect, flavor companies must become small scale food processors.

## 14.2 THE PEOPLE

Flavor applications laboratories tend to be staffed by Food Science graduates. They generally will have completed a four-year degree at a university offering this program. They know food systems (food chemistry) and how foods are processed (food engineering/processing). Since few Food Science programs offer courses in flavor technology, these graduates typically lack a thorough knowledge of flavors and

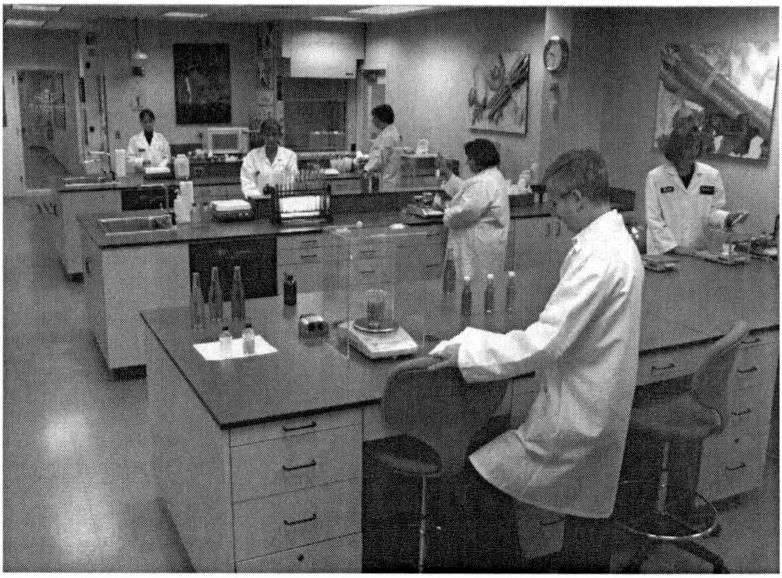

**FIGURE 14.1** Basic flavor applications laboratory facility. (Courtesy of Robertet Flavors, Inc., Piscataway, NJ. With permission.)

flavoring systems, but this is learned on the job. Individuals often come from segments of the food industry that a given flavor company serves. This provides substantial product knowledge for the company and enhances customer:flavor company interactions.

## 14.3 THE LABORATORY

The laboratories vary greatly in staffing expertise and food processing equipment, depending upon the food industry they wish to service. They will all have a laboratory-like space for the formulation of the product (Figure 14.1). This means ingredient storage, weighing facilities, and the appropriate area for packaging, and formal and informal evaluation of the ultimate product. They must also have a food processing facility. If they are selling flavors to a culinary market, they would have food service equipment such as stoves and ovens (a variety of them). If they wish to sell flavorings to an aseptic dairy, they will need to have an aseptic milk processing facility. Thus, one can appreciate that the needs vary greatly across the industry. It is noteworthy that the industry generally must work with specialized, small-scale equipment that is very expensive.

## 14.4 SPECIFIC FLAVORING APPLICATIONS

If we consider market data, the flavor industry has four large food product areas (shown in Table 14.1 [5]). Beverages (all beverage types) are the largest segment of the business followed by confectionery products. It is no surprise that dairy follows,

**TABLE 14.1**
**Distribution of Flavoring Sales Across the Industry by Major Product Area**

| | |
|---|---|
| Beverages | 31.5% |
| Confectionery | 20.0% |
| Dairy, fats, oils | 15.0% |
| Culinary products | 14.5% |
| Oral hygiene | 8.0% |

*Source:* From *Perfumer and Flavorist*, 1999 Worldwide flavor market, http://www.perfumerflavorist.com. With permission.

but it is of interest that culinary products have such a prominent position. The recent growth in the food service, restaurant, and institutional businesses likely make this segment of the market even larger than indicated in this figure (1999 data).

The sections that follow consider the flavoring of some of the major food flavor applications areas. Much of this information still comes from Henry Heath's writing in this subject area [1]. Additional information comes from Leora Hatchwell (Chicago, IL), essentially her teaching at short courses once given at the University of Minnesota. Substantial credit must be given to these individuals.

### 14.4.1 Culinary and Meat Products

Culinary products embrace all types of soups, bouillons, sauces, gravies, marinades, and other convenience savory items supplied to the food industry. Historically there was little need for such products since these foods were prepared at home using basic food ingredients. However, the fact that most households have both adults working outside the home has changed our eating habits. There is no time (or desire) to prepare meals at home from scratch. The large scale use of convenience foods (meals for home preparation, takeout meals from supermarkets, restaurants, delis or fast food outlets, as well as eat-in meals at restaurants) has resulted in culinary related food flavorings experiencing very rapid and continuing growth in the industry. It is interesting that today the culinary products group in a larger flavor company will generally include several commercial chefs. The chefs are the equivalent of a flavorist in other flavoring areas. A chef leads the creation and he/she, or another group, will turn their creations into commercial products. At one time, this was a need serviced by a few small, specialized companies, but today nearly all of the major flavor companies have groups focused on this market.

In addition to the rapid growth of the food service industry, we also find a vast array of seasonings, gravies, marinades, and sauces available to the consumer for home use (e.g., grill aids, hamburger seasonings, and mesquite marinades). Thus, these types of flavorings not only find a commercial market but a home market as well. The flavoring of meat and meat products is also included under this heading due to the similarity in flavorings used. Very commonly, the culinary products discussed above contain, or more likely are based on, meat flavorings.

#### 14.4.1.1 Soups and Stocks

The first commercial prepared soups were produced in 1810 and sold in sterilized glass jars. They were not prepared for home use but for voyages where the preparation of food would be problematic. Bouillons date from the time of meat extracts (1861) and on a truly commercial scale with the development of bouillon cubes based on Hydrolyzed Vegetable Protein (HVPs) (1886). HVPs bring to mind the names of Julius Maggi and Carl Knorr. Dry soups are more recent, dating from the 1930s [6]. Certainly today's products have evolved greatly from these early renditions to improve quality, variety, and nutritional content. The most recent addition of frozen soups has added a very high quality product to the market.

Manufactured soups occur in three forms: (i) canned, either single strength ready to eat or concentrates, (ii) dry mix for reconstitution as required with or without the need for cooking, and (iii) frozen. Each of these calls for a different approach to flavoring so as to achieve quality and consistency in the end product.

##### *14.4.1.1.1 Canned Soups*

Canned soups are generally composed of some combination of vegetables, meats (and meat-like flavorings that bring meat character plus flavor enhancers), starch (or modified starch), fat, seasonings (salt, herbs, spices, and derivatives thereof), and noodles or rice (of some type). They may be single strength or concentrated (double strength). The concentrated products are usually formulated at higher solids levels as opposed to undergoing any concentration process. The ingredients are blended and then preheated before packaging (assuming a batch process) for sterilization. Most soups are still retorted (metal or polymer cans, or pouches), but continuous aseptic processing is possible in some cases. Retorting exposes the product (and flavoring) to a very detrimental heat treatment. Aseptic processing lessens the heat treatment, but soups with particulates are problematic in processing. Aseptic processing using direct steam injection requires a vacuum cooling step that again is extremely detrimental to flavor (stripping of flavor). Thus, any flavorings used in canned soups must be designed through formulation and delivery system (encapsulation) to be as heat stable as possible.

Since these products are liquid, it is most common to use liquid flavorings and seasonings, although dispersed spices blended with other dry ingredients are quite satisfactory. Controlled release (heat stable) flavorings add cost and do not perform very well under the harsh processing conditions. The flavoring materials used in soups (and culinary products in general) have been discussed in detail in other chapters of this text (Chapter 6, Chapter 9, and Chapter 11) and will not be discussed in any additional detail here or subsequent sections of this chapter.

##### *14.4.1.1.2 Dry Soups*

Dehydrated soups may be either instant or require cooking. Instant soups are formulated using thickeners that perform at lower temperatures (e.g., pregelatinized starch), and problematic ingredients are agglomerated to facilitate dissolution. Regular dry soups do not have these requirements. The ingredients used are similar to those used in the formulation of liquid soups but of course are in the dry form. A dry soup is typically made by dry blending the desired ingredients and packaging.

In terms of flavorings, dry process flavorings (containing flavor enhancers and natural or synthetic top notes), dry vegetables, ground spices, and herbs as well as encapsulated or plated spices serve as the foundation of the flavor systems. Plated spice oils are not particularly stable and may be a problem in terms of oxidation or evaporative losses. The use of encapsulated essential oils offers greater stability, but they do not have very much intrinsic smell in the dry form and may be misjudged by the consumer in consequence.

##### *14.4.1.1.3 Frozen Soups*

These products offer the finest quality to the consumer. Due to their price, they are generally made from fresh vegetables and premium ingredients. Many of the same types of flavorings are used in the formulation of these products as the canned or dry soups. The low processing temperatures and frozen storage conditions result in a high quality product.

##### *14.4.1.1.4 Stocks or Bouillons*

Stocks or bouillons are basically made in the same manner as the soups. The primary difference is that the stocks and bouillons are designed to be used as an ingredient of a further preparation while a soup is typically the final product. They may be sold as liquids, cubes, powders, or pastes. Most commonly these products are simpler in composition than soups and are designed to carry primarily a meaty character as opposed to a complete product (such as soup). Thus, there are no vegetables, noodles, rice, or other components in this product. Flavor development and overall system have been discussed previously in Chapter 9 and Chapter 11.

#### **14.4.1.2 Sauces, Seasonings, and Marinades**

Sauces serve the function of adding lubrication and flavor to some item of the meal typically lacking in these attributes such as meat, potatoes, pasta, or rice. They are based either on meats (e.g., gravies) or vegetables (e.g., tomato sauce). At one time, there was little need for meat sauces for they were a byproduct of meat preparation. Today the meat portion is often precooked, and thus there is little opportunity to prepare gravy. In other cases, the use of a prepared sauce is a time saver both at home and in commercial operations.

Sauces are available in both liquid and dry forms. The liquid forms are generally shelf stable (sterilized) but can be frozen. The frozen sauces are typically a part of a frozen meal as opposed to simply selling the sauce in that form. Most sauces have been thermally processed in creation, and thus the additional mild thermal processing involved in making a shelf stable form does little damage to the product. Thus, the market is dominated by shelf stable product forms.

Sauces use many of the same basic ingredients as soups: process flavors (meaty notes), comminuted vegetables or extracts, thickening agents (e.g., starches, flour, gelatin, and hydrocolloides), ground spices or derivatives thereof (e.g., essential oils or oleoresins), salt, fat, sugar, and some specialty ingredients associated with a particular sauce (e.g., wine, sherry, etc.). One could consider them to be thick soups.

**Liquid sauces** — As noted earlier, liquid sauces are typically sold in shelf stable forms. They generally have a low pH that minimizes the heat treatment required for microbial stability. They may be lower fat versions that are thickened by starches and/or hydrocolloides, or high fat products (e.g., hollandaise-based sauces) that obtain their viscosity from the fat emulsion. Each type of sauce has its own unique formulation requirements for retaining its physical properties during both thermal processing and storage.

**Dry sauces** — This sauce form is produced using similar ingredients as the dry soups. The larger proportion of thickening agents may dictate the use of a bulking agent (maltodextrin, lactose, etc.) to facilitate reconstitution. The issue of physical stability during storage is irrelevant and is a concern only at the time of preparation. Typically dry forms of the desired ingredients are dry mixed and packaged. It is a simple process.

#### 14.4.1.3 Meat Products

The prepared meat products include fresh, semi-dried, dried, fermented, deep-frozen, and canned meats that may be eaten directly either cold, after reheating, or after some further preparation or cooking at home or in a food service operation. The opportunities for using added flavorings are almost limitless, but the diversity of product types, many of which are now traditional, imposes considerable constraints on the nature of any added seasoning or flavoring materials. In addition, in many developed countries the meat industry is controlled by legislation that is either separate from, or supplements, food legislation so that special regulations may govern the nature and quality of any additives.

In savory foods the prime taste adjunct is salt, and its level influences the shelf-life of many products and the total flavor profile as it affects product palatability. Additional flavoring effects may be achieved by using seasonings (marinades or topical powders). A *seasoning* is an inclusive term applied to any ingredient which by itself or in combination adds flavor to a meat product [1]. Most seasonings are blends of natural herbs and spices, or products thereof, often admixed with other flavoring ingredients such as MSG, the ribonucleotides, process flavors, hydrolyzed vegetable protein, and/or yeast autolysates, all of which may enhance or impart characteristic meaty notes to the product.

The basic technology of meat processing is complex and often specific to a country, a manufacturer, or, indeed, to an individual product line. Whatever the process, the following factors must be taken into account when selecting an appropriate seasoning:

1. The nature of the raw materials used, particularly the lean:fat:water ratio
2. The nature of any pretreatment, particularly the use of curing agents
3. The stage and method of incorporating the seasoning
4. The degree and nature of any comminution stage

5. Postmixing treatment, particularly that involving heat (i.e., cooking, smoking, drying or retorting)
6. The temperature and times involved at any stage, particularly in an open system
7. The nature of any added preservatives, particularly sulfur dioxide
8. Methods of packaging, particularly if this involves exposing the product directly to a vacuum
9. Postpackaging handling and storage, particularly refrigeration or deep freezing

A precise knowledge of the end product and its method of manufacture is required in order to develop a flavoring suitable for a given meat product.

The principal aim in incorporating seasonings into meat products is to provide added flavoring notes that will enhance the natural meat flavors developed during cooking, maybe modifying them to suit individual tastes but not overwhelming them. Traditionally, seasonings have been prepared from herbs and spices ground to varying degrees of fineness. For most purposes such products should pass through United States standard sieves of No. 20–No. 60 mesh.

Increasingly, the industry has adopted seasonings based on spice extractives — the dispersed or soluble spices, which give a standardized flavoring effect. Such commercially compounded seasonings are generally in the form of dry powders comprising not only the appropriate herbs and spices but also other permitted additives that may include flavor enhancers, hydrolyzed vegetable protein, yeast extracts, salt, phosphates, and colorants. They are frequently supplied preblended in unit packs to facilitate addition to a single processing bulk at the chopping stage. To overcome volatile losses from such products, several spice houses commonly offer seasonings based on spice extractive encapsulated by spray drying. These have a considerably longer shelf life and in many uses are preferred.

With an increase in automated processing (computer-controlled dosing of ingredients), the use of seasonings in the form of liquid emulsions is rapidly gaining in popularity. These have all the advantages of the dry processed products but can be accurately metered in-line and almost instantly and uniformly dispersed into the prepared meat emulsion. They are more concentrated than powdered seasonings and many are designed for use at 2 g/kg of meat mix. Precise usage level should be established with the supplier. The presence or absence of added preservatives will dictate whether or not the product is supplied in a single unit or multi-dose pack.

### 14.4.2 Baked Goods and Bakery Products

This product group includes such widely different products as bread and rolls, sweet yeast dough products, biscuits, cookies and crackers, pies and pastries, cakes, and breakfast cereals. These products are generally composed of some combination of: flour (9 to 12% protein with the remainder being primarily carbohydrate), liquid (eggs, milk, or water), leavening (yeast or chemical — pH), shortening (fat), sugar (many choices), flavoring (optional), and salt (0.2 to 1%). Matz [4] is recommended for more detailed discussion of this product area.

#### 14.4.2.1 Problems in Flavoring Baked Goods

When one designs a flavoring for these products, one has to consider the potential for flavor/ingredient interactions as well as the effects of the baking process (see Chapter 6). Proteins and starch offer substantial opportunity for flavor binding both in the baking process as well as later during storage (staling). Losses during baking are usually determined experimentally and an additional quantity of flavor is added or perhaps the flavor formulation itself is adjusted to give the desired flavor balance and level in the end product. The dosage level of any added flavoring should be adjusted carefully as over-flavored baked goods are much less attractive than under flavored. Baking losses are reduced by using a lipid-soluble flavoring (added to the shortening in the recipe if possible) since volatiles will have reduced vapor pressure in an oil solvent providing better retention during this process. If the flavoring contains essential oils, they are often folded oils since they again offer better heat stability. Chemically leavened products have a higher pH (than yeast leavened products), which may also be detrimental to some flavorings, e.g., almond, butterscotch, caraway, cinnamon, ginger, honey, and vanilla. Also, encapsulated/controlled release flavor forms may be used as is discussed later in an example.

#### 14.4.2.2 Flavoring Baked Goods

Within this product group, flavorings may be incorporated in one of four ways: (a) mixing into the dough or batter prior to baking; (b) spraying onto the surface of the product as it emerges from the oven; (c) dusting on to the surface after cooking and oiling; and (d) introduction into the cooked product as a cream filling, glaze, or coating [1,7].

Flavors added to the dough before baking generally are in the form of ground or whole spices, essential oils, cheeses, or an encapsulated/controlled release form. In yeast leavened products, care must be taken to not inhibit yeast growth during the proofing stage. Many flavorings are good microbial inhibitors, and loaf volume can suffer from the addition of some flavorings. Graf and Soper [8] have advocated the use of a coacervated flavor (e.g., garlic oil) to minimize this problem since the flavoring would not be released until later in the baking process.

For some products, a flavoring may be dispersed or slurried in an oil (liquid and dry forms, respectively) and then sprayed onto the finished good. This works only for thin goods and has the negative effect of adding oil to the product (label declaration). The flavoring may also take the form of a dry seasoning and be dusted onto the product. Again this is best suited for thin goods.

It is fairly common to flavor baked goods with a fillings or glazes. The common fillings are presented in Table 14.2. Glazes typically are simple mixtures of liquid and sugar but may be more complex mixtures, including fat, stabilizers, and foaming agents. Within these flavor application methods, encapsulated/controlled release flavorings, plated flavorings, baker's emulsions (previously discussed), liquids, and paste forms may be used.

**TABLE 14.2**
**Fillings Used to Flavor Baked Goods**

| Filling Type | Ingredients | Comment |
|---|---|---|
| Dairy and synthetic whipped creams | Base: 40% cream or 18–20% cream (half and half) | May get destabilization of cream foam from citrus fruit pastes. Use citrus extracts to advantage |
| Butter cream | a. Beaten mix of butter and sugar<br>b. Blend gelatin (marshmallow) and butter<br>c. Fondant and NFDM plus butter | Sweet flavors most appropriate: gelatin or protein based |
| Fondant | Dry fondant base plus water and corn syrup | |
| Custard | a. Traditional starch-based custard mix is boiled and then cooled to 60–70°C prior to adding flavor or b. Citrus pectin-based made as starch-based product | |

*Source:* From Heath, H.B., *Flavor Technology: Profiles, Products, Applications*, AVI Pub. Co.: Westport, CT, 1978. With permission.

### 14.4.2.3 Heat Resistant Flavorings

Heat resistant dry flavors made by multistage encapsulation (typically a fat coating is applied to some water soluble encapsulated flavoring) are commercially available and are very effective under these harsh conditions [9]. The secondary capsule, being water insoluble, ensures that the flavoring does not dissolve during preparation of the dough and melts to expose the primary capsule only during the later stages of the baking cycle. At this stage the water content of the product is too low to result in a full release of flavor, which only occurs slowly during the baking process.

Heiderich and Reineccius [9] have provided some data on the use of controlled release flavorings in baked cookies. They added a series of esters (to simulate a flavoring) to sugar cookie dough in various forms (in alcohol, spray dried in both gum acacia and chemically modified starch, and then lipid coated versions of each of the spray-dried forms) and then baked the cookies in a conventional oven. After baking, they analyzed the cookies for residual esters (Figure 14.2). As one might expect, the esters dissolved in alcohol (simulates a typical water soluble flavoring) were retained the poorest. The esters spray dried in gum acacia and a chemically modified starch fared better, with gum acacia being the better of the two. The reason for gum acacia retaining more esters during baking is not known. When the spray-dried powders were fat coated (sample designations "CR" in Figure 14.2), retentions improved again. The fat coating would retard dissolution of the spray-dried flavoring until later in the baking process thereby affording protection against evaporation.

The decision regarding the use of controlled release encapsulated flavorings in an application is largely dictated by cost considerations. If it is more economical to increase the usage level of a liquid flavoring (or a simple encapsulated flavoring)

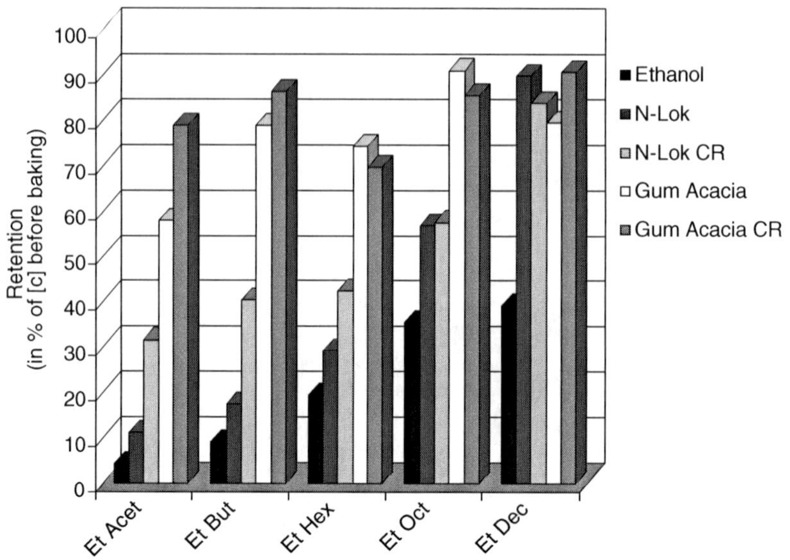


**FIGURE 14.2** The effect of flavor delivery system on the retention of esters added to cookie dough after baking it into cookies. (Ethanol: esters added in ethanol solvent; Mod starch: esters spray dried in a modified starch and then added to cookie dough; Mod starch: CR — esters spray dried in a modified starch and then coated with a high melt triglyceride before adding to cookie dough; Gum acacia: same as Mod starch but gum acacia was used as flavor carrier.) (From Heiderich, S. and G. Reineccius, *Perfum. Flavorist*, 26, 6, 2001. With permission.)

than to use a sophisticated controlled release encapsulated flavoring, the less expensive flavoring will be used. The only situation which would potentially justify the use of the more expensive flavor form is if some unique property could be given to the product that would justify the cost. For example, perhaps the use of a spray-dried, fat-coated flavoring could deliver the very volatile aroma constituents (e.g., fresh notes), which may not be possible in any other flavor form. This may give the baked product a value that would justify the added cost. Otherwise, it is more cost effective to simply increase the dosage level of a cheaper flavoring.

There are generally few problems with the application of flavorings after baking, assuming one can obtain the desired initial flavor profile in the baked good. The creation of suitable flavorings to apply to doughs is problematic for the reasons discussed. The short shelf life of baked goods generally limits any shelf life issues other than moisture loss or microbial spoilage.

### 14.4.3 Snack Foods

The snack food market is very broad and, depending upon the classification, may be comprised of the products listed in Table 14.3. However, for our purposes, we will discuss only savory snack foods as highlighted in this table. For more detail in this product area, Matz [10] is recommended.

**TABLE 14.3**
**Snack Foods Categories and Examples of Products in Each Category**

| Category | Products |
|---|---|
| Hot Snacks | Minipizzas; Pizza Baguettes; etc. |
| | Toasts au Gratin; Spring Rolls; filled croissants |
| Cold Snacks | |
| Milk and dairy | Yogurts, plain or fruit; Minicheese cubes |
| Products | Frozen ice cream novelties |
| Bakery products | Cake bars; Mini Tarts; Cookies; Biscuits; etc. |
| Bars | Granola/muesli bars; Energy bars; Breakfast cereal bars; Chocolate bars; Mini-break bars |
| Confectionery products | Candies |
| Fruits and vegetables | Fresh carrot sticks; celery sticks; apples; oranges; Dry fruit rolls |
| Savory products | Chips and sticks; Extruded products; Crackers; Pretzels and salt sticks; Nuts and nut mixtures |

*Source:* Tettweiler, P., Snack foods worldwide, *Food Technol.*, Feb., p. 58, 1991.

### 14.4.3.1 Problems in Flavoring Snack Foods

The two primary problems in flavoring snack foods are the initial interactions of the flavoring with the snack ingredients and then flavor loss during processing (extrusion, baking, or frying). In terms of interactions, there are numerous interactions that may occur between the flavorings and food ingredients, complicating flavor formulation. The waning trend towards lower fat products and the current low carb trend make one painfully aware of how important these interactions can be and the implications on both flavor formation during processing as well as the performance of added flavorings. While flavor loss from baked goods was previously discussed, snacks generally receive a greater heat treatment than baked goods, so flavor loss during thermal treatment is exacerbated.

Heath [1] has provided a tabulation of typical heat treatments given snacks (Table 14.4). It should be obvious from this table that flavor losses are extremely problematic for most snack foods especially considering that snacks are typically thin products, which greatly favors flavor loss.

### 14.4.3.2 Snack Flavorings

Kuntz [12] has noted that the flavorings applied to snack foods typically comprise 6 to 12% the weight of the snack. As noted earlier in this text, the flavorings applied to snacks are typically called seasonings in the industry. The seasoning may contain:

1. A carrier (or diluent) that can serve some function as a binder and free flowing agent
2. Spices (whole or ground) and/or herbs (often only for appearance)
3. Dairy products (fermented products)

**TABLE 14.4**
**Thermal Processing of Snack Foods Produced by Various Methods**

| Thermal Process | Temperature | Time |
|---|---|---|
| Baking | 150–260°C | 10–30 min |
| Deep fat frying | 162–182°C | 2–3 min |
| Extrusion | | |
| Low pressure | <100°C | Dependent on product Followed by drying |
| High pressure | 110–150°C | |

*Source:* From Heath, H.B., *Flavor Technology: Profiles, Products, Applications*, AVI Pub. Co.: Westport, CT, 1978. With permission.

4. Artificial or natural flavors/extracts
6. Process flavors
7. Additional adjuncts (e.g., antioxidants, binders and/or colorants)
8. Flavor enhancers — MSG, 5′-nucleotides
9. Yeast extracts and/or HVPs

Finished seasonings are created by simply dry blending, or alternatively, slurring and then spray drying some combination of these ingredients [13]. The latter approach produces flavorings which offer better uniformity in flavor character and application on the snack than the dry blended product since the flavoring cannot fractionate (separate) either during application or subsequent shipping. The negative is the cost since adding water to make a slurry and then redrying is more costly than simply dry blending ingredients.

#### 14.4.3.3 Flavoring Materials

**Salt** — It is generally considered that salt application can either be on the surface or added internally to the product; however, the placement will make a difference in perception and product flavor. van Osnabrugge [7] has suggested the use of fine salt to give greater salt perception (faster dissolution and greater surface area prior to dissolution). He has also suggested topically applying salt, i.e., put the salt where it will make the quickest and greatest sensory impact — on the product surface. However, omitting salt from a dough and putting it on the surface can alter flavor formation in that dough during subsequent thermal processing (it alters the course of the Maillard reaction). If salt is applied to the surface, it must be a fine, crystal pure, vacuum dried salt to facilitate adherence to the surface.

**Herbs, spices, and pieces of vegetables (e.g., onions, caraway seed, sesame seed, etc.)** — Herbs or pieces of vegetable material are generally used for

**TABLE 14.5**
**Sample Formulation for a Taco Seasoning**

| Component | % | Purpose | Component | % | Purpose |
|---|---|---|---|---|---|
| Yeast | 30 | Binder | Salt | 20 | Taste |
| MSG | 5 | Taste enhancer | Corn flour | 16 | Binder |
| Onion powder | 10 | Flavor | Garlic powder | 5 | Flavor |
| Cumin; ground | 3 | Flavor | Oregano; ground | 2.5 | Flavor |
| Caramel color | 3 | Color | Chili pepper | 3 | Flavor/color |
| Oleo paprika | 0.5 | Color | Citric acid | 1 | Taste |
| Anticaking agent | 1 | Anticaking | | | |

*Source:* From Heath, H.B., *Flavor Technology: Profiles, Products, Applications*, AVI Pub. Co.: Westport, CT, 1978. With permission.

appearance (sesame seeds or pieces of onions or green herbs). Spices may or may not be used for flavoring depending on the spice that is used. A ground spice such as turmeric or paprika may be used solely to provide a natural color to the product. The seasoning formula provided in Table 14.5 illustrates the use of spices and spice derivatives.

**Seasonings based on dairy products** — Cheeses and sour creams are commonly used in snack seasonings. Natural cheeses are often used for label or familiarity purposes. Enzyme modified cheeses or creams are typically used to carry the flavor due to both flavor strength and cost issues. The dairy character of the seasoning may be fortified with other natural flavorings, e.g., lactic acid, butyric acid, diacetyl, etc. Examples of dairy-based snack seasonings are presented in Table 14.6 and Table 14.7.

**Essential oils and oleoresins** — These flavor forms offer the advantages of being natural flavoring materials, carrying a large amount of flavor, can be blended to offer uniformity, and are readily released on eating (as opposed to the whole spice). Their main problem is a lack of stability to heating.

**Natural and artificial compounded flavorings (including smoke flavorings)** — These flavoring materials are used to carry or reinforce a flavoring. They seldom carry the entire flavor profile. An example of a seasoning using these components is presented in Table 14.8. Pszczola [14] has provided an overview of smoke flavorings available on the market.

**Process flavors** — Process flavors are generally used to create meaty flavors (e.g., chicken) for snacks. As discussed in an earlier chapter (Chapter 9), they are created based on process flavor technology from amino acid/sugar reactions plus other ingredients such as flavor enhancers (MSG and/or 5′-nucleotides), antioxidants, emulsifiers, flavoring (e.g., smoke), and anticaking agents (since they will be sold as powders).

**Flavor enhancers** — There is very broad use of MSG in snack seasonings. MSG is a very effective flavor enhancer of savory foods and thus usage

**TABLE 14.6**
**Sample Formulation for a Sour Cream and Onion Seasoning**

| Component | % | Component | % |
|---|---|---|---|
| Nonfat milk solids | 10 | Sour cream solids | 1 |
| Whey solids | 6 | Salt | 17 |
| Onion powder | 8 | Toasted onion powder | 13 |
| MSG | 4 | Parsley flakes | 2 |
| Flavor (nat/art) | 1 | HVP | 5 |
| Sugar/dextrose | 15 | Celery seed; ground | 1 |
| Paprika; ground | 1 | Garlic powder | 1 |

*Source:* From Heath, H.B., *Flavor Technology: Profiles, Products, Applications*, AVI Pub. Co.: Westport, CT, 1978. With permission.

**TABLE 14.7**
**Sample Formulation for a Nacho Seasoning**

| Component | % | Purpose | Component | % | Purpose |
|---|---|---|---|---|---|
| Romano cheese | 16 | Label, flavor, appearance | Cheddar cheese | 7 | Label, appearance |
| Parmesan cheese | 5 | Label, flavor, appearance | Whey solids | 12.4 | Binder |
| Buttermilk solids | 4 | Label | Salt; fine | 18 | Taste |
| MSG | 5 | Taste enhancer | Dextrose/sugar | 5 | Taste |
| Tomato powder | 5 | Appearance | Ribotide | 0.1 | Taste enhancer |
| Onion powder | 5 | Flavor | Garlic powder | 1 | Flavor |
| White pepper; ground | 0.5 | | Wheat flour | 15 | Binder |
| Anticaking agent | 1 | | Color | — | Appearance |

*Source:* From Heath, H.B., *Flavor Technology: Profiles, Products, Applications*, AVI Pub. Co.: Westport, CT, 1978. With permission.

levels in the seasoning often reach 5%. It is also common to see seasonings containing the 5′-nucleotides in combination with MSG due to their synergistic effects.

**Yeast extracts/HVP** — Normally one uses bakers yeasts or extracts thereof in seasonings. They serve both as a flavorant and bulking agent. HVPs are typically used in meaty flavorings. These materials will add some flavor of their own typically considered savory or meaty in character. They also contribute flavor enhancers — MSG (HVP and yeast extracts) and 5′-nucleotides (yeast extracts).

**TABLE 14.8**
**Sample Formulation for a Barbecue Seasoning**

| Component | % | Purpose | Component | % | Purpose |
|---|---|---|---|---|---|
| Salt; fine | 20 | Taste | Filler (flour or soy grits) | 19.5 | Binder |
| Sugar | 18 | Taste | Torula yeast | 15 | Binder |
| Citric acid | 1 | Taste | Onion powder | 4 | Flavor |
| Garlic powder | 1 | Flavor | Red pepper | 1 | Taste |
| Mustard seed; ground | 10 | Flavor | Smoke flavor | 1 | Flavor |
| Tomato powder | 3 | Color | Oleo paprika | 0.5 | Color, flavor |
| Anticaking agent | 1 | Anticaking | Chili powder | 4 | Flavor, color |
| MSG | 2 | Taste enhancer | | | |

*Source:* From Heath, H.B., *Flavor Technology: Profiles, Products, Applications*, AVI Pub. Co.: Westport, CT, 1978. With permission.

## 14.4.3.4 Means of Applying Flavorings

Snack foods may be flavored by either incorporating a flavoring into the dough prior to thermal processing or topically applying the seasoning after heat treatment. Due to the severe heat treatment given most snack foods, topical seasoning is most common.

### *14.4.3.4.1 Topical Flavoring*

Topical flavorings may be applied either in the liquid or the dry form. Liquid, oil-soluble flavorings are dissolved in a carrier such as vegetable oil (may be partially hydrogenated) and then sprayed onto the surface of the hot snacks (e.g., savory crackers) as they emerge from the baking process. The surface application of oil-based flavorings reduces the dry sensation in the mouth associated with baked (as opposed to fried) snacks. A disadvantage of liquid flavorings is that they offer no protection against evaporative losses during storage or degradation due oxidation or light induced reactions.

Dry flavorings are applied directly to fried or baked snacks after spraying with hot oil or a gum acacia solution. The oil (or gum solution) serves as an adhesive for the dry flavoring. This is done in a seasoning drum [15]. The use of dry seasonings offers ease of application (economics), control of flavor separate from snack (produce one base snack and flavor as desired), and provides an immediate impact (the flavor is on the surface). While this method of seasoning is very common, it has several disadvantages including: (1) anticaking agents required in dry seasonings result in very dusty plant conditions (also, the anticaking agent will potentially add a mouth drying note to the snack); (2) the presence of hygroscopic ingredients such as dehydrated onion, and hydrolyzed vegetable protein may pick up sufficient moisture that the seasoning forms balls in application that creates difficulties in obtaining a uniform application; (3) the seasoning may fractionate in the hopper resulting in nonuniform flavor character and application rate; (4) they are messy for consumer and processor; (5) there is no flavor in the snack, only on the surface; (6) under the best of conditions, uniform application is problematic; and (7) high flavor usage levels are required to impart the desired sensory properties.

#### *14.4.3.4.2 Internal Flavoring*

It may be desirable to put an internal flavor into a snack either to provide a boost to the flavor or to flavor the product more uniformly. Unfortunately, this is generally not economical (major losses during processing) and often is impossible due to flavor degradation associated with the high heat treatment. It is common to lose 98% of the more volatile flavor components during an extrusion process. Heat stable flavorings such as smoke, meat, sweet brown, and some spices may fare better and may have some application [16]. Also, there is no issue with the stability of capsaicin or piperine flavoring in this process. Williams [16] mentions other issues with internal flavoring as well. He notes that loss of the flavoring to the frying oil limits the use of that fryer (oil) to only that flavor (this has manufacturing implications) and some essential oils are detrimental to fryer oil stability.

## 14.4.4 Sugar-Based Confectionery Products and Chewing Gum

Other than chocolate and chewing gum, most confectionery products are based on sugar (sucrose), and the addition of other ingredients differentiates the products. The additional ingredients may include gums or starch, marshmallow, nuts, cocoa powder, pectin, or fat, for example. While the thermal breakdown of sugars (and additional ingredients) during manufacture gives some flavor, these products typically have high flavor usage levels.

### 14.4.4.1 Hard Candies

Hard candies are essentially made by adding water to sugar to dissolve it with heat and then removing the water in a manner that produces a clear, amorphous mass. The water is removed by boiling in the open air, under vacuum, with high pressure steam, or in a thin film evaporator. The sugar mass typically is cooked to temperatures of 110 to 130°C and this yields products ranging in moisture content from 1 to 4%. Due to the very high temperatures, it is necessary to add the flavoring as late in the process and at the lowest possible temperature to minimize flavor loss. The data presented in Figure 14.3 show the effect of temperature of flavor addition on flavor retention. As anticipated, the lower the temperature, the better the retention of flavor. The lower limit is the viscosity of the molten sugar mass since as temperature decreases, the viscosity increases greatly thereby complicating the flavor addition process. One will notice in the figure that flavor losses reached ca. 95% for limonene — this is immense.

Flavor losses are also influenced by the flavor solvent. Volatile solvents such as ethanol rapidly flash off from the hot mass, carrying flavor along with it. Thus, less volatile solvents, e.g., propylene glycol, are preferred.

Typically the acid component of the flavor is also added late in the process. The issue here is not acid loss but acid induced hydrolysis of the sucrose. Acid levels generally range from 0.5 to 2.5%, and citric, malic, or tartaric acids are most com-

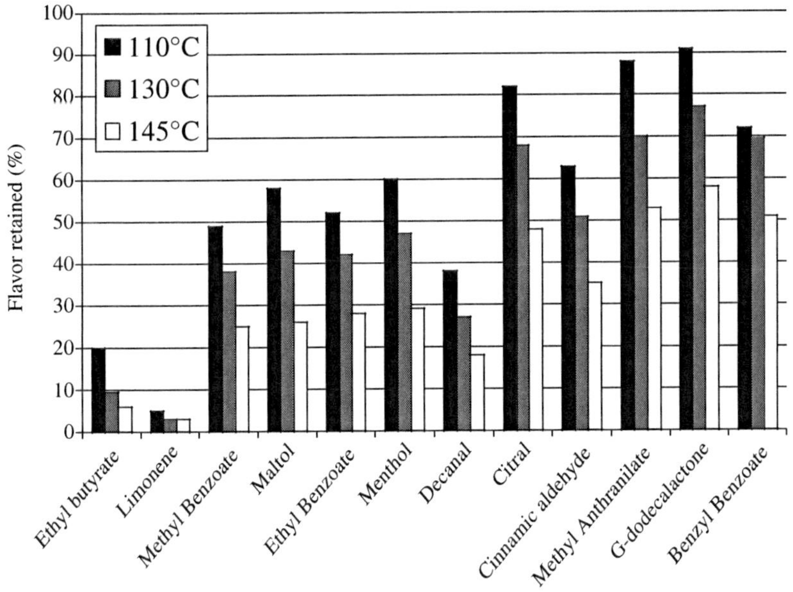


**FIGURE 14.3** The influence of addition temperature on the retention of flavorings during the manufacture of hard candies.

monly used. The acid chosen influences the sensory character of the product: malic enhances cherry, tartaric enhances grape, and citric enhances citrus. Tartaric acid is very expensive, which limits its use.

The primary flavoring issue is flavor loss (function of flavor formulation, solvent choice, product composition, and temperature). However, it is well known that methyl anthranilate (grape flavor) will undergo browning during manufacture, the product may be cloudy (wrong solvent [17] or too much solvent), oxidation may occur (particularly troublesome for citrus flavored candies), or the product may be tainted by the environment.

### 14.4.4.2 Caramels (Toffees)

Caramels are made from some combination of sugar, glucose syrup, milk solids (unsweetened condensed milk, evaporated milk and/or spray-dried milk), whey, and fats (often butter). Additional ingredients may be used such as molasses, salt, malt, licorice, emulsifiers, and other flavorings. Flavor is formed from these base ingredients during the cooking process (120 to 132°C). If flavorings are used, they are added as late in the process as possible to avoid flavor loss due to evaporation (when cooled to 60 to 65°C). The final product contains 8 to 12% moisture.

Dairy flavors may be used to compensate for the use of nondairy fats or simply to enhance the dairy notes. Product differentiation may be achieved through the use of additional flavorings such as rum, butterscotch, and cream flavorings. The primary flavor issues are adequate dispersion in the product during manufacture and fat rancidity during storage.

#### 14.4.4.3 Pressed Tablets

Pressed tablets are made from a wide variety of ingredients and processing conditions. Ingredient choice and processing conditions determine the rate of dissolution thereby determining the rate of flavor release. Products that dissolve more rapidly require less flavor than those that dissolve slowly. Also, products that are small in size (small surface area) require higher flavor dosage than large products since there is less material to dissolve.

Since these products are made with little or no heat, one would think that this eliminates the problem of flavor loss during manufacture, but it does not. The simple addition of a liquid flavoring to a dry blender may result in the loss of up to 50% of the flavor. This loss is reduced through minimizing mixing time, using closed systems, and the use of encapsulated flavorings. Oxidation during storage may also be an issue, depending upon the flavoring type used and expected shelf life.

#### 14.4.4.4 Starch-Deposited Chews

This diverse group of products is grouped together due to their being deposited in a starch mold where they are given form, lose moisture, and cool to solidify. They are based on gelatin, gum acacia, agar, starch, or pectin each producing a slightly different product. Agar, pectin, starch, and low levels of gelatin produce soft gels. Gum acacia and higher levels of gelatin generally produce harder, more chewy gels. These products are generally cooked to a lower temperature than the hard candies or caramels, and thus, have around 20% moisture.

An issue in the flavoring of these products is the addition of acidulant. If the acidulant level is too high, or cooking time/temperature too long/high (when acid is present), the gelling properties of the hydrocolloid are reduced (hydrolysis) and some portion of the sucrose is inverted. These problems limit the amount of acid that may be used in formulation. Typical acid levels used in different chews are listed in Table 14.9. It is reasonable that the acidulants are added as late in the process as possible.

These products generally require high levels of flavoring (180 to 240 ml essential oil/50 kg batch) because they do not give up their flavor readily on eating. They do not dissolve like a tablet or hard candy, and thus, flavor only is given up at the interface created during mastication. Pectin is generally considered to most readily release its flavor followed by agar and gelatin. Flavor loss occurs at all levels of manufacture and storage, and it can be very significant.

Fruit juice concentrates, pastes, or powders are commonly used in these products particularly in pectin or agar-based products. One has to be concerned with the Maillard browning that can occur with the use of juice-based flavorings. Gelatin is often flavored with more intense flavorings since it can be very firm and thus not give up its flavor on eating. Oxidation is common in these products since they are minimally packaged, present no oxygen barrier themselves (they are not in a glassy state), and often a two-year shelf life is expected.

**TABLE 14.9**
**pH and Acid Levels Used in the Manufacture of Different Chews**

| Gelling Agent | Added Acid (expressed as % citric ) | pH of Product |
|---|---|---|
| Agar-agar | 0.2–0.3 | 4.8–5.6 |
| Pectin | 0.5–0.7 | 3.2–3.5 |
| Gelatin | 0.2–0.3 | 4.5–5.0 |
| Starch | 0.2–0.3 | 4.2–5.0 |
| Gum acacia | 0.3–0.4 | 4.2–5.0 |

*Source:* From Heath, H.B., *Flavor Technology: Profiles, Products, Applications*, AVI Pub. Co.: Westport, CT, 1978. With permission.

#### 14.4.4.5 Chewing Gum

The diversity of this product makes it deserving of a book in itself but little is available in the public domain other than patent literature or sections of other book titles (an exception is Suck [18]). The variations in formulation, sweetening system, manufacturing process as well as interactions between all of these variables make this a most complicated product. An example to provide an idea of a typical gum formulation may be: 43% sweetener system (corn syrup 21.14%, glycerin 7.05%, lycasin 18.02%, sorbitol 53.69%); 29% sucrose; 25% gum base; 1+% flavor and a high potency sweetener [19].

The primary flavoring issues come from the long time in the mouth (long-time flavor delivery) and the difficulties of flavor release from the gum base. The gum base is hydrophobic and thus captures most flavoring materials and only slowly gives them up on chewing. As discussed earlier in this text, we now appreciate that in order to support the flavor of a food, we must also continue to deliver sweetness and acidity, not just aroma [20]. The task of developing flavor systems for chewing gums has received more patent applications than the flavoring of any other food product. A search of the U.S. Patent Office turned up 2,418 patents with the keywords *chewing gum* and *flavor.*

A secondary issue has been the interaction of some flavorings with aspartame. Chewing gums provide a good opportunity to undergo such reactions due to their long shelf life and the mobility of substrates within it. Chewing gums depending upon aldehyde-based flavorings (e.g., cinnamic aldehyde, benzaldehyde, or citral) cannot be sweetened with aspartame because the aspartame reacts with these aldehydes, resulting in the loss of both the flavorings and aspartame sweetener. The loss of both flavor and sweetener is illustrated in Figure 14.4 [19,21]. One can see that the flavorings will degrade in the absence of aspartame but much faster in its presence. It is also obvious that aspartame is lost at the same time. Thus, chewing

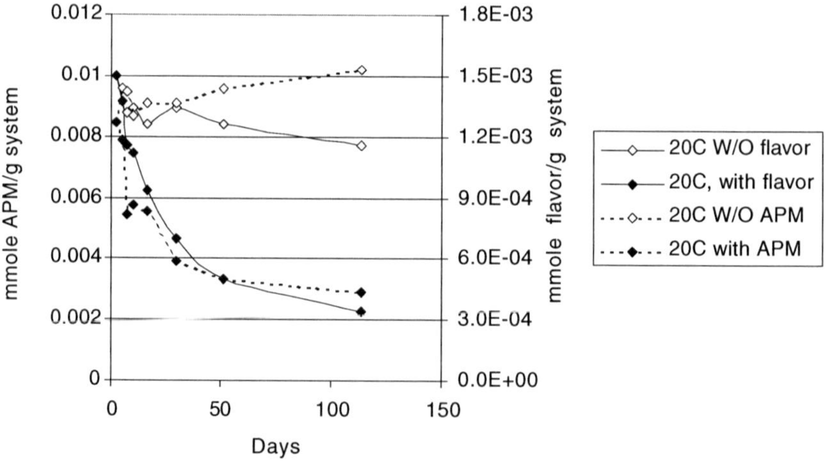


**FIGURE 14.4** The loss of flavor (aldehydes) and aspartame (APM) during storage of chewing gums. (Dashed lines are aldehyde concentrations; solid lines are APM concentrations.) (From Schirle-Keller, J.P., G.A. Reineccius, L.C. Hatchwell, in *Food Flavor Interactions,* R.C. McGorrin, J.V. Leland, Eds., Amer. Chem. Soc., Washington, D.C., 1996. With permission.)

gums with aldehyde-based flavorings are made with other high-potency sweeteners such as acesulfame k or sucralose.

### 14.4.5 Dairy Products

Dairy products represent a major market for flavoring materials. Extremely large volumes of flavored milks, yogurts, and frozen dairy desserts are sold in the U.S. These products have very little inherent flavor, and thus represent a large and growing market for the flavor industry. The market for flavored milks, particularly single serving sizes, has grown greatly in recent years. A few years ago, milks used to be flavored with chocolate, vanilla, or strawberry flavorings. Today one sees a very diverse range in flavors including root beer, coffee, banana, cappuccino, blueberry, coffee, and some fantasy flavors [22]. The ice cream industry has seen a somewhat similar trend away from the old standby flavorings to a host of creative flavor concoctions that have become extremely successful. The area of flavored coffee creamers presents another area of growth, for again new flavorings have been created for those that do not particularly like coffee flavor but enjoy a low calorie, hot beverage. We will consider the flavoring of these mainstream products.

#### 14.4.5.1 Flavored Milks

While milk has an indigenous flavor of its own that is altered by its thermal processing, it is still a very bland product. Its growth in the marketplace depends upon the addition of new and exciting flavors. The products being flavored are primarily low in fat, single service, and extended-life products. Extended-life products receive a higher heat treatment than normal pasteurization but less than shelf stable milks.

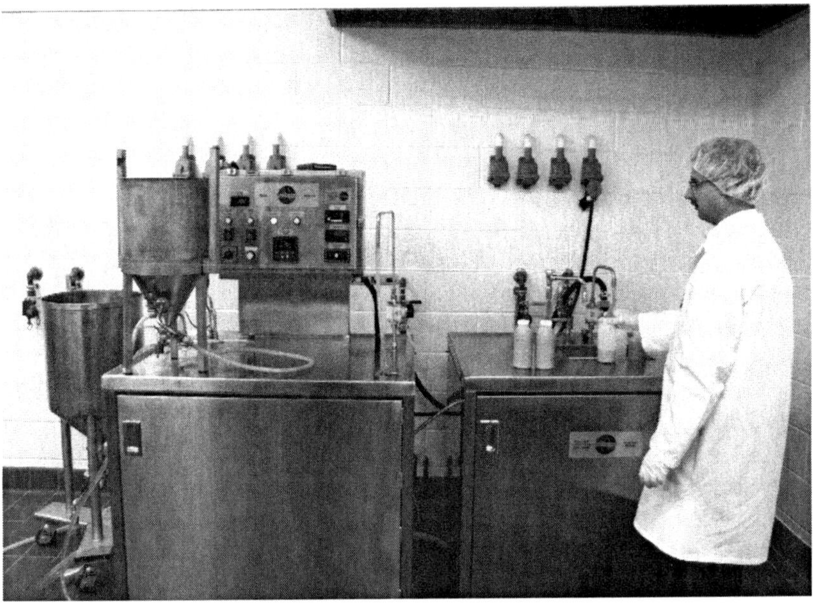

**FIGURE 14.5** Processing of flavored milks in a Microthermics unit. (Courtesy of Robertet Flavors, Inc. With permission.)

Process times/temperatures of about 80°C for 20 seconds or alternatively, higher temperatures for shorter times if the equipment permits are generally used (Figure 14.5). Often they are processed in steam injection systems, which are very detrimental to the flavor system. In steam injection systems, steam is injected into the milk for heating, it is held the desired amount of time and then cooled by evaporative cooling (flashed into a vacuum chamber to remove the added water and cool the product). This cooling (evaporation) step tends to strip much of the added flavors from the milk. Thus, any flavoring added to milk processed in this manner must be robust.

Flavored milks also have added gums, sugar, and vitamins. The gums impart viscosity and stability to the milk system. They influence flavor through their interaction with the flavorings and imparting viscosity. Vitamins may add off-notes associated with vitamin degradation.

Flavorings used in extended milks must be formulated to survive the thermal process and be compatible with the added vitamins and gum systems. Issues in storage are the loss of flavor quality. This is due to some combination of the loss of desirable flavoring (chemical reactions) and the formation of off-flavors probably related to traditional milk spoilage.

### 14.4.5.2 Flavored Yogurts

Whole milk, partially skimmed milk, skim milk, or cream may be used as a base component for yogurt. In addition to the milk base, other ingredients may be used which include: other dairy products (concentrated skim milk, nonfat dry milk, whey, or lactose), sweeteners (glucose or sucrose, high-intensity sweeteners such as aspartame), stabilizers (gelatin, carboxymethyl cellulose, locust bean, guar, alginates, carrageenans, whey protein concentrate), flavors, and fruit preparations (which may include natural and artificial flavoring and color). In manufacture, the milk plus other ingredients are blended together, homogenized and pasteurized (30 min at 85°C, or 10 min at 95°C). When cooled to the desired fermentation temperature, starter cultures are added (*S. thermophilus* and *L. bulgaricus*) and the desired fermentation carried out. The desired flavoring (including a fruit preparation) may be added, mixed and the product packaged (Stirred Set Yogurt). Alternatively, the inoculated milk may be added on top of a fruit preparation in the final carton and allowed to ferment to the desired acidity (Set Style Yogurt). (Note: The University of Guelph has excellent web pages on yogurt, see http://www.foodsci.uoguelph.ca/dairyedu/yogurt.html)

The primary flavor issues in this product category are obtaining the desired flavor profile initially (acidity/sweetness/flavor) and then maintaining it in a "living" product. As the product ages, the sweetness and acidity change, thereby changing the flavor balance. Furthermore, there are living microorganisms in the yogurt that may metabolize flavor components or enzymatically convert them to other less desirable components. This changes the volatile portion of the flavor system.

#### 14.4.5.3 Flavored Dairy Desserts

Ice creams are made from milk fat (usually in the form of cream), milk solids (concentrated skim milk, milk, buttermilk solids, condensed whey, or NFDM), sweetener (usually sucrose, corn syrup solids, or high fructose corn syrup (HFCS)), 0.2–0.5% stabilizers, and flavoring. These ingredients are formulated to meet the desired Standard of Identity (Table 14.10). These Standards of Identity dictate compositional limits as well as some aspects of flavoring.

The desired ingredients are mixed in a tank, pasteurized, homogenized, cooled and then allowed to age (4+ h). After aging, the mix is frozen to the desired overrun (this relates to the amount of air whipped into the product during freezing). (For an excellent website covering most aspects of ice cream making and why, see: http://www.foodsci.uoguelph.ca/dairyedu/icecream.html).

The sales of ice creams in the U.S. by flavor are shown in Table 14.11. Vanilla is number one but this is likely due to its wide usage as a base for many other ice

**TABLE 14.10**
**Standards of Identity for Dairy Desserts**

| | Ice Cream | Ice Milk | Sherbet |
|---|---|---|---|
| Min./max. milk fat | 10% min. | 2–7% | ½% |
| Min. MSNF | 10% unless milk fat >10% | min. 9% unless milk fat >2% | not <1% |
| Total milk solids | must = 20% | must = 11% | not < 2% or > 5% |
| Bulky flavors | If used, may be 8% fat and 16% total milk solids | Characterized by fruit must have titratable acidity of 0.35% | |

**TABLE 14.11**
**U.S. Supermarket Sales of Ice Cream by Flavor, 2002 (% of Volume)**

| | | | |
|---|---|---|---|
| Vanilla | 28.4 | Chocolate | 8.0 |
| Nut flavors (including butter pecan @ 4.3%) | | | 10.4 |
| Neapolitan | 7.4 | Cookies and bakery | 5.8 |
| Fruit flavors (including strawberry @ 3.2%; cherry @ 2.0%) | | | 7.6 |
| Candy flavors | 3.4 | Chocolate chip | 3.7 |
| Mint chocolate chip | 3.0 | Coffee/mocha | 2.2 |
| | | All other | 19.7 |

*Source:* From http://www.foodsci.uoguelph.ca/dairyedu/icecream.html. Original Source: The Latest Scoop, International Dairy Foods Association.

cream desserts (sundaes, floats, banana splits, etc.). There is a fairly uniform distribution of flavors after vanilla and chocolate. The newest members of this group are the cookie-based and bakery flavorings.

The time and flavor addition in the manufacturing process depends upon the flavor being used. Flavoring materials that can withstand the manufacturing process (pasteurization, homogenization, and freezing) are added to the mix tank. This would include liquid flavorings, syrup bases, fruit purees, and colors. When one creates variegations, this is done with a special variegating pump after the freezing step. Since particulates (nuts, cookie doughs, candy pieces, fruits, etc.) cannot survive the agitation involved in the freezing process, they are added after freezing using a fruit feeder.

The flavoring system determines labeling as specified by law and is:

- Category I: Contains no artificial flavoring (e.g., vanillin/ethylvanillin)
- Category II: Contains both natural characterizing flavor and artificial flavor but natural flavor predominates
- Category III: Flavored only with artificial flavor or a combination where artificial flavoring predominates

In terms of flavorings, as shown in Table 14.11, vanilla and chocolate make up over 40% of sales. The use of these two flavorings in ice cream is covered well on the Guelph website and thus, the reader is encouraged to go there for detail (http://www.foodsci.uoguelph.ca/dairyedu/icflavours.html). Fruit preparations for this purpose have undergone change in the last few years. While fresh, frozen, and canned fruits still find considerable use, today there is a wide variety of aseptically processed fruits and fruit preparations available for ice creams. These products are of very good quality and find wide usage.

In selecting flavorings and their usage levels, a few guidelines should be kept in mind. First, ice creams are eaten cold and thus usage level will be higher than expected if one bases usage level on flavoring a sample of the mix (refrigeration temperature). Second, as the overrun increases, so does the flavor usage level. One is delivering less product in a spoon of high overrun ice cream than a low overrun ice cream. Third, increasing the level of fat in an ice cream increases the flavor need (nearly a linear relationship between flavor dosage needed and product fat level). Fourth, stabilizers increase the flavor level needed as well. For example, the use of carboxymethyl cellulose permits one to eat ice cream colder, which means higher flavor levels are needed. Guar gum increases chewiness, which again increases the need for flavor. Emulsifiers can add plastic, soapy, or bitter notes. Finally, increasing the milk solids (protein) again increases the need for flavor.

A last point is ice creams will deteriorate in flavor with time. They have a finite shelf life. They will oxidize from light or simply enough time. They are subject to tainting from the packaging. Also flavor binding, particularly vanilla, occurs over time as well.

### 14.4.6 Soft Drinks

#### 14.4.6.1 Introduction

This product group includes carbonated beverages, both clear and cloudy, noncarbonated beverages (still products), which may or may not be concentrated requiring dilution by the consumer, and crystal beverages which are in powder form and require reconstitution with water [23]. Flavorings for such a diverse product group must (a) impart the characteristic profile implied by the name; (b) be technologically compatible; (c) be stable to heat, light, acids, and preservatives, particularly sulfur dioxide; (d) impart the correct physical appearance to the end product; (e) be free from spoilage organisms; and (f) comply with existing legislation.

Their application is, of course, dictated by the nature of the beverage as consumed. Many flavoring materials are either insoluble or only very sparingly soluble in water, so that special techniques have to be employed to ensure a uniformly flavored, stable product. Two widely used methods are the following:

- Flavorings may be used that are soluble in the concentrated bottling syrup (e.g., extracts, or terpeneless isolates) and thus are stable when diluted with water for bottling.
- They may be emulsified to produce a stable emulsion. Such flavorings are generally used to produce a cloud in a drink that would otherwise be clear (see emulsions discussion in Chapter 13).

Fruit and berry juices are widely used as flavor bases for soft drinks, and most of these are concentrated by the removal of water under vacuum to give a commercial product, which is between four and six times stronger than the original juice. The flavor value of these concentrates depends not only on the degree of concentration but on the precise processing conditions used in their manufacture. From an application point of view, fruit products may be offered in sealed containers that have been pasteurized, in which case the whole contents must be used once the container has been opened, or they may be in multidose containers the contents of which contain a permitted preservative, usually sodium benzoate. All fruit-based flavorings are best stored under refrigeration or in a cold store. Fruit essences also may be used to flavor clear waters. The essences are produced by alcoholic infusion or distillation of the fruit.

Crystal (dry) beverages offer few problems in flavor application. They fall into two broad categories: (i) high quality products made from spray-dried fruit and natural flavorings; and (ii) cheaper, lower quality products, based on imitation flavorings. A typical formulation contains sugar (62%), dextrose (24%), citric acid (7.5%), sodium citrate (1%), ascorbic acid (0.2%) together with a dry flavoring and an appropriate colorant [1]. The use of encapsulated flavors is recommended as these give a long shelf life. Where a cloudy product is desired, a spray-dried vegetable oil or terpene-based cloud may be incorporated into the above formulation at about 4%.

### 14.4.6.2 Carbonated Beverages

#### *14.4.6.2.1 Formulation/Manufacturing*

Carbonated beverages make up the largest portion of the beverage market. The manufacturing process for carbonated beverages begins with making a Bottlers Syrup. This product is often manufactured by the parent company (who then controls the formulation) and is sold to local bottlers. The Bottlers Syrup is composed of: 1. Some or all of the sugar; 2. Fruit juice solids (optional), 3. Acidulants; 4. Flavoring; 5. Coloring; and 6. Preservatives — Na Benzoate at a max of 0.1%. The bottler uses this syrup to prepare a finished bottled beverage by first preparing a sugar syrup, adding the Bottlers Syrup to it, and making a uniform blend thereof. The blend is homogenized, an aliquot added to a bottle, filled with carbonated water and then capped/finished.

One has to pay strict attention to:

1. Water quality — Concern is for hardness and undesirable aroma. Hard water will result in cloudiness in the beverage over time due to mineral precipitation. Any off odors in the water may taint the product yielding an undesirable product.
2. Acidulants — Acidulant choice and level affect flavor. Citric acid is the most commonly used acidulant although malic acid may be used in grape and phosphoric acid is common to colas. Typically, the pHs of colas are the lowest being about 2.5, with most other beverages ranging from 2.5–3.0 except for cream soda which has a pH of about 5.0.
3. Sweeteners — Sweeteners have a strong influence on flavor. While sucrose is still used in some high-end beverages, high fructose corn syrups are the primary bulk sweetened used today. Diet products may use acesulfame K, aspartame, saccharin, or sucralose (typically some combination thereof).
4. Emulsifiers — Gum acacia or a modified starch is used as an emulsifier for the flavoring (when needed). This helps impart physical stability to the flavoring.

#### *14.4.6.2.2 Flavoring*

As noted above, the flavoring may be soluble or insoluble. An example of a soluble flavoring is a washed lemon extract used in clear lemon-lime beverages. The washed extract is used at a level of 0.625 to 3.125 L/50 L Bottlers Syrup. While the washed extract imparts a fresh flavor to the beverage, the alcohol in the product is not permitted in some countries due to religious practices, and thus, a deterpenated oil may be used (generally produced by distillation). Cloudy beverages are made from emulsions: this topic has been thoroughly discussed in Chapter 13 and a recent review is available [24].

#### *14.4.6.2.3 Flavor Deterioration*

Carbonated beverages are expected to have a shelf life of several months. During this time, numerous chemical changes may occur in the beverages. For example, terpenes and aldehydes may oxidize, lactones polymerize, esters hydrolyze, and various flavor/solvent reactions occur (e.g., acetal formation) [25]. These reactions result in flavor deterioration either due to the loss of desirable flavor components or the formation of off-flavors. Of these changes, acid catalyzed changes (sweetener or terpene degradation) are most common. In terms of acid catalyzed changes in the sweetener, sucrose is readily hydrolyzed at beverage pHs. This changes the level of sweetness as well as the character of the sweetness. The prevalent high intensity sweetener, aspartame, is most stable at pH pHs 4–5 and degrades at lower pHs. Aspartame degradation results in the loss of sweetness, but more importantly, it results in the formation of off-flavors. This is exacerbated by temperature abuse that often happens during shipping and distribution. Thus, the approval of acesulfame k and more recently sucralose has resulted in their frequent use in place of aspartame.

Off-flavors in citrus flavored beverages due to acid catalyzed terpene degradation is extremely common. The fresh citrus (particularly orange or lemon) flavor is short lived, and off-notes characteristic of fragranced household cleaners quickly predominate. In lemon flavored beverages, citral is often nearly completely degraded only hours after bottling [26,27], and thus the characterizing lemon flavor disappears. At the same time, the degradation of α-pinene, β-pinene, myrcene, and α-terpinene occurs, resulting in additional off-notes of ρ-cymene, α-terpineol, fenchyl alcohol, methylacetophenone, ρ-cresol, and 1-terpinen-4-ol [26,28–30].

## 14.5 SUMMARY

Each product group within the food and related industries poses its own specific flavoring problems to a point where generalizations are of limited value. Within each segment of the industry there are technologists whose responsibility it is to ensure that the end products are of a consistent quality and correctly flavored to satisfy customer expectations, technological needs, and legal requirements. Within the flavor industry there are also technical service experts available to give advice on the choice of flavorings and their optimum application. The value of their advice depends on a knowledge of the manufacturer's precise flavoring problems. The better the communication between the development technologist, the creative flavorist, and the flavorings applications technologist, the better the chances of success for the end product.

## REFERENCES

1. Heath, H.B., *Flavor Technology: Profiles, Products, Applications*, AVI Pub. Co.: Westport, CT, 1978, p. 542.
2. Ashurst, P.R., *Food Flavorings*, 2nd ed., Blackie Academ. Prof., New York, p. 332, 1995.
3. Ziegler, E. and H. Ziegler, *Flavourings: Production, Composition, Applications, Regulations*, Wiley-VCH, New York, 1998, p. 710.
4. Matz, S.A., *Cookie and Cracker Technology*, 3rd ed., Van Nostrand Reinhold, New York, 1992, p. 404.
5. *Perfumer and Flavorist*, 1999 Worldwide flavor market, http://www.perfumerflavorist.com, accessed 04/22/05.
6. Olsman, H. and S.C.E. Romkes, Flavouring of bouillons, soups and sauces, in *Food Flavourings*, E. Ziegler, H. Ziegler, Eds., Wiley-VCH, New York, 1998, p. 492.
7. van Osnabrugge, W., How to flavor baked goods and snacks effectively, *Food Technol.*, 27, 1, p. 74, 1989.
8. Graf, E., J.C. Soper, Flavored flour containing allium oil capsules and method of making flavored flour dough product, U.S. Patent 5536513, 1996.
9. Heiderich, S. and G. Reineccius, The loss of volatile esters from cookies, *Perfum. Flavorist*, 26, 6, p. 14, 2001.
10. Matz, S.A., *Snack Food Technology*, 3rd ed., Pan-Tech International, McAllen, TX, p. 450, 1992.
11. Tettweiler, P., Snack foods worldwide, *Food Technol.*, Feb., p. 58, 1991.
12. Kuntz, L.A., Flavoring systems for savory snack, *Food Prod. Des.*, May, p. 59, 1997.
13. Anon., New flavor technology separates blend flavors, *Food Process.*, 53, 11, p. 44, 1992.
14. Pszczola, D.E., Tour highlights production and uses of smoke-based flavors, *Food Technol.*, Jan., p. 70, 1995.
15. Matz, S.A., *Snack Food Technology*, 1st ed., Chapman Hall, New York, p. 349, 1976.
16. Williams, D., Flavors for snack-food applications, *Perfum. Flavorist*, 24, p. 29, 1999.
17. Anon., Encapsulated spice, seasoning extracts add flavor, color to dry-mix products, *Food Engineer.*, 53, 8, p. 69, 1981.
18. Suck, A.H., *Chewing Gum: History and Development, Raw Materials, Production, Packaging*, Haarmann Reimer, Holtzminden, Germany, 1988, p. 60.
19. Schirle-Keller, J.P., G.A. Reineccius, L.C. Hatchwell, The loss of aspartame in chewing gums during storage, in *Food Flavor Interactions,* R.C. McGorrin, J.V. Leland, Eds., Amer. Chem. Soc., Washington, D.C., 1996, p. 143.
20. Davidson, J.M., R.S.T. Linforth, T.A. Hollowood, A.J. Taylor, Effect of sucrose on the perceived flavor intensity of chewing gum, *J. Agric. Food Chem.*, 47, p. 4336, 1999.
21. Schirle-Keller, J.-P., *Flavor Interactions with Fat Replacers and Aspartame,* Ph.D. thesis Food Science and Nutrition, University of Minnesota, St. Paul, p. 237, 1995.
22. Zimmerman, C.M., A wide variety of flavors take milk far beyond chocolate, *School Foodservice Nutrition*, Aug., p. 44, 1995.
23. Downer, A.W., The application of flavors in the soft drinks industry, *Flavour Ind.*, 4, p. 488, 1973.
24. Tan, C.T., Beverage Emulsions, in *Food Emulsions*, 3rd ed., S.E. Friberg, K. Larsson, Eds., Marcel Dekker, New York, p. 485, 1997.

25. Sinki, G., R. Assaf, J. Lombardo, Flavor changes: a review of the principal causes and reactions, *Perfum. Flavorist*, 22, 4, p. 23, 1997.
26. Freeburg, E.J., B.S. Mistry, G.A. Reineccius, Stability of citral-containing and citral-less lemon oils in flavor emulsions and beverages, *Perfum. Flavorist*, 19, 4, p. 23, 1994.
27. Ikenberry, S., The effect of temperature and acidity on the stability of specific lemon flavor constituents, II. Kinetics, in *Food Flavors, Ingredients and Composition*, Charalambous, Ed., Elsevier, New York, 1993, p. 355.
28. Schieberle, P. and W. Grosch, Quantitative analysis of important flavour compounds in fresh and stored lemon oil/citric acid emulsions, *Lebens. Wiss. u. Technol.*, 21, p. 158, 1988.
29. Ueno, T., H. Masuda, S. Muranishi, S. Kiyohara, Y. Sekiguchi, C.-T. Ho, Inhibition of the formation of off-odour compounds from citral in an acidic aqueous solution, in *Flavour Research at the Dawn of the Twenty-First Century*, J.-L. Le Quere, P.X. Etievant, Eds., Tec. & Doc., Paris, 2004, p. 128.
30. Liang, C.P., M. Wang, J. Simon, C.-T. Ho, Stabilization of Citral Flavor and Fish Oil Fatty Acids by Phenolic Extracts, in 228th ACS National Meeting, Philadelphia, PA, U.S., Amer. Chem. Soc., Washington, D.C., Aug. 22–26, 2004.

# 15 Flavor Legislation and Religious Dietary Rules

## 15.1 INTRODUCTION

Legislation impacts the flavor industry in many ways. This ranges from controlling the materials it may use in formulations in a given country, to costs of materials (e.g., taxes on alcohol), to disposal of waste materials, conditions in the workplace environment, and even the shipping of finished materials. The impact of legislation is profound and broad in effect. It is rather impossible to cover all aspects of legislative effects on the industry and certainly some legislative actions have a greater impact on the industry than others. Two areas of legislation that have an important impact are those that limit the use of flavor compounds and those that determine how a product is labeled (e.g., natural vs. artificial). Thus, this chapter will focus on these two legislative topics. One may make some of the same statements about the impact of religious dietary rules: certainly these rules control the selection and manufacture of food flavorings. They also impact cost in the sense that religious oversight adds cost to the ingredients used in foods. Thus, a brief summary of these rules will also be included in this chapter.

The reader is strongly encouraged to read Salzer and Jones [1] for details in both of these topic areas. These authors have provided an excellent overview of these topics on a global basis.

## 15.2 LEGISLATION LIMITING THE USE OF FLAVOR COMPOUNDS

This is an extremely problematic area for multinational companies since the laws governing the use of flavor chemicals may be unique to a given country. Thus, what is a legal flavor formulation in one country is not necessarily legal in another country. This situation is getting better due to the formation of various trade agreements which standardize food laws across numerous countries. Examples include the formation of the European Union (EU) (Austria, Belgium, Cyprus, the Czech Republic, Estonia, Finland, France, Germany, Greece, Hungary, Ireland, Italy, Latvia, Lithuania, Luxembourg, Malta, the Netherlands, Poland, Portugal, Slovakia, Slovenia, Spain, Sweden and the U.K.), NAFTA (Mexico, U.S., and Canada), and the Arabian Standardization and Metrology Organization (ASMO; Jordan, Iraq, Bahrain, Kuwait, Quatar, Osman, Saudi Arabia, and the United Arab Emirates). Unfortunately, large population centers such as Asia, Central and South America, and Africa do not have such agreements, and the laws of individual countries must be consulted.

Simply keeping up on global legislation adds substantial costs to doing business on a global basis.

Due to the shear volume of information in this topic area, space permits only discussing the legislation governing the use of flavor compounds in the U.S. As noted earlier, the chapter authored by Salzer and Jones [1] is recommended. For more current information, the Leffingwell [2] website is also recommended. The flavor related legislation for Japan, Australia, and the EU is linked to the Leffingwell home page.

### 15.2.1 U.S. Flavor-Related Legislation

The 1958 Food Additives Amendment outlines the laws regarding the use and labeling of food additives. This amendment shifted the emphasis from the government (the Food and Drug Administration, i.e., the FDA) having to demonstrate that a food additive was harmful to a policy that the food industry had to show that a food additive was safe. It was recognized that it would be impossible to require that all food additives in use at that time be withdrawn from use until tested and proven safe. The solution was that all food additives that had prior sanction (approval of their use prior to the amendment) would not be considered a food additive and therefore be exempt from petition and regulation (grandfathered approval). A second grandfather clause stated that food additives which were Generally Recognized As Safe (GRAS) by the scientific community would also be exempt from food additive status. These materials became known as GRAS substances and are listed in the CFR Title 21 Part 170–180. The concept that a food additive could be exempt from petitioning and the required safety testing if deemed safe by the scientific community opened the door for the flavor industry to establish its own panel of safety experts to approve the use of flavoring substances.

The Flavor Extract and Manufacturers Association (FEMA) formed an expert panel in 1959 composed of toxicologists, pathologists, organic chemists, biochemists, and oncologists for this purpose [3]. The panel initially evaluated the GRAS status of several hundred flavor compounds used in the industry and published the first GRAS list (in 1965) containing over 1100 flavoring substances. Since that time, additional FEMA GRAS lists have continued to be published expanding this list to over 2000 compounds. Approximately 125 compounds are being added to this list each year in an effort to make the FEMA list consistent with the approved use of flavoring compounds in the EU. The evaluation procedures and criteria used in this evaluation include anticipated exposure in foods, natural occurrence in foods, chemical identity, metabolic and pharmacokinetics, available toxicological data, and the application of principles outlined in the FDA's Red Book on the safety evaluation of food additives [3]. These criteria and procedures are discussed in detail by Woods and Doull [4]. The data used in the evaluation of flavor chemicals are published as Scientific Literature Reviews (SLR) by FEMA for open evaluation.

The concept that a flavoring material might be used without petitioning and rigorous safety testing has scientific merit for several reasons. Flavoring materials are unique among the food additives in that their use is self-limiting. It is virtually impossible to substantially overdose a food with a given food flavor component since the food would not taste good. Thus, the issue of regulation of quantity is not relevant. Flavor compounds are also unique in the sense that they typically are the same chemicals that are found in nature in the food flavor being imitated. It stands to reason that the most true flavor representation will be the result of using the same aroma compounds as nature used (also at the same levels). Literally, only a handful of aroma chemicals in use today are not found in nature, therefore, we have a long history of consuming these compounds. When a new flavor compound is being evaluated for potential approval for food use, an important consideration is whether it exists in nature and if the approval of this aroma compound will result in a significant increase in its consumption. Most of the studies on consumption document that the flavor industry adds a rather insignificant amount of aroma chemicals to man's diet vs. what we consume as a part of our natural foods [5–8]. A final consideration is that flavoring materials are generally used at extremely low levels in foods often in the ppb or lower levels. This results in very low exposure to any given flavor chemical.

In addition to these safety considerations, there are some practical considerations to be noted. The flavor industry, even collectively, is very small but has a very diverse market for fine chemicals. If the same level of animal testing were required for the approval of a food flavor component as other food additives, the cost would be prohibitive. In essence, no new flavor chemicals would be added to the approved list. This would greatly limit the industry's ability to improve the sensory quality of our food supply. In the flavor business, the global market for a given flavor chemical may be 100 kg a year. Assuming that animal testing would cost $10 million, it would be impossible to justify the cost of safety testing for the small annual sales. While this is not justification for *not* doing animal testing, it is practical consideration.

As a result of legislation, we have a positive list of flavor chemicals that may be used in a food: if a chemical is not on this list, it cannot be used. Some countries have a negative list, a list of compounds that cannot be used, or a mixed system, a positive and negative list. The FEMA Expert Panel has continued to serve as a scientific body which is advisory to the FDA. Over the years, the FDA has accepted the decisions of this committee. It has gained a very credible record both in the U.S. and around the world.

## 15.3 RELIGIOUS DIETARY RULES

The two main religions that have strict rules regarding the preparation of foods are Judaism and Islam. Of these two religions, approximately 1.3 billion people are Muslims and 14 million are Jews [9]. Interestingly, kosher foods are well established in the marketplace. This obviously is not due to the number of Jews in the world (or Seventh Day Adventists and some Muslims who may also purchase kosher foods to meet their needs) but the belief that kosher foods are made with greater care and are in some ways premium products, thus they appeal to people who are not Jewish as well. Halal foods have existed since the birth of Islam but only in recent times have gained importance outside of the Muslim countries. While the expanding presence of Muslims throughout the world is partially responsible for recognition, there is also an impression of better quality. In this section of this chapter, the rules of Judaism and Islam regarding foods and flavors will be outlined.

### 15.3.1 Kosher Dietary Laws

Kosher law is based on the Torah: the first Five Books of Moses (ca. 1275 B.C.) While many of the laws make sense in terms of health considerations, they are based on religious considerations. The primary rules are that: 1) All plant materials are kosher; 2) Flesh and Blood: man is permitted to eat these because of his desire for it; 3) Milk and Meat: these two food groups must be separated; and 4) Only certain species of animals are suitable to eat, and these must be slaughtered and processed in a specified manner. To elaborate further on the rules, only animals that both chew their cud and have cloven hooves may be eaten (these are herbivores). The restrictions on birds are more elaborate. The birds cannot be birds of prey or have a front toe. They must have a craw and catch food thrown in the air, place it on the ground and then tear it with their bill before eating it. For fish, only fish with fins and removable scales may be eaten. All shellfish are prohibited. The slaughtering procedure is also defined. It must be quick and lead to maximal bleeding (use of a sharp knife and cutting of the carotid artery, jugular vein, and windpipe). This must be done by a trained individual called a "schochet." The meat is also salted to remove any remaining blood.

There are different categories of foods: meat (fleishig) and dairy (milchig: dairy and all dairy products derived thereof. The milk cannot be from a nonkosher animal, e.g., a camel or pig), neutral (parve or pareve,) and unacceptable (traif, nonkosher). Parve products can be used in either meat- or milk-based foods, but meat and dairy can never be mixed. Also, although fish as defined above is considered parve, it cannot be mixed with meat. This classification is done by a rabbi.

There are strict requirements for the food equipment and process of making kosher foods. A primary equipment consideration is that it is either nonporous (glass) or easily cleanable (e.g., stainless steel or other metal). The cleaning process depends on what was made in the equipment previously. It minimally involves leaving the equipment empty for 24 hours and then filling it with boiling water until it over flows. However, the cleaning process and subsequent manufacturing is done under rabbinical supervision and thus must meet the requirements of the rabbi present.

There are numerous rabbis and rabbinical organizations involved in the certification of kosher foods. Circle U is the largest of these organizations. Circle U certification is administered by the Union of Orthodox Jewish Congregations [10]. This symbol originated in 1923 and the circle represents the letter O for Orthodox, and the U inside stands for Union (see figure of circle U symbol).

OUKOSHER.ORG

Circle K is also a major certification symbol and has been in existence since 1935. It is administered by Organized Kashruth Laboratories [11]. The *Kashrus* magazine publishes an index to all the kosher symbols and the organizations representing them. This magazine can be obtained from Kashrus, P.O. Box 204, Brooklyn, NY 11204 (request the Kosher Supervision Guide issue).

Kosher certification has an impact on the flavor industry in that there are numerous flavor ingredients that are derived from animal sources. For example, gelatin may be used in flavor encapsulation (coacervation) or process flavors to give body. While gelatin from a beef animal (skin and bones) may be kosher if processed under kosher certification, there is no kosher commercial source of gelatin today [3]. Emulsifiers such as stearates, mono and diglycerides, and glycerol may be derived from animal sources. Animal fats may be used in process flavors to give species notes. Starter distillates (or natural diacetyl) may be made from dairy ingredients making them a dairy ingredient that must not be mixed with anything meat. The same is true of casein, lactose, lactic acid, whey, enzyme modified cheeses, or any fermented product made from a dairy material. Hydrolyzed proteins may be of either dairy or meat origin, limiting their use. The hydrolyzed animal proteins must have been processed in a kosher manner to be used at all. Grape and grape products (e.g., wines, cognacs, etc.) must be made by Jews to be acceptable. This could have implications in the use of fusel alcohols (made from a grape source) for biotechnological conversions to other flavor compounds. Alcohols from other sources (not made by Jews) are acceptable if they have been pasteurized.

There are additional requirements for foods for Passover. Wheat, rye, barley, spelt, corn, legumes, rice, and mustard and any derivatives thereof are not permitted. This would, for example, include any alcohol, beer, dextrose, maltodextrin, or corn syrup solids (from corn or wheat), sorbitol, etc. made from these products [3].

One can see that these rules can very broadly affect the use of flavoring materials by the industry. The flavor industry must be vigilant in making sure that any of the products they produce are kosher and that their customer knows that they are present in the flavoring.

### 15.3.2 Halal Rules

Like many religions, Islam has established dietary rules for believers to follow. The Quran (Surah 5:3–4) provides the canonical basis for these dietary laws that were explained by the Prophet Mohammed and codified by the various schools of Islamic law [12]. The rule is that all foods are halal that are not haram (prohibited). The imposition of these laws was to "promote good hygiene, prevent intoxication and make believers mindful of their debt to God for His sustenance" [12]. Prohibited foods include:

- All pork and pork byproducts
- Alcohol and other intoxicants
- Blood and blood byproducts
- Any food over which the name of any deity other than God has been invoked

- Birds of prey, animals with claws and fangs, and almost all insects and reptiles
- The flesh of dead animals or improperly slaughtered animals [12]

The slaughtering of animals is done in a specified manner that kills the animal as humanely as possible, prevents contamination of its meat, and, ideally, is slaughtered by a "pious man" using a sharp knife and a single cut that severs the windpipe and the jugular vein, draining the animal of its blood. The animal is also blessed before it is killed (the animal's face is turned toward the holy city of Mecca and a prayer is said over it). The process is very similar to that of the kosher slaughtering of animals.

The primary considerations for the flavor industry are the use of any meat or meat byproduct in process flavors, gelatin in process flavors or coacervation processes, and the use of alcohol as a component on any basis in a flavoring are all prohibited. There are various organizations that oversee halal certification of foods. Unlike the kosher certification, halal certification is younger and thus less well organized. Two organizations that come up on a web search are IFANCA (The Islamic Food and Nutrition Council of America, 5901 N. Cicero Avenue, Suite 309 Chicago, IL, 60646) and Halal Transactions, Inc., P.O. Box 4546, Omaha, NE, 68104).

## 15.4 LABELING OF FOOD FLAVORINGS

The labeling of foods is done for informational purposes and safety. In terms of information, if a significant segment of the population is interested in knowing if a food contains a given ingredient (e.g., MSG, or an artificial flavor or flavors), then this ingredient may be required to be listed on the food product label. Information as to religious law is also regulated in the sense that FDA requires that foods labeled as kosher or halal must have been produced under the requirements of that designation (21.CFR 101.29, [3]). The requirement of listing an ingredient does not imply inferiority, lesser quality, or a lack of safety. It simply is there for those that want to know. In terms of safety, the primary issue is of allergens. It is critical that foods generally recognized as being potential allergens are listed clearly for the consumer.

### 15.4.1 Bulk Labeling Requirements

The purpose of bulk labeling is to provide adequate information to the food processor so that the company may properly label its products for the consumer. There is no requirement to label flavor components other than their natural status (natural, natural/artificial, artificial, with other natural flavors (WONF), natural type flavor [none of the named product is in the flavor], or artificial but noncharacterizing flavor). All nonflavor ingredients must be listed as well as any allergens and religious status if relevant.

### 15.4.2 Labeling for the Consumer

There are two label areas on a food product where flavor is relevant: the Principle Display Panel (PDP) and the Ingredient Statement (IS). The PDP informs the consumer of: 1) what the product is; 2) its flavor (e.g., cherry, orange, etc.), 3) natural status of the flavors suggested by name and/or picture; and 4) provide any statement as to meeting any religious laws (e.g., kosher or halal) [3]. The ingredient statement lists in detail the ingredients in the food as defined by law. The labeling laws for flavor are reasonably clear and are found in the CFR Title 21 Section 101.22 Sec. 101.22 "Foods; labeling of spices, flavorings, colorings and chemical preservatives." This section of the CFR is reprinted below, with personal comments inserted in italics. Certain words or sections are bolded for emphasis, and some sections not related to flavoring have been omitted.

## TITLE 21 VOLUME 2 — FOOD AND DRUGS

### Chapter I — Food and Drug Administration, Department of Health and Human Services (continued): Park 101 — Food Labeling — Table of Contents

### Subpart B — Specific Food Labeling Requirements

#### Sec. 101.22 Foods; labeling of spices, flavorings, colorings, and chemical preservatives.

(a) (1) The term artificial flavor or **artificial flavoring** means any substance, the function of which is to impart flavor, which is not derived from a spice, fruit or fruit juice, vegetable or vegetable juice, edible yeast, herb, bark, bud, root, leaf or similar plant material, meat, fish, poultry, eggs, dairy products, or fermentation products thereof. Artificial flavor includes the substances listed in Sec. 172.515(b) and Sec. 182.60 of this chapter except where these are derived from natural sources. ***(Artificial is in effect defined as not being natural.)***

(2) The term **spice** means any aromatic vegetable substance in the whole, broken, or ground form, except for those substances which have been traditionally regarded as foods, such as onions, garlic and celery; whose significant function in food is seasoning rather than nutritional; that is true to name; and from which no portion of any volatile oil or other flavoring principle has been removed. Spices include the spices listed in Sec. 182.10 and part 184 of this chapter, such as the following: Allspice, Anise, Basil, Bay leaves, Caraway seed, Cardamon, Celery seed, Chervil, Cinnamon, Cloves, Coriander, Cumin seed, Dill seed, Fennel seed, Fenugreek, Ginger, Horseradish, Mace, Marjoram, Mustard flour, Nutmeg, Oregano, Paprika, Parsley, Pepper, black; Pepper, white; Pepper, red; Rosemary, Saffron, Sage, Savory, Star aniseed, Tarragon, Thyme, Turmeric. Paprika, turmeric, and saffron or other spices which are also colors, shall be declared as "spice and coloring" unless declared by their common or usual name.

(3) The term **natural flavor** or natural flavoring means the essential oil, oleoresin, essence or extractive, protein hydrolysate, distillate, or any product of roasting, heating or enzymolysis, which contains the flavoring constituents derived from a spice, fruit or fruit juice, vegetable or vegetable juice, edible yeast, herb, bark, bud, root, leaf or similar plant material, meat, seafood, poultry, eggs, dairy products, or

fermentation products thereof, whose significant function in food is flavoring rather than nutritional. Natural flavors include the natural essence or extractives obtained from plants listed in Sec. 182.10, Sec. 182.20, Sec. 182.40, and Sec. 182.50 and part 184 of this chapter, and the substances listed in Sec. 172.510 of this chapter. ***(Part 1 and Part 3 above essentially define artificial and natural flavor components. Spices are defined in Part 2. Deleted Part 4.)***

(5) The term **chemical preservative** means any chemical that, when added to food, tends to prevent or retard deterioration thereof, but does not include common salt, sugars, vinegars, spices, or oils extracted from spices, substances added to food by direct exposure thereof to wood smoke, or chemicals applied for their insecticidal or herbicidal properties.

(c) A statement of **artificial flavoring**, artificial coloring, or chemical preservative shall be placed on the food or on its container or wrapper, or on any two or all three of these, as may be necessary to render such statement likely to be read by the ordinary person under customary conditions of purchase and use of such food ***(must appear on the PDP unless it meets an exception noted later)***.

(d) A food shall be exempt from compliance with the requirements of section 403(k) of the act if it is not in package form and the units thereof are so small that a statement of artificial flavoring, artificial coloring, or chemical preservative, as the case may be, cannot be placed on such units with such conspicuousness as to render it likely to be read by the ordinary individual under customary conditions of purchase and use.

(e) A food shall be exempt while held for sale from the requirements of section 403(k) of the act (requiring label statement of any artificial flavoring, artificial coloring, or chemical preservatives) if said food, having been received in bulk containers at a retail establishment, is displayed to the purchaser with either (1) the labeling of the bulk container plainly in view or (2) a counter card, sign, or other appropriate device bearing prominently and conspicuously the information required to be stated on the label pursuant to section 403(k). ***(deleted Section f)***

(g) A flavor shall be labeled in the following way when shipped to a food manufacturer or processor (but not a consumer) for use in the manufacture of a fabricated food, unless it is a flavor for which a standard of identity has been promulgated, in which case it shall be labeled as provided in the standard ***(Bulk flavor labeling requirements)***:

(1) If the flavor consists of one ingredient, it shall be declared by its common or usual name.

(2) If the flavor consists of two or more ingredients, the label either may declare each ingredient by its common or usual name or may state "All flavor ingredients contained in this product are approved for use in a regulation of the Food and Drug

Administration." Any flavor ingredient not contained in one of these regulations, and any nonflavor ingredient, shall be separately listed on the label. ***(These can be flavor carriers, solvents, emulsifiers, preservatives, etc.)***

(3) In cases where the flavor contains a solely natural flavor(s), the flavor shall be so labeled, e.g., "strawberry flavor," "banana flavor," or "natural strawberry flavor." In cases where the flavor contains both a natural flavor and an artificial flavor, the flavor shall be so labeled, e.g., "natural and artificial strawberry flavor." In cases where the flavor contains a solely artificial flavor(s), the flavor shall be so labeled, e.g., "artificial strawberry flavor."

(h) The label of a food to which flavor is added shall declare the flavor in the **statement of ingredients** in the following way:

(1) Spice, natural flavor, and artificial flavor may be declared as "spice," "natural flavor," or "artificial flavor," or any combination thereof, as the case may be.

(2) An incidental additive in a food, originating in a spice or flavor used in the manufacture of the food, need not be declared in the statement of ingredients if it meets the requirements of Sec. 101.100(a)(3).

(3) Substances obtained by cutting, grinding, drying, pulping, or similar processing of tissues derived from fruit, vegetable, meat, fish, or poultry, e.g., powdered or granulated onions, garlic powder, and celery powder, are commonly understood by consumers to be food rather than flavor and shall be declared by their common or usual name.

(4) Any salt (sodium chloride) used as an ingredient in food shall be declared by its common or usual name "salt."

(5) Any monosodium glutamate used as an ingredient in food shall be declared by its common or usual name "monosodium glutamate."

(6) Any pyroligneous acid or other artificial smoke flavors used as an ingredient in a food may be declared as artificial flavor or artificial smoke flavor. No representation may be made, either directly or implied, that a food flavored with pyroligneous acid or other artificial smoke flavor has been smoked or has a true smoked flavor, or that a seasoning sauce or similar product containing pyroligneous acid or other artificial smoke flavor and used to season or flavor other foods will result in a smoked product or one having a true smoked flavor.

(7) Because protein hydrolysates function in foods as both flavorings and flavor enhancers, **no protein hydrolysate used in food for its effects on flavor may be declared simply as "flavor," "natural flavor," or "flavoring."** The ingredient shall be declared by its specific common or usual name as provided in Sec. 102.22 of this chapter. ***(HVPs must be listed as well as their source protein.)***

(i) If the **label, labeling, or advertising of a food** makes any direct or indirect representations with respect to the primary recognizable flavor(s), **by word, vignette, e.g., depiction of a fruit, or other means**, or if for any other reason the manufacturer or distributor of a food wishes to designate the type of flavor in the food other than through the statement of ingredients, **such flavor shall be considered the characterizing flavor** and shall be declared in the following way:

(1) If the food contains no artificial flavor that **simulates, resembles, or reinforces the characterizing flavor** ***(this wording shows up in the regulations and is controversial. Would a flavor manufacturer add a flavor chemical that does NOT simulate, resemble, or reinforce the characterizing flavor? It seems that any flavor chemical that one would use in a flavoring would do that. This wording permits artificial flavors to be added to foods and not have to be declared on the PDP. [They must be listed in the Ingredient Statement.] In extreme cases, a food may be flavored entirely with artificial flavorings and not have to be declared as being artificially flavored on the PDP!),*** the name of the food on the principal display panel or panels of the label shall be accompanied by the common or usual name of the characterizing flavor, e.g., "vanilla," in letters not less than one-half the height of the letters used in the name of the food, except that:

(i) If the food is one that is commonly expected to contain **a characterizing food ingredient** ***(The ingredient must be in an adequate quantity to "characterize" the flavor of that product)***, e.g., strawberries in "strawberry shortcake," and the food contains natural flavor derived from such ingredient and an amount of **characterizing ingredient insufficient to independently characterize the food**, or the food contains no such ingredient, the name of the characterizing flavor may be immediately preceded by the word "natural" and shall be immediately followed by the word "flavored" in letters not less than one-half the height of the letters in the name of the characterizing flavor, e.g., "natural strawberry flavored shortcake," or "strawberry flavored shortcake."

(ii) If none of the natural flavor used in the food is derived from the product whose flavor is simulated, the food in which the flavor is used shall be labeled either with the flavor of the product from which the flavor is derived or as "artificially flavored." ***(No one opts for this choice. They take the following alternative).***

(iii) If the food contains both **a "characterizing" flavor from the product whose flavor is simulated** and other natural flavor which simulates, resembles or reinforces the characterizing flavor, the food shall be labeled in accordance with the introductory text and paragraph (i)(1)(i) of this section and the name of the food shall be immediately followed by the words "with other natural flavor" in letters not less than one-half the height of the letters used in the name of the characterizing flavor. ***(This paragraph is again controversial. If the food contains a flavor from the named source that "characterizes" the product, the flavor can be called a WONF instead of artificial. What is "characterizing"? The method to determine if a flavor***

***ingredient "characterizes" the product is not defined. So is flavor "characterizing" to whom? Me, you, or the company selling the product?)***

(2) If the food contains any artificial flavor that simulates, resembles or reinforces the characterizing flavor, the name of the food on the principal display panel or panels of the label shall be accompanied by the common or usual name(s) of the characterizing flavor, in letters not less than one-half the height of the letters used in the name of the food and the name of the characterizing flavor shall be accompanied by the word(s) "artificial" or "artificially flavored," in letters not less than one-half the height of the letters in the name of the characterizing flavor, e.g., "artificial vanilla," "artificially flavored strawberry," or "grape artificially flavored."

***(The remainder of this section outlines the requirement that there is a written and signed document from an officer of the flavor company that their natural flavors do not contain any artificial components. The means of validation are spelled out as well.)***

(4) **A flavor supplier shall certify, in writing, that any flavor he supplies which is designated as containing no artificial flavor does not, to the best of his knowledge and belief, contain any artificial flavor, and that he has added no artificial flavor to it.** The requirement for such certification may be satisfied by a guarantee under section 303(c)(2) of the act which contains such a specific statement. A flavor user shall be required to make such a written certification only where he adds to or combines another flavor with a flavor that has been certified by a flavor supplier as containing no artificial flavor, but otherwise such user may rely upon the supplier's certification and need make no separate certification. All such certifications shall be retained by the certifying party throughout the period in which the flavor is supplied and for a minimum of three years thereafter, and shall be subject to the following conditions:

(i) The certifying party shall make such certifications available upon request at all reasonable hours to any duly authorized office or employee of the Food and Drug Administration or any other employee acting on behalf of the Secretary of Health and Human Services. Such certifications are regarded by the Food and Drug Administration as reports to the government and as guarantees or other undertakings within the meaning of section 301(h) of the act and subject the certifying party to the penalties for making any false report to the government under 18 U.S.C. 1001 and any false guarantee or undertaking under section 303(a) of the act. The defenses provided under section 303(c)(2) of the act shall be applicable to the certifications provided for in this section.

(ii) **Wherever possible, the Food and Drug Administration shall verify the accuracy of a reasonable number of certifications made pursuant to this section, constituting a representative sample of such certifications, and shall not request all such certifications.**

(iii) Where no person authorized to provide such information is reasonably available at the time of inspection, the certifying party shall arrange to have such person and the relevant materials and records ready for verification as soon as practicable: Provided, That, whenever the Food and Drug Administration has reason to believe that the supplier or user may utilize this period to alter inventories or records, such additional time shall not be permitted. Where such additional time is provided, the Food and Drug Administration may require the certifying party to certify that relevant inventories have not been materially disturbed and relevant records have not been altered or concealed during such period.

(iv) The certifying party shall provide, to an officer or representative duly designated by the Secretary, such qualitative statement of the composition of the flavor or product covered by the certification as may be reasonably expected to enable the Secretary's representatives to determine which relevant raw and finished materials and flavor ingredient records are reasonably necessary to verify the certifications. The examination conducted by the Secretary's representative shall be limited to inspection and review of inventories and ingredient records for those certifications that are to be verified.

(v) Review of flavor ingredient records shall be limited to the qualitative formula and shall not include the quantitative formula. The person verifying the certifications may make only such notes as are necessary to enable him to verify such certification. Only such notes or such flavor ingredient records as are necessary to verify such certification or to show a potential or actual violation may be removed or transmitted from the certifying party's place of business: Provided, That, where such removal or transmittal is necessary for such purposes the relevant records and notes shall be retained as separate documents in Food and Drug Administration files, shall not be copied in other reports, and shall not be disclosed publicly other than in a judicial proceeding brought pursuant to the act or 18 U.S.C. 1001.

## 15.5 SUMMARY

There is little question that food laws and regulations have profound impact on the flavor industry. Clearly laws and regulations are required for the protection and safety of the consumer; this was the basis for food laws coming into existence. In the flavor industry this results in some aroma chemicals being prohibited from use in flavorings and other materials being clearly labeled (e.g., potential allergens). With time, additional laws have emerged regarding labeling that serve the function of being informative, e.g., natural, religious status, organic, genetically manipulated, contain MSG, etc. These labeling requirements are not necessarily based on safety but a desire of a segment of the population to know if certain ingredients are present in a food. While these latter requirements may have little foundation based on scientific knowledge, they are laws and the industry must abide by them. Each of these laws adds cost to a product and imposed restrictions on the flavoring materials being used. Often they will result in decreased quality or performance of a flavoring.

While there may be some differences in the flavoring ingredients restricted in various countries due to safety concerns, the largest variations exist in the "desire to know" category. The population of one country may be quite concerned about genetic manipulation (GM) of a food ingredient while there is no concern for GM in another. Thus, issues of labeling laws, as well as legally permitted ingredients, become extremely problematic for flavor companies that are multinational. This results in every multinational company employing a staff of individuals who simply must keep up on the regulations of their market countries and insure that every flavoring sold meets the regulations of that country. This adds substantial overhead to the business. The movement toward larger common markets (e.g., the EU, NAFTA, and ASMO) is helping to ease this task. Strong efforts to harmonize regulations across common markets are also underway. Conceivably, one day we will be able to reach much of the world with a common product.

## REFERENCES


1. Salzer, U.J. and K. Jones, *Legislation/Toxicology* in *Flavourings,* E. Ziegler, H. Ziegler, Eds., Wiley-VCH, New York, 1998, p. 645.
2. Leffingwell & Assoc., http://www.leffingwell.com, accessed 04/22/05.
3. Manley, C., Flavor Laws and Regulations, presented at the IFT sponsored Flavor Interactions in Food workshop, Orlando, FL, March, 2004.
4. Woods, L.A. and J. Doull, GRAS evaluation of flavouring substances by the expert panel of FEMA, *Regul. Toxicol. Pharmacol.*, 14, 1, p. 48, 1991.
5. Stofberg, J., The future of artificial flavoring ingredients, *Perfum. Flavorist*, 5, p. 17, 1980.
6. Stofberg, J., Setting priorities for the safety evaluation of flavoring materials, *Perfum. Flavorist*, 6, 4, p. 6, 1981.
7. Stofberg, J., Safety evaluation and regulation of flavoring substances, *Perfum. Flavorist*, 8, p. 53, 1983.
8. Stofberg, J. and J. Stoffelsma, Consumption of flavoring materials as food ingredients and food additives, *Perfum. Flavorist*, 5, 7, p. 19, 1980.
9. Major Religions of the World Ranked by Number of Adherents, http://www.adherents.com/Religions_By_Adherents.html, accessed 04/22/05.
10. Orthodox Union, 11 Broadway, 14th Floor, New York NY, 10004, http://www.oukosher.org, accessed 04/22/05.
11. OK Kosher Certification, 391 Troy Avenue, Brooklyn, NY, 11213, http://www.okkosher.com, accessed 04/22/05
12. Diet and Food: Dietary Rules: Halal and Haram, http://mediaguidetoislam.sfsu.edu/dailylife/02a_diet.htm, accessed 10/30/04.

# 16 Quality Control

## 16.1 INTRODUCTION

A flavor house must provide a good quality, consistent flavoring material for the consumer. It must also be certain that the flavor meets legal restrictions on ingredients and their qualities. The Quality Control/Assurance (QA) department meets these needs by performing both analytical tests and sensory evaluation on incoming ingredients and finished flavors.

The following chapter will outline in varying levels details that are performed broadly across the flavor industry. Since each flavor company may differ in the products they purchase as ingredients and products they produce, any given flavor company will only run some the tests to be presented.

The methods used in the industry generally come from sources such as the AOAC International, American Spice Trade Association (ASTA), American Oil Chemists Society (AOCS), U.S. Pharmacopea/National Formulary, or other organizations that have activities in establishing standard methods. However, some methods are sufficiently unique that they are generated within companies and may relate to a given product or customer. Also, standard methods are unquestionably accurate and transferable across companies but they often suffer from being generations behind in speed and accuracy. Thus, a company may choose to use newer methods in place of the published standard methods, e.g., near infrared to determine moisture as opposed to a vacuum oven. A textbook on general food analysis such as *Food Analysis* by Nielsen [1] is recommended for background reading in this area.

## 16.2 ANALYTICAL TESTS

### 16.2.1 Overview of Physicochemical Tests

This first section will simply provide a list of analytical procedures used in the flavor industry for the analysis of a given flavor material. This listing makes it clear that a flavor company that is a producer of essential oils, for example, would run a given set of analytical procedures (Section 16.2.1.2) and this set of procedures would be quite different from that run by a company that largely buys its ingredients and manufactures finished flavorings (Section 16.2.1.6). Thus, the overview that follows should be considered as an overall listing, one that any given flavor company regardless of its size, would not run in its entirety. This list was originally compiled by Henry Heath [2].

#### 16.2.1.1 Natural Plant Materials

A sample may represent whole unground, broken, or previously ground material and should first be judged on overall appearance and visual quality. In most cases, whole samples must first be reduced to a uniform, moderately fine powder before subjecting them to any of the following tests.

#### *16.2.1.1.1 General Tests*

*Ash.* Total, acid insoluble, water soluble.
*Crude fiber.*
*Extractive matter.* Alcohol soluble, other solvents, nonvolatile ether extractive. The test may be followed by an examination of the resulting extract after removal of the solvent.
*Extraneous matter and filth.* Microscopy.
*Moisture content.* Oven drying, Infrared drying, Karl Fischer method, Near Infrared.
*Volatile oil content.* Distillation — may be followed by an examination of the recovered essential oil by GC and/or other test methods (USP).

#### *16.2.1.1.2 Tests of Limited Application*

*Copper-reducing substances.*
*Starch content.* Followed by microscopic examination of the starch grains.
*Sieve analysis of particle size distribution* (ground material).
*Microscopic examination* — for insect parts, undesirable plant parts or other foreign matter.
*Limit tests as specified.*

#### *16.2.1.1.3 Additional Specific Tests*

With specified limits in many countries.

*Allspice.* Quercetannic acid.
*Capsicum* (cayenne, red pepper). Capsaicin content, Scoville pungency, color content.
*Clove.* Quercetannic acid, clove stems, and other foreign matter.
*Curry powder.* Sodium chloride.
*Ginger.* Gingerine content (or other measure of pungency), starch, lime (CaO), cold-water extractive, water-soluble ash.
*Marjoram.* Stems and other foreign matter.
*Mustard.* Allyl isothiocyanate, sodium chloride, starch, and other farinaceous matter.
*Paprika.* Color index, iodine value of fixed oil.
*Pepper, black and white.* Piperine content (Kjeldahl nitrogen or instrumental method), pepper-starch content.
*Saffron.* Yellow styles and other foreign matter.
*Tumeric.* Curcumin content (or color index), starch.
*Vanilla beans.* Vanillin content.

### 16.2.1.2 Essential Oils

Essential oils are liquids, some of which may crystallize and solidify on cooling. Such samples should be gently warmed until completely liquid before carrying out any tests. Samples should first be evaluated for appearance (i.e., color, clarity), preferably being compared with a reserve sample, followed by an evaluation of the odor and the flavor in water or a diluted sugar syrup.

###### 16.2.1.2.1 General Tests

*Specific gravity* (normally at 25°C or corrected for any other temperature) or weight per ml at 20°C.
*Refractive index* (RI) at 20°C.
*Optical rotation* (OR) at 20°C.
*Solubility* in alcohol of a specified dilution.

###### 16.2.1.2.2 Tests of Limited Application

*Boiling range* (C).
*Melting point* (C).
*Congealing point* (C).
*Flash point.* Open or closed cup (F or C).

###### 16.2.1.2.3 Instrumental Tests

*Gas chromatography* (to establish identity/origin, detect adulteration, determine compositional changes, and for production quality control).
*Infrared spectroscopy.*
*Ultraviolet spectrophotometry.*
*Mass spectrometry.*

###### 16.2.1.2.4 Specific Tests for Constituents

*Acetals.*
*Acid value.*
*Alcohols.* Aceto-formylation method, total alcohols, tertiary terpene alcohols, citronellol by formylation.
*Aldehydes and ketones.* Bisulfite method, neutral sulfite method, phenylhydrazine method, hydroxylamine method.
*Chlorinated compounds.*
*Esters.* General, high-boiling solvent method.
*Heavy metals.*
*Phenols.* Free, total.

###### 16.2.1.2.5 Tests Specific for Citrus Oils

*UV absorbance.*
*Evaporation residue.*
*Peroxide value.*
*Foreign oils.*

###### 16.2.1.2.6 Specific Tests and Procedures

The special tests applicable to all the major essential oils are described in the Standards and Specifications of the Essential Oil Association of U.S. Where standards are set for essential oils, it is usual for the test method to be described. Variations in the test method may lead to different test results so that the method prescribed should be adhered to in detail.

#### 16.2.1.3 Oleoresins

In physical form, oleoresins range from thin to viscous fluids, from semisolids to waxy or sticky pastes. Samples should be well mixed to ensure uniformity before testing. Initially, the sample should be evaluated for appearance (often characteristic), odor and flavor after dilution in a suitable medium.

##### *16.2.1.3.1 General Tests*

*Volatile oil content,* followed by an examination of the recovered essential oil.
*Solvent residues.*
*Solubility* in a specified solvent.

##### *16.2.1.3.2 Specific Tests*

*Capsicum (cayenne, red pepper) oleoresins.* Capsaicin content, Scoville pungency index, color index.
*Ginger oleoresin.* Gingerine content (or other pungency index).
*Paprika oleoresin.* Color index, pungency test.
*Pepper* (black and white) oleoresin. Piperine content.
*Turmeric oleoresin.* Curcumin content (or color index).

#### 16.2.1.4 Plated or Dispersed Spices

Manufactured flavorings that comprise oleoresins and/or essential oils plated or spread on an edible carrier (e.g., salt, dextrose, flour). If stored in clear glass, the sample may show surface color fading. Samples should be well mixed before carrying out tests.

##### *16.2.1.4.1 General Tests*

*Volatile oil content* — distillation.
*Extractives* — soluble in a nominated solvent.
*Carrier base constituents* — specific tests as appropriate.
*Sieve classification* or test for particle size distribution.

##### *16.2.1.4.2 Tests of Limited Application*

*Microbiological examination.* Total plate count, freedom from coliform, salmonella, etc.

##### *16.2.1.4.3 Specific Tests*

See preceding Oleoresins.

#### 16.2.1.5 Synthetic Chemicals

Samples may be liquids or crystalline solids. The nature of the testing is often determined by the intended end use. Routine physicochemical tests are necessary for control of purity, but odor and flavor evaluations are of more importance when the chemical is intended for use in a flavoring or fragrance composition.

*16.2.1.5.1 General Tests: Liquids*

*Specific gravity* (normally at 25°C or corrected for any other temperatures) or weight per ml at 20°C.
*Refractive index* (RI) at 20°C.
*Optical rotation* (OR) at 20°C.
*Solubility* in alcohol of stated dilution or in other specified solvent.
*Boiling point/range* (C) at specified pressure (mm Hg).
*Freezing/congealing point* (C).
*Flash point* (F or C). Open or closed cup method.

*16.2.1.5.2 General Tests: Solids*

*Melting point* (C).
*Congealing point* (C).
*Solubility* in specified solvent, freedom from insoluble matter.

*16.2.1.5.3 Specific Tests for Chemical Identity and Purity-Instrumental Methods*

*Gas chromatography.*
*Infrared spectroscopy.*

#### 16.2.1.6 Finished Flavorings

Flavorings may be either liquid, emulsions, or dry powders. Samples should first be evaluated for appearance (i.e., color, clarity, opalescence) then for odor and flavor after dilution in an appropriate medium. They should preferably be compared directly with a standard sample that has been maintained under optimum conditions and changed at regular intervals.

*16.2.1.6.1 General Tests: Liquid Flavorings*

*Specific gravity* (SG, normally at 25°C or corrected for other temperatures) or weight per ml at 20°C.
*Alcohol or other specified solvent content* — gas chromatography or distillation and SG.
*Refractive index* at 20°C.
*Color intensity* or shade (as appropriate).
*Flash point* (F or C). Open or closed cup method (if required).

*16.2.1.6.2 General Tests: Emulsions*

*Specific gravity* (25°C) or weight per ml (20°C).
*Volatile oil content.* Steam distillation method.
*Microscopic examination.* Particle size distribution.

*Microbiological examination.* Total plate count, yeasts, molds.
*Ringing test* (for beverage emulsions).
Specific tests for stabilizers, antioxidants, etc.

#### *16.2.1.6.3 General Tests: Encapsulated Dry Flavorings*

*Particle size analysis.* Laser light scattering, Coulter Counter, or microscopic method.
*Moisture content.* Oven drying, Infrared drying, Karl Fischer method, Near Infrared.
*Volatile oil content.* Steam distillation method or organic solvent method.
*Microbiological examination.* Total plate count, coliforms, salmonella, molds, and yeasts.

### 16.2.1.7 Vanilla Extract

Natural vanilla extract is subjected to a wide range of tests to ensure its authenticity. These include:

*Specific gravity* (25°C) or weight per ml at 20°C.
*Alcohol content.*
*Glycerin content.*
*Vanillin content.*
*Neutral lead number* (Wichmann).
*Total solids.*
*Ash.* Total, water soluble, water insoluble (followed by test for alkalinity of ash).
*Acidity* (pH).
*Coloring matter* soluble in amyl alcohol.

### 16.2.1.8 Fruit-Based Products

Fruit juices and fruit-based flavorings should be kept chilled unless the soluble solids content exceeds 65°C Brix, or they should be suitably preserved. Samples should be made uniform and should be subjected to sensory evaluation before applying any of the physicochemical tests.

#### *16.2.1.8.1 General Tests*

*Alcohol content.*
*Total solids.*
*Ash.*
*Fruit and sugar content.*
*Sucrose content.*
*Pectins.*
*Microbiological examination.* Total plate count, molds, yeasts.

*16.2.1.8.2 Special Tests*

*Detection of β-ionone in fruit extracts* — adulteration.
*Detection of strawberry aldehyde* — adulteration.

### 16.2.2 Generally Used Analytical Testing Methods

The following discussion will highlight some of the more commonly performed tests in the typical mid- to large flavor house.

#### 16.2.2.1 Density/Specific Gravity

One of the most commonly performed analyses on ingredients and finished flavors is measuring density (specific gravity). Density is a physical property of a raw material (liquid) or finished flavor which is dependent on composition. Density traditionally was determined using a pycnometer or hydrometer. However these methods are tedious and have been replaced by rapid instrumental methods. Today Mettler Toledo and Paar make automated flow through systems for the rapid density determination (Figure 16.1). To determine density, an empty sample tube is initially electromagnetically excited to vibrate at its natural (resonance) frequency. The tube is then filled with sample and the tube is again excited to its new resonance frequency. The difference in frequency between the tube loaded and empty is related to the mass difference. Therefore, mass and subsequently density is calculated by the instrument. Some advantages to this technique are that neither volume nor weight of sample need be measured. Also, the determination is independent of sample viscosity. Instruments are available that are sensitive to up to five decimal places, which is excellent for QC purposes. Also, these instruments are available with

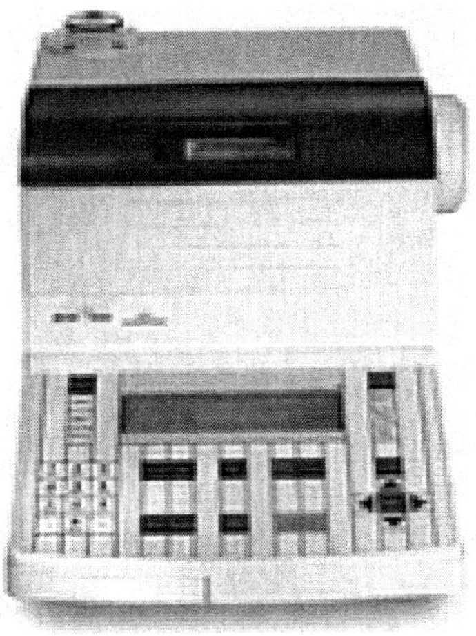

**FIGURE 16.1** Combined refractive index and density meter from Mettler Toledo (Model DR45 Combined density meter and refractometer). (Source http://www.mt.com/mt/product_-detail/product.jsp?m=t&key=Uz_Tc4NjM1NT)

automatic sampling systems and dual functionality (e.g., determines Refractive Index [RI] at the same time) to permit unattended operation. The data output is compatible with most laboratory data systems thereby facilitating easy data handing.

#### 16.2.2.2 Refractive Index

The second analysis that is performed on most liquid ingredients and finished flavors is refractive index (RI). Refractive index is typically done using an Abbe refractometer in a smaller company or by automated systems in mid to large flavor companies. The Abbe refractometer is accurate to four decimal places, covers the RI range typically encountered in flavoring materials, and is rapid. The RI of a flavor is a function of all the components and their proportions. Therefore, this determination is very sensitive to mixture composition. RI and SG will detect many of the compounding errors made in the formulation area.

The system shown in Figure 16.1 is a combined (density and refractive index) system that provides two measurements at one time. This system uses a total reflection method to obtain refractive index. This permits the determination of refractive index of turbid or dark samples.

#### 16.2.2.3 Optical Rotation

Optical rotation is commonly used as in QC for essential oils and oleoresins. The most important flavor constituents in these oils very often are optically active. Therefore, optical rotation may be used to quantify specific flavor components in the oleoresin or essential oil (e.g., AOAC Official Method 920.142 for orange and lemon oils).

#### 16.2.2.4 Alcohol Content

Alcohol (ethanol) content is determined on all flavors that use alcohol as a significant part of the flavoring. That is, where alcohol may be one of the flavor solvents. Flavor companies typically use large volumes of ethanol in manufacturing. They must pay tax on the alcohol (to the Bureau of Alcohol, Tobacco and Firearms, the ATF) unless the flavor company can document that the alcohol was used in a product which was not potable as is. Since most flavorings are not potable at production concentrations, the flavor house can get a drawback (tax rebate) on the alcohol tax. This amounts to a substantial dollar value in most flavor houses so alcohol usage is monitored very closely. Alcohol is typically determined by gas chromatography. The official method (AOAC) involves directly injecting a sample of the flavor into the GC (AOAC Official Method 973.23).

#### 16.2.2.5 Residual Solvent

Residual solvents must be determined on any flavoring material that is obtained by solvent extraction. Oleoresins and absolutes compromise the largest group of flavorings in this category. Table 16.1 presents the legal limits on residual solvents in oleoresins. Note that residual solvents are permitted only in the low ppm levels. Gas

**TABLE 16.1**
**Permitted Levels of Residual Solvents in Oleoresins**

| | | |
|---|---|---|
| 173.210 | *Acetone* | A tolerance of 30 ppm is established for acetone in spice oleoresins when present therein as a residue from the extraction of spice |
| 173.230 | *Ethylene dichloride* | A tolerance of 30 ppm is established for ethylene dichloride in spice oleoresins when present therein as a residue from the extraction of spice; *Provided, however,* that if residues of other chlorinated solvents are also present, the total of all residues of such solvents shall not exceed 30 ppm |
| 173.240 | *Isopropyl alcohol* | Isopropyl alcohol may be present in the following foods under the conditions specified: (a) In spice oleoresins as a residue from the extraction of spice at a level not to exceed 50 parts per million and (b) In lemon oil as a residue in production of the oil at a level not to exceed 6 parts per million |
| 173.250 | *Methyl alcohol* | Methyl alcohol may be present in the following foods under the conditions specified: (a) In spice oleoresins as a residue from the extraction of spice at a level not to exceed 50 parts per million |
| 173.255 | *Methylene chloride* | Methylene chloride may be present in food under the following conditions: (a) In spice oleoresins as a residue from the extraction of spice at a level not to exceed 30 parts per million; provided, that if residues of other chlorinated solvents are also present, the total of all residues of such solvents shall not exceed 30 parts per million |
| 173.270 | *Hexane* | Hexane may be present in the following foods under the conditions specified: (a) In spice oleoresins as a residue from the extraction of spice at a level not to exceed 25 parts per million |
| 173.290 | *Trichloroethylene* | Tolerances are established for residues of trichloroethylene resulting from its use as a solvent in the manufacture of foods as follows: Spice oleoresins 30 parts per million (provided that if residues of other chlorinated solvents are also present, the total of all residues of such solvents in spice oleoresins shall not exceed 30 parts per million) |

*Source:* From 2002, 21CFR173 — Subpart C — Solvents, Lubricants, Release Agents and Related Substances

chromatography is the method of choice for residual solvents (e.g., AOAC Official Method 969.29, Ethylene Dichloride and Trichloroethylene in Spice Oleoresins). An oleoresin is often subjected to an initial distillation step to isolate the volatiles (including residual solvents) from nonvolatile extractives. This distillate is then analyzed for residual solvent.

### 16.2.2.6 Particle Size of Emulsions

As has been pointed out earlier in this text (Chapter 13), emulsions are inherently unstable. Since a primary determinant of stability is particle size, the QC department routinely determines particle size distributions of emulsions.

The method used to determine particle size depends both upon the size of the company and upon the importance of emulsions to the business. A small company may not have the resources to invest $30,000 to $50,000 in sophisticated particle size measurement instrumentation. Even a larger company which may have such resources may choose not to invest this amount of money in this instrumentation if the business does not support it.

The simplest direct method to determine particle size is light microscopy. Typically, a 0.1 to 1.0% dilution of the emulsion in water (or glycerin) is made and then a couple of drops of this dilution are placed on a glass slide. The emulsion slide is covered with a cover slip, then mineral oil and viewed under the oil immersion lens. The microscope is fitted with a stage micrometer so that particle size can be estimated. A quality control decision at this point becomes rather subjective due to the data one obtains in this manner. The data are compromised for several reasons. First, since the particles are not fixed but will move across the field, and only one layer of particles is in focus, it is difficult to obtain a good view of the emulsion. Second, to obtain an accurate estimate of the average droplet size distribution, *several hundred particles* must be sized. Time constraints do not permit this effort — generally, one gets views of a few fields that hopefully are representative of the overall emulsion. Third, the data/view one gets are incomplete in that the resolution of an optical microscope is only about 0.5 μm, which is above the particle sizes that one may find in a good flavor emulsion [3]. Thus, one sees only a portion of the overall size distribution (albeit, one is most interested in the presence of larger particles, i.e., those visible with an optical microscope). In practice, a trained technician typically examines several microscope fields and either accepts or rejects the emulsion. While this would appear to be a rather questionable approach, practical constraints often require that this approach is taken. Contrary to expectations (and the weaknesses mentioned above), this method can be rather effective if the individual has adequate training and experience.

A relatively simple indirect method to estimate emulsion quality is to determine its ability to scatter light using a simple spectrophotometer (spectro-turbidimetric method, [3,4]). In this analysis, an emulsion is diluted to ca. 0.1%, placed in a cuvette, and then light is passed through the sample. The wavelength of the light is chosen so there is no absorption of the light due to emulsion components (ranges from 300 to 800 nm but 400 nm is common), and the reduced light signal is due only to scattering effects. The average particle size can be calculated from light scattering theory or a standard curve (absorbance vs. average particle size) can be generated in-house over time via correlation with other methods. While this method lacks sophistication, it is quite simple and is reasonably accurate.

Alternatively, there are several instrumental methods to determine the particle size distributions of emulsions. The most commonly used techniques involve light scattering or electrical resistance measurements. Common but less used techniques include spectrophotometry and Zeta potential.

Static light (laser) scattering is used to determine particles ranging in size from 0.1–1000 μm [3]. In this technique, a dilute sample of emulsion is placed in a cuvette and a laser beam is directed through the sample. The laser beam is scattered by the emulsion droplets based on their size. The scattering pattern is detected by an array

**FIGURE 16.2** A diagram of the sample cell used to measure particle size. (From Beckman Coulter, Multisizer™ 3 COULTER COUNTER®, www.beckman.com/products/Discipline/LifeScienceResearch/prdiscgenpartchar.asp?bhcp=1. With permission.)

of photodiodes and analyzed by a computer program that fits a size distribution profile to the observed scattering pattern. This technique is rapid, automated, and easy to use. The primary disadvantages are that the emulsion must be transparent and very dilute [5]. There is some danger that dilution may disrupt emulsion flocculation and thereby be in error.

Methods based on electrical resistance (electrozone sensing) are more than 50 years old and thus have a long history of use in the field. This technique involves drawing an emulsion through an orifice and measuring the change in electrical resistance as emulsion particles (emulsion is diluted in electrolyte) are drawn through an orifice. As is seen in Figure 16.2, one electrode is placed on one side of an orifice while the second electrode is placed on the opposite side of the orifice. Since the only path for electrical conductivity is through this orifice, the passage of an emulsion particle through the orifice changes electrical current flow. The degree of change in current flow is proportional to the volume of the particle. Therefore, this technique both counts and sizes each particle drawn through the orifice. Particle size distributions can be obtained in only a few minutes. The primary drawbacks of this method are: the particle size range is limited (0.4 to 400 μm according to McClements, [3] but 0.4 to 1200 μm according to the manufacturer [6]); different orifices are required for different particle sizes; the emulsion must be conductive (generally have to add an electrolyte which may change emulsion properties); the emulsion must be diluted (may change flocculation state) and the very small orifices are prone to plugging [3,7]. It is this author's experience that the method is excellent for typical flavor or cloud emulsions, but it can be problematic when trying to measure these emulsions due to the need to use a small orifice (15 μm orifice). This orifice readily plugs when using natural materials (e.g., gum acacia).

The final method to be mentioned is Zeta potential (may also be referred to as an electrophoretic method). In this method, the diluted emulsion is placed in a measurement cell, and a static electric field is applied. Emulsion droplets carrying an electrostatic charge will move in the field at a velocity that is a function of several factors, including viscosity of the emulsion, net charge on the particle, and particle size. The velocity of movement can be determined using a light scattering method or optical microscopy (measures how far a particle moved in a given time). This method is liked by some [8] since it measures both electrostatic charge on the emulsion particles as well as their size. Since charge is related to droplet stability (repulsive forces), this determination adds another dimension related to stability. It is this author's observation that while the method has theoretical merit, its cost and time requirements are not justified.

#### 16.2.2.7 Volatile Oil

The volatile oil content of various botanicals is a measure of flavor strength. It is typically the volatile components which carry the flavor of a spice or herb (exceptions being those materials that have a "bite"). Therefore, the QC department will run volatile oil content on many botanicals used as flavoring materials.

Volatile oil is also run on some spray dried flavorings [9,10]. Spray-dried citrus oils, mint oils, spice oils, etc., are commonly analyzed for volatile oil to determine the efficiency of the encapsulation procedure. Volatile oil is generally measured using a distillation method (e.g., AOAC Official Method 962.17). The Clevenger trap (Figure 16.3) is designed to volumetrically measure volatile oil content of water insoluble flavorings. Sufficient material is placed in the sample flask to yield 10 to 15 ml of volatile oil. Water and antifoam are added to the sample flask, and then it is heated under reflux for 2 to 4 h. The volume of oil present in the sample can be read directly from the graduated trap.

An alternative method for measuring volatile oil is the Fosslet method. This method involves placing the sample (a botanical or spray-dried powder) into a sample holder, which has a sliding hammer inside it. Perchloroethylene is added, and then the sample holder is placed in the vibrating reactor. The vibrating reactor is similar to a paint shaker and subjects the sample to vigorous shaking plus the smashing action of this hammer. Extraction is completed in 1 to 2 min. The sample holder is then opened, a dessicant added, and the perchloroethylene/essential oil mixture is

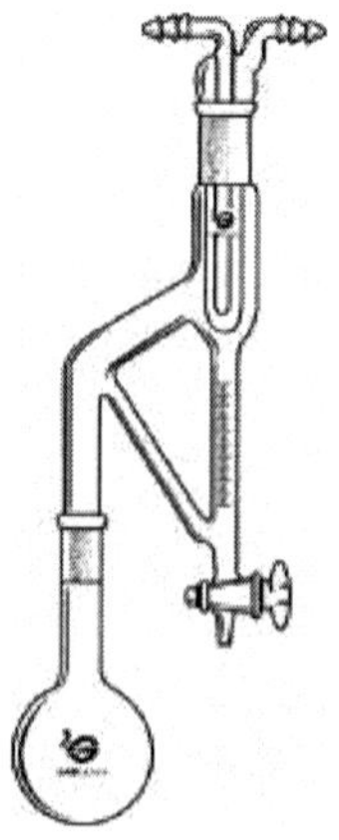

**FIGURE 16.3** Clevenger trap for the analysis of volatile oils lighter than water. (Source: http://www.lab-glass.com/html/nf/DSTA-LG-6655.html. With permission.)

filtered into an electronic device for the measurement of specific gravity of the filtrate. Since perchloroethylene has a very high specific gravity, specific gravity will be lowered depending upon the concentration of essential oil in solution. The instrument may be calibrated to read out the percentage of extractable oil. This is a very rapid method (5 to 10 min per test) that is gaining use in the industry.

#### 16.2.2.8 Surface Oil

Surface oil refers to the flavoring oil that is on the surface of an encapsulated flavoring material. This surface oil is of particular interest since it is exposed to light, oxygen, and package environment and will readily become oxidized and potentially produce an off-flavor in the product.

Surface oil is typically determined only on encapsulated essential oil products. This is because some of the essential oils (e.g., citrus oils) are extremely labile to oxidative deterioration. This analysis is not performed on a routine basis but only occasionally to monitor the spray drying process.

Surface oil is most commonly determined gravimetrically. This involves placing the encapsulated product in an Erlenmeyer flask and adding organic solvent. The flask is shaken for several minutes, the powder filtered from the solvent, and then the solvent removed by evaporation. The residue, or surface oil, would be determined by weighing the residue. Residue may also be quantified by gas chromatography [10]. In this procedure, the organic solvent would contain an internal standard. Evaporation of the filtrate would be stopped at a concentration suitable for gas chromatographic determination.

#### 16.2.2.9 Moisture Content

Moisture content is of concern in dry flavorings (e.g., dry whole spices and encapsulated flavors). Since flavorings are volatile, one has to be careful in the application of drying methods. Several methods are available for this purpose [10].

The most common method for the determination of moisture in botanical materials and dry flavorings is the *Karl Fischer*. The Karl Fischer method is based on the titration of water in the sample with the Karl Fischer reagents as noted below [11]. The reagents are formulated such that 1 ml of reagent will titrate 3.5 mg of water. This method may use an optical endpoint (appearance of a dark red-brown color) or a conductivity endpoint that permits automated operation.

$$C_5H_5N\bullet I_2 + C_5H_5N\bullet SO_2 + H_2O \rightarrow C_5H_5N\bullet HI + C_5H_5N\bullet SO_3$$

$$C_5H_5N\bullet SO_3 + CH_3OH \rightarrow C_5H_5N(H)\bullet SO_4\bullet CH_3$$

If the samples are dry, it is necessary to extract the water from the sample typically with hot anhydrous methanol. The extracted water is then titrated to determine the water in the extracting solvent. Errors may enter the determination due to: (1) incomplete water extraction, (2) contamination with atmospheric moisture, and (3) interferences from some food constituents. Food constituents that may interfere with the method are ascorbic acid (oxidized to dehydroascorbic acid — yields higher values), carbonyl compounds (may be many flavor components — carbonyls react with methanol to form acetals thereby releasing water. Also cause fading endpoints), and fatty acids (react with iodine so overestimate moisture content). Even considering these potential interferences, the method is widely used in the industry.

Despite the earlier note that drying methods must be used with caution, simple drying methods may be applied to moisture determination. Very commonly, aroma constituents are trapped within an encapsulated flavor or botanical material and are not readily lost even on heating. Moisture in encapsulated flavorings may be determined by placing a sample in an oven at 100°C and allowing the sample adequate time to reach a constant weight.

Historically moisture has often been determined using a distillation technique (AOAC Official Method 986.21. Moisture in Spices). In this method, a sample is placed in a standard taper flask and covered with an adequate amount of solvent (toluene for most spices and encapsulated flavorings and hexane for capsicums, onions, garlic, and other spices containing large amounts of sugar), a Bidwell-Sterling trap attached (Figure 16.4), and then the sample is heated to bring the solvent to boiling. The solvent vapor carries some water vapor with it into a condenser where both the solvent and water vapor are condensed [11]. The water and solvent are immiscible, and the water is heavier so they separate with the water being trapped in the collection finger of the Bidwell-Sterling trap. The solvent flows back into the sample flask to reboil and entrain more water. At some point, there is no more water in the sample and the distillation is stopped. The water is read from the graduated side trap. While this method is simple and reasonably accurate, the desire to remove solvents from the workplace and costs of disposing of used solvents make this method less popular today.

There are several other methods that a flavor house may use to determine moisture content. They include microwave or infrared ovens (may be used in a spray drying operation for rapid analysis for process control) and refractive index (liquid

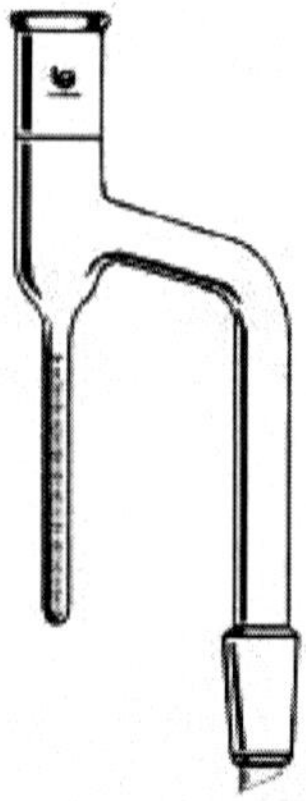

**FIGURE 16.4** A reflux distillation (Bidwell-Sterling) trap for the determination of moisture in flavorings. (Source: http://www.lab-glass.com/html/nf/DSTA-LG-6655.html. With permission.)

streams, e.g., a fruit juice, or a spray-drier infeed emulsion), but these methods have very specific uses. We are seeing near infrared coming into use due to its wide applicability and speed of analysis.

### 16.2.2.10 Gas Chromatography

Gas chromatography (GC) is generally performed on all incoming (volatile) raw materials and some finished flavorings. Incoming ingredients are analyzed by GC to determine flavor quality, purity, or detect adulteration in the ingredient. Many flavoring materials are prone to deterioration during storage. Citrus oils are exceptionally susceptible to oxidative degradations. Even pure chemicals may undergo undesirable reactions when stored. Benzaldehyde is readily oxidized to benzoic acid, and limonene both oxidizes and polymerizes. Therefore a GC profile may be used as one method to determine quality of the ingredient. Needless to say, GC is well suited to purity measurements and may be used in detecting ingredient adulteration.

Historically, GC was not routinely performed on finished flavorings. GC was slow, and the sample load did not permit this analysis on all outgoing products. However, the advent of automated operation and fast GC has resulted in GC being done on most finished flavorings. The purpose of running GC on a finished flavor may be more for insurance than to catch errors in formulation. Most compounding errors are caught by the simple RI, density, and sensory analyses.

#### *16.2.2.10.1 Incoming Ingredients*

GC is well suited to determining the composition of flavoring materials whether they are pure chemicals or natural plant materials. GC analysis of pure chemicals insures that one has purchased the chemical intended and the primary component is present at the expected level. One has to recognize that a food grade chemical is not necessarily pure. For example, natural linalyl acetate is only 80+ % pure [12] while the corresponding synthetic chemical (Kosher synthetic) is available at 97+ % purity [13]. Food grade is an assurance that the method of synthesis has not left any toxic components in the material at levels considered hazardous. GC analysis of the material assures the purchaser that the proper active component is present at the levels anticipated. Impurities present in the chromatogram are of limited interest unless they have sensory significance. Then one would monitor not only the primary component, but selected impurities, i.e., those with sensory significance that may change the sensory character of the ingredient.

Plant materials (botanical extracts such as essential oils, oleoresins, absolutes, etc.) are also analyzed by GC. The QA laboratory will develop standardized GC protocols that permit the generation of standard chromatograms that can be over laid on GC runs of previously acceptable incoming ingredient. While one will never obtain an identical GC chromatogram from one sample to another, with time, the operator develops a feeling for what variation in the chromatogram is acceptable and what is not. This is where laboratory data systems have become very useful. Previous chromatograms can be stored and retrieved to make chromatographic overlays of many lots of a given ingredient. One can see the variability and note changes over time in the material.

### *16.2.2.10.2 Finished Flavors*

As was mentioned earlier, only recently has it been practical to obtain chromatograms of most if not all of the outgoing flavorings. The primary purpose of obtaining these chromatograms is for in-house use. This author does not recommend that chromatograms be provided of flavorings to their customers. The reason is that the chromatographic profile of a given flavor may vary considerably and yet have the same sensory properties. This variation will typically result from using a different supplier of an ingredient (different impurities in a pure chemical or composition of a botanical extract). The variation in the raw material may be evident in the chromatogram but have no sensory impact whatsoever. Yet, the customer may look at the chromatogram and be concerned that the flavor is different and a problem between the supplier and customer may occur needlessly.

The in-house use of GC chromatograms of finished flavors relates to documenting the composition of a flavoring. This documentation becomes useful when there is a problem with the flavoring, e.g., the sensory analyst rejects a new production lot of a flavoring. The QA manager can readily retrieve the GC chromatogram of the new production lot, chromatograms of past production lots, plus all other reference data (e.g., density, refractive index, UV profile, etc.) and have a good idea of the problem. While the density, refractive index and UV profile etc. will suggest there is a problem, only the GC chromatogram will suggest the *source* of the problem. Perhaps there is too much solvent, too little solvent, some component missing, or too much of some component(s), or unusual peaks appearing in the new production lot relative to previous lots. The GC chromatogram is useful in problem solving.

### *16.2.2.10.3 Flavor Duplication*

The QA lab may be called on to help in the duplication of a competitor's flavoring, and GC is the major tool used in this task. While much of this work has traditionally been done by the creative flavorist, instrumental data can be very helpful in duplication.

The difficulty of obtaining a reliable analysis of a competing flavor depends upon the level of information requested and the flavoring being analyzed. If the flavorist just wishes to know what components are present in the target flavoring, the task is relatively easy. However, if the flavorist asks for a quantitative estimate of the components present in a flavoring, the task becomes much more difficult. Also, it is easier to analyze flavors that have simple solvent systems and only pure chemicals as opposed to those that contain nonvolatile solids or complex botanical extracts. An idea of how this task is approached follows.

The first question in analysis is how to introduce a sample into a GC. If the flavoring is present in a simple solvent system, e.g., alcohol, triethyl citrate, triacetin, or propylene glycol, it can be directly injected into a GC. However, if the flavoring contains nonvolatile components (e.g., vegetable oil solvent, fruit solids, or emulsifiers), this is not possible and some sample preparation method must be initially applied to the sample before GC analysis. Unfortunately, the sample preparation method will bias the analytical profile (perhaps even miss some compounds) and greatly complicate any quantitative efforts (refer to Chapter 3). Sample preparation may include steam distillation (suitable for either oil or water soluble flavoring systems), solvent extraction (suitable for water based flavoring systems), dynamic headspace or SPME/Stir Bar analysis.

The GC analysis of the flavor, or suitable flavor isolate, should be done using two different polarity GC columns (typically Carbowax and methyl silicone phases). One should also do GC olfactometry (GCO) analysis since this helps to identify compounds, tells one if there are compounds with very low sensory thresholds being used (perhaps no GC peak there), and also if there is a compound coeluting with the solvent. The next step is compound identification (this has been discussed in Chapter 3).

If quantitative data are desired, one has several approaches available. One could use the raw GC peak areas, but this will result in substantial error even if the sample is analyzed by direct injection. Normalized GC peak areas would yield much more accurate values, but this requires calibration of the instrument and has little value if any sample isolation methodology was used. Other methods for quantification include standard addition, the development of standard curves, or the use of stable isotopes. Whatever method is used, one needs to formulate the flavor according to the analytical data and analyze it in the same manner as the target sample. One can then compare results between the formulated duplicate and the target sample. The sample formulation will likely have to be adjusted, reanalyzed, and reformulated again. There is little question that reasonably good analytical data can be obtained if there is time for this task. Unfortunately, time is usually limited and the data are less reliable than desired (or possible).

### 16.2.2.11 Spectroscopic Analysis

UV-Vis spectroscopy is used in a similar manner as density and refractive index. This determination is performed on most incoming and finished flavorings (liquids). The UV-Vis spectrum ranges from 200–700 nm (UV 200–350 and Vis 350–700 nm). This determination measures specific components of a material, i.e., those that absorb light in the noted regions, as opposed to overall composition as is measured by density and refractive index. Thus, one has a quick and simple method to measure specific components of a mixture, i.e., those that contain double bonds.

UV-Vis may be used in the QA laboratory to either determine the concentration of a particular component in an incoming ingredient (quantitative analysis) or to obtain an overall spectrum (scan from 200–700 nm). An overall spectrum has value in providing a profile for future reference of either new batches of an incoming ingredient or a newly made finished flavor. These determinations are often made on dilutions that are used for sensory testing and require only 30 sec to obtain.

QA accept/reject decisions are aided by being able to recall the UV-Vis spectra of the last several batches of an ingredient or flavoring, and overlay their spectra much like one does with GC chromatograms. The objective of this testing is to assure that a new flavor formulation or incoming ingredient is acceptable in the quickest manner possible. Thus, methods such as this that are quick, cheap, and *automated* are exceptionally valuable.

#### 16.2.2.12 Microbiological Analysis

The majority of flavorings produced in a flavor house will not support the growth of microorganisms. As mentioned earlier, they are toxic to most microorganisms. Therefore, the QC lab does not typically run a large number of microbiological assays. They will run microbiological analyses on spray dried products, emulsions, fruit preparations, and some of the natural flavorings. The microbiology lab may be set up to run standard plate count, yeasts, and molds, and lactobacilli. Salmonella and other pathogens are generally sent out to service laboratories for analysis.

#### 16.2.2.13 Electronic Noses

Electronic noses have been discussed in Chapter 3 of this text. As was discussed, it is this author's opinion that electronic noses have little application in research but may have application in QA applications. An electronic nose offers another rapid technique to perhaps determine if a new ingredient or formulated flavoring differs from a previously accepted lot. Similar to other analytical methods employed in the QA laboratory, this technique is able to detect certain errors and thereby may help accomplish the mission of the QA laboratory.

## 16.3 SENSORY ANALYSIS

Sensory analysis is performed on all raw materials on receipt and periodically during their life within the company. The commitment to retesting *prior to use* is very much a key to meeting the primary objective of the QA laboratory, i.e., to insure that flavorings are formulated properly using approved, quality ingredients. Inherent in the approved, quality ingredient concept is that they are of the proper quality not only when received but more importantly, when used. Sensory analysis is also performed on all finished flavors before they go to the customer. Sensory is the final word on acceptance of an ingredient or finished flavor. The following description of sensory analysis as a QC function is reasonably typical of the industry as a whole. There will be some differences in operation from company to company; however, the basic approach is similar.

### 16.3.1 Incoming Raw Ingredients

An appropriate sample (one that accurately represents the material received) of the ingredient is brought to the QA laboratory and properly logged into the system (chain of custody). Samples are taken for analytical testing and a sample is prepared for sensory evaluation. The sensory sample is typically prepared at a dilution appropriate for that given material (prepared in water as the final tasting medium). While the samples are commonly prepared at low ppm levels, some materials are extremely potent and must be prepared at ppb (or lower) levels. A retention sample of the last *accepted* lot of the same material is prepared as well. The new sample is then evaluated relative to the retention sample for appearance, odor, and flavor. This evaluation is most commonly done by one experienced individual. While paneling would be more desirable, the large number of samples that come through the QC lab make this virtually impossible. It is common for over 50 samples a day to be evaluated in a larger flavor house. If on evaluation, the individual feels there is a difference between the new sample and the retention sample, a second individual would be asked to evaluate the two samples. If there is agreement that the samples differ in some respect, then a second sample may be taken, being *absolutely sure* it is representative of the lot, and additional retention samples may be retrieved for comparison as well. At this point, it is critical to mention that the status of the retention samples is critical.

Retention samples must be stored under conditions that permit easy retrieval and minimal change over time. Commonly the last three to four samples of accepted lots of raw materials are retained. This sample history may cover a significant time period, and inappropriate storage conditions may result in sample deterioration such that the retention samples do not represent the new fresh material. The first key to dealing with this issue is storing samples in colored bottles, full, in the dark, and at the appropriate temperature. While one may be inclined to use a very low temperature (refrigeration or frozen storage), some samples may be adversely affected by the low temperature, e.g., some components may precipitate. Thus, a storage temperature of ca. 55°F is common. This slows chemical reactions but would not have an adverse affect on quality. The second key is to have an experienced individual(s) doing the evaluation. An experienced person would know what changes to expect in a retention sample over time. For example, a citrus oil may lighten in color, become slightly more viscous, and take on a terpeny note. A vanilla sample may show signs of Schiff's base formation (color formation). A juice may show signs of browning both in terms of color and the appearance of a pruney flavor. In each of these cases, the newly received sample should not be rejected if it lacked these signs of aging.

Ease of sample retrieval was mentioned earlier because this becomes an issue considering the large number of samples that must be retained. While only three to four samples of raw materials may be retained, it is common that the last six samples of a finished flavoring may be retained. This may cover up to two years time. The total number of samples retained is formidable, and the proper cataloging and retrieval becomes problematic. Storage facilities such as shown in Figure 16.5 become a tremendous asset. This is a climate controlled, vertical stacking tray assembly with computer-based filing and retrieval.

### 16.3.2 Finished Flavors

#### 16.3.2.1 Sensory Evaluation

The evaluation of finished flavors is quite similar to that used in the evaluation of incoming ingredients. Samples come to the QA laboratory directly from the compounder in production. Along with each sample is a form with the name and number of the flavor, the amount produced, date of production, compounder's name, and customer's name. (Today this may be replaced by a bar code that references this data in a central database.) This information allows QA to retrieve information on that particular flavor.

**FIGURE 16.5** Sample storage/retrieval system at Robertet Flavors, Inc. (Photo courtesy of Robertet Flavors, Inc.)

Similar to raw materials, the standard for comparison on finished flavors is a sample of the most recent lot of the flavor that was accepted by that customer. The samples (new and retention) are set up in a side-by-side comparison in order to evaluate appearance, odor, and flavor. While a company may chose to evaluate all flavorings in a water solution, it is common to evaluate certain flavors in bases that enhance the flavor being evaluated. This may mean that a flavor company uses water, sweetened water, or sweetened, acidified (citric acid) water in this evaluation. Water may be used for tasting flavors such as meat, onion, garlic, butter, coffee, and most spicing compounds. The sweetened water may be used to taste flavors such as mint, spearmint, peppermint, wintergreen, vanilla, and banana flavors. The sweet/acidic water solution may be used to evaluate fruit flavors. The level of flavor used for evaluation depends upon the strength of the given flavor. This would be specified in the testing protocols and commonly ranges from 0.1 to 1%.

#### 16.3.2.2 Changes in Finished Flavors with Age

Finished flavorings undergo the same types of reactions as their raw materials plus some caused by reactions with other materials not found in a given raw material or with the solvent. Examples of flavor reactions with two common flavor solvents (ethanol and propylene glycol) are illustrated in Figure 16.6 and Figure 16.7. Figure 16.6 illustrates the reaction of aldehydes (acetaldehyde in this example) with an alcoholic solvent (ethanol or propylene glycol) to form the corresponding acetals. The acetals are generally fairly unpleasant and are seldom desired in a flavoring. This reaction takes place fairly quickly. A second flavor:solvent reaction is shown in Figure 16.7. This reaction yields esters. The point being that a flavor changes with age and a freshly made flavor may change with time (from the time initially made to even one day later). One may in fact chose *not* to evaluate some flavors until they have been aged a given amount of time. Likewise, an older retention sample may have undergone some of these reactions and pick up an acetal note or increased in fruitiness (ester formation). In either of these cases, one would not reject the freshly made flavor because it lacked the acetal note or was slightly less fruity than the older sample. This insight comes from experience evaluating flavors.

Thus, if the retention sample is substantially older than the new sample, any one or combination of changes may have occurred in the older sample. The discrepancy between the retained sample and new sample then is not necessarily cause for rejection of the new sample. If, however, the new sample shows such signs of color or flavor changes, this may be cause for concern.

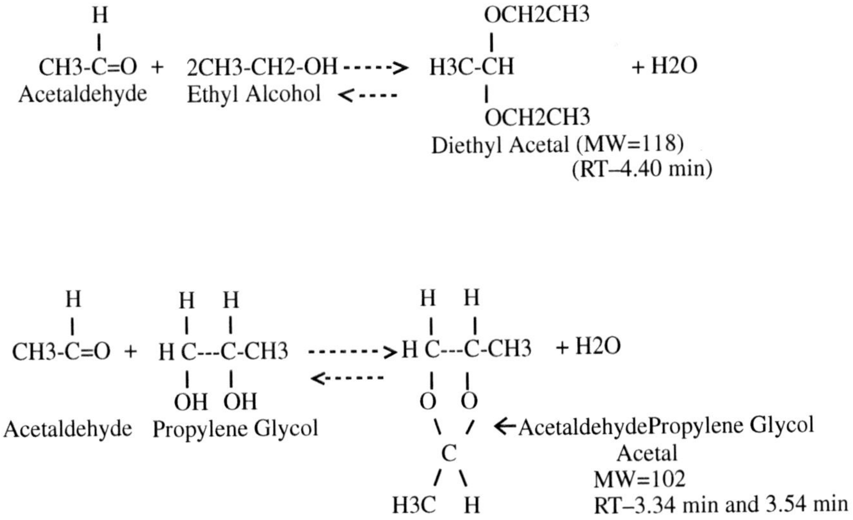


**FIGURE 16.6** The formation of acetals in flavorings from aldehyde:ethanol reactions. (From Vollaro, F., Physical/Chemical Evaluation: Methods of Detection of Changes, Presentation at an IFT sponsored flavor workshop titled, Flavor Interaction with Food, Orlando, FL, March, 2005. With permission.)

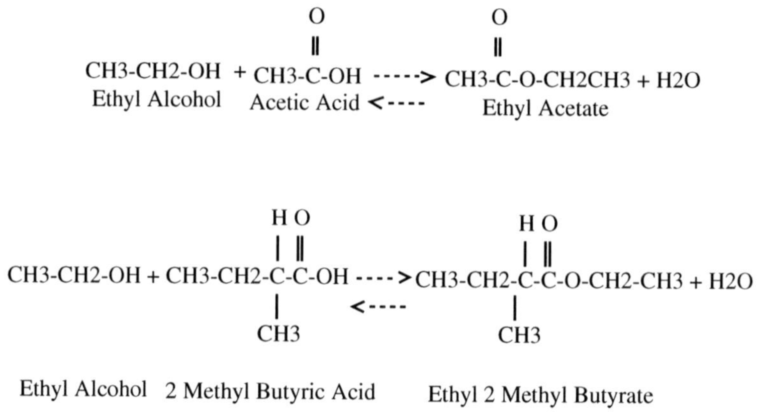


**FIGURE 16.7** The formation of esters in flavorings due to alcohol:acid reactions. (From Vollaro, F., Physical/Chemical Evaluation: Methods of Detection of Changes, Presentation at an IFT sponsored flavor workshop titled, Flavor Interaction with Food, Orlando, FL, March, 2002. With permission.)

### 16.3.2.3 Sample Rejection

Assuming that the retention sample has undergone little deterioration, the criteria for rejection of a flavor is a difference in color and/or flavor between the newly produced flavor and the older retention sample. Flavor character and strength are most important. Slight color differences are often tolerated, depending on what the finished product will be. If the newly made flavor is different from the retention sample, the first step is to pull the second most recent retention sample and resample the newly made flavoring. It is always possible that this first retention sample was off for some reason or the sample of newly made flavor was not representative. If this does not seem to be the case, then the new flavor must be rejected.

There are several reasons why the new flavoring may be rejected. The first is that the compounder might have improperly sampled the batch or even sent the wrong sample. That is why a second sample is always taken. A second cause of rejection may be the use of old stock materials. If old juice concentrates or citrus oils have been used, the finished flavors would exhibit flavor and color changes characteristic of browned or oxidized stock material. Since the compounders keep very extensive data on each flavor they make, it is easy to determine the age of the stock materials used through lot numbers. Also the use of computerized formulations can tie into stock records and expected life of a raw material and tell the compounder that a given raw material needs to be retested prior to use. This helps greatly to eliminate this type of formulation error.

A third alternative would be that the compounder used an incorrect ratio of aromatics to solvents. This would result in a difference in flavor strength. For example, if too much alcohol solvent were added, the flavor would be weaker than the stock sample. An excess of alcohol would make the specific gravity lower than the specified range. This problem can be solved by calculating the amount of flavor materials needed to bring the finished flavor back to the proper ratio of aromatics to solvent. The result is a larger batch of properly proportioned material.

The fourth alternative is one in which a compounder has used a wrong flavor component(s). In this case, it is necessary to reject the material, throw it away and begin again. This is a costly error since flavor materials are expensive to purchase and dispose of. The use of automated compounding stations where the desired flavoring materials are electronically selected and metered into a mixing vessel minimizes this type of error.

### 16.3.3 Colorings

Some items are used in flavors primarily to impart color. Two examples of these are caramel coloring and special grape extract. Quality control on these items is a color check both sensorially and instrumentally. The sensory evaluation is done in a 50-ml Nestler tube. The tube is filled with 50 ml of water to which 0.1 ml of the caramel color is added. This sample is compared to a retention sample for color, intensity, and clarity. Special grape extract is set up the same way, except that three drops of 50% citric acid are added to the Nestler tube. Special grape extract is made from grape skins and gets its color from anthocyanins. Color from anthocyanins is pH dependent, therefore, a small amount of citric acid must be added to get the characteristic reddish, purple color. In a basic solution, the special grape extract would appear blue. It is usually used in fruit flavors that are acidic or that will go into a finished product that is acidic. Instrumental analysis would be done either by a UV-Vis scan or a Hunter reflectance measurement. Again, the instrumental data are compared to retention samples for acceptance or rejection.

### 16.3.4 Scoville Heat Units

One of the few situations requiring a QC panel evaluation is the determination of Scoville heat units. While there are analytical methods available for the determination of capsaicins [15], a sensory approach [16,17] is often still used.

The Scoville heat unit is a measurement of the pungency or "biting" of products such as oleoresin red pepper. This test involves making an initial dilution of the flavoring material in alcohol. This solubilized material is then diluted with water (5% sucrose) to a standard dilution (that would give a Scoville value of 240,000). The solution is then sequentially diluted to determine the dilution at which three of five tasters can detect pungency. A one-to-one dilution of the standard dilution gives the flavoring 480,000 Scoville units while a one-to-two dilution yields 720,000 Scoville units, etc.

### 16.3.5 Summary

The QA laboratory ultimately combines both analytical and sensory evaluations to make a final decision on the acceptability of an incoming ingredient or outgoing flavor. Any questionable data point, sensory or instrumental, is checked most carefully. The QA laboratory must assure that good quality ingredients are used in manufacturing and that finished flavors meet the specifications demanded by the customer. The QA laboratory is a key part of a typical flavor house since the flavor house will maintain an inventory of somewhere between 2,000 and 4,000 raw materials and at least the same number of finished flavors [18]. The task of guaranteeing that each raw material is good and each flavoring made properly is a most demanding task.

## 16.4 ADULTERATION TESTING

The QA laboratory serves functions beyond assuring the quality of manufactured products. It is involved in any problem solving regarding flavor quality, supports research (analytical capabilities), and may check incoming ingredients not only for quality but also adulteration. The responsibility for troubleshooting is obvious. The QA laboratory has all of the data that were used in the acceptance of a given production run and is the place to start the search for the problem. They also have the most experience in what types of reactions/problems can occur in a given flavor. Thus, this responsibility is quite logical.

Its function in supporting research is also obvious. Larger companies do considerable synthesis of new chemicals, analysis of new raw materials, and processes for their recovery. The very strong analytical capabilities of this laboratory are again logically utilized in these activities.

Checking incoming materials for adulteration is primarily done by instrumental methods, placing this responsibility also in the hands of the QA laboratory. Adulteration has been a substantial problem for many years. Some natural flavoring ingredients will sell for several hundred dollars per pound. The dilution of an expensive ingredient with a cheaper ingredient is unfortunately too common.

Monitoring materials for adulteration can be extremely complicated or relatively simple. The complexity involved depends upon the ingredient and the level of assurance of authenticity desired. Due to the potential complexity, this activity is usually only pursued in depth by mid- to large flavor houses. A brief description of this activity follows.

The detection of adulteration is a particularly challenging task since adulteration is typically accomplished by adding the same chemical component(s) to an ingredient as is (or, as are) present there normally only the addition is from a cheaper source — perhaps a synthetic chemical as opposed to a natural chemical, or a cheaper essential oil vs. an expensive oil. Thus, we are generally trying to distinguish between sources of a given chemical as opposed to the presence of a unique chemical (a purity or deterioration issue).

In discussing new isotopic methods to detect flavor adulteration, Martin et al., [19] have provided a table summarizing methods used to detect adulteration in flavoring materials (Table 16.2). This table has been edited to be somewhat more inclusive.

### 16.4.1 GAS CHROMATOGRAPHY/MASS SPECTROMETRY

GC by itself has very limited capabilities to detect adulteration. One obtains a profile that may be used to monitor consistency of the material, but that is about it. One

**TABLE 16.2**
**Instrumental Methods Used to Detect Adulteration in Flavoring Materials**

| | |
|---|---|
| | **Gas Chromatography** |
| Information obtained | Chromatographic profile<br>Quantitative determination |
| Aim | Monitor the consistency of supplies (after identification using GC-MS) |
| Limitations | Cannot differentiate between natural and synthetic flavors |
| | **GC-MS** |
| Information obtained | Chromatographic profile with identification and relative quantification |
| Aim | Closer monitoring of the consistency of supplies; detection of compounds that do not conform to their designation and/or are synthetic (nonnature identical) |
| Limitations | Although different constituents can be identified; the technique cannot differentiate between natural and nature identical (NI) flavors |
| | **Enantioselective MDGC-MS (Multidimensional GC-MS)** |
| Information obtained | Enantiomeric distribution of optically-active molecules |
| Aim | Detection of the addition of chiral NI molecules<br>Authentication of chiral molecules |
| Limitations | Only able to conclude for optically-active constituents of a flavor<br>Biosynthetic flavors can be difficult to classify |
| | **$^{14}C$ Dating** |
| Information obtained | Determine the level of radioactivity present in flavor compound |
| Aim | Determine the age of the carbon in the sample |
| Limitations | Can only determine of carbon is from recent time period<br>Requires large amount of pure sample |
| | **GC-IRMS** |
| Information obtained | $^{13}C/^{12}C$ isotope ratio of main peaks in chromatogram |
| Aim | Check the authenticity (NI or natural) of components that can be separated |
| Limitations | Using ratios it is not always possible to discriminate between all natural and synthetic sources |
| | **SNIF-NMR** |
| Information obtained | Isotope ratios $^{13}C/^{12}C$ (D/H) of different sites of flavor molecules extracted from a product |
| Aim | Check the authenticity (natural or NI) |
| Limitations | Requires > 100 mg of pure single material extraction often necessary |

*Source:* From Martin, Y.-L., G.G. Martin, G. Remaud, *Food Testing Anal.*, 3, 6, 1998. With permission.

needs the ability to identify components to have more value. If one knows what is present and how much, the monitoring of adulteration is facilitated greatly. Three examples of applying GC-MS to this task are presented.

The simplest and surest application of GC-MS methodology is to search for nonnatural compounds in a flavor. For example, ethyl vanillin, allyl caproate, and ethyl maltol do not exist in nature so their presence in a flavoring (e.g., ethyl vanillin in a vanilla flavor, allyl caproate in a pineapple flavor, and ethyl maltol in a sweet flavor) are guarantees that the flavor has been adultered with a synthetic component(s). Likewise the identification of a solvent commonly associated with artificial flavors (such as triethyl citrate, triacetin, or propylene glycol) would raise questions about their origin in a flavoring. While adulteration with these solvents (and often complete artificial flavors) was commonplace at one time, adulteration has become much more sophisticated, and thus one seldom finds the task so easy.

The adulteration of expensive essential oils with cheaper oils traditionally has been detected by obtaining GC-MS profiles of the oil. The analyst would search for components in the expensive oil that are not commercially available and are unique to this oil. An example is β-selinene in oil of celery. Authentic oil of celery should contain 7 to 7.5% β-selinene [20]. Oils containing less than 7% are suspected of being adulterated. A second example would be the adulteration of oil of peppermint with oil of cornmint. Oil of peppermint contains sabinene hydrate at a level of 1% while cornmint does not contain this component. Thus peppermint oils low in sabinene hydrate would be suspect as being adulterated. Unfortunately, more of these markers are becoming commercially available to correct the GC profile, so this approach has lost much of its value. Also, the oils today are often adulterated with pure components (e.g., synthetic linalool is added to oil of neroli, or cinnamic aldehyde is added to oil of cassia) as opposed to another essential oil. Thus a GC profile of an adulterated oil may appear to be high in certain components (e.g., linalool in neroli or cinnamic aldehyde in oil of cassia) but natural variation in the oil or method of processing make these quantitative differences difficult to interpret. One has to determine the source of these given chemicals to prove adulteration.

A final approach is to look for trace impurities associated with the synthesis of a given chemical, i.e., one added as an adulterant. One would expect to find very unique impurities associated with an organic synthesis pathway as opposed to typical enzymatic pathways used in nature. This is illustrated for the detection of synthetic cinnamic aldehyde in oil of cassia [21]. Synthetic cinnamic aldehyde contains traces of phenyl pentadienal (400–500 pm). Thus, finding phenyl pentadienal in a suspect oil can strongly indicate the addition of synthetic cinnamic aldehyde to the oil. As is shown in Figure 16.8, the combination of GC-MS operated in the Selected Ion Monitoring (SIM) mode provides the ability to detect 2% addition of synthetic cinnamic aldehyde to oil of cassia. The way to avoid detection using this method is to obtain extremely pure cinnamic aldehyde, i.e., one that contains very little of the marker impurity. Similarly, one might look for dihydro linalool in synthetic linalool (present at 0.5–2% in synthetic linalool) [21].

### 16.4.2 Chiral Compounds

Enantioselective GC-MS or MDGC-MS (multidimensional GC-MS) has value in determining the chiral composition of a flavoring which is useful in detecting adulteration [22]. Natural aroma chemicals are produced enzymatically, and thus have strong chiral preference. Compounds produced by organic syntheses, at least those involving a chiral center, will produce a racemic mixture. Thus one can differentiate the natural status of some components of a flavoring in this manner. The process may involve a simple chromatographic run on a enantioselective (chiral) column or

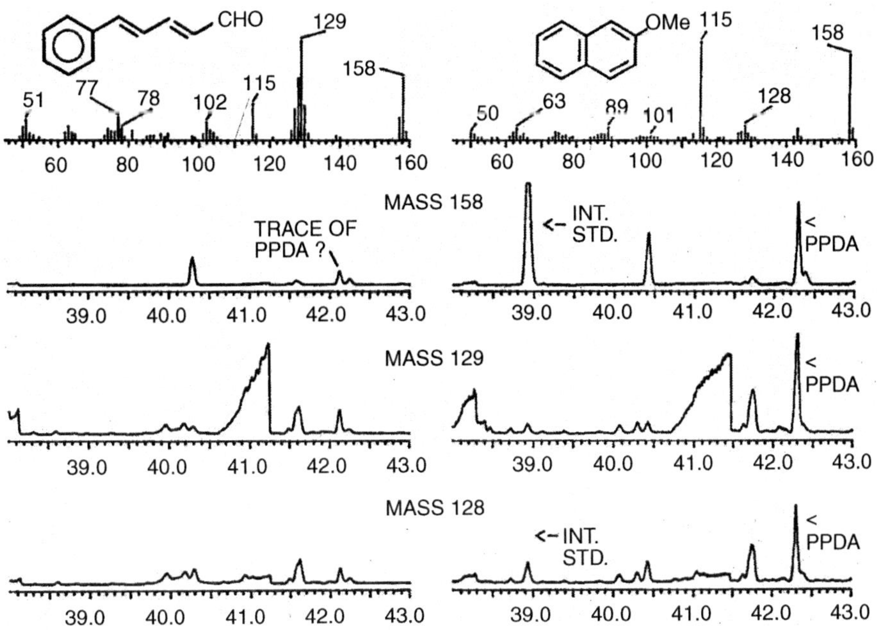


**FIGURE 16.8** SIM MS traces of pure cassia oil (left) and cassia oil adulterated with 2% synthetic cinnamic aldehyde (right). Masses 158, 128, and 129 are unique to phenyl pentadienal (PPDA). (From Frey, C., in *Flavors and Fragrances: A World of Perspective*, B.M. Lawrence, B.D. Mookerjee, B.J. Willis, Eds., Elsevier, Amsterdam, 1988. With permission.)

require heart cutting and rechromatography on a chiral column if the sample is very complex. One has to be aware that there may be racemization of the sample due to nonenzymatic reactions (e.g., photooxidation or autoxidation), or during processing and storage [23]. Thus, the chiral profile of a processed flavoring may not be identical to that of the starting materials.

This method has been applied to many flavorings. For example, if one suspects that synthetic linalool has been added to neroli or bergamot oil, linalool from the oil may be isolated and analyzed by chiral GC. One would expect a given chiral distribution for a natural linalool from bergamot but a racemic mixture for the synthetic counterpart.

### 16.4.3 HPLC

High Performance Liquid Chromatography (HPLC) has traditionally found little use in most flavor companies. The typical flavor house has focused on volatile substances, and thus GC was the primary tool. However, HPLC finds application in flavor houses that work with nonvolatile components such as capsaicins, colorants, savory enhancers (MSG or the 5′-nucleotides), salts, or other flavor adjuncts. It can also be used to advantage for flavor chemicals that are difficult to analyze by GC, i.e., those that are low in volatility, thermally unstable, or difficult to resolve from the matrix or other flavor components.

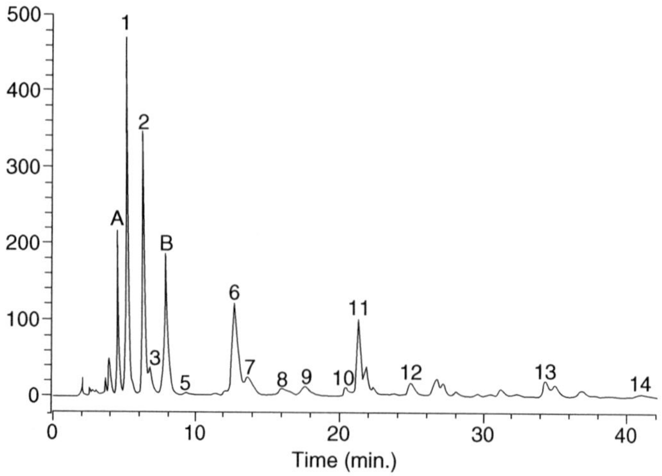


**FIGURE 16.9** HPLC chromatogram of adultered lemon oil. A is ethyl p-dimethylaminobenzoate and B is auraptene + 5-isopentyloxy-7-methoxycoumarin. (From McHale, D. and J.B. Sheridan, Detection of the adulteration of cold-pressed lemon oil, in *Flavours and Fragrances*, B.M. Lawrence, B.D. Mookherjee, B.J. Willis, Eds., Elsevier Science Publishers, Amsterdam, 1988. With permission.)

There has been some use of HPLC for detecting adulteration in flavor ingredients. McHale and Sheridan [24] reported on using HPLC to detect the adulteration of lemon oil. Adulteration was done by diluting lemon oil with lower quality steam stripped oil. In order to have the adulterated oil pass UV standards, various UV absorbers were added to the oil. These absorbers include menthyl salicylate and anthranilate, ethyl p-dimentylaminobenzoate, and cold pressed lime and grapefruit oils. HPLC examination of the oil can show the presence of these adulterants (Figure 16.9). It was rather remarkable that 25% of the commercial, supposedly authentic lemon oils on the U.K. market were found to contain ethyl p-dimethylaminobenzoate. A significant number of other oils were found to contain grapefruit oil pigments. Adulteration is a problem in the flavor industry.

### 16.4.4 Carbon 14 Dating

Another approach to the detection of adulteration has been to determine whether a pure chemical has been synthesized from petroleum-based chemicals or from recent plant metabolism [25]. All recent plant tissues contain some $^{14}C$ due to $^{14}C$ in the $CO_2$ of the atmosphere (nuclear testing). The only way for a flavor compound to show no $^{14}C$ is if it has been synthesized from petroleum sources. $^{14}C$ has a half-life of about 5,730 years and thus is depleted in crude oil. An obvious weakness of this method is it does not distinguish synthetic chemicals made from plant starting materials from truly natural materials.

Bricout and Koziet [26] demonstrated how $^{14}C$ analysis could be used to determine the natural status of citral. On $^{14}C$ dating, they reported 20.1 dpm/g $^{14}C$ for citral from lemon grass and 0.25 dpm/g $^{14}C$ for citral synthesized from petrochemical sources. As would be expected, they did not find a significant difference in $^{14}C$ values between synthetic citral synthesized from pinene, for example, or natural citral isolated from lemon or lemon grass. A further limitation of this technique is that it is suitable only when relatively large (g quantities), pure samples are available. Thus, this technique is most useful for the analysis of pure ingredients, not finished flavors or flavors in food products.

### 16.4.5 Stable Isotope Ratio Analysis

Stable isotope ratio analysis (SIRA, GC-IRMS [Gas Chromatography-Isotope Ratio Mass Spectrometry], and Site Specific Isotope Fractionation — Nuclear Magnetic Resonance [SNIF-NMR]) have proven useful in many adulteration situations. In nature, $^{13}C$ and $^{12}C$ exist at relative proportions of 1.11:98.89 [27]. The photosynthetic process selectively enriches the plant in $^{12}C$ dependent upon the type of photosynthetic process used by the plant. Plants using the Hatch-Slack pathway (e.g., corn, sugar cane, millet, and lemon grass) give $\delta^{13}C$ values the closest to the standard, i.e., $\delta^{13}C$ values of ca. -10. The $\delta^{13}C$ value is calculated as:

$$0/00 = [(R_{sample}/R_{reference}) - 1] * 1000$$

where $R_{sample}$ = the ratio $^{13}C/^{12}C$ in the sample and $R_{reference}$ = $^{13}C/^{12}C$ in the reference (same equation applied to the D/H calculation).

Plants using the Crossulacean Acid Metabolism (CAM) pathway (e.g., vanilla, pineapple, and agave) have $\delta^{13}C$ values of ca. -20. Plants using the Calvin cycle (e.g., trees [source of lignin], wheat, sugar beets, grapes and citrus fruits) have the greatest $^{12}C$ enrichment giving $\delta^{13}C$ values near -30. Thus, this method is useful in determining the type of plant used as the source material to make a chemical.

One of the earliest applications of stable isotope analysis was to detect adulteration in vanilla (CAM pathway). The $\delta^{13}C$ value of vanillin from authentic vanilla (-20.9) is significantly higher than in vanillin that has been synthesized from lignin (-27.8) or guaiacol (-31.5) [26]. Lignin and guaiacol are the primary starting materials for the synthesis of synthetic vanillin and use the Calvin pathway. This method of monitoring vanilla adulteration worked until adulterators synthesized a $^{13}C$ enriched vanillin (aldehyde and methoxy groups) which, when added in trace amounts, boosted the $^{13}C$ level up to that normally found in natural vanillin [19]. This methodology became of limited value, i.e., limited to catching only the less sophisticated adulterers, until researchers found that this was being done. The chemists monitoring adulteration then developed methodologies to specifically determine the level of $^{13}C$ in the aldehyde and methoxy groups, which then was useful in monitoring adulteration.

The next development in stable isotope monitoring was the measurement of deuterium in a flavor component [24]. Deuterium exits at a ratio of about 0.015 to 99.9 to hydrogen [27]. Minor differences in this ratio exist due to the physical and chemical processes occurring in the synthesis of a compound. For example, deuterium is depleted in water as it evaporates and condenses into rain. This changes the

**TABLE 16.3**
**Stable Isotope Abundances in Cinnamic Aldehyde**

| Source | $^{14}C$ dpm/g C | $\delta\,^{13}C$ | $\delta D$ |
|---|---|---|---|
| Natural | 16.3 | –27.2 | –116 |
| Synthetic (toluene oxidation) | 0.1 | –25.0 | 464 |
| Synthetic (other natural precursors) | 14.0 | –27.1 | 72 |

*Source:* From Culp, R.A., J.E. Noakes, *J. Agric. Food Chem.*, 38, 5, p. 1249, 1990. With permission.

level of deuterium with geographical origin. Isotopic fractionation also occurs in a given plant. These factors can be used to detect the source of a natural chemical.

Originally, the total $\delta D$ value was used in this determination [25]. This number comes from the total deuterium in the sample (2–5 mg of sample required in analysis). Culp and Noakes [25] showed how this determination can be used to detect synthetic cinnamic aldehyde (Table 16.3). The first row in Table 16.3 is for a natural cinnamic aldehyde. It has the desired $^{14}C$ demonstrating it is from recent plant metabolism, the proper $\delta\,^{13}C$ value based on its normal plant source (photosynthetic pathway) and a $\delta D$ indicating geography and plant type. The second row provides these values for a cinnamic aldehyde made from petrochemicals. The lack of $^{14}C$ clearly demonstrates its origin as petrochemical as does the $\delta D$. The third row provides values for cinnamic aldehyde made by other synthetic routes, clearly starting with the chemical transformation of a related plant component. While the $^{14}C$ and $^{13}C$ values are in agreement with the natural authentic material, the $\delta D$ values indicate this material has been chemically altered. Thus, this method has value in detecting chemically altered plant materials. But again, the inventive adulterers have managed to find economic means of balancing the total isotope ratios thereby minimizing the utility of the method. This has called for innovation on the part of the individuals who monitor this activity.

The next development has been to do site specific stable isotope analysis (SNIF-NMR). While total $^{13}C$ or D can now be done by combined GC-MS analysis, which makes the method both sensitive and rapid, site specific analysis requires the isolation and purification of components and nmr time [28]. This adds complexity to the procedure, and it must be recognized that the analysis potentially will alter the isotope distribution/composition of the component being monitored [29]. However, Remaud et al. [28] have demonstrated how this analysis can be applied to the detection of adultered vanilla.

Remaud et al. [28] reported that they could detect as little as 5% adulteration of vanillin with vanillin ex-guaiacol (petrochemical source) and 10% adulteration with vanillin ex-lignin (wood source) using SNIF-NMR. These detection limits were established using the D/H data at all sites (as opposed to any one site). Fronza et al. [30] demonstrated that using SNIF-NMR they could distinguish between raspberry ketone that came from authentic botanical origin, microbial conversions, and other nonnatural sources. The method appears to be extremely difficult to circumvent. The major drawback of this method is it is extremely tedious and thus impossible to routinely apply on incoming materials. One would also like to have detection limits of <10%.

### 16.4.6 Comments on Adulteration

The government does not do adulteration testing (except in a limited sense for vanilla) but focuses on safety issues rather than ethical or economic issues associated with adulteration. Thus, the job of adulteration monitoring rests on the industry itself. While the industry does this job partially for noble reasons, there often is an economic incentive. One may find that a company cannot match the quality, character, or price of a competitor's flavoring. This raises a question as to whether the competitor is using some materials that are adulterated (lower in cost or unavailable in natural form) and inspires the flavor company to do some testing for authenticity. If the company finds that the flavoring is adulterated, the company may choose to tell the client company about this and hope the client will change the source to the more ethical company — the one who caught the adulteration.

Adulteration is unfortunately more common in the industry than it should be. This is illustrated by the finding at one time that there was virtually no authentic natural benzaldehyde available in the world or that the vast majority of lemon oils sold in the U.K. were adultered. This does not mean that all of the flavor companies using and selling adultered benzaldehyde or lemon oil were unethical: the industry often has few sources of any given material and one unethical manufacturer can taint the supply of a large number of flavor companies who use these materials. QA labs in larger companies devote considerable effort to ensuring that they get authentic flavor ingredients; however, the magnitude of this task makes it impossible to assure that all materials are in fact authentic. Unfortunately, adulterers have the same instrumentation as the "police" and thus the adulterer plays a game with the police always trying to stay one step ahead of enforcement. The ultimate goal is to make adulteration so expensive to accomplish that the economic incentive is removed.

## REFERENCES


1. Nielsen, S.S., Ed., *Food Analysis*, 3rd ed., Kluwer Academic Plenum Publishers, New York, 2003, p. 557.
2. Heath, H.B., *Source Book of Flavors*, 2nd ed., AVI Pub. Co., Westport, CT, 1981, p. 863.
3. McClements, J., Analysis of food emulsions, in *Food Analysis,* 3rd ed., S.S. Nielsen, Ed., Kluwer Academic Plenum Publishers, New York, 2003, p. 571.
4. Tse, K.Y. and G.A. Reineccius, Methods to predict the physical stability of flavor-cloud emulsion, in *The Chemical and Physical Stability of Flavors in Foods*, C.T. Ho, C.-T. Tan, C.-H. Hsiang, Eds., Amer. Chem. Soc., Washington, D.C., 1995, pp. 172–182.
5. Malvern Instruments, I want to measure particle size, www.malvern.co.uk/malvern/rwmalvern.nsf/vwaareas/Lab%20Products, accessed 04/22/05.
6. Beckman Coulter, Multisizer™ 3 COULTER COUNTER®, www.beckman.com/products/Discipline/LifeScienceResearch/prdiscgenpartchar.asp?bhcp=1, accessed 04/22/05.

7. Rawle, A., Basic principles of particle size analysis, Surface Coatings Intern., Part A: *Coatings J.*, 86(A2), p. 58, 2003.
8. Tan, C.-T., Beverage emulsions, in *Food Emulsions,* 3rd ed., K. Larssen, S. Friberg, Eds., Marcel Dekker, New York, 1997, p. 491.
9. Kernik, K., G.A. Reineccius, J.P. Scire, An evaluation of methods for the analysis of encapsulated flavors, in *Progress in Flavour Research*, J. Adda, Ed., Elsevier Publishers, Amsterdam, 1985, p. 477.
10. Anandaraman, S. and G.A. Reineccius, Analysis of encapsulated orange peel oil, *Perfum. Flavor.*, 12, 2, p. 33, 1987.
11. Bradley Jr., R.L., Moisture and Total Solids Analysis, in *Food Analysis*, 3rd ed., S.S. Nielsen, Ed., Kluwer Academic Plenum Publishers, New York, 2003, p. 119.
12. Sigma-Aldrich Co., Natural linayl acetate, http://www.sigmaaldrich.com/cgi-bin/hsrun/Suite7/Suite/HAHTpage/Suite.HsViewHierarchy.run?Detail=Product&ProductNumber=ALDRICH-W263613&VersionSequence=1, accessed 04/22/05.
13. Sigma-Aldrich Co., Kosher synthetic linayl acetate, http://www.sigmaaldrich.com/cgi-bin/hsrun/Suite7/Suite/HAHTpage/Suite.HsViewHierarchy.run?Detail=Product&ProductNumber=ALDRICH-W263605&VersionSequence=1, accessed 04/22/05.
14. Vollaro, F., Physical/Chemical Evaluation: Methods of Detection of Changes, Presentation at an IFT sponsored flavor workshop titled, Flavor Interaction with Food, Orlando, FL, March, 2005.
15. Johnson, E.L., R.E. Majors, L. Werum, P. Reiche, The determination of naturally occuring capsaicins by HPLC, in *Liquid Chromatographic Analysis of Foods and Beverages*, G. Charalambous, Ed., Academic Press, New York, 1979, p. 17.
16. ASTA, *Official Analytical Methods*, Vol. 3, American Spice Trade Assoc., New York, 1985.
17. Gemert, L.J., L.M. Nijssen, H. Maarse, Improvements of the Scoville method for the pungency determination of black pepper, in *Flavour '81*, P. Schreier, Ed., de Gruyter, Berlin, 1981, p. 211.
18. Dorland, W.E. and J.A. Rogers Jr., *The Fragrance and Flavor Industry*, Wayne Dorland, New Jersey, 1977.
19. Martin, Y.-L., G.G. Martin, G. Remaud, New isotopic possibilities for flavor authentication, *Food Testing Anal.*, 3, 6, p. 10, 1998.
20. Strauss, D.A. and R.I. Wolstromer, The examination of various essential oils, *Congr. Essential Oils*, 1, 1974.
21. Frey, C., Detection of synthetic flavorant addition to some essential oils by selected ion monitoring GC/MS, in *Flavors and Fragrances: A World of Perspective*, B.M. Lawrence, B.D. Mookerjee, B.J. Willis, Eds., Elsevier, Amsterdam, 1988, p. 517.
22. Mosandl, A., Analytical authentication of genuine flavor compounds, in *Flavor Chemistry: Thirty Years of Progress*, R. Teranishi, E.L. Wick, I. Hornstein, Eds., Kluwer Academic Plenum Publisher, New York, 1999.
23. Mosandl, A., R. braunsdorf, G. Bruche, A. Dietrich, U. Hener, T. Kopke, P. Kreis, D. Lehman, B. Maas, New methods to assess authenticity of natural flavors and essential oils, in *Fruit Flavors*, R.L. Rouseff, M.M. Leahy, Eds., Amer. Chem. Soc., Washington, D.C., 1995.
24. McHale, D. and J.B. Sheridan, Detection of the adulteration of cold-pressed lemon oil, in *Flavours and Fragrances*, B.M. Lawrence, B.D. Mookherjee, B.J. Willis, Eds., Elsevier Science Publishers, Amsterdam, 1988, p. 525.
25. Culp, R.A., J.E. Noakes, Identification of isotopically manipulated cinnamic aldehyde and benzaldehyde, *J. Agric. Food Chem.*, 38, 5, p. 1249, 1990.

26. Bricout, J., J. Koziet, Characterization of synthetic substances in food flavors by isotopic analysis, in *Flavor of Foods and Beverages*, G. Charalambous, G.E. Inglett, Eds., Academic Press, New York, 1978, p. 199.
27. Faure, G., *Principles of Isotope Geology*, 2nd ed., John Wiley & Sons, New York, 1977, p. 589.
28. Remaud, G.S., Y.S. Martin, G.G. Martin, G.J. Martin, Detection of sophisticated adulterations of natural vanilla flavors and extracts: application of the SNIF-NMR method to vanillin and p-hydroxybenzaldehyde, J. *Agric. Food Chem.*, 45, p. 859, 1997.
29. Moussa, I., N. Naulet, M.-L. Martin, G.J. Martin, A sitespecific and multielement approach to the determination of liquid-vapor isotope fractionation parameters: the case of alcohols, *J. Phys. Chem.*, 94, p. 8302, 1990.
30. Fronza, G. and C. Fuganti, Natural abundance 2H nuclear magnetic resonance study of the origin of raspberry ketone, *J. Agric. Food Chem.*, 46, p. 248, 1998.

# Index

## B

## C

## D

## E

## F

## G

## H

## I



## J



## K

## L



## M

## N



## O

# Q

## T

## U



## V

## W



## X



## Y



## Z

Printed by Publishers' Graphics Kentucky